Digital Elevation Models for Environmental Studies

Adriano Rolim da Paz

Digital Elevation Models for Environmental Studies

Springer

Adriano Rolim da Paz
Federal University of Paraíba
João Pessoa, Brazil

ISBN 978-3-032-04522-5 ISBN 978-3-032-04523-2 (eBook)
https://doi.org/10.1007/978-3-032-04523-2

Translation from the Portuguese language edition: "Modelos Digitais de Elevação para Estudos Ambientais" by Adriano Rolim da Paz, © ABRHidro owns the copyright to the book in Portuguese. All other rights stay with author. 2024. Published by ABRHidro—Associação Brasileira de Recursos Hídricos. All Rights Reserved.

The original submitted manuscript has been translated into English. The translation was done using artificial intelligence. A subsequent revision was performed by the author(s) to further refine the work and to ensure that the translation is appropriate concerning content and scientific correctness. It may, however, read stylistically different from a conventional translation.

This Springer imprint is published by the registered company Springer Nature Switzerland AG
The registered company address is: Gewerbestrasse 11, 6330 Cham, Switzerland

Dedicated to my children, Mateus and Luísa

Foreword

It is with great pleasure that I write this Foreword for *Digital Elevation Models for Environmental Studies* prepared by my colleague and friend Dr. Adriano Rolim da Paz.

Although developing maps goes back to the times of the Babylonian and Greek ancient civilizations, those first maps were rather inaccurate as they were produced with very rudimentary methods and tools, and with limited data. Maps continued being created with somewhat inadequate resources until, in the late XVIII century, the Carte Géométrique de la France was first published. This set of maps constituted a breakthrough in cartography because it used a network of triangles to increase horizontal accuracy and, also, because it covered a large area such as the entire country of France. Still, despite the remarkable improvement these maps meant in the field of cartography, they did not include elevation information. Up to that point, relief in maps was represented by drawings of mountains without numeric elevation values.

Basic elevation maps, in the form of contour lines, were first developed in France in the mid XIX century and, soon after, became a standard format for topographic maps in other European countries and the United States. Still, it was not until the 1960s–1970s that topographic maps were converted from analog to digital format, which was part of an overall and worldwide conversion that also affected, for example, music and photography, to mention two apparently unrelated types of information. These topographic maps, now in raster format, were termed Digital Elevation Models (DEM) and, since then, DEMs of different resolutions and accuracies have been developed using a combination of surveying data, stereo photography, satellite data, and LiDAR and IfSAR data, among others. In this XXI century, it has become obvious that DEMs are fundamental data to study a wide range of problems, such as water flow, exposure to wind and/or light, slope stability, and many more. That is why it is essential to develop accurate DEMs and research the many different ways in which these data can be used.

In this book, Adriano walks us over a comprehensive list of very necessary raster-data management tools, fundamentals of projections and spatial resolution of raster data, and raster-data operations. This book also discusses, in detail, practical applications of DEM-based terrain analysis for environmental studies such as hydrologic modeling, slope stability evaluation, reservoir capacity estimation, to mention a few.

I first met Adriano in 2007, when I was invited to the Federal University of Rio Grande do Sul to teach a short course on integrating Geographic Information Systems (GIS) and hydrologic modeling. I recall being impressed by the caliber of the research conducted at their Institute of Hydraulic Research and encouraging them to publish their findings so that the world could benefit from their work. At that time, Adriano was a Ph.D. student and a fundamental member of this research team. Fast forward to 2025, and Adriano is a Professor of the Department of Civil and Environmental Engineering of the Federal University of Paraiba. He does research on water resources engineering, the effect climate change on the hydrologic cycle, sustainable urban drainage, and surface hydrologic processes. Additionally, Adriano's work benefits from his unique approach of accounting for location of the different elements of the hydrologic system through the use of GIS. This book is the result of years of dedicated work to expand the frontiers of knowledge and takes advantage of his many journal publications, and personal experience teaching classes and training courses to students and professionals in DEM processing.

Reading a book is a trip to the unknown. Enjoy the journey.

July 2025

Francisco Olivera, Ph.D., P.E., F. ASCE
Associate Professor of Civil
and Environmental Engineering
Texas A&M University
College Station, USA

Preface

In 2000, the space shuttle Endeavour orbited the Earth as part of the Shuttle Radar Topography Mission (SRTM), collecting near-global topographic data with an unprecedented level of detail: initially released at a spatial resolution of 90 m, later released with 30 m resolution. Before this mission, global or near-global topographic data were only available at much coarser resolutions—1 km or more.

The SRTM mission represented a revolution in studying topography through Digital Elevation Models (DEMs). Since then, dozens of other global DEMs have been, and continue to be, made available, some derived from new orbital surveys, others from advanced processing techniques applied to previously collected data. More recently, efforts have focused on improving the removal of surface artifacts from these global products, such as vegetation and buildings, to produce DEMs that approximate a Digital Terrain Model (DTM), rather than remaining as a Digital Surface Model (DSM).

Raster-based DEMs—structured grid images composed of pixels—offer a practical representation of topography. Alongside advances in acquiring and refining these global products, significant progress has been made in developing software and algorithms for processing this data, with easy-to-use graphical user interfaces. Consequently, such processing has increasingly been undertaken by professionals across diverse fields of application, including water resources, ecology, forest engineering, geology, and many others.

However, after working in this field for over 20 years, I have observed firsthand the widespread difficulties many professionals face in mastering and conducting DEM processing studies. Significant gaps exist in conceptual understanding, comprehension of how software commands work, and critical interpretation of the resulting outputs.

One reason for these challenges is the limited availability of textbooks specifically focused on Digital Elevation Models. This is the primary contribution that this book seeks to offer: to provide professionals and students, both at undergraduate and graduate levels, from various disciplines, with a comprehensive resource for understanding the fundamentals and operations involved in DEM processing. This book can be adopted in courses related to Geographic Information Systems,

Geoprocessing, and Geospatial Analyses, at both undergraduate and graduate levels, with content depth adaptable to different learning needs.

Importantly, this is not a software manual. Instead, it was conceived to reveal what lies behind software commands. The aim is to describe the relevant operations, explain the rationale and implications of different methodological choices, justify parameter values, etc. Complementing this, real-world examples are presented for each operation, using areas and river basins within Brazilian territory.

These practical examples are intended to help readers strengthen their understanding of theoretical and conceptual aspects and illustrate different methods for analysis and result presentation. When readers perform similar processing tasks, they can compare their results with those presented in the book to check for consistency. Using examples from Brazil is simply due to the author's familiarity with these regions. However, care has been taken to ensure that the analyses derived from these examples are generalizable and reproducible in any other area.

The depth of the concepts and explanations, and how the examples are explored, have been carefully designed to maximize the reader's comprehension. The extensive use of schematic diagrams, illustrative figures, and visual analysis is intentional: graphical representations are expected to reinforce understanding in ways that purely textual descriptions cannot.

The effort invested in transforming this motivation and these ideas into the concrete form of this book was considerable. Yet it was undertaken with great pleasure. The most excellent satisfaction, however, comes from knowing that this work will help students and professionals continue engaging with DEM processing and derived products, now with higher confidence in their methods and analyses.

I wish you all an excellent reading experience and a deep and enduring understanding!

João Pessoa, Brazil
May 2025

Adriano Rolim da Paz

Acknowledgements To all my students (graduate and undergraduate advisees, students from graduate and undergraduate courses, and extension programs): your questions, difficulties, and doubts challenged me to evolve in my lecturing style, provided me with valuable lessons, and motivated me to conceive this book. I hope new students benefit from this material, and may their future questions contribute to improving subsequent book editions!

To Prof. Walter Collischonn, from Federal University of Rio Grande do Sul (southern Brazil), for his teachings and exchanges of ideas on processing Digital Elevation Models, which laid a foundation of knowledge and encouraged me to deepen work on this topic, and for gently writing the foreword of the Portuguese edition of this book.

To Prof. Francisco Olivera, from Texas A&M University, for kindly writing the Foreword for the English edition of this book.

To Prof. Rafael Mendonça, from Federal University of Paraíba (northeastern Brazil), for his clarifications regarding concepts of Cartographic Engineering and for reviewing parts of the Portuguese version of this text.

To Environmental Engineer Alice Dantas, for her kindness in reviewing the Portuguese version of this manuscript.

To my wife, Marina, and my children, Mateus and Luísa, for their understanding and support in sharing my time during the long work sessions dedicated to the development of this book.

To my parents, Alisson (in memoriam) and Ilma, and to my brother Alexandre and my sister Andréa, for their support and encouragement to always strive to do my best.

To Robison dos Santos, Senior Editor at Springer, who believed, advised, and encouraged me to continue developing this book.

To Mayra Castro, Senior Editor at Springer, and the whole edition team for their support and hard work on transforming the manuscript into this edited format.

To Springer, thank you for the partnership in publishing this book.

To everyone who, upon learning about the idea or during the development of this book, expressed support and offered encouragement to continue this journey.

Competing Interests The author has no competing interests to declare that are relevant to the content of this manuscript.

Contents

Chapter 1
Introduction

1.1 Why Digital Elevation Models for Environmental Studies?

The motivation for analyzing, processing, and working with Digital Elevation Models (DEMs) in developing environmental studies is broad. However, it can be tailored to specific fields of knowledge, such as hydrology, ecology, agronomy, or renewable energy. At the most general level of environmental sciences, this motivation can be summarized by establishing a logical reasoning based on three fundamental arguments (Fig. 1.1):

(i) Spatial analysis is essential for understanding the environment
(ii) Topography underlies environmental relationships and processes
(iii) The world is under intense monitoring.

The following subsections discuss these arguments in detail, creating a narrative that guides the reader toward establishing a central line of reasoning that connects Digital Elevation Models as a foundational element for environmental studies.

1.1.1 Spatial Analysis as a Foundation for Understanding the Environment

The environment is spatial. Everything that occurs at a certain point in the environment results from its surroundings. All processes, phenomena, characteristics, and environmental factors at a point are explained, influenced, and governed by what occurs in the surrounding space. This space can be considered an area of influence around the point. However, this area of influence does not have a clearly defined boundary, as the environmental aspects related to the point act on different spatial scales, interacting with each other and variably over time at different temporal scales.

A. Rolim da Paz, *Digital Elevation Models for Environmental Studies*,
https://doi.org/10.1007/978-3-032-04523-2_1

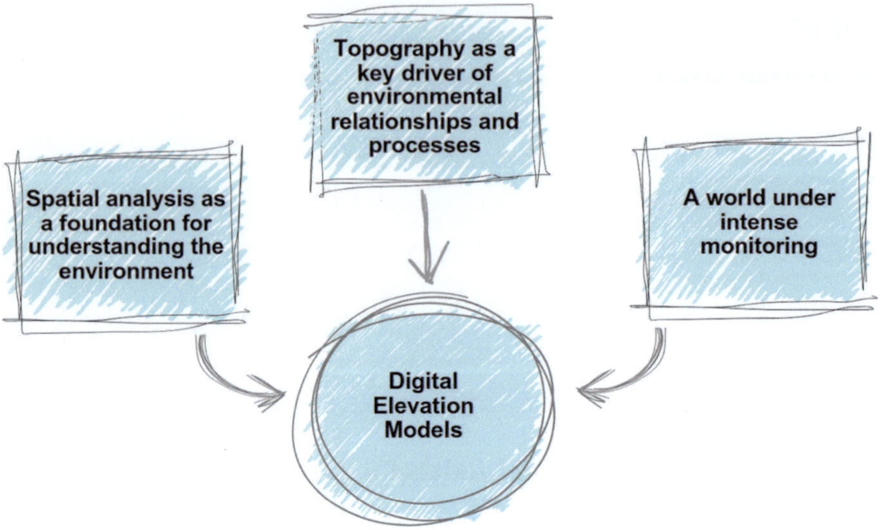

Fig. 1.1 Three key aspects driving the application of Digital Elevation Models in environmental studies

The spatial and temporal scales of analysis are specific and not spatially uniform for each process, phenomenon, or environmental characteristic one wants to analyze. Despite these variations and indefiniteness, the core issue is that the multiple space–time scales of numerous factors condition the environment as it presents itself.

Knowing and understanding how the environment works requires unconditionally studying the spatial variability of several variables and identifying the associated spatiotemporal patterns. This is aligned with the concepts established in the so-called Science of Where, a term coined by Jack Dangermond, founder of the well-known GIS software company Environmental Systems Research Institute (ESRI).

The central idea of the Science of Where is that spatiality must be incorporated to adequately analyze and understand complex phenomena, situations, and problems in various areas, such as the environment, logistics, and urban planning. Spatial analysis allows the identification of patterns, relationships, and trends that other analysis procedures might not readily perceive.

1.1.2 The Topography as a Key Driver of Environmental Relationships and Processes

From the perspective of the need to study the environment spatially, the topography of the Earth's surface stands out as a fundamental and essential condition for various relationships, phenomena, and environmental characteristics.

The word topography originates from the Greek term *topographía*, which is formed by combining two words: '*topos*' and '*graphen*'. The term '*topos*' means '*place*', while '*graphen*' refers to '*description*'. The word topography has a literal origin, meaning a '*description of a place*'. It can be understood more broadly that topography refers to the representation of a limited portion of the Earth's surface.

For example, the spatial variability of topography is a driving factor for differences in air temperature, humidity, and the degree of exposure to solar radiation and wind action. The variation of these characteristics contributes to the formation of microclimates, which in turn influence the spatial distribution of species and the soil characteristics, for instance.

As influenced by topographic characteristics, exposure to solar radiation and winds affects the potential for harnessing solar and wind energy. The influence of topography on wind action also significantly affects the dispersion of atmospheric pollutants.

Soil characteristics are also influenced by other topography-related factors, such as slope, landform shape in terms of convexity or concavity, and the convergence of contributing flow areas. These factors affect soil moisture and runoff drainage and influence soil properties such as organic matter content, sand and silt composition, depth, texture, and nutrient availability.

In addition to the effect of microclimates, the characteristics of soils related to topography also influence ecological aspects such as net primary productivity, richness, and composition of plant species, as well as spatial patterns of vegetation cover. However, topography also acts in other ways on ecology. For example, the topographic characteristics influence the difficulty of movement of species in the environment, the ability of the habitat to provide shelter against extreme weather conditions, and the ability of the habitat to facilitate escape from predators or to facilitate predation, which also contributes to the spatial variability of species diversity.

In the agronomic context, beyond microclimates and soil characteristics, topography also influences aspects of agricultural suitability related to crop management, such as irrigation, fertilization, and mechanized harvesting.

The topographic issue is also paramount for hydrological studies since water movement on the surface is primarily governed by gravitational forces. These water flow dynamics are closely linked to other topics, such as water quality in water bodies, surface erosion processes, and sediment transport, which are highly influenced by topographic characteristics.

Topography is also a key conditioning factor in natural disaster risk assessments. Slope gradients and the amount of drainage area influence landslides. Several topography-related factors also play a role in the occurrence and intensity of floods.

Different connections between topography and environmental issues have been outlined here, but they represent only a tiny sample of the vast number of associations that could be established. Topographic characteristics are intrinsically linked to other environmental attributes and play an essential role in understanding the functioning and spatial diversity of landscapes, relationships, and environmental processes.

1.1.3 A World Under Intense Monitoring

Since the Cold War period in the 1950s and 1960s, satellites have been launched into Earth's orbit, and data acquisition has been carried out remotely for Earth observation, configuring what is understood by orbital remote sensing. The evolution of orbital remote sensors, data acquisition capacity, the quality of these data, and the diversification of variables observed in different areas of knowledge have been expressive since then and continue to increase.

There is a true wealth of data obtained by remote sensing, which continues to expand. More variables and processes are being monitored with higher spatial and temporal detail. For example, data about atmospheric and climatic conditions, land use and land cover, vegetation cover, oceans, continental water bodies, and urban settlements have been collected and made available. Never has the world been under such an intense level of environmental monitoring. And this level continues increasing.

One type of data obtained by orbital remote sensing is topography. Since the 2000s, various products have been available with topographic data covering almost the globe and a spatial resolution of less than 100 m. This sparked a true revolution regarding the ease and speed of characterizing the topography of areas in any geographic location, including regions with difficult access.

1.1.4 Digital Elevation Models: Spatiality, Topography, and Digital Representation

Topographic data acquired by orbital remote sensors is usually made available as a raster-type Digital Elevation Model (DEM). DEMs with almost global coverage have been available, especially those produced since the 2000s. Numerous advantages drive the use of these DEMs.

On one hand, there are intrinsic advantages to this type of topographic representation, the raster-based Digital Elevation Model (DEM): the relative ease of processing the data, automating this processing, overlaying with other raster data, and generating analyses with the development of specialized computational algorithms available in GIS software. These software programs have increasingly enhanced graphical user interfaces, making operations easier, while there is also a growing availability and dissemination of supporting instructional materials.

On the other hand, DEMs originating from global orbital surveys have particular advantages: free and easy access; global coverage (or almost); data standardization; ease of handling; extensive product description; results in the literature with validation, precision analyses, and other studies from these data; and acceptance by the scientific community.

A DEM combines the advantages of representing the spatiality necessary to study the environment (consisting of a representation of a key variable to understand the environment, which is topography) with the convenience and usefulness of being a geographically referenced digital representation in matrix format. This fully justifies the broad applicability of DEMs in environmental studies in various areas of knowledge.

1.2 Book Organization

This book is organized into 14 chapters. Chapter 1 presents conceptual foundations that justify studying Digital Elevation Models for those working in the environmental field.

Chapters 2 and 3 show concepts and fundamentals in geoprocessing necessary for adequately learning about the concepts and methods of processing Digital Elevation Models, which are addressed in the remaining chapters. Chapter 2 deals with the characteristics of geographic data, while Chap. 3 describes the primary operations involving data in raster format, which is the primary format of DEMs.

It is not intended to cover all the essential aspects of geoprocessing with these two preliminary chapters. The reader is encouraged to consult the books proposed for this specific aim. Here, a synthesis of the essence of the theme is offered, giving the reader the opportunity for directed learning and serving as a source of consultation for an objective review. For the reader not so familiar with geoprocessing but interested in learning about DEMs, Chaps. 2 and 3 establish a foundation of minimum content for understanding the concepts and methods addressed in the following chapters, explicitly related to DEM.

Even the reader with some experience in geoprocessing is invited to study Chaps. 2 and 3 before diving into the theme of DEMs. This type of reader can benefit from these two chapters for one or more of these four reasons, at least: (i) reinforcement of contents that they already work with; (ii) deepening in specific items that they were unaware of; (iii) eventual correction of understanding, when the reader thought to have understood but has superficial or mistaken knowledge; (iv) terminology adjustment, since there are readers with different backgrounds and there are various materials in the field of geoprocessing with some variation between terms used by different authors.

From Chap. 4 onwards, the theme addressed directly involves the DEM. In each chapter, concepts and methods are described, and various real-world applications are presented, taking Brazilian areas or watersheds. The theoretical knowledge, understanding of how each DEM processing operation works, and examples of interpreting real-world results are combined.

Chapter 4 introduces the reader to the main fundamentals of the DEMs, such as involved concepts, characteristics, and descriptions of various global or near-global data sources. This chapter describes the basic operations of preparing a DEM, such as windowing, concatenation, and spatial filtering.

Chapter 5 focuses on how a DEM can fundamentally characterize the relief. Issues such as the elaboration of contour lines, hypsometric curves, and transversal topographic profiles are addressed. Other DEM-derived products, like surface slope, aspect, shading, and curvature, are also described.

Chapter 6 deals explicitly with flow directions, which indicate the pixel-by-pixel flow path that water follows from upstream to downstream. This product is derived from the DEM and is considered the foundation for various other products. The primary methods used to generate flow directions and the related challenges are described, mainly due to removing terrain depressions and treating flat areas.

Chapter 7 focuses on the river drainage networks. Concepts and methods for obtaining accumulated drainage areas and defining the river drainage networks are presented. It describes how to derive characteristics such as length, slope, and sinuosity of river reaches, in addition to the longitudinal topographic profile. In the final part of this chapter, two special topics are discussed. The first is incorporating an existing vector drainage network to improve DEM processing, a procedure known as stream burning. The second topic is how to change the scale of the river drainage networks to obtain this data layer in spatial resolutions coarser than the resolution of the DEM.

Chapter 8 covers watershed characterization, delineation, discretization into subbasins by different methods, and determining area, perimeter, drainage density, and other information.

Chapter 9 addresses a more specific issue: using DEM to estimate the area and volume of a lake formed by the construction of a reservoir. It also shows how to determine the area-volume-elevation curve of this reservoir.

Chapter 10 addresses the accounting of distances related to water flow, such as distances to the river drainage network and distances along the river drainage network, with different focuses.

Chapters 11, 12, and 13 present various topographic indices, which are attributes derived from DEM processing that seek to relate the spatial variation of topography to help explain different patterns and processes in distinct areas, such as hydrology, ecology, geography, and renewable energies.

Chapter 11 brings together ten topographic indices with two focuses: the analysis of elevations relative to the neighborhood of each pixel and the issue of terrain roughness. Chapter 12 describes eight topographic indices involving flow convergence and/or the influence of topography on soil moisture and flooding risk. The highlight of this chapter is the HAND index, with an example of applying this attribute to estimate flood-prone areas. Chapter 13 describes seven topographic indices that characterize the terrain regarding protection and exposure to wind action.

Chapter 14 ends the book by dealing with the issues of errors and uncertainties associated with DEM processing. It discusses how to evaluate the quality of a drainage network and describes different methods and metrics used for this purpose. It also addresses the types of impacts that errors in some products derived from the DEM can cause in further hydrological studies.

Chapter 2
Geographic Data

Working with Digital Elevation Models and their derived products means dealing with geographic data. But what characterizes geographic data? What defines it? This is the primary focus of this chapter, which also describes the formats in which such data are represented, discusses key aspects of their symbology, and introduces related concepts such as temporality, projection systems, and spatial referencing. Particular attention is given to data format conversion operations, including transformations from vector to raster and vice versa, along with the implications involved.

These foundational concepts are essential knowledge for anyone working with GIS and geoprocessing. Understanding them is crucial to avoid errors and misunderstandings when processing Digital Elevation Models and in analyses that combine them with other environmental data.

2.1 Formats

Geographic data can be understood as a subtype of spatial data, that is, data to which a spatial dimension is associated, and this spatiality is related to geographic location on the Earth's surface. But how can we make a spatial representation of geographic data? This is a fundamental characteristic of this type of data: how it visually presents itself, how we see it in terms of its geometry, and how it is positioned over its location.

There are two main ways to store and spatially represent geographic data: the vector and matrix formats. In the vector format, geometric features are of the point, line, and polygon types (Fig. 2.1). Points are the most essential elements of the vector format, and from them, it is possible to construct representations as lines and polygons.

Lines are straight segments defined by the vertices (points) that join them. Polygons are straight segments forming closed-area elements. Each polygon is a set of lines whose spatial arrangement configures an area element. A vector's points, lines,

A. Rolim da Paz, *Digital Elevation Models for Environmental Studies*,
https://doi.org/10.1007/978-3-032-04523-2_2

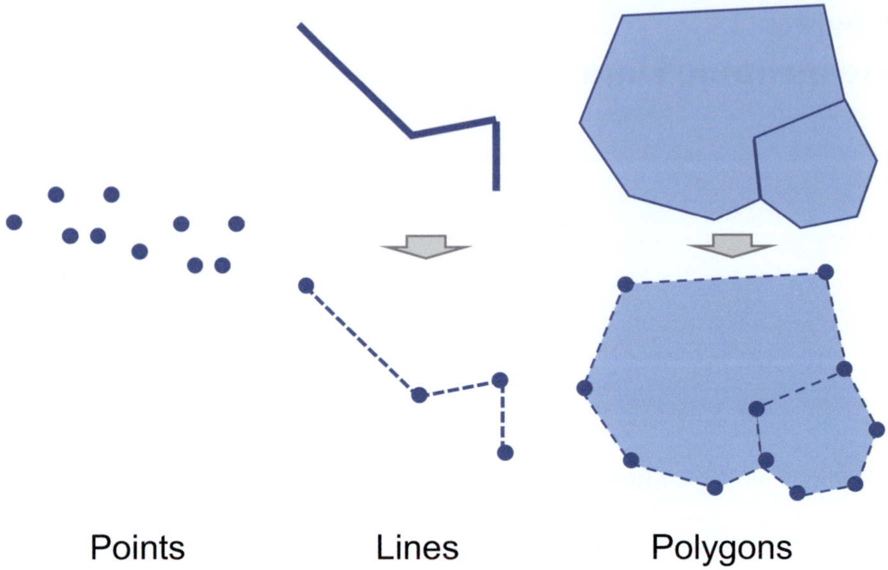

Fig. 2.1 Geographic data types in vector format: points, lines, and polygons

or polygons are also generically referred to as features of the vector data layer. In the example of Fig. 2.2, there is a point vector with ten features, a line vector with one feature (this feature has four vertices and three straight segments), and a polygon vector with two features (one feature with eight vertices and another with six vertices, with three spatially coincident vertices between the two features).

As the name suggests, the matrix format is a data representation through a matrix. It is more commonly referred to as a raster or simply an image. Each element of this raster is called a pixel, with dimension (d) along the image, which is the so-called spatial resolution of the image (Fig. 2.2).

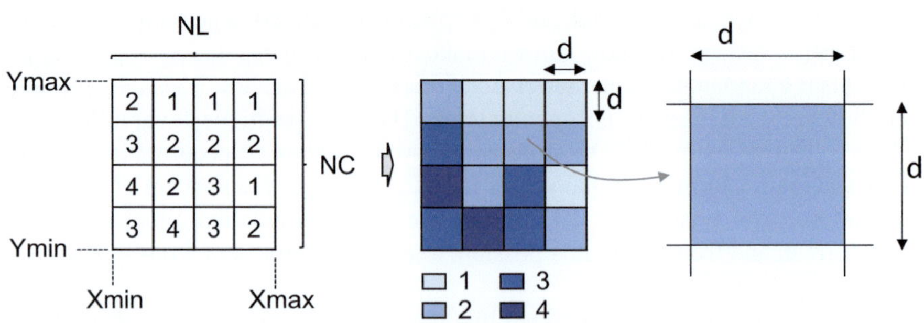

Fig. 2.2 Main characteristics of a raster: number of lines (NL), number of columns (NC), spatial resolution (*d*), and geographic coverage defined by the coordinates $X_{\min}$, $X_{\max}$, $Y_{\min}$, and $Y_{\max}$, with zoom detail for a pixel

In a raster, the pixels are organized in lines (NL) and columns (NC), defining the matrix's size. In the area occupied by each pixel, the stored information is only one, which is the value of the matrix in this element. In other words, it is assumed that the variable represented in the raster is spatially uniform along the area of the pixel. In the example of Fig. 2.2, the highlighted pixel has a value of 2, so in this raster, this is the value represented for the entire area covered by this pixel.

The matrix needs to have geographic coordinates to become geographic data. One way to store this information is simply by keeping the coordinates of the vertices of the matrix ($X_{min}, X_{max}, Y_{min}, Y_{max}$). It is enough to know the coordinates of one of the vertices, and from this information, the number of lines and columns, and the pixel size, one can deduce the coordinates of the other vertices of the image. Similarly, by this procedure, one knows the coordinates of the vertices of any image pixel.

2.2 Attributes

Geographic data can be a vector or a matrix. But what is being represented in these data? They are the data's attributes, representing the main non-spatial characteristics of geographic data. In the vector format, the attributes are associated with the geometric elements that comprise the data (Fig. 2.3). For example, a point type vector may refer to soil sampling sites with attributes such as pH, clay content, texture, water infiltration in the soil, soil type, and data acquisition date (Table 2.1). In this table, each line represents a point of the vector. This vector has eight attributes (fields) associated with each of the five points. A remarkable characteristic of vector data is its ability to store and work with several attributes simultaneously, unlike matrix data, as seen later.

Attributes can be quantitative variables, such as pH, clay content, and water infiltration in the soil in Table 2.1. In the case of these three fields, the attribute value is numeric and indicates the quantity or intensity of that attribute on a given magnitude scale. A clay content of 40% represents twice a clay content of 20%, for example. Quantitative numeric attributes can follow the format of integer numbers (*integers*) or numbers with decimal places (*floats*).

However, in some cases, numeric attributes may not represent values on a magnitude scale. In this case, these numbers function merely as labels, meaning the distinction between classes. For example, a field refers to the soil texture code in the point vector of the previous example. The values 1, 2, or 3 are associated with clayey, sandy, or loamy soils, respectively. This is a qualitative attribute where the values represent classes. The value 3 of this attribute does not represent triple the value 1, for example, but only indicates that they are of different classes. In this case of qualitative data, one should opt for integer numbers, avoiding numbers with decimal places.

The first field in Table 2.1 is a qualitative numeric attribute that refers to the point identifier. Every vector should have this field type, with values that serve as a unique code that identifies and distinguishes each point relative to the others.

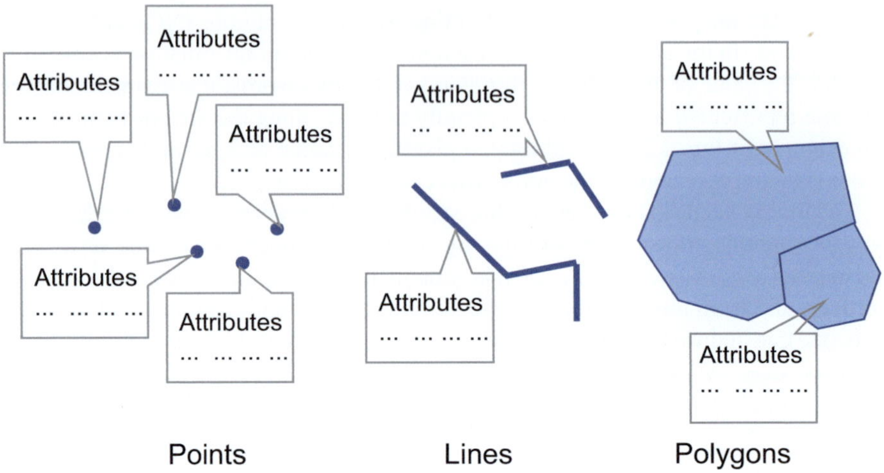

Fig. 2.3 Attributes associated with a point vector with five features, a line vector with two features, and a polygon vector with two features

Table 2.1 Example of an attributes table associated with a vector of soil sampling points

Identifier	pH	Clay content (%)	Soil texture (class id)	Soil texture (class name)	Soil water infiltration (mm/h)	Soil type	Data acquisition date
1	6.4	39	1	Clayey	58	Ultisol	23/03/2000
2	6.6	12	2	Sandy	121	Oxisol	27/03/2000
3	6.5	41	1	Clayey	46	Vertisol	12/04/2000
4	6.8	25	3	Loamy	110	Oxisol	19/05/2000
5	6.6	11	2	Sandy	149	Oxisol	25/05/2000

In addition to numeric attributes, vector data can also contain textual attributes. In the example of soil sampling points, the fifth attribute refers to the term that designates the soil texture: “clayey”, “sandy”, “loamy”. Another type of attribute is the one that follows the date format, like the data acquisition date attribute in the example. Given this range of formats and types of data allowed, the expression alphanumeric data is used to relate the data that can be worked on in vector format.

In the line vector, the attributes are specified for each sequential set of geometrically connected line segments constituting an individual line element or feature. In the example of Fig. 2.3, there are two line elements, each with its attribute values. Similarly, the polygon vector has attributes specified for each polygon element.

In contrast, the matrix format is limited to a single attribute. This attribute can assume different values for each pixel and is exclusively in numeric format. It can be a quantitative or qualitative quantity. For example, a raster of the soil’s clay content could be generated by interpolating the values sampled at points. In this case, the raster has a quantitative numeric attribute, with data in a numeric format with decimal places. Another example of quantitative data is a raster whose attribute is the average annual rainfall in each element.

What if the user wants to represent the soil textures in a raster—that is, to have this information recorded in each pixel of an image? The solution is to have a raster whose attribute is the value of the texture class to which it belongs (values 1, 2, or 3, for example). This is because a raster cannot store textual data (string data) like the names of each soil texture class ("sandy", "clayey", "loamy"). The raster whose attribute is the numeric identification of the soil texture class is an example of qualitative numeric data.

As commented from the example shown in Fig. 2.2, the raster representation is also limited to a single attribute value for each pixel. The value assigned to each pixel in a raster is defined in various ways, according to the characteristics of the information acquired by a remote sensor or according to the geoprocessing operation performed to generate this data layer.

In general, the value assigned to the pixel of a raster follows one of these three principles (Fisher, 1997): (a) represents the value of the variable in most of the pixel area; (b) represents the value of the variable at some systematically defined location within the pixel (for example, the centroid of the pixel); (c) is a value considered representative of the variable by some other specific criterion.

The consideration of a single value for each pixel implies assuming that the variable represented by the raster is spatially uniform throughout the area of each pixel. This means that, for convenience and necessity, a division of the geographic space is imposed into a grid of regular elements. The variable represented has different values in each element of this grid, but is uniform internally in the space occupied by each one. This is a considerable simplification of reality since, in the real world, the variable or phenomenon represented can vary continuously throughout the space occupied by the pixel. How much the simplified representation is closer or further from reality and how much such representation is acceptable varies according to various factors, including the variable or phenomenon worked on, the adopted spatial resolution, the characteristics of the study area, and the purpose of using this information.

Four main incompatibility types can occur between the representation in pixels and reality (Fisher, 1997). The first situation is the existence of sub-pixel information (Fig. 2.4a). In this situation, there are elements of reality with dimensions smaller than the pixel size, such as a tree or a residence, if the raster has tens or hundreds of meters of spatial resolution.

The second situation is when the pixel is located at the transition of a variable value to other values (Fig. 2.4b). For example, a pixel whose area covers the change from one type of land use and land cover to another (example: forest to pasture, urban area to lake) or from one kind of soil to another (example: Oxisols to Cambisols). The third type of situation is the presence of linear elements in the real world running through the extension of the pixels (Fig. 2.4c). Typical examples are roads or transmission lines in rasters that represent land use and land cover.

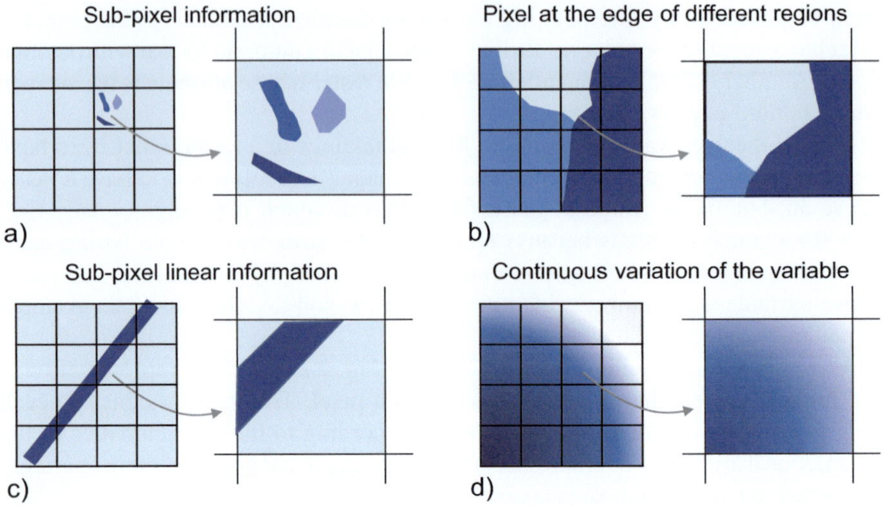

Fig. 2.4 Situations of incompatibility between the raster representation and reality (based on Fischer, 1997): **a** Sub-pixel information; **b** Pixel at the edge of transition between regions; **c** Sub-pixel linear information; **d** Continuous variation of the variable

The fourth situation is the representation in the raster of a variable whose value continuously changes throughout the area of the pixel, like a gradient (Fig. 2.4d). A classic example is the representation of topography in the raster format. Each pixel stores a single elevation value of the terrain, assuming this value represents the entire area of the pixel. However, there is a gradual variation in topography across the area occupied by the pixel.

2.3 Symbology

A vector can have its appearance or symbology edited to reflect one or more specific attributes stored in it, thus enabling the visual identification of patterns. This is one of the most prominent utilities of geographic data. For example, the points referring to the soil sampling from the previous example could have colors and even sizes or shapes varying according to the clay content of the soil (Fig. 2.5a), as it is a quantitative magnitude.

In a line vector, like the rivers of a watershed, each river reach could assume a color according to a gradient (or ramp) of colors related to the average flow value of this reach, also because it is a quantitative magnitude. In a line vector representative of roads, the symbology could reflect the distinction between paved and unpaved roads (Fig. 2.5b), as it is qualitative information.

In a polygon vector, like watersheds, each polygon could have a specific color to distinguish the situation of basins with and without an established basin committee,

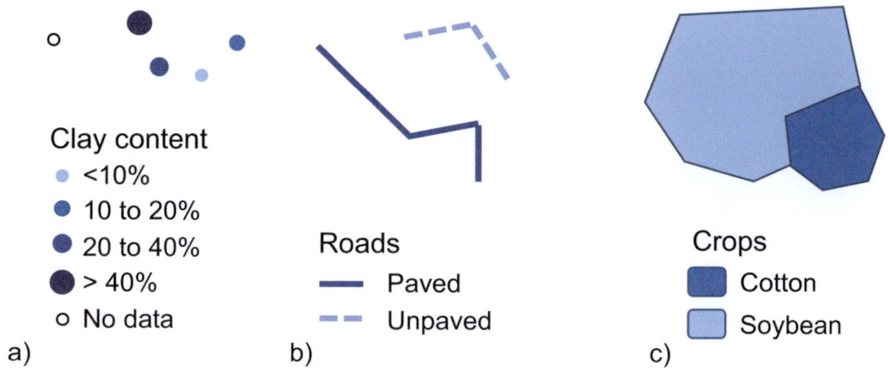

Fig. 2.5 Example of symbologies associated with **a** a point vector referring to soil samples with an attribute related to clay content; **b** a line vector referring to roads with an attribute relative to classes of paved or unpaved roads; **c** a polygon vector referring to crops with an attribute relative to cotton or soybean cultures

for example. The symbology could vary to differentiate existing crops in a polygon vector representative of cultivated species (Fig. 2.5c). Both examples refer to qualitative variables.

It should be noted that, as the vectors' attributes can be alphanumeric and have multiple fields, the symbology can simultaneously indicate more than one attribute, including combining quantitative and qualitative magnitudes. For example, the vector of points referring to soil samples could maintain the size graduation to refer to the clay content and vary the symbol between circles and squares to distinguish between points corresponding to samples collected in two years.

In the raster-type geographic data, the matrix values can be associated with a color scale, quantitative or qualitative, to interpret the image and identify spatial patterns. For a quantitative magnitude, when the attribute varies continuously in space, such as the terrain's slope, the soil's clay content, or the average annual rainfall, the most suitable symbology is a color scale according to a continuous ramp (Fig. 2.6a).

A quantitative magnitude could be categorized into classes, such as ranges of average annual rainfall values (Fig. 2.6b). In this case, the raster of the classes has a qualitative attribute, which is the class number (value 1–5), but the classes are presented according to the increasing order of the value ranges. The adopted symbology uses specific colors for each class but follows a color gradient that reflects the rainfall variation indicated by the classes.

In the example of Fig. 2.6c of a raster with soil types, the order of the classes has no meaning. Class 1 was for Luvisols and 3 for Oxisols, but they could be inverted, for example, without any detriment to the purpose of interpretation and use of information. In this type of raster, the interpretation is purely qualitative, so the assigned colors are most suitable to vary randomly according to the classes to avoid incorrect interpretation. However, it would be possible to use tones of colors close to

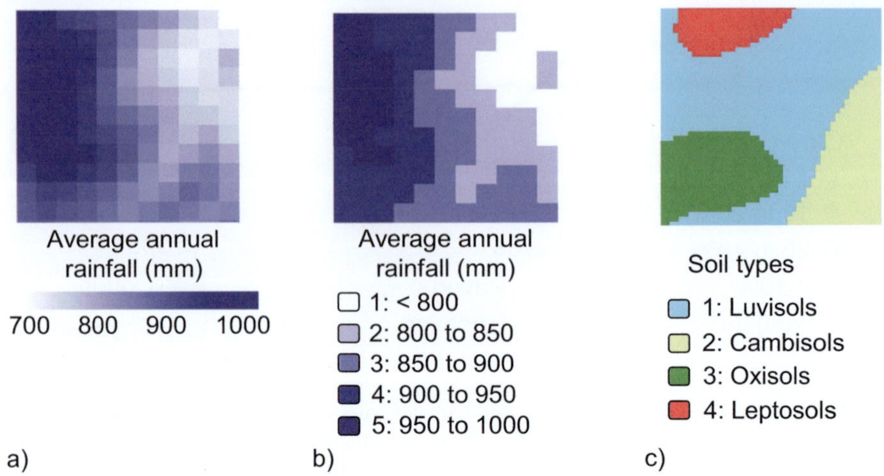

Fig. 2.6 Example of symbologies associated with **a** a raster with an attribute given by the average annual rainfall; **b** a raster with an attribute given by classes of different ranges of average annual rainfall; **c** a raster whose attribute is the soil type

each other for more similar soils, which would facilitate the identification of patterns of spatial variability of soils for some specific purpose.

2.4 Temporality and Metadata

Another very relevant characteristic of geographic data is temporality, the characterization in time of the geographic information. For example, when the soil sampling occurred, or to which year or period the rainfall value represented in the raster refers. In the vector case, this information can be stored as one of the attributes, as in the example illustrated in Table 2.1. In this case, the temporal information is explicitly stored as an attribute of the vector (data acquisition date).

Alternatively, the temporal issue can be recorded in the metadata for both the vector and the raster. The metadata represents the documentation about the geographic data, like a kind of label (Fig. 2.7). Other information, such as the name of the technician or the institution responsible for the data acquisition, the type of instrument used, the method used for acquisition or estimate, the unit to which the quantitative data refers, the acquisition time, and other descriptions about the attributes, can also be present in the metadata. This is helpful information for better managing and using geographic data, avoiding confusion between different versions of the data and incorrect interpretations.

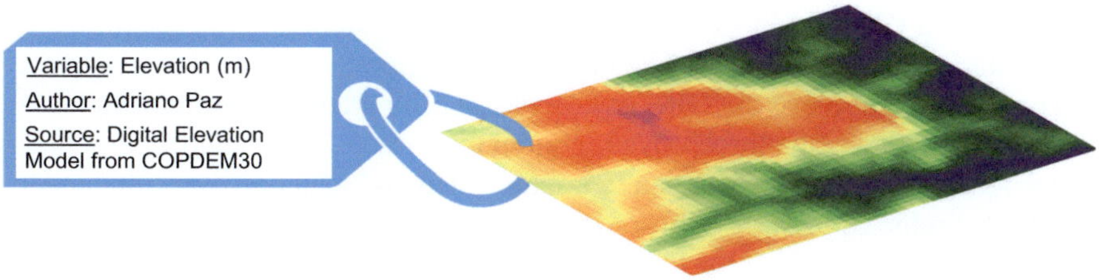

Fig. 2.7 Illustration of raster metadata as the dataset's label, in this case for a raster whose attribute is terrain elevation

2.5 Geodetic Reference System

This section presents a synthesis of concepts and fundamentals associated with a geodetic reference system, including issues of datum, coordinate system, and representation of the Earth's surface. It seeks to show the essence of knowledge related to geodetic reference systems, which is considered relevant in the context of those who work with Digital Elevation Models applied to environmental studies. The scope and terminology adopted are aimed at this target audience. They may differ from those that would be strictly the most appropriate from the point of view of cartographic engineering. A complete and in-depth description of these elements is beyond the scope of this chapter, which can be found in specialized books, like Lu et al. (2014) or Silva and Segantine (2015), the latter specifically for Portuguese readers.

2.5.1 Definition and Purpose of a Geodetic Reference System

When working with geographic data, a Geodetic Reference System (GRS) is adopted to accurately, consistently, and standardly represent this data, defining its relationship in a Geographic Information System, for example, with the real world.

Adopting a GRS allows spatial data to be stored, manipulated, and analyzed coherently, facilitating the overlap and relationship between different geographic datasets. This also allows these data to be projected onto maps, converting the three-dimensional representation of the Earth's surface into a two-dimensional format.

For a GRS to be used, it must be defined and materialized. The definition of the GRS involves the adoption of an ellipsoid of revolution to represent the Earth's surface, the definition of datum, and the establishment of various parameters. For example, the origin and orientation of the GRS relative to the Earth must be defined.

A GRS can be of geocentric origin when this origin is the center of mass of the Earth, or not geocentric. In the first case, the proposal is that the GRS adequately represents the Earth's surface globally. In the second case, the origin of the GRS is a chosen point located on the Earth's surface. This point is usually called a geodetic vertex; such a GRS is topocentric. The topocentric system aims to maximize the representation of the Earth's surface in a specific region.

The materialization of the GRS requires establishing a set of points with known coordinates that describe the Earth's topographic surface. This network of points is known as the geodetic network and is implemented and materialized with geodetic coordinates referred to as the ellipsoid adopted in the GRS.

2.5.2 *Representing the Earth's Surface: Geometric and Geoidal Models*

The Earth's surface, also known as the topographic surface, is highly irregular and contains numerous details describing the local topography fully. However, this topographic surface is unfeasible to serve as the basis for a Geodetic Reference System, which requires something simpler, with a certain degree of abstraction and generalization.

The geometric and geoidal surfaces are simplified ways of representing the Earth's surface. The geometric surface refers to an idealization and simplification of the Earth's surface in the geometric format of a sphere or an ellipsoid of revolution. In the case of the ellipsoid, a geometric figure known as an ellipse is defined. This ellipse is equivalent to a flattened circle at the top and bottom and is rotated around its minor semi-axis (Fig. 2.8). A geometric solid called an ellipsoid of revolution is formed, like a sphere flattened at the poles. For this reason, the geometric surface is also called an ellipsoidal surface.

Due to its well-defined nature, an ellipsoid of revolution allows one to determine mathematically its surface in a relatively simple way and, consequently, the location of a point on it. An ellipse is defined mathematically by the major (a) and minor (b) semi-axes. From these two quantities, an ellipse is also usually characterized by

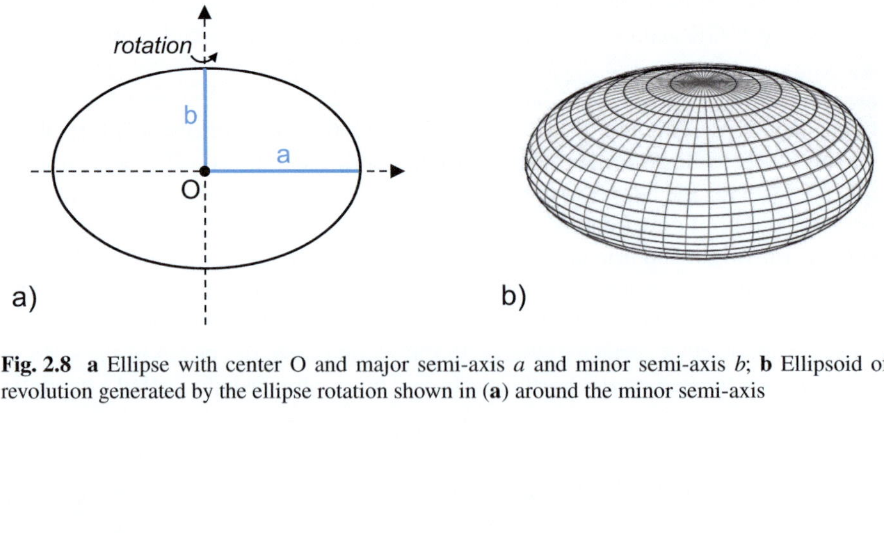

Fig. 2.8 **a** Ellipse with center O and major semi-axis a and minor semi-axis b; **b** Ellipsoid of revolution generated by the ellipse rotation shown in (**a**) around the minor semi-axis

flattening (f), defined by the relationship between the difference between the semi-axes and the extension of the major semi-axis (Eq. 2.1). The higher the difference between the major and minor semi-axes of the ellipse, the higher the flattening. On the contrary, the smaller this difference, the smaller the flattening. In the limiting case, when the semiaxes are equal, the flattening is zero, forming a circle.

$$f = \frac{a - b}{a} \tag{2.1}$$

Several ellipsoids have been proposed to represent the Earth's surface, defined based on major semi-axis and flattening parameters, as exemplified in Table 2.2.

The geoidal surface represents the Earth and is determined based on its gravitational field. The surface of a geoid seeks to represent the average sea level, disregarding the effects of tides, winds, and other factors that affect the movement of oceans, except for the Earth's gravitational field. This surface is considered equipotential in terms of the gravitational field and is a more realistic representation of the Earth's surface than the ellipsoidal surface.

The geoidal surface has a highly irregular shape and its determination is extremely complex, making it impossible to define through a mathematical formulation. This limits the practical use of the geoid as a reference surface.

There are several geoids proposed to represent the Earth. One of the most used types of geoid is the geopotential model made available by the National Geospatial-Intelligence Agency (NGA) of the USA, called the EGM (Earth Gravitational Model). At the time of the preparation of this book, there were three versions of this type of geoid: EGM84, EGM96, and EGM2008, whose acronyms refer to the abbreviation of the expression Earth Gravitational Model, followed by the years 1984, 1996, and 2008, respectively. These versions were developed in this chronological sequence and differ from each other by the increasing number of parameters involved in their determination, which has increased the accuracy of the versions.

Table 2.2 Overview of some ellipsoids of revolution used in Geodetic Reference Systems

Name	Abbreviation	Geometric characteristics	
		Major semi-axis a (m)	Flattening f
Clarke ellipsoid 1866	Clarke 1866	6,378,206	1/294.9786998214
Hayford ellipsoid 1924	Hayford 1924	6,378,388	1/297
Geodetic Reference System 1967	GRS67	6,378,160	1/298.25
Geodetic Reference System 1980	GRS80	6,378,137	1/298.257222101
World Geodetic System 1984	WGS84	6,378,137	1/298.257223563

2.5.3 Geodetic Datum

In Geodesy, a datum or geodetic datum can be understood as the reference from which the positions of geographical elements on the Earth's surface are determined (Silva & Segantine, 2015). The definition of this datum requires information that includes the representation of the Earth's surface, the definition of a coordinate system, and the establishment of the origin and orientation of this system.

The datum can be horizontal or vertical. The horizontal datum is used as a reference to define the position of points in a planimetric way on the Earth's surface. An ellipsoid of revolution is usually defined as the horizontal datum, given the mathematical ease of conceiving the geometric surface and determining the coordinates of points on it. As the previous item shows, the several ellipsoids differ by the geometric characteristics that define the ellipse of origin.

The vertical datum is used as a reference to define the altitudes of the points. As a vertical datum, one can have an ellipsoidal or geoidal surface. When the ellipsoidal surface is adopted, the altitudes are called geometric or ellipsoidal altitudes. In Brazil, the technical standard NBR 13.133 of 2021 modified this nomenclature, referring to the altitude measured with respect to the ellipsoidal surface as the geodetic altitude. We have the orthometric altitude when the geoidal surface is adopted as a vertical datum.

At a certain point on the Earth's surface, the difference between the geodetic altitude and the orthometric altitude is called geoidal undulation (Eq. 2.2).

$$h = H + N \tag{2.2}$$

where h is the geodetic altitude, H is the orthometric altitude, and N is the geoidal undulation.

The surfaces of an ellipsoid and a geoid intersect, alternating above or below (Fig. 2.9). Thus, the geoidal undulations can be positive when the geoid is above the ellipsoid or negative when it is below the ellipsoid.

Figure 2.10 illustrates the spatial distribution of the geoidal undulation, which refers to the differences between the WGS84 ellipsoid and the EGM96 geoid for the entire globe. The areas in shades of orange and red have negative undulations, which indicate that the geoid is below the ellipsoid. In shades of green and blue, areas with positive undulations indicate the opposite: the geoid is above the ellipsoid.

For instance, the geoidal undulation between the WGS84 ellipsoid and EGM96 geoid along the Brazilian territory varies from −33.9 to 28.7 m (Fig. 2.11). This means that topographic data obtained concerning the WGS84 ellipsoid and used in conjunction with data referring to the EGM96 geoid may present an elevation discrepancy of more than 30 m between them for the Brazilian territory. This difference has a tremendous spatial variation, as the earlier figure illustrates. For example, in the region of the municipality of Belém, this magnitude varies between −25.2 and −24.4 m. In contrast, in Porto Velho, the values are positive and have a more extensive range of variation in the municipality area, varying between 7.2 and 22.3 m.

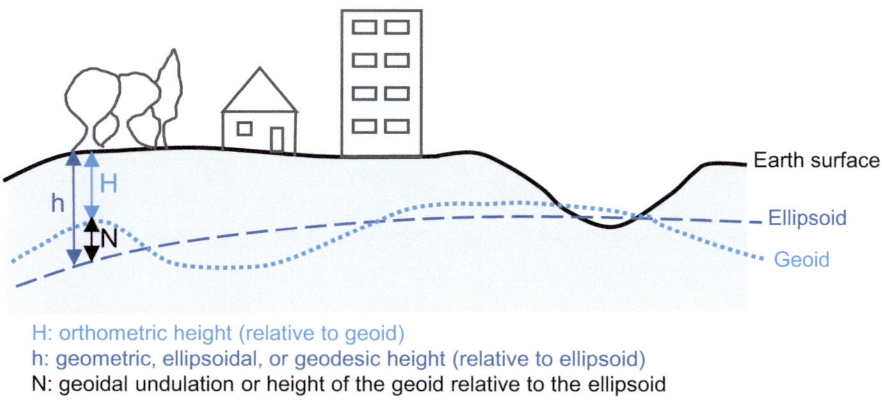

Fig. 2.9 Illustrative scheme of the altitudes measured concerning the geoid and the ellipsoid and the difference between them

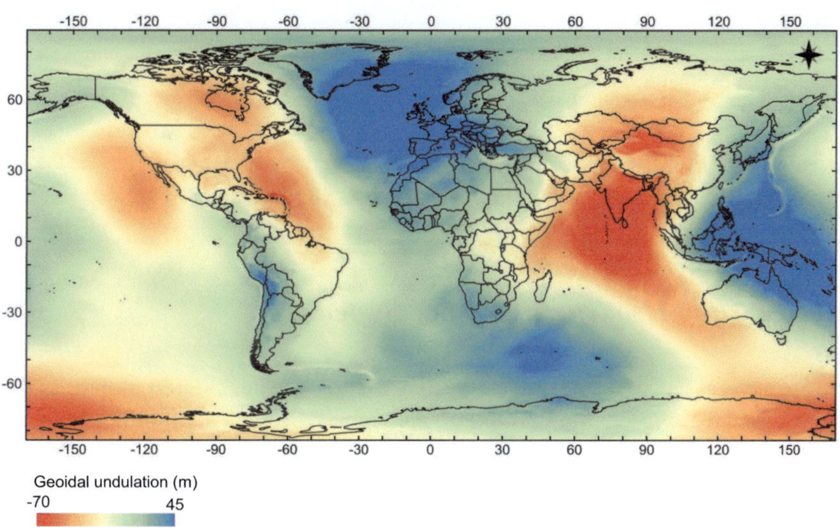

Fig. 2.10 Global geoidal undulation concerning the difference between the ellipsoid WGS94 and the EGM96 geoid

To use topographic data measured with respect to an ellipsoid and a geoid together, it is necessary to convert the elevations from one reference to another (Eq. 2.2). For example, for the municipality of Porto Alegre, the altitudes measured concerning the EGM96 geoid should be added to a value between 4.5 and 5.3 m, depending on the specific location to which the data refers, to transform them into altitudes relative to the WGS84 ellipsoid. For the municipality of João Pessoa, the data measured

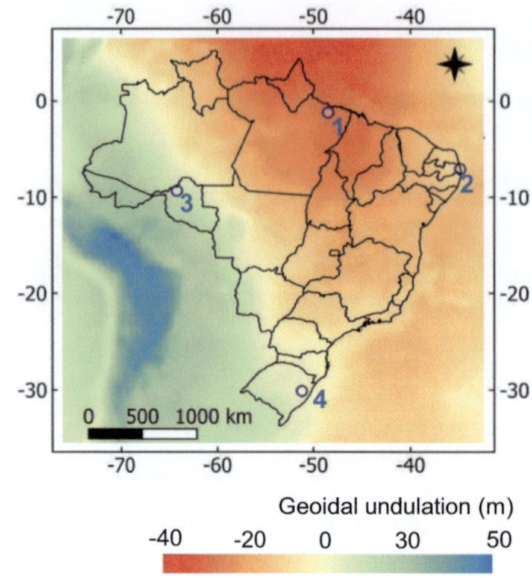

Municipality	Geoidal undulation	
	Minimum (m)	Maximum (m)
Belém	-25.2	-24.4
João Pessoa	-7.4	-4.4
Porto Velho	7.2	22.3
Porto Alegre	4.5	5.3

Fig. 2.11 Geoidal undulation referring to the difference in elevation between the WGS84 ellipsoid and the EGM96 geoid for the Brazilian territory and surroundings, with an indication of the location of the municipalities of Belém, João Pessoa, Porto Velho, and Porto Alegre, and the corresponding minimum and maximum values of the geoidal undulations

concerning the EGM96 geoid need to be subtracted by a value between 4.4 and 7.4 m to correspond to altitudes relative to the WGS84 ellipsoid.

2.5.4 Existing Geodetic Reference Systems

There are several Geodetic Reference Systems. Some are considered global, as they were proposed to represent the Earth's surface globally. The most prominent example is the World Geodetic System 1984 (WGS84), the most widely used worldwide. This geocentric system has the ellipsoid of revolution, also called WGS84, as its horizontal datum. In North America, two examples of GRS are NAD27, topocentric, and NAD83, geocentric, designed to replace the former.

In Brazil, between the 1950s and 1970s, the official GRS was the topocentric system centered in Córrego Alegre, with the Hayford 1924 ellipsoid as the reference. Until 2015, two other systems were considered official in the country: one based on SAD69 (South American Datum 1969) and another based on SIRGAS2000 (Geocentric Reference System for the Americas 2000).

Based on SAD69, the Brazilian GRS is topocentric, originating at a vertex called Chuá, marked in Uberaba, Minas Gerais state. It considers the International Reference Ellipsoid 1967. Since 2015, only the system based on SIRGAS2000 has been regarded as an official GRS in Brazil, also referred to as the Brazilian Geodetic System (SGB).

Table 2.3 Listing of some Geodetic Reference Systems

Name	Acronym	Coverage	Origin	Reference ellipsoid	Horizontal datum
World Geodetic System 1984	WGS84	Global	Center of mass of the Earth	WGS84	WGS84
European Terrestrial Reference System 1989	ETRS89	Europe	Center of mass of the Earth	GRS80	ETRS89
North American Datum 1927	NAD27	North America	Meades Ranch (Kansas, EUA)	Clarke 1866	Meades Ranch
North American Datum 1983	NAD83	North America	Center of mass of the Earth	GRS80	NAD83
Brazilian Geodetic System Córrego Alegre	SGB-CA	Brazil	Córrego Alegre	Hayford 1924	Córrego Alegre
South American Datum 1969	SAD69	South America and Brazil	VT-CHUÁ (Uberaba-MG, Brazil)	GRS67	VT-CHUÁ
Geocentric Reference System for the Americas	SIRGAS2000	South America and Brazil	Center of mass of the Earth	GRS80	SIRGAS2000

The SGB SIRGAS2000 is geocentric and represents the Earth's surface using the surface of the ellipsoid GRS80. The system based on SAD69 and the one based on SIRGAS2000 use the mean sea level, based on observations made in the bay of Imbituba, on the coast of Santa Catarina state, as their vertical datum.

Due to the enormous similarity in mathematical definition between the ellipsoids GRS80 and WGS84, in practical terms, within the scope of environmental studies, it is not necessary to perform transformations of horizontal coordinates between the WGS84 and the GRS SIRGAS2000 (Table 2.3).

2.5.5 Geographic Coordinate System

Several coordinate systems are commonly adopted in GIS. One of the most well-known and used is the geographic coordinate system, which works with coordinates as degrees of latitude and longitude. For this reason, this system is colloquially referred to as the latlong system.

In a strict sense, from the point of view of cartographic engineering, the expression "geographic coordinates" refers to coordinates determined on a reference spherical surface, using angular arc values measured with respect to such a surface, as explained in Silva and Segantine (2015). According to the same authors, when the geometric reference surface is an ellipsoid, the correct term would be "geodetic coordinates" or "geodetic geographic coordinates," with other authors using the expression "ellipsoidal coordinates". Similarly, latitude and longitude should be referred to as geodetic latitude and geodetic longitude, respectively.

However, for convenience and ease of recognition by the general community working with Digital Elevation Models applied to environmental studies, in this book, we choose to keep the expressions "geographic coordinates" and "geographic coordinate system" to refer to what is defined in cartographic engineering as "geodetic coordinates" and "geodetic coordinate system," respectively. We also chose to mention only latitude and longitude, which would be more correct to refer to as geodetic latitude and geodetic longitude, respectively.

In the geographic coordinate system, the lines drawn along the ellipsoidal surface joining the North and South poles are called meridians (Fig. 2.12). The lines drawn perpendicular to the meridians and parallel to the equatorial plane are called parallels. Since the ellipsoidal surface is obtained by the revolution of the reference ellipse around its minor semi-axis, located in the north–south direction, it is understood that geometrically, the meridians form elliptical surfaces and that the parallels constitute circles (Fig. 2.13).

Zero latitude is defined according to the parallel located in the equatorial plane. The north latitudes are to the north of this parallel, conventionally considered positive and assuming values up to 90°. To the south of the equatorial parallel are the south latitudes, which assume negative values by convention, up to −90°.

Zero longitude refers to the meridian that passes through the Royal Observatory in the city of Greenwich, in southwest London, in the United Kingdom. To the east of this meridian are the east longitudes, which assume positive values by convention, up to 180°. The western longitudes are west of the Greenwich meridian, assuming negative values up to −180°.

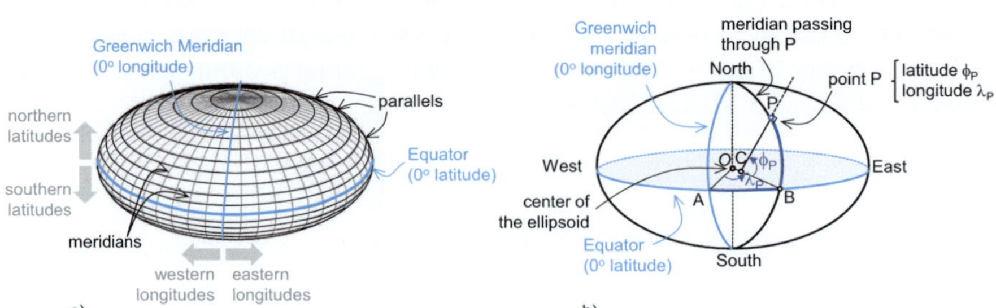

Fig. 2.12 **a** Indication of the meridians, parallels, the Equator, and the Greenwich Meridian in a 3D view of a reference ellipsoid, according to the geodetic coordinate system; **b** Illustration of the determination of the latitude and longitude of a point P in the geodetic coordinate system

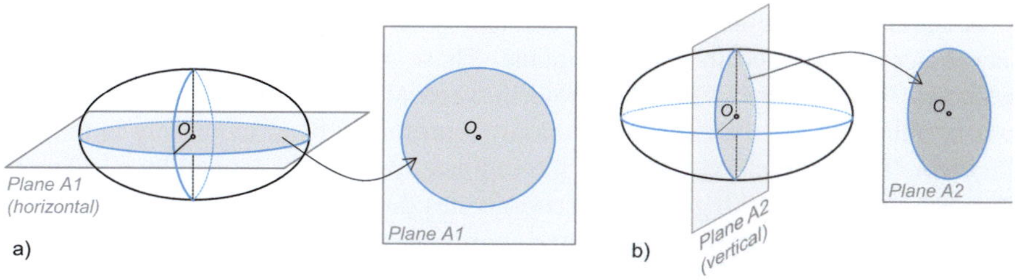

Fig. 2.13 **a** Cross-sectional cut of a ellipsoid of revolution with the center of mass at point O by a horizontal plane at the level of the Equator (plane A1) and the resulting circle; **b** Cross-sectional cut of the same ellipsoid by a vertical plane (plane A2) passing through point O and the ellipse resulting from this cut

To illustrate how geographic coordinates are defined at a certain point on the Earth's surface, represented by the ellipsoidal surface, consider the point P indicated in Fig. 2.12b. Initially, identify the meridian along which P is located. The crossing of this meridian with the equatorial parallel is point B. Draw a line from the center of mass of the ellipsoid (point O) to B and a line from O to the point of crossing of the Greenwich meridian with the equatorial parallel (point A). The longitude of P (λ_P) is defined as the angular measure corresponding to the arc A–B, located along the equatorial parallel. This angular measure is the angle formed between the lines O–A and O–B.

To determine the latitude of P (φ_P), a line is drawn perpendicular to the meridian that passes through P. The intersection of this line with line O–B, located in the plane of the Equator, is point C. Point C does not coincide with the center of mass of the ellipsoid (point O) because of its flattened geometry at the poles. If the Earth's surface representation were a sphere, point C would always be at its center of mass for any point P on its surface. The angle between the lines C–B and C–P represents the latitude of P, which is the angular measure of the arc B–P located along the meridian that passes through P.

The geographic coordinate system is the most easily understood and widely used system for multiple purposes. However, some studies are interested in working with the Earth's surface in a flat format or projected onto a flat surface. In this situation, projected coordinate systems are used, as discussed in the following item.

2.5.6 UTM Coordinate System

When you want to work with the Earth's surface in a flat form, you must project it onto the plane. Essentially, this operation corresponds to transforming the position on the Earth's surface, defined in terms of latitude and longitude, into a position according to the coordinates of a Cartesian system.

A Cartesian coordinate system is defined by two geometric axes orthogonal to each other and located in the same plane. These axes are usually oriented in the west–east (horizontal axis) and south-north (vertical axis) directions. The horizontal axis is called the abscissa and is commonly represented by *x*, increasing in the east direction. The vertical axis is called the ordinate axis and is defined by *y*, rising to the north. A point P, with geographic coordinates ϕ_P (latitude) and λ_P (longitude), is projected onto the Cartesian plane to the coordinates x_P (abscissa) and y_P (ordinate).

Different cartographic projection systems perform the projection of the Earth's surface onto a plane. Regardless of the projection system used, distortions are inherent to the concept of projection. Despite this, projection is widely performed due to the convenience and ease of working with the projected surface on the plane.

Among the projected coordinate systems, the system referred to as UTM (Universal Transverse Mercator) is the most used, based on a cartographic projection called the Transverse Mercator Cylindrical Projection. According to this projection system, the Earth is divided into 60 vertical zones, sequentially numbered from 1 to 60, and each 6° of longitude wide (Fig. 2.14). Each zone has this constant width in degrees and extends from one pole to the other of the Earth. For example, eight zones partially cover different areas of the Brazilian territory—zones 18 to 25 (Fig. 2.15).

In the UTM system, within each zone, the coordinates are represented on two perpendicular axes, called *x* (west–east direction) and *y* (south-north direction), with units in meters and having as origin the center of the zone. For each zone of the UTM system, specific coordinates are measured concerning the zone's origin. Therefore, when working with coordinates in the UTM system, one must specify to which zone it refers (the number of this zone). For example, to work in areas located in Santa Catarina state, one should use the UTM system with zone 22. One should also specify whether the area is in the Northern or Southern hemisphere because the rules

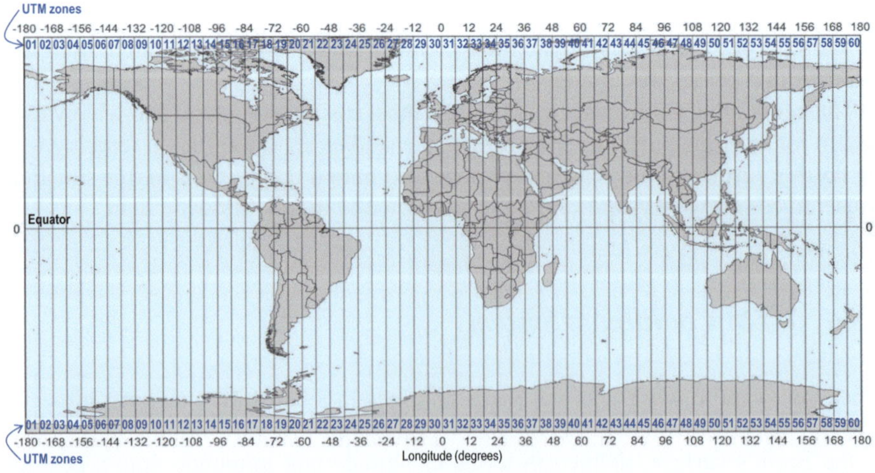

Fig. 2.14 The 60 zones that make up the reference coordinate system UTM (Universal Transverse Mercator), each zone corresponding to a longitude interval of 6°, without overlap

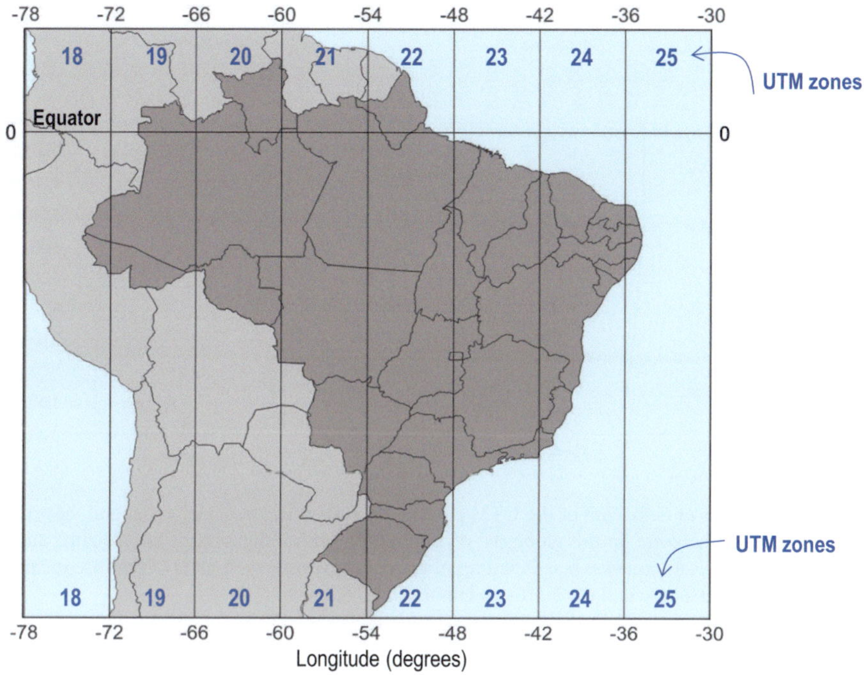

Fig. 2.15 The Brazilian territory's UTM zones (18–25)

for defining the origin and calculating coordinates within the zone differ for each hemisphere. For Santa Catarina, the complete specification is UTM zone 22 South, or with the abbreviation UTM-22S, as it is commonly used.

More specifically, the coordinates of the UTM system are measured concerning the zone's central meridian for the x-axis and the Equator for the y-axis. To avoid negative coordinates, the coordinate 500,000 m is assigned to the central meridian of each zone on the x-axis (Fig. 2.16). Moving from the central meridian to the east, this x coordinate increases from the value of 500,000. Conversely, as you move from the central meridian to the west, the x coordinate decreases from 500,000. This coordinate scheme on the x-axis is identical in both the Northern and Southern hemispheres.

Different schemes are adopted between the hemispheres on the y-axis of the UTM coordinate system. For coordinates in the Northern Hemisphere, the value zero is assigned to the Equator and taken as the origin. In this case, the coordinates on the y-axis increase as you approach the North Pole. For coordinates in the Southern Hemisphere, a y coordinate value of 10,000,000 m is assigned to the Equator. This value decreases as you move away from the Equator toward the South Pole.

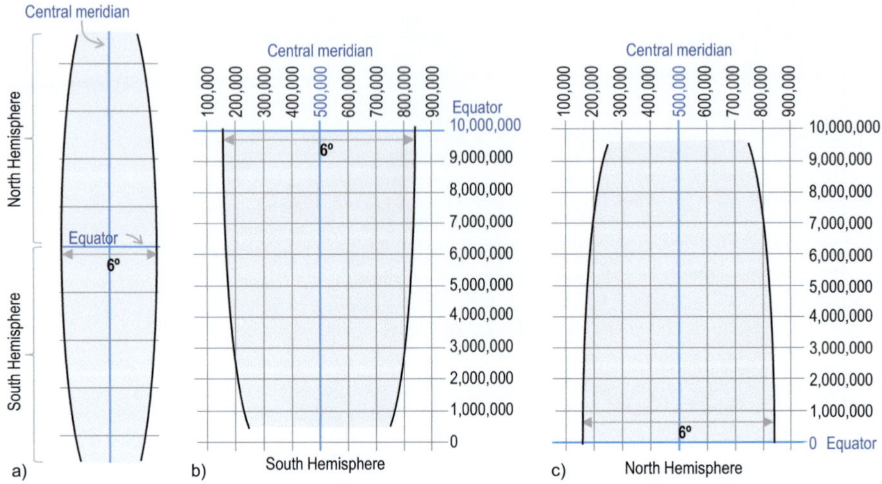

Fig. 2.16 Scheme of definition of the UTM coordinate system in any zone: **a** General view of the area covered by the zone on the globe (6° of longitude), with indication of the Equator and the Central Meridian as references; **b**, **c** Detailing of the coordinates for the parts of the fuse located in the Southern Hemisphere and the Northern Hemisphere, respectively

2.5.7 Coordinate Systems and Digital Elevation Models

In the context of Digital Elevation Models, the predominance is generally the use of geographic coordinate systems (latlong system) when working with relatively extensive areas that cross areas corresponding to more than one UTM zone. The UTM system has been preferred for areas that do not fit this condition, have less geographical coverage, and are fully inserted within the same UTM zone.

Other factors peculiar to each study and application may require the definition of a particular reference coordinate system, such as the need to overlay with other data layers, standardization with existing geographic databases, requirements from mathematical models or other tools for which the DEM can serve as input data, and even user preferences.

2.6 Operations for Converting Data Format (Vector-Raster)

Changing the geographic data format is typical, going from vector to raster or from raster to vector. This is an essential operation available in almost any GIS software.

In the case of vector data of points being taken to the raster format (Fig. 2.17), each point would correspond to a pixel of the raster if the spatial resolution is sufficient so that there is not more than one pixel located inside the same raster element (Fig. 2.17a, b). These pixels would have the value of one of the numerical attributes of the point vector as an attribute chosen by the user in the operation configuration. For example, in the case of soil sample points, the soil pH attribute could be the information taken to the raster. If there is interest in having other attributes of the points in raster format, the operation could be repeated, generating a raster for each attribute.

Being a matrix, the raster generated from the conversion of a point vector is not formed only by the pixels corresponding to the points but also by a certain number of lines and columns that spatially and continuously cover the location of all pixels (Fig. 2.17c). The pixels that do not correspond to any point of the vector are usually assigned the value zero, a constant value informed by the user, or an identifier of the non-existence of information (a flag like 'null data,' for example).

The user can change the spatial coverage of the raster while converting points to a raster to cover an area more or less extensive than the location of the points, according to their interest. In the first case, when the raster is expanded relative to the original coverage of the vector, lines and columns with zero-value pixels would be created to meet the desired extension for the raster (Fig. 2.17d). In the second case, where the raster has coverage less than the vector, one or more points of the vector would not have corresponding pixels in the generated raster (Fig. 2.17e).

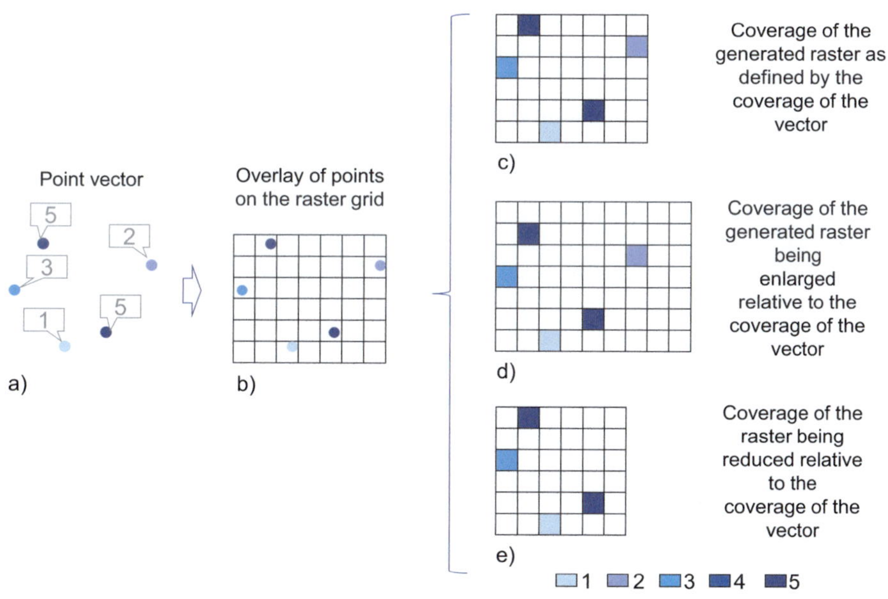

Fig. 2.17 Conversion of a point vector to raster format: **a** Point vector; **b** Overlaying the point vector to the raster grid; **c** Raster generated from the points, with spatial coverage defined by the location of the points; **d** Raster generated from the points, with increased spatial coverage concerning the location of the points, with the addition of a line and two columns of pixels; **e** Raster generated from the points, with reduced spatial coverage concerning the location of the points, with the elimination of a column of pixels—in this case, note that the point farthest to the right is no longer represented in the raster

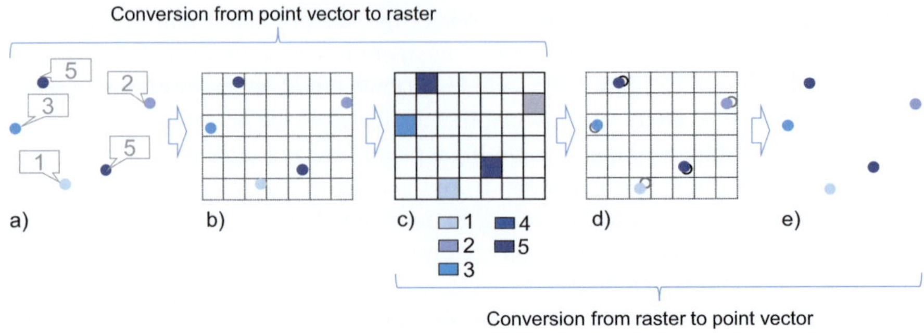

Fig. 2.18 Conversion of a point vector to raster format and conversion back to points: **a** Point vector; **b** Overlaying the point vector to the raster grid; **c** Raster generated from the points, with each pixel having attribute corresponding to the value of the equivalent point; **d** Point vector generated from the raster (solid circles), with superposition to the raster grid and the points of the original vector (hollow circles); **e** Point vector generated from the raster with indication of the corresponding attributes

In the conversion in the opposite direction, passing the information from a raster to a vector of points, the standard operation consists of creating a point corresponding to each pixel whose attribute is non-null, with the point located at the centroid of the pixel. Figure 2.18 shows the transformation of points to raster, which is already shown in Fig. 2.17, but now also makes the inverse transformation, taking the resulting raster to vector. As highlighted in Fig. 2.18d, the points originating from the raster correspond to the centroids of the pixels, which do not coincide with the original location of the points before the conversion to raster.

The automatic correspondence between each vector point and a raster pixel may not exist if two or more points are located in the area covered by the same pixel (Fig. 2.19). In this situation, the pixel can receive the average values associated with the points as an attribute, an interpolation procedure, or another type of decision, such as the highest or lowest values, the mode (value that repeats the most), etc. Attention should be paid to the type of data involved. If it is of quantitative magnitude, operations such as the average or interpolation are generally valid. However, if it is qualitative data, operations such as the average or interpolation should not be performed, while the mode would be more appropriate.

The conversion of a line vector to a raster involves more decisions about which pixels receive the attributes of each line than the conversion from point to raster. Different rules and methods exist for defining the pixels representing each line, mainly due to the various degrees of belonging that can be established between each line and the set of pixels along which the line follows its path.

The line can cross right through the center of the pixel or pass lightly at one of the corners of the pixel, for example. A simple rule could be that all pixels "touched" by the line would represent it in the raster, receiving the value of its attribute. Another alternative would be to seek the pixels most related or representative of the main path of each line.

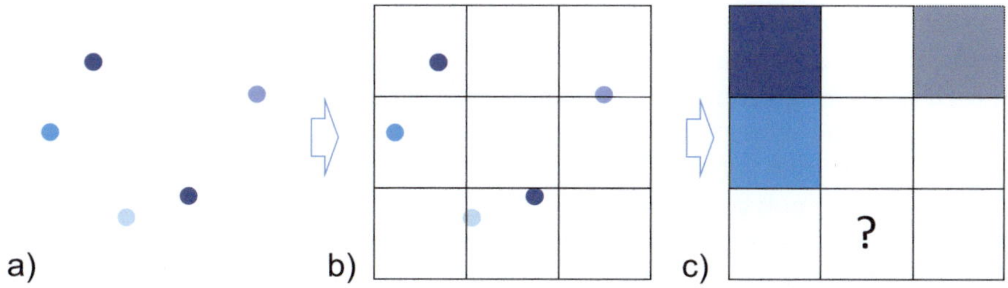

Fig. 2.19 Conversion of a point vector to raster format: **a** Point vector; **b** Superposition of the point vector to the raster grid; **c** Raster generated from the points, with each pixel having attribute corresponding to the value of the equivalent point (an undefined pixel is identified, to which two points simultaneously correspond)

Figure 2.20 illustrates the conversion of lines to raster using approaches according to these two ideas. The first approach leads to more pixels representing each line in the raster (Fig. 2.20c, d). In contrast, the second approach induces fewer pixels, with a higher occurrence of pixels aligned diagonally (Fig. 2.20e, f).

The conversion in the opposite direction, from raster to line vector, also involves decisions, and different methods exist for this; it is not the aim of this book to detail them. However, Fig. 2.21 illustrates such a conversion. In this example, a simplistic rule was considered to generate straight line segments for neighboring pixels orthogonally to each other and, once these possibilities are exhausted, generate straight line segments corresponding to the pixels aligned diagonally—always having

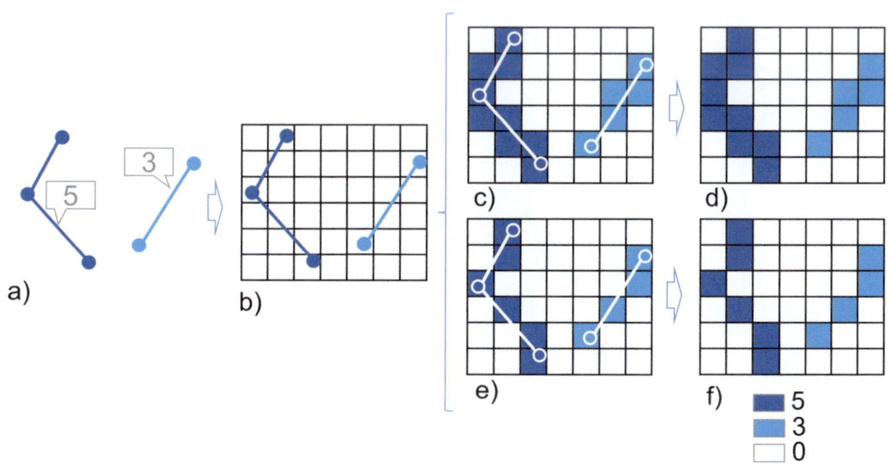

Fig. 2.20 Conversion of a line vector to raster format: **a** Line vector; **b** Overlay of the line vector on the raster grid; **c**, **d** Raster generated from the lines, with a procedure to include in the raster representation all pixels touched by the lines; **e**, **f** Raster generated from the lines, with a more restrictive procedure, to include in the raster representation the pixels whose alignment follows the main path of the lines

the start and end points of the straight line segments located at the centroids of the pixels. By this approach, the figure of this example illustrates how the conversion of each raster (originating from the same vector) leads to new vectors quite distinct from the original vector and different from each other, considering the two methods used to convert from vector to raster.

Converting a polygon vector to a raster involves decisions and criteria, with different methods available. For example, a criterion related to including the pixel as representative of the polygon may exist if the centroid of the pixel is included in the polygon or if a certain minimum fraction of the pixel area is part of the polygon. Different approaches with distinct tolerance levels for including pixels lead to different results (Fig. 2.22).

The opposite direction of the conversion, from raster to polygon, also allows variations of methods (Fig. 2.23). A widely used approach aims at obtaining polygons with simplified geometries, minimizing the number of straight line segments needed to represent the raster pixels (Fig. 2.23b, c). An alternative to this method is for the polygon to be traced strictly following the external vertices of the pixels of the raster border that the polygon represents.

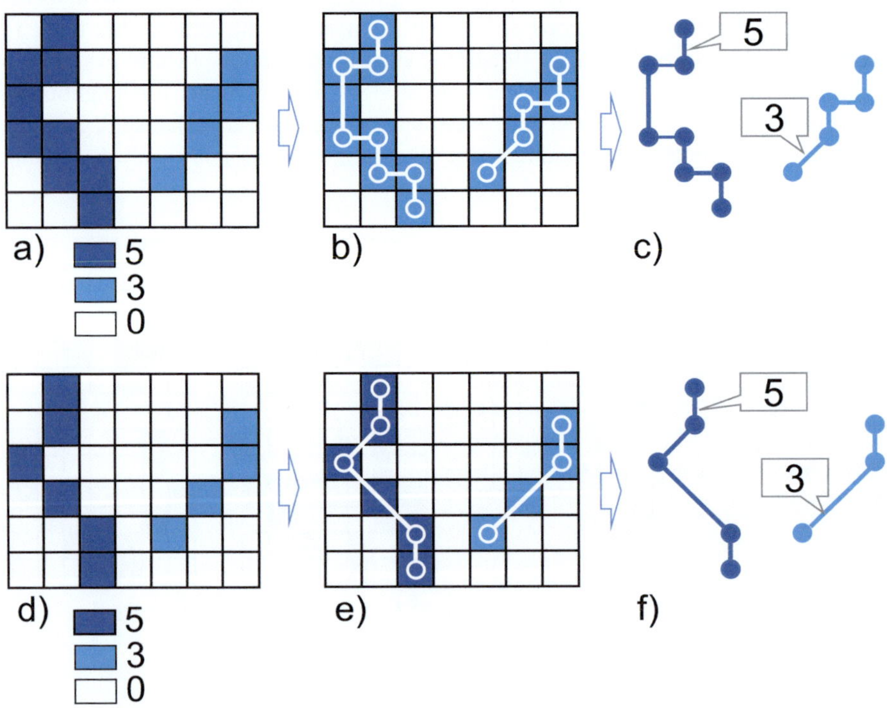

Fig. 2.21 Conversion of a raster into a line vector: **a, d** Rasters to be converted; **b, e** Overlay of the raster and the corresponding line vectors generated; **c, f** Line vectors generated from the rasters, with an indication of the attributes

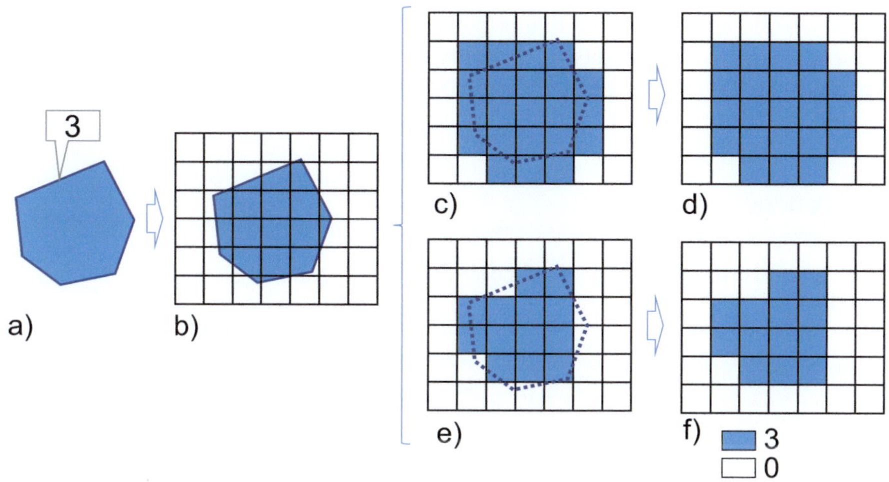

Fig. 2.22 Conversion of a polygon vector to raster format: **a** Polygon vector; **b** Overlay of the vector on the raster grid; **c**, **d** Raster generated from the polygon considering the inclusion in the raster representation of all pixels 'touched' by the polygon; **e**, **f** Raster generated from the polygon considering the inclusion in the raster representation only the pixels whose centroid was covered by the polygon

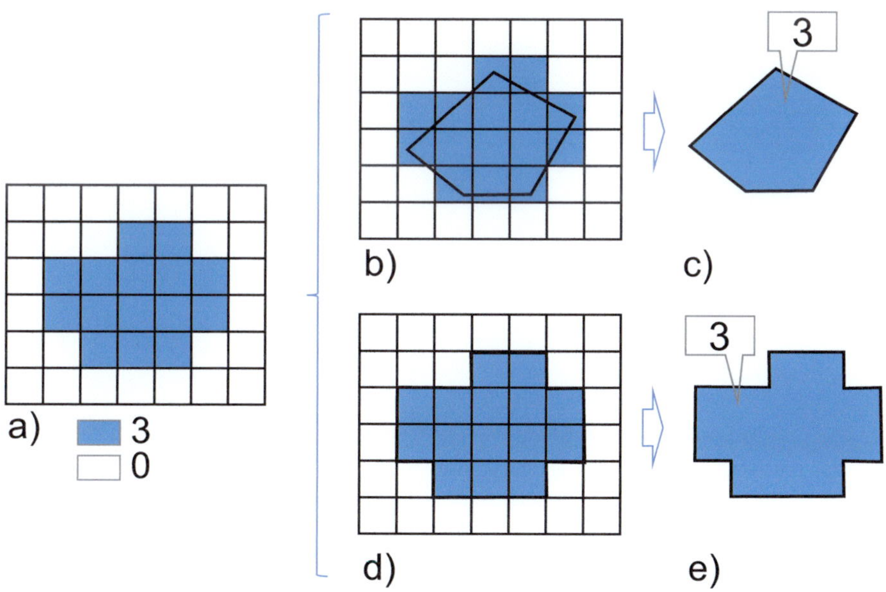

Fig. 2.23 Conversion of a raster into a polygon vector: **a** Raster to be converted; **b**, **c** Polygon vector generated from the raster, considering a procedure to obtain a simplified geometric shape; **d**, **e** Polygon vector generated from the raster, considering a procedure to follow the vertices of the pixels of the raster border

The second approach leads to a polygon with a pixelated edge shape (Fig. 2.23) due to the edges of the pixels. If this polygon is converted back to a raster format (Fig. 2.23-d), it results in a raster identical to the original (Fig. 2.23-e).

References

Fisher, P. (1997). The pixel: A snare and a delusion. *International Journal of Remote Sensing, 18*, 679–685.

Lu, Z., Qu, Y., & Qiao, S. (2014). *Geodesy: Introduction to geodetic datum and geodetic systems* (p. 401). Springer.

Silva, I., & Segantine, P. C. L. (2015). *Topografia para Engenharia* (1st ed.). Elsevier.

Chapter 3
Basic Operations with Raster Data

The primary way to digitally represent topography is in raster format. Whether a raster represents a Digital Elevation Model or a raster that represents another variable, such as average annual precipitation or concentration of a particular atmospheric pollutant, there are a series of standard operations usually performed in GIS software. Procedures such as windowing, statistics generation, frequency histogram, arithmetic operations, and spatial filtering, among others, are detailed in this chapter.

The procedure for each operation involving raster data is didactically described. Examples allow the reader to understand what is behind the operation when using GIS software.

3.1 Cropping Raster Data

It is common for the user to have an image with spatial coverage much larger than the area of interest. One can crop the image to avoid storing and unnecessarily processing a large amount of data. An image is generated with the exact spatial resolution but with fewer rows and columns, and consequently, a smaller spatial extent. Considering the raster as a matrix, the cropping operation involves removing specific numbers of rows and/or columns from the original matrix to generate a second matrix. This operation is also referred to as raster windowing in some software.

In Fig. 3.1's example, the original raster has seven rows and nine columns. A crop was performed to eliminate two columns at the left end, a column at the right end, and a row at the top and bottom. The resulting raster from this cut has five rows and six columns.

Cutting a raster is a trivial operation, usually available in any geoprocessing software. It is common for the user to be able to specify the geographical coordinates that define the window of interest or identify the position of the cut in terms of rows and columns of the original image (for example, from row 240 to row 1020 and from

A. Rolim da Paz, *Digital Elevation Models for Environmental Studies*,
https://doi.org/10.1007/978-3-032-04523-2_3

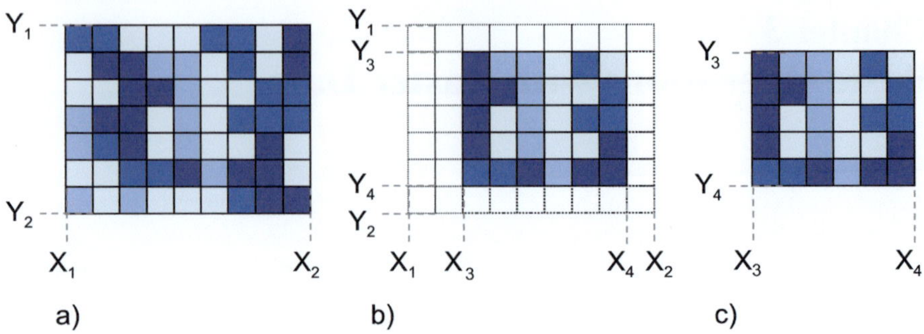

Fig. 3.1 **a** Raster of spatial extent defined by the coordinates X_1, X_2, Y_1, Y_2; **b** Cut of this raster according to the specification of the new spatial extent (coordinates X_3, X_4, Y_3, Y_4), indicating the elimination of three columns and two rows; **c** Cut raster, with the same attributes as the original image within the extent of the coordinates X_3, X_4, Y_3, Y_4

column 815 to column 2321). Some software has convenient options for the user, such as cutting a raster to the area shown on the screen; that is, the user adjusts the zoom of the raster view to be cut, and the area on display defines the spatial extent of the cut to be generated. Another approach is to indicate the spatial extent of the cut by drawing a rectangle on the viewing screen of the original raster.

When there are already other images cut for a specific window of interest, some software allows for extracting information that defines the window from one of these images, that is, determining the extent of the cut by mirroring a pre-existing image (Fig. 3.2). This procedure is interesting in avoiding errors or imprecision related to typing coordinates or rounding and, in this way, guarantees cuts of identical spatial extent between several images. This can be a fundamental prerequisite for a series of analyses that may be done involving two or more of these images or for these images to serve as input for some mathematical modeling software.

3.2 Raster Concatenation

Another essential operation in geoprocessing is image concatenation, which consists of deriving a new image from the spatial combination of two or more images. This is very common when the user's area of interest is partially inserted in more than one scene (or tile) from a particular data source. The user obtains the individual scenes and combines them to compose a single image sufficient to cover the entire area of interest.

In the pure concatenation process, no arithmetic operation is performed between the attributes of the pixels from the individual images. In areas where only one of the images provides pixel coverage, the attributes from that image are carried over to compose the resulting concatenated image. In the example shown in Fig. 3.3, the

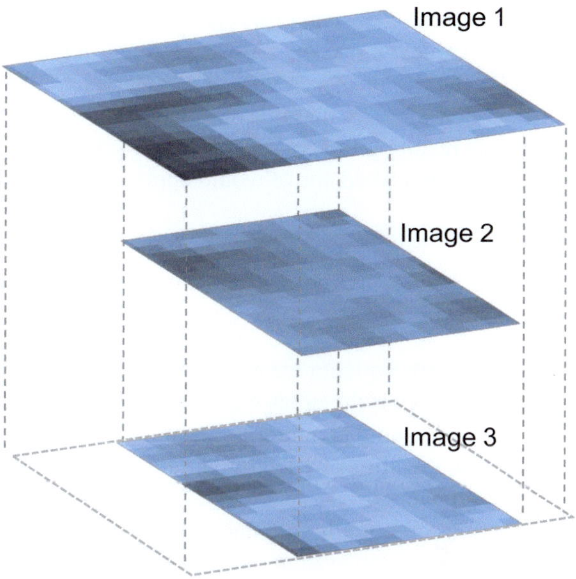

Fig. 3.2 Windowing a raster (image 1) from the spatial extent defined by another raster (image 2), resulting in image 3, composed of pixels with the same attributes as image 1, but eliminating the pixels in areas not covered by image 2

area marked as A contains pixels whose attributes come solely from Image 1. In contrast, the area marked as B corresponds to attributes exclusively from Image 2.

If areas are simultaneously covered by two or more of the individual images, the typical procedure is to define a hierarchical priority among the images. The attributes of the pixels from the image with the highest priority are used to compose the resulting image. In the example in Fig. 3.3, the area marked as C would contain pixel values from either Image 1 or Image 2, depending on the user-defined image priority.

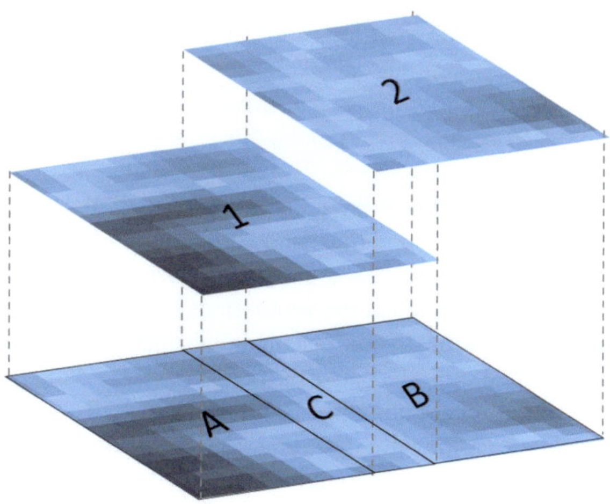

Fig. 3.3 Concatenation of images 1 and 2 generating a third image, which is composed by a part that only existed in image 1 (part A), whose attributes are identical to this image; a part that only existed in image 2 (part B), whose attributes are precisely those existing in this image; and a part that existed in both images 1 and 2 (part C), whose attributes are either equal to image 1 or image 2, according to the decision rule that is applied

Alternatively, a decision rule can be applied to overlapping areas that ignore zero or invalid values in one or more images and retain the valid values from the other image. In area C of the previous example, each pixel would be checked for valid values in both images, and if a valid value is found in only one, that value would be used in the resulting image. If both images contain valid values for a given pixel, the priority rule described above would apply.

3.3 Raster Areas

For a raster with a qualitative attribute, one of the most demanded fundamental analyses is quantifying the areas occupied by each attribute value. This operation can be performed simply by counting the number of pixels of the same value and multiplying these quantities by the individual pixel area to obtain the area quantification in planar area units.

Adopting a unique individual area for all pixels is valid when the raster is in a projected coordinate system, such as the UTM system. In this case, there is a flat representation of the Earth's surface, with linear units of measure for the coordinates—usually in meters, and all pixels have the exact dimensions and area in square meters.

Suppose the raster is in a geographic coordinate system, in which a three-dimensional geometric surface, such as an ellipsoid, is used to locate each point on Earth in terms of latitude and longitude. In that case, the unit of the coordinates is the degree. For example, a raster with a spatial resolution of 0.001°. In this case, each pixel has the exact dimensions in degrees but has variable dimensions in terms of equivalence to distances in meters or kilometers, depending on the location of this pixel on the Earth's surface. This is due to the variation of equivalence between measures in degrees and arc lengths.

For example, 1° of longitude at the Equator equals about 111.32 km. In comparison, 1° of longitude at 60° latitude corresponds to 55.80 km, using the WGS84 ellipsoid of revolution as a reference for representing the Earth's surface. Chapter 4 (item 4.6) details how to make this equivalence between angular units and distances in kilometers.

Figure 3.4 shows the delineation of the Tapajós River watershed in the Midwest and North of Brazil in the raster format with a spatial resolution of 0.002°. This basin has an elongated north–south shape, corresponding to a latitude variation of almost 13°. A pixel located at the southern end of the basin has an approximate surface area of 0.047945 km^2. In contrast, at the northern end, the area of a pixel is 0.049529 km^2, a difference of 3.3% between them.

An example of quantifying areas of a raster is illustrated in Fig. 3.5. There is a land use classification with three classes (forest, agriculture, and pasture) and unclassified pixels with a zero attribute. The image has a spatial resolution of 30 m; therefore, each pixel has an area of 900 m^2 ($= 30 \times 30$). There are 42 pixels in this image (6 rows $\times$ 7 columns), with the quantities of pixels belonging to each class as indicated

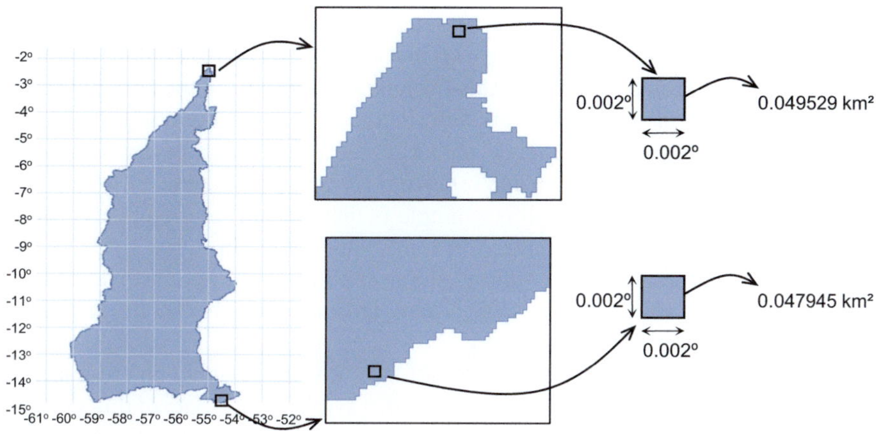

Fig. 3.4 Delineation of the Tapajós River watershed in raster format, with a spatial resolution of $0.002° \times 0.002°$, and indication in the details of a pixel located near the northern limit of the basin and another pixel situated near the southern end of the basin, with their respective areas in planar units of these pixels

in the figure. To obtain the areas occupied by each class, such as forest, agriculture, and pasture, multiply the count of pixels belonging to each class by the individual area of a pixel (900 m^2). For example, the class called forest has 12 pixels, whose area is equivalent to $12 \times 900 = 10{,}800$ m^2.

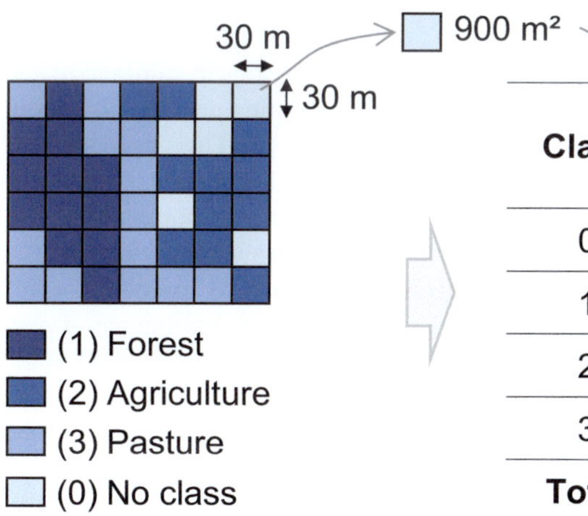

Class	Area	
	Pixels	m²
0	6	5,400
1	12	10,800
2	11	9,900
3	13	11,700
Total	**42**	**37,800**

Fig. 3.5 Raster of land use classification with four classes and spatial resolution of 30 m, with an indication of the number of pixels belonging to each class and the corresponding total areas of these pixels in planar units; thin arrows indicate the example for the forest class, whose area of 10,800 m^2 is obtained by multiplying 12 pixels by 900 m^2 (area of each pixel)

In the example of Fig. 3.5, it should be noted that the pixels belonging to the forest class are all contiguous, forming a single set in which all pixels are neighbors of others of the same class. However, some pixels form isolated groups in the agriculture and pasture classes. Even so, all of them were included in the accounting of the area occupied by the class. What matters is the basic rule: quantify the pixels of the same value, regardless of their location and relative position among them.

Another point to highlight in this example is that the unclassified pixels, which have a zero attribute in the image, were also quantified. In some software, the command for area quantification allows the user to mark an option that indicates whether or not to ignore pixels with a zero attribute, which can be interesting depending on the purpose of the survey.

A distinct situation is when pixels do not have a valid attribute due to a lack of data. For example, the user treats the image with a symbolic attribute of this situation for such pixels, which is known as a flag of the type null value. In quantifying the areas occupied by each class, the invalid pixels can also be ignored or quantified, depending on the user's interest and the software used.

When no pixels are disregarded, either because they have a zero attribute or are considered invalid, the quantification of raster areas must necessarily result in areas whose sum corresponds to the total number of pixels in the image. In the example of Fig. 3.5, the sum of the forest, agriculture, and pasture classes with the unclassified pixels equals 42, which is the dimension of the raster in pixels (6 rows $\times$ 7 columns).

3.4 Descriptive Statistics of a Raster

For a given dataset, users are often interested in applying statistics to summarize and describe the data through descriptive measures. These statistics provide an overall view and can serve as a basis for comparison with other datasets. This approach falls under descriptive statistics and includes measures of central tendency, dispersion, skewness, kurtosis, and positional measures such as percentiles and quantiles.

Determining these descriptive measures for a raster of p rows and q columns is relatively simple, analogous to any other dataset. A straightforward approach for this is by treating the values of the raster pixels as a one-dimensional sequence of data (Fig. 3.6), from the first element of the raster (positioned in row 1, column 1) to the nth element (located in row p, column q). Standard mathematical formulations are applied to this unidimensional dataset to compute statistical parameters such as the mean, mode, and median. The main parameters are listed in Table 3.1

In the standard form, the statistics are determined for the entire set of pixels of the raster, that is, considering the whole image. However, the user may be interested in having a statistical characterization for one or more specific regions of the raster. Depending on the software used, this operation can be done by defining these areas of interest in the vector format of polygons or as a second raster. In the first case, the statistics are generated for each polygon or set of polygons with the same attribute. Generally, a numeric identifier of the areas of interest is used for this purpose.

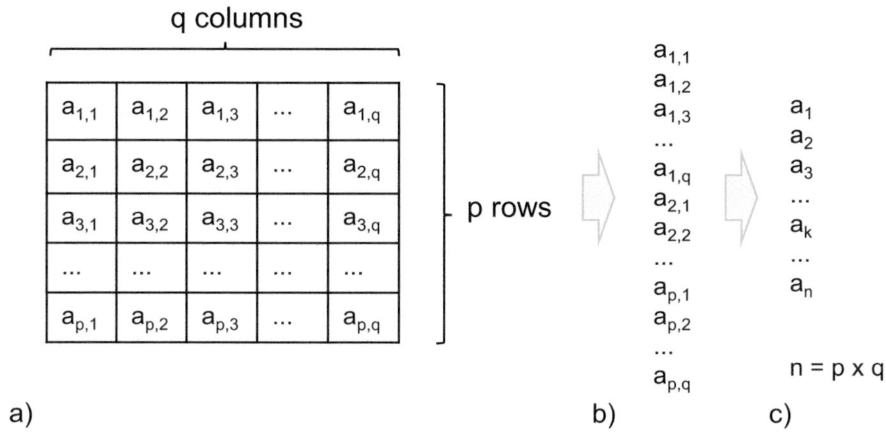

Fig. 3.6 **a** Matrix structure of a raster with p rows and q columns, formed by elements named by a_{ij}, where i indicates the row position and j the column position; **b** Ordering of all the elements a_{ij} of the raster into a unidimensional sequence of values; **c** Renaming of the raster elements to a unidimensional sequence a_k of values, where k is the position in the sequence, which varies from 1 to n, being n the total number of elements of the raster ($n = p \times q$)

In the second case, the statistics are determined by separately analyzing the pixels of the image according to the correspondence with the pixels of the raster that defines the areas of interest. Statistics are generated for the image under analysis only by taking the pixels referring to a specific area indicated in the second raster. Then, the procedure is automatically repeated for each of the other regions.

The procedure for obtaining the average, maximum, and minimum statistics of a raster of terrain slope, both in analyzing the entire image together and separately based on specific regions defined by a second raster, is illustrated in Fig. 3.7. In the second option, notice that all pixels representative of a given region have an attribute equal to the value of the region identifier. This characterizes which pixel enters the computation of the statistics of each region.

3.5 Reclassifying a Raster

Reclassifying a raster, also known as segmentation or slicing, consists of generating a new raster with attributes determined by a set of reclassification rules of the attributes of an existing raster. This existing raster can have a quantitative or qualitative attribute type. The resulting raster has the same characteristics as the reclassified raster regarding spatial coverage, number of rows and columns, and spatial resolution. The attribute of the output raster can be generated with quantitative or qualitative information.

Table 3.1 Relationship and description of the main types of measures obtained in the descriptive statistics analysis applied to a raster

Type of measure	Statistical parameter	Definition
Central tendency	Mean	Average value of the raster pixels
	Mode	The value that appears with the highest frequency among the raster pixels
	Median	The value that occupies the central position among the ordered values of the raster pixels, from the lowest to the highest
Dispersion	Maximum	Maximum value among all raster pixels
	Minimum	Minimum value among all raster pixels
	Range	Difference between the maximum and minimum values among the raster pixels
	Standard deviation	Measure of the degree of dispersion of the raster pixel values relative to the mean, obtained by the square root of the sum of the squared deviations divided by the number of pixels
	Variance	Measure of the degree of dispersion of the raster pixel values relative to the mean, obtained by squaring the standard deviation
	Coefficient of variation	Relative measure of dispersion of the raster pixel values, given by the ratio between the standard deviation and the mean
Quantiles	Quartile	Values that divide the set of ordered raster pixel values into four equal parts. The 1st quartile indicates that 25% of the pixels have a value less than or equal to it. The 2nd quartile indicates that 50% of the pixels have a value less than or equal to it. The 3rd quartile indicates that 75% of the pixels have a value less than or equal to it
	Decile	Values that divide the set of ordered raster pixel values into ten equal parts. The 1st decile indicates that 10% of the pixels have a value less than or equal to it, while the 2nd decile represents 20%. The 9th decile indicates that 90% of the pixels have a value less than or equal to it
	Percentile	Values that divide the set of ordered raster pixel values into one hundred equal parts. The 1st percentile indicates that 1% of the pixels have a value less than or equal to it, while the 2nd percentile represents 2%. The 99th percentile means that 99% of the pixels have a value less than or equal to it
Skewness	Pearson's skewness coefficient	Indicates the shape of the distribution of raster pixel values relative to the mean
Kurtosis	Kurtosis measure	Indicates the degree of flatness or peakedness of the distribution of raster pixel values relative to the mean

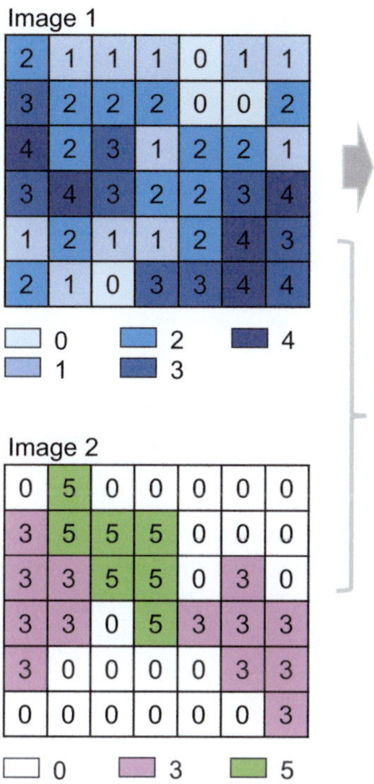

Statistics of the entire image 1		
Maximum	Minimum	Average
4	0	2.02

Region	Statistics of image 1 by region defined in image 2		
	Maximum	Minimum	Average
0	4	0	1.5
3	4	1	3
5	3	1	1.86

Fig. 3.7 Obtaining descriptive statistics average, maximum, and minimum for a raster (image 1) considering all its elements and the same type of statistics generated considering regions defined by another raster (image 2); in the second case, for each region defined by a set of pixels with the same attribute, the corresponding statistics are generated

The key issue of this procedure is the appropriate establishment of the reclassification rules, which must satisfy the user's interests and meet logical issues. Each rule is individually composed of two parts. The first part defines a range of attribute values to which this rule applies. The second part indicates the new value to be adopted when the corresponding rule is met.

All the pixels of the image are analyzed to see if they fit into some reclassification rule. If a pixel has an attribute between the lower and upper value established by a rule, that pixel has the value indicated by that rule in the resulting image. If the pixel does not fit into any rule, its original attribute goes to the resulting image. Some software allows the user to specify that, in this situation, the pixel has a zero or invalid value.

In the example of Fig. 3.8, a raster that presents the average annual rainfall as an attribute is reclassified. Five reclassification rules are adopted, and the resulting attribute of each rule is merely the number of the corresponding class, a qualitative variable. The reclassified image no longer has the information of average annual rainfall as an attribute, but rather the value of the classes of values of this variable. In this case, all the resulting image's pixels have an attribute varying from 1 to 5. Based

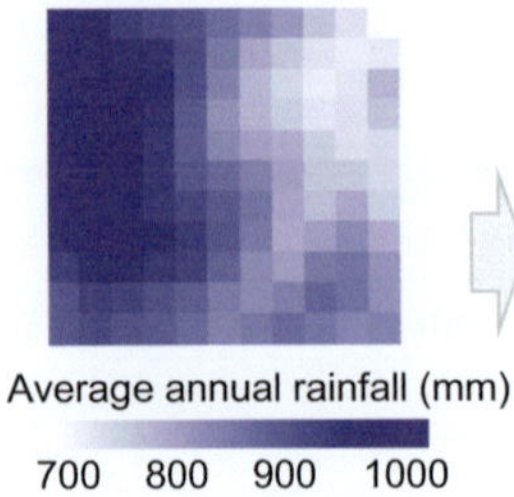

Reclassification

Lower limit (mm)	Upper limit (mm)	New value
750	800	1
800	850	2
850	900	3
900	950	4
950	1000	5

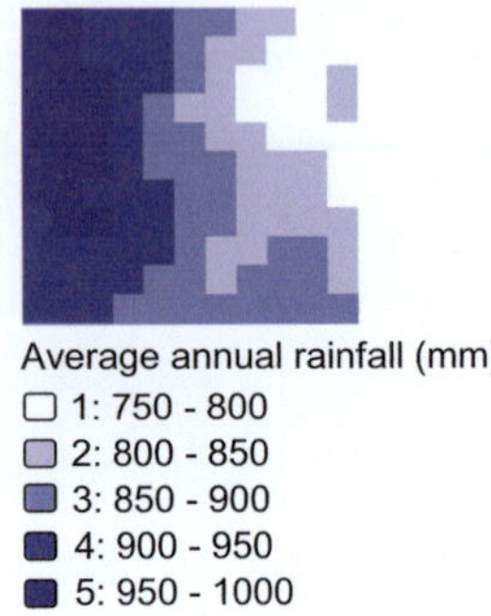

Fig. 3.8 Reclassification of a raster whose attribute is the average annual rainfall, considering five classes defined based on the lower and upper limits indicated and the corresponding resulting raster

on the legend associated with the raster, it is possible to interpret the spatial gradient of precipitation in this area. However, the reader should realize that there is no longer a distinction in how the rainfall varies among the same class's pixels. There was a simplification in the representation of the phenomenon's spatial variability. This simplification is more intense the wider the intervals adopted for each class.

A common mistake is the user not realizing that the established rules do not cover all the values of the image to be reclassified, resulting in non-reclassified pixels. For example, if there were pixels with an attribute lower than 750 mm in the image, there would be pixels in the resulting image that did not belong to any of the classes, which would remain with the original rainfall value in mm. The resulting image would be weird, with pixels with class values (1–5) and others with an annual rainfall value in mm.

Having pixels that do not fit into any reclassification rule may not be a mistake if this is done intentionally, with the user's interest in not changing some of the original values of the image. For example, suppose that the land use classification of Fig. 3.9 initially differentiated areas of low and high urban density. However, during the study, it was decided to treat them equally, as just a broader class reported as an urban area. In this case, the reclassification operation needs a single rule to convert the pixels of class 2 to class 3 (or vice versa) without changing the other pixels of the image.

Another frequent error is to have an intersection between the reclassification rules when specific values of the input raster fit into more than one rule. In this case, one of the rules will prevail, depending on how the software operates. For example, some software may consider the last rule the pixel fits into as the one to be applied. However, it is not expected that reclassification rules with intersections between them are intentionally established.

Reclassification

Lower limit	Upper limit	New value
2	2	3

Fig. 3.9 Reclassification of a raster whose attribute is land use type, to convert pixels classified initially as class 2 ('high-density urban area') into class 3; in the resulting raster, class 3 was renamed in the legend simply as 'urban area', as it now includes both the 'high-density urban area' pixels and the 'low-density urban area' pixels from the original image; note that pixels from other classes were not modified, since the reclassification rule only applied to class 2

3.6 Frequency Histogram

In the previous item, we saw how reclassifying a raster works. Earlier in this chapter, the logic behind calculating areas of a raster was shown. Determining a frequency histogram of a raster's attributes combines concepts used in the reclassification and area determination processes. The idea is the same as that used to obtain a histogram of any tabular dataset. The intervals of the classes to construct the histogram should be defined, and then the number of pixels whose attribute belongs to each class should be counted.

In the image shown in Fig. 3.10, we have the average monthly rainfall in millimeters over a particular region. Note that the image's attribute is of quantitative magnitude with decimal values. Five classes were established to elaborate the frequency histogram of this image, with ranges of 10 mm amplitude each. For each established interval, the number of corresponding pixels was counted (2nd column of the table). For example, there are 16 pixels with an average monthly rainfall between 180 and 190 mm. However, does this represent a high or a small portion of the region?

Notice that using absolute occurrence counts makes understanding the distribution of values harder. Moreover, it's not the most effective way to present results in a technical or scientific report, let alone compare them with other studies' findings. Other analyses may have been done for the same region and period, but with images of different spatial resolutions from the example. This would prevent comparison in terms of absolute quantities per class. It is more interesting to work with relative frequencies in percentages of the total pixels of the image (3rd column). In the 180 to 190 mm class, 38.1% of the image's pixels (or 38.1% of the region) fit into this class. It provides more complete information and can be interpreted and cross-referenced with other data.

Image 1

212.3	203.5	199.3	189.9	186.2	180.1	181.9
214.1	200.1	188.9	177.6	174.5	181.6	183.0
217.0	206.7	195.0	175.8	177.3	180.7	185.8
209.4	194.8	178.1	175.5	182.4	188.6	182.5
189.1	177.2	171.5	182.0	181.6	194.5	190.9
172.6	163.8	150.3	174.8	179.0	189.3	195.2

Class	Quantity per class		Accumulated quantity	
	Number of pixels	Percentage of the total (%)	Number of pixels	Percentage of the total (%)
150 – 160	1	2.4%	1	2.4%
160 – 170	1	2.4%	2	4.8%
170 – 180	11	26.2%	13	31.0%
180 – 190	16	38.1%	29	69.0%
190 – 200	13	31.0%	42	100.0%
Total	42	100.0%	42	100.0%

Fig. 3.10 Quantification of the frequency of occurrence of pixels of a raster (image 1, whose attributes are average monthly rainfall in mm) according to ranges of attribute values (classes), in absolute terms (number of pixels) and regarding the percentage of the total image; it also shows the accumulated quantities along the sequence of classes, both in absolute terms and in percentages of the total

This example also determined the accumulated quantity of pixels along the classes (4th column). The class 180–190 mm includes 29 pixels in lower classes or inside this range ($= 1 + 1 + 11 + 16$). More understandable information is obtained when converting these quantities to percentages of the total (5th column): 69% of the image's pixels have an attribute less than or within the 180–190 mm range.

The graphical construction of the frequency histogram allows for a better visual understanding of the data distribution and a better identification of patterns than the direct observation of the numbers in the table. For the example discussed earlier, Fig. 3.11 shows the graphs of the histograms in terms of both absolute and relative quantities. Each shows the histogram of the amounts per class and the curve of the accumulated quantities.

If an image has a qualitative magnitude as an attribute, such as types of soils, for example, the frequency histogram usually involves classes of unique values instead of ranges. In the example of Fig. 3.12, the image's attribute assumes values 0, 1, 2, 3, or 4. The corresponding histogram was elaborated by counting the occurrence of pixels with an attribute equal to each value. The rest of the procedure concerning the

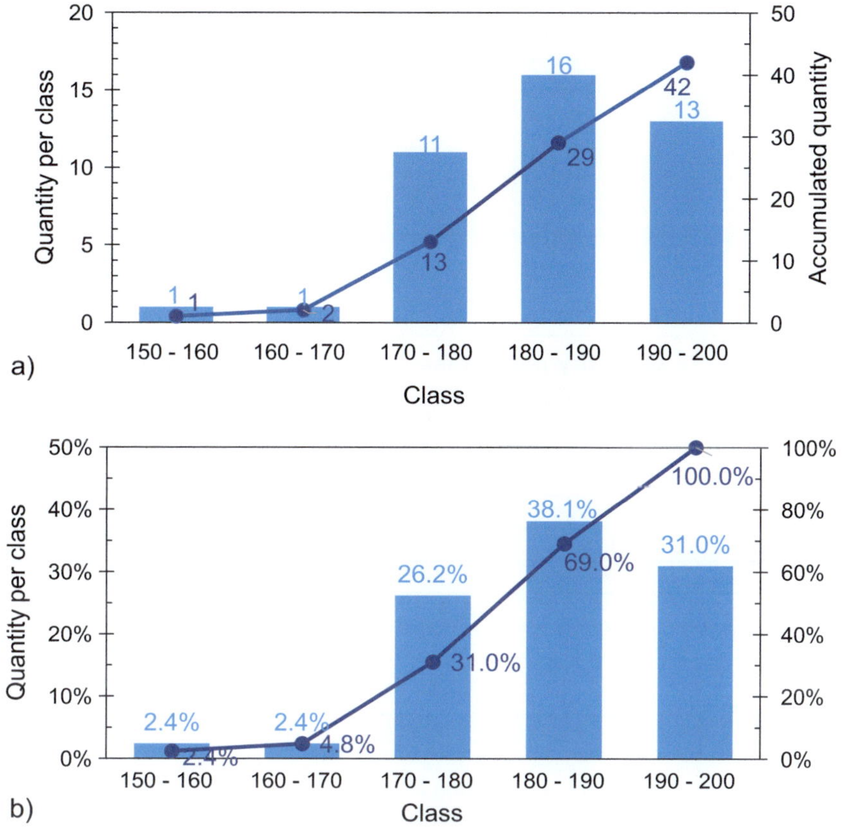

Fig. 3.11 Frequency histogram of pixel occurrences in a raster (image 1 in the previous figure; the attribute represents average monthly rainfall in mm), based on ranges of attribute values (classes), presented in absolute terms, as pixel counts (**a**), and in relative terms, as percentages of the total (**b**)

calculation of absolute, relative, and even cumulative quantities is analogous to what was described for the previous example.

3.7 Changing Spatial Resolution

A fundamental characteristic of a raster is its spatial resolution. For reasons such as overlaying with other datasets, reducing data storage space, or reducing processing cost, the spatial resolution of a raster can be changed. The most common approach is to coarsen the spatial resolution, that is, to increase the pixel size. For example, a raster that has a 30 m spatial resolution can be changed to a spatial resolution of 120 m. In this case, there is a coarsening factor equal to 4 (=120/30), which indicates that the raster resulting from the operation has reduced the number of rows and the

Image 2

2	1	1	1	0	1	1
3	2	2	2	0	0	2
4	2	3	1	2	2	1
3	4	3	2	2	3	4
1	2	1	1	2	4	3
2	1	0	3	3	4	4

Class	Quantity per class		Accumulated quantity	
	Number of pixels	Percentage of the total (%)	Number of pixels	Percentage of the total (%)
0	4	9.5	4	9.5
1	11	26.2	15	35.7
2	13	31.0	28	66.7
3	8	19.0	36	85.7
4	6	14.3	42	100.0
Total	42	100.0	42	100.0

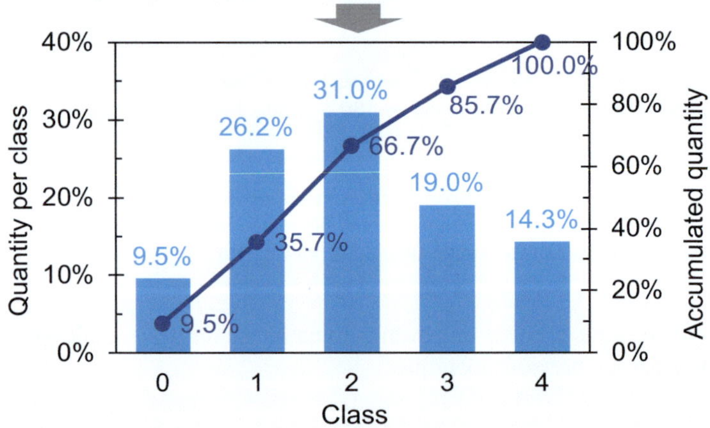

Fig. 3.12 Quantification of the frequency of occurrence of pixels of a raster (image 2, whose attributes are types of land use) in each class, in absolute terms (number of pixels) and regarding the percentage of the total image; it also shows the accumulated quantities along the sequence of classes, both in absolute terms and in percentages of the total, and the frequency histogram in terms relative to the total

number of columns by 4 times. This means a reduction in the number of pixels equal to 16 times.

In some cases, it is desired to refine the spatial resolution, reducing the pixel size. For example, from a raster with a 30 m resolution, another raster with a resolution of 10 m is generated. In this case, there is an expansion factor equal to 3 (=30/10). Changing the resolution with a coarsening or expansion factor that is not an integer value is also possible, such as going from 30 to 100 m (coarsening factor equal to 3.33) or 30 to 20 m (expansion factor equal to 1.5).

The coarsening of spatial resolution consists of generating a new raster in which each pixel has an attribute calculated based on a set of pixels from the original image. Suppose the operation corresponds to a coarsening factor of an integer value. In that case, if there is a multiplicity between original and coarsened resolutions, this calculation is generally done by the average, mode, minimum value, maximum value, or central pixel (Fig. 3.13). In the example, one pixel of the coarsened image corresponds to nine pixels of the original image.

The average method assigns the arithmetic average value of the attributes of the corresponding pixels in the original image to the coarsened pixel. It is a recommended approach for quantitative magnitudes, such as average annual rainfall, where it makes sense to calculate an average, but not for qualitative magnitudes, such as soil types. In the example of Fig. 3.13b, the arithmetic average of nine pixels is the value assigned to the pixel of the coarsened image.

In the mode method, also known as *pixel majority*, the coarsened pixel receives the value that repeats the most among the refined pixels to which it corresponds

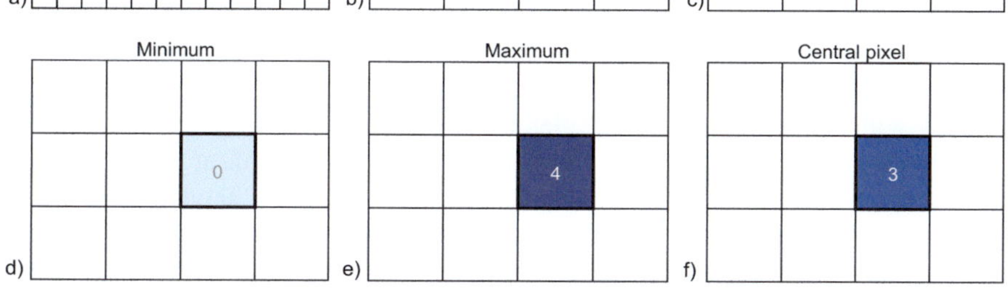

Fig. 3.13 Example of methods of spatial resolution coarsening of a raster, by a coarsening factor equal to 3: **a** Raster to be coarsened, with the indication of the attributes of some pixels; **b–f** Rasters generated by the coarsening of the spatial resolution of the raster indicated in (**a**), via methods of the average, mode, minimum, maximum and central pixel, respectively; the value indicated in the rasters **b–f** refers to the pixel of the coarsened image corresponding to the nine pixels highlighted in the original image

(Fig. 3.13c). In this example, attribute 1 is the value that repeats the most and, therefore, is considered the attribute of the corresponding pixel in the coarsened image. This operation of spatial resolution coarsening by the mode method is not recommended for a raster with a quantitative magnitude, especially if the values are decimal numbers.

The methods of minimum and maximum assign, respectively, the minimum or maximum value among the pixels of the original image to the coarsened pixel (Fig. 3.13d, e). The pixel thinning method establishes that the attribute of the coarsened pixel is the pixel's value located systematically in a given position in the corresponding window in the original image. For example, it may choose the central pixel of the original image regarding the spatial extent covered by the coarsened pixel or the original image's pixel in the lower right corner of such an extent. When the central pixel is chosen, the pixel thinning approach is called the central pixel method.

In the example of Fig. 3.13f, the pixel of value 3 of the refined image is located in the central position relative to the window that defines the pixel of the coarsened image. Suppose there is no central pixel (for example, a coarsened pixel covers four rows and four columns of pixels of the original image). In that case, additional rules exist in the software to choose which pixel to take the attribute from. An example is the identification of the central pixel in the original image as the one whose centroid is closest to the centroid of the coarsened pixel. In this case, the central pixel method is also known as the nearest neighbor method.

Depending on the purpose of the operation and the user's interest, the methods of minimum value, maximum value, and central value can be applied to both quantitative and qualitative magnitudes. Interestingly, the mode, minimum, maximum, and central pixel methods are examples of approaches that preserve original values in the resulting image. In other words, these methods result in a coarsened image in which all pixel attribute values are exactly found in the original image. The average method, on the other hand, generates values different from those that exist in the original image. Some authors point out this as a disadvantage of applying this method, owing to artificially creating values that do not exist in the original information.

Another approach to coarsening an image's spatial resolution is using interpolation methods, such as bilinear interpolation, cubic convolution, or spline interpolation. Depending on the technique used, the attribute value of the pixel in the coarsened image is derived from a broader set of pixels in the original image than those formed by those with the exact spatial equivalence.

Interpolation is understood as estimating the value of a variable at a point in space based on the value of the points around that point, which is the neighborhood. As a result, the surface interpolated by methods such as bilinear, cubic convolution, and spline is smoother than that obtained by methods like the nearest neighbor method, and the calculated values do not necessarily exist in the original image. Therefore, this type of interpolation cannot be used for a qualitative magnitude. For a quantitative magnitude, any of these methods can be used in principle, according to the purpose of the operation, the studied variable, and the user's interest.

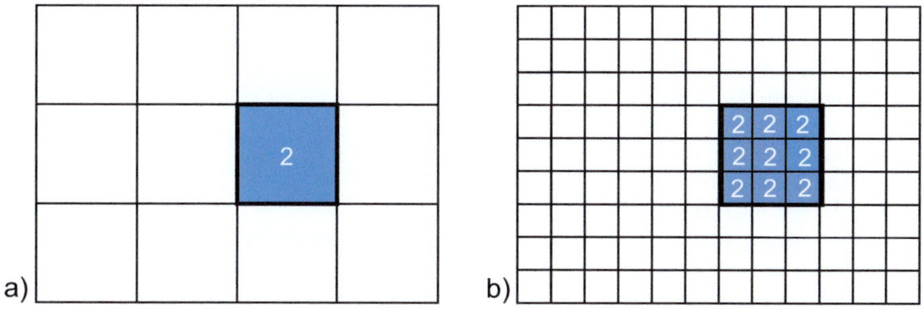

Fig. 3.14 Example of a method of refining the spatial resolution of a raster by an expansion factor equal to 3: **a** Raster to be refined, with an indication of the attribute of a pixel for exemplification; **b** Raster generated by the refinement of the spatial resolution of the raster indicated in (**a**), via a simplified method of repeating the pixel attribute of the original image for all corresponding pixels of the refined image

The inverse operation of refining the spatial resolution has, as its most usual method, simply the repetition of the attribute of the original image's pixel as the attribute of the pixels of the refined image. In the example of Fig. 3.14, the coarsened pixel has an attribute equal to 2, and nine pixels are generated in the refined image, all with this exact attribute value. This method considers that, as there is no sub-pixel information in the original image, no additional information is created. Only the quantity of rows and columns has been multiplied, but the same quality of information remains. In the mentioned example, the entire area of the coarsened image's pixel with attribute equals 2 also continues with this value, despite now being presented by 9 pixels instead of a single element.

Analogous to the case of spatial resolution coarsening, mathematical interpolation methods can be used to refine the spatial resolution of an image. Due to their mode of operation, these methods generate smoother surfaces. Visually, these surfaces appear more pleasing to the user, who sees more significant spatial variability and gradual transitions of values throughout the image. However, one must be cautious in this operation, as it artificially generates more detailed information than initially. The more pleasing visual effect may be merely aesthetic, not representing the reality of that variable in the raster.

3.8 Arithmetic Operations Between Rasters

Geographic data in raster format is handy for overlaying and combining different layers. Ideally, this requires that the rasters to be crossed have the exact spatial resolution and the same number of rows and columns, that is, the same spatial coverage. As a raster has a matrix structure, arithmetic operations such as addition, subtraction, multiplication, and division between two rasters involve pixel-to-pixel calculations:

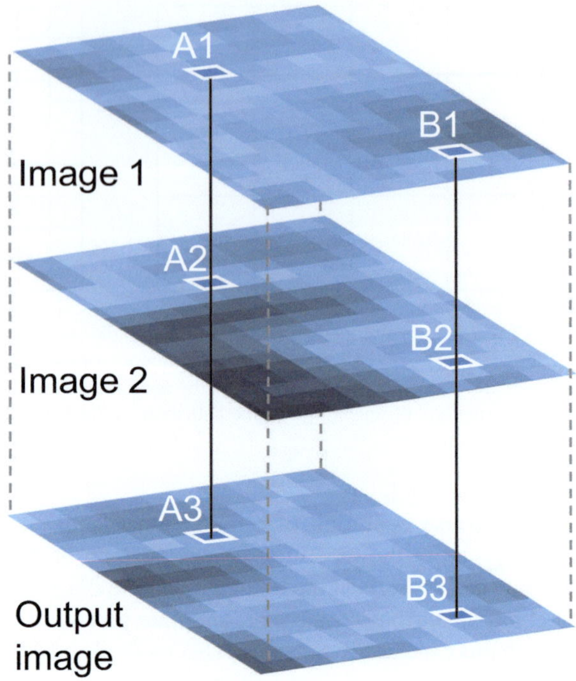

Fig. 3.15 Example of arithmetic operation between two rasters, which involves applying the operation pixel by pixel between the corresponding elements of image 1 and image 2, to generate the specific pixel in the resulting image: pixel A1 with pixel A2, to obtain pixel A3; pixel B1 with pixel B2, to obtain pixel B3

each pixel of one image with its corresponding pixel in the other image (Fig. 3.15). This is the standard procedure available in almost all GIS software.

For example, consider images 1 and 2 shown in Fig. 3.15. When performing an addition operation between these images, pixel A3 of the resulting image is determined by A3 = A1 + A2. Similarly, for pixel B3, we have B3 = B1 + B2, and so on, for all pixels in the image. The resulting image has the same number of rows and columns as the images used in the operation and the same spatial coverage and spatial resolution.

Some software, however, allows this type of operation between two images, even if there is a difference in spatial resolution between them or differences in spatial coverage. In other words, they allow operating rasters without an exact equivalence of one pixel from one image for one pixel of the other. Internally, in executing the mathematical operation, the software performs procedures to make decisions and ensure compatibility. For example, it identifies the correspondence between the pixels of the two images by the geographical location of each element. It decides not to perform the operation when a correspondence is not found.

In some cases, the user has access to options to configure this procedure. For example, the user can choose whether the resulting image should have spatial extension according to one of the two images used in the operation, or even if this coverage is the union of the coverages of the two rasters. Another example is the user deciding whether the spatial resolution of the resulting image should be equal to one of the two input images.

As governed by the properties of mathematics, the order of the images used in addition and multiplication operations between two rasters does not influence the result. In subtraction and division operations, however, the order of the images matters and changes the result.

The multiplication operation between rasters can be used as a trick so that a raster has pixels with a valid attribute only within a specific region of interest (Fig. 3.16). For this, there must be a raster defining this area of interest, composed of pixels with an attribute value of 1 within this area and pixels with a value of zero (or *null data*) outside this area. Image 2 in the earlier figure is an example of a raster used for this purpose.

This type of raster, like that of image 2, is also known as a mask. By multiplying the raster whose attributes you want to restrict to a particular region (image 1) by this mask (a raster of attribute 1 or 0, like image 2), you get a resulting image (image 3) with: (i) pixels of the same attribute as image 1, when these pixels are located in the area of interest (defined by the attribute value 1 in image 2); or (ii) pixels with zero attribute, when these pixels are not part of the region of interest (defined by the zero attributes in image 2).

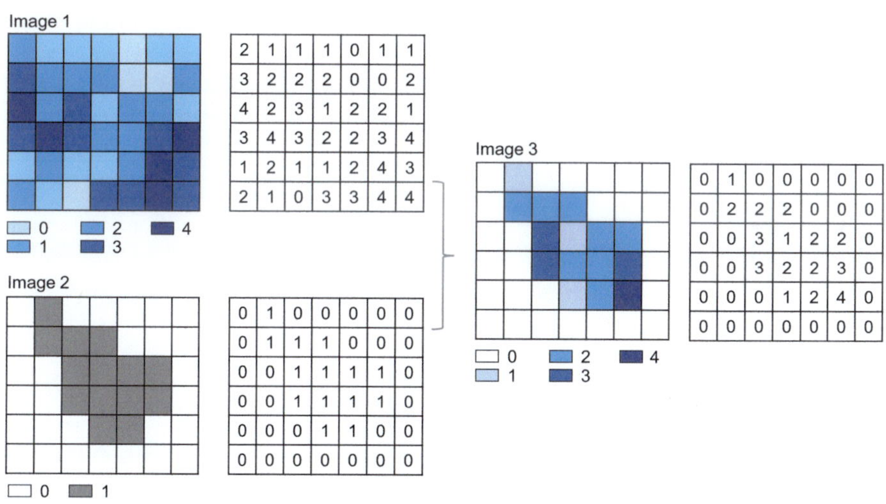

Fig. 3.16 Example of multiplication operation of a raster (image 1) by another (image 2), aiming to obtain a result (image 3) that consists of image 1 with valid pixels only corresponding to the pixels with attribute 1 in image 2, which defines a region of interest, or mask

3.9 Arithmetic Operations Between Rasters and Scalars

In some situations, the user wants to perform an arithmetic operation between a raster and a scalar quantity (a number). A typical example of this need is converting a raster's attribute unit. Suppose a raster represents the shortest distance from each point to some road and was obtained in meters (Fig. 3.17). To convert the unit of the attribute of this raster of distances to kilometers, divide the raster by 1000. The operation divides the value stored in each pixel by 1000 and stores the result as the attribute of a new raster. This operation can be seen as the operation between two rasters, replacing the scalar with a raster whose attribute is uniform for all pixels and equal to the scalar value (Fig. 3.18). The operation proceeds pixel by pixel, as already discussed in the previous item about the arithmetic between two rasters.

It should be noted that the arithmetic operations of addition, subtraction, multiplication, and division between a raster and a scalar change the value of the pixels but do not change the pattern of spatial variability of the variable represented in the raster. This happens because it involves a value (the scalar) constant for all pixels. In the example of Fig. 3.17, by maintaining the same color gradient adjusted for the range between the minimum and maximum value of the raster, both image 1 and the resulting image appear to have the same spatial patterns. When this operation was represented equivalently as the operation between two rasters in Fig. 3.18, it is evident that the spatial variability of image 1 is carried over to the resulting image since image 2 is spatially uniform.

Another important observation is to remember that storing the unit in the raster is impossible, but this can be recorded in the metadata of this layer to avoid confusion in subsequent data use.

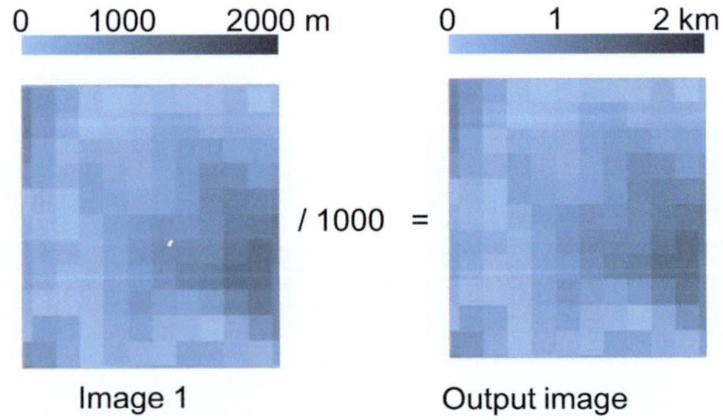

Fig. 3.17 Example of arithmetic operation of division between a raster (image 1) and a scalar (value 1000), resulting in a raster in which each pixel is the value of the corresponding pixel in image 1 divided by 1000 (resulting image)

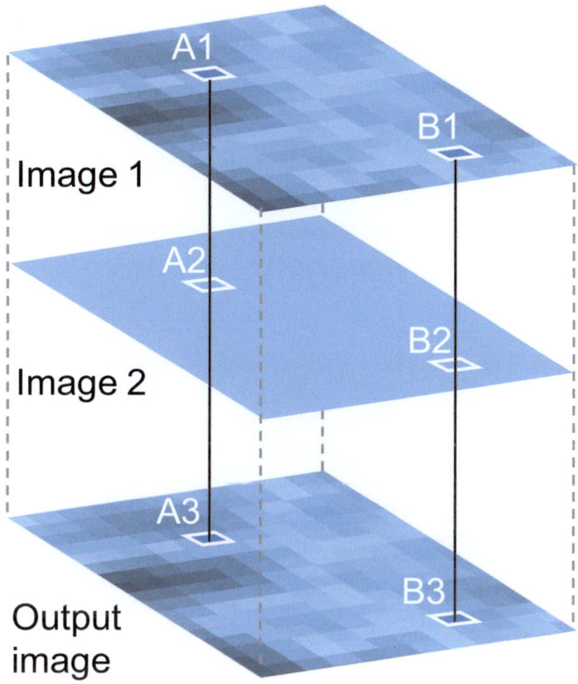

Fig. 3.18 Example of arithmetic operation between two rasters, one of them (image 2) equivalent to a scalar, since all pixels have the same attribute: the operation occurs pixel by pixel, as between two any rasters (pixel A3 is the result of the operation between pixels A1 and A2, while pixels B1 and B2 result in pixel B3)

3.10 Applying Mathematical Functions to Rasters

In addition to the arithmetic operation between rasters or between a raster and a scalar, mathematical functions can be applied directly to a raster. Among the most used mathematical functions, trigonometric (sine, cosine, and tangent), inverse trigonometric (arc sine, arc cosine, and arc tangent), exponential, and logarithmic functions stand out.

Applying these functions to a raster works analogously to what is described for arithmetic operations, with the calculation for each pixel to determine the pixel's value in the resulting image. In each pixel, the argument of the mathematical function is its attribute in the input image, and the result is the attribute to be recorded in the output image (Fig. 3.19).

Other functions that can be developed involve descriptive statistics, such as calculating the maximum, minimum, or average between rasters. For the operation of the average between two images, each pixel of the resulting image is given by the sum of the values of the corresponding pixels in the two images divided by 2, that is, the arithmetic average of the corresponding pixels. In the case of the maximum or minimum function applied to two images, each pixel in the resulting image is determined as being the highest or lowest of the values of the two pixels of the input images of the operation, respectively.

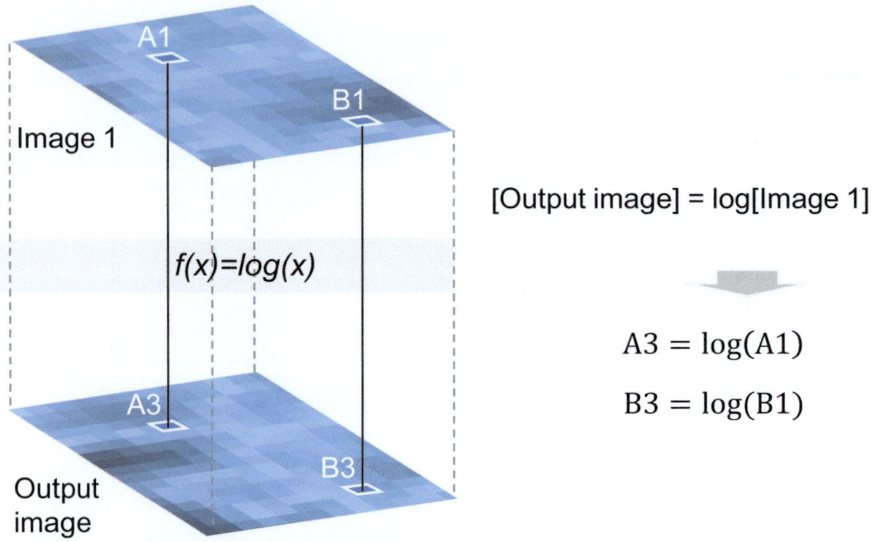

Fig. 3.19 Example of applying a mathematical function (logarithmic function) to a raster (image 1) to obtain another raster (resulting image), in which each pixel has as an attribute the value of the logarithm of the attribute of image 1

In general, one can think of the combined operation between raster arithmetic, scalar arithmetic, and the application of mathematical functions. The procedure follows as described individually for each type of operation, and now, multiple operations constitute the established mathematical expression. One must consider the order of priority of operations and the presence of parentheses, brackets, and braces, as are usually practiced in numerical expressions in mathematics. What is inside the parentheses should be solved before what is inside the square brackets, and these should be solved before what is inside the curly braces. Similarly, power or square root operations are solved first, followed by multiplication and division operations, and then addition and subtraction operations.

Figure 3.20 illustrates an example of a combined mathematical calculation with several operations involving two rasters. Initially, the tangent is calculated for each pixel of image 1. Then, the square root of each pixel of image 2 is computed. After that, the result of the initial operation on image 2 is multiplied by the scalar 1.5, and this result is then added, pixel by pixel, to the result of the initial tangent operation applied to image 1.

Suppose the user defines the entire mathematical expression in some software. The calculation is performed without them being aware of the order in which it is done or the preliminary results of each part of the expression. Even so, understanding that there are rules regarding the priorities of operations is essential for the user to correctly write the mathematical expression to be calculated, especially regarding the placement of parentheses, brackets, and braces when necessary.

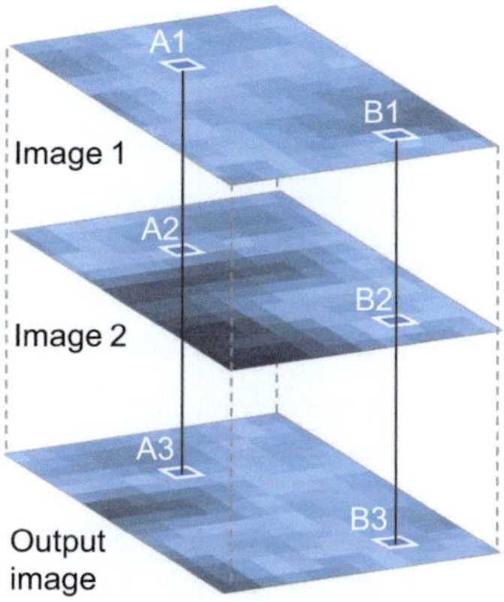

[Output image] = tan [Image 1] + 1.5 x sqrt[Image 2]

$$A3 = \tan(A1) + 1.5 \text{ x } \sqrt{A2}$$

$$B3 = \tan(B1) + 1.5 \text{ x } \sqrt{B2}$$

Fig. 3.20 Example of applying a mathematical expression with multiple operations between two rasters (image 1 and image 2) to obtain another raster (resulting image), in which each pixel has as an attribute the value received by the mathematical expression calculated with the attributes of the corresponding pixels in images 1 and 2

3.11 Profile Over Space

A raster represents the variation in two-dimensional space (the *xy* plane) of a particular phenomenon or variable *z*. Sometimes, there is interest in observing this variability of *z* along a specific direction. In this case, a procedure of generating a profile along a space is performed, and it is usually available in any GIS software.

This procedure requires specifying a vector line that defines the path along which the raster will be profiled. This line is generally referred to as a transect. The pixels of the raster crossed by the transect have their attributes recorded, along with the position of each pixel along the line. This position is indicated as the distance calculated from the initial vertex of the line. Knowing the initial and final points of the line used to generate the profile is necessary to avoid oppositely interpreting the result, as if the end point were the initial and vice versa.

Each pixel identified to be part of the profile corresponds to a pair of distance and attribute coordinates, which are usually plotted on the *x* and *y* axes, respectively. The result is a graph that indicates the spatial variation of the variable represented in the raster along the transect.

In Fig. 3.21, the raster represents the average annual precipitation in a region with a spatial resolution of 25 km. The profile of this raster along a given line, in the direction from A to B, is shown graphically regarding the pairs of distance and precipitation coordinates. A strong increasing precipitation gradient is observed from north to south, going from less than 800 mm to almost 1000 mm over 260 km.

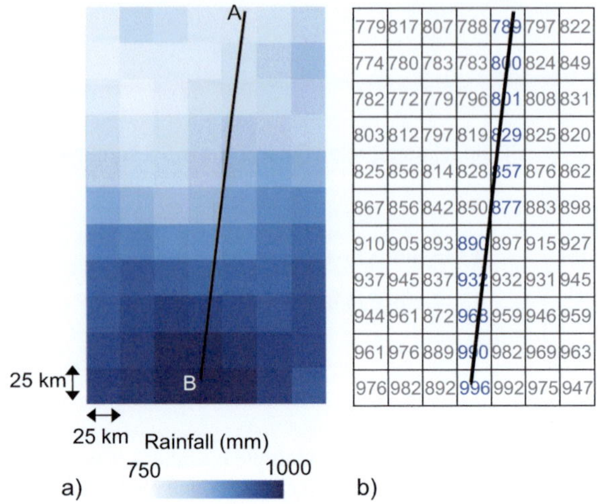

c)

Distance from A (km)	Rainfall (mm)
0	789
25.00	800
50.00	801
75.00	829
100.00	857
125.00	877
160.36	890
185.36	932
210.36	968
235.36	990
260.36	996

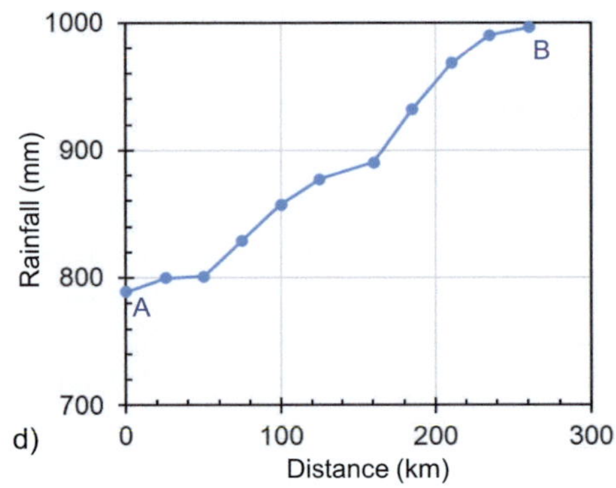

Fig. 3.21 **a** Raster of average annual precipitation with indication of transect A–B; **b** Numerical matrix of the raster, with emphasis in blue on the pixels located along the transect; **c** Pairs of distance and precipitation coordinates along the transect A–B; **d** Profile of the precipitation raster along the transect A–B

3.12 Raster Overlay

Overlaying rasters is an operation between rasters that does not involve mathematical calculation with their attributes. This type of analysis aims to evaluate the combination of occurrences between the attributes of the corresponding pixels of the two rasters, like a cross-tabulation.

In the example of Fig. 3.22, image 1 shows the types of soils in the region (Oxisols, Cambisols, and Luvisols) and, in image 2, a land use classification in the Forest and Agriculture categories. Overlaying the two images can result in up to six classes, the maximum possible combinations between the three types of soils and the two classes of land use. In the example, there are five combinations as indicated in the legend since there was no occurrence of Forest combined with Luvisols. Pixel A3, for example, is in the Oxisols-Agriculture category in the resulting image, as it is the result of overlaying pixel A1 (Oxisols, in image 1) with pixel A2 (Agriculture, in image 2).

The reader should remember that non-numeric attributes cannot be stored in a raster. Therefore, in image 1, each soil class has a numeric value representative of its

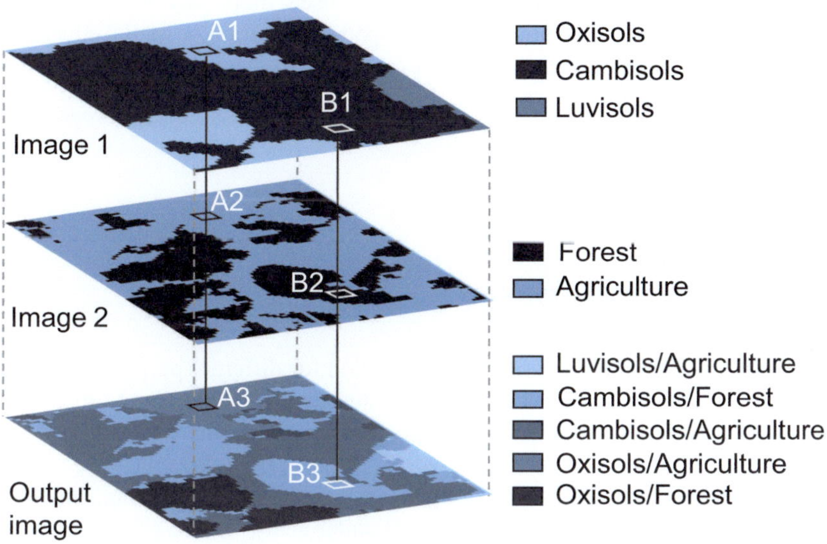

Soil type (image 1)		Land use (image 2)		Cross-tabulation of soil type and land use (output image)	
id	Name	id	Name	id	Name
1	Luvisols	1	Forest	1	Luvisols/Forest
1	Luvisols	2	Agriculture	2	Luvisols/Agriculture
2	Cambisols	1	Forest	3	Cambisols/Forest
2	Cambisols	2	Agriculture	4	Cambisols/Agriculture
3	Oxisols	1	Forest	5	Oxisols/Forest
3	Oxisols	2	Agriculture	6	Oxisols/Agriculture

Fig. 3.22 Example of overlaying a raster of soil type (image 1) and a raster of land use (image 2), obtaining a raster (resulting image) given by the combination of soil type and land use in the corresponding pixels

code (the identifier 'id'), and the same occurs analogously for image 2. The resulting image also has numeric values distinct for each class.

Conceptually, the overlaying operation of images requires the existence of pixel-to-pixel correspondence between the two images; that is, both have the same spatial coverage, number of lines and columns, and, therefore, the exact spatial resolution. Images 1 and 2 from the previous example meet these criteria, resulting in an output image having the same pixel size and spatial coverage as the two images used in the operation. However, depending on the software used, there may be flexibility to allow overlaying rasters that do not coincide in spatial resolution or spatial coverage. In this case, the software performs internal procedures to manage these differences in the overlaying operation.

3.13 Spatial Filtering

3.13.1 Principles and Concepts

An image's spatial filtering operation can be applied for different purposes, such as spatial improvement, increasing the image's visual quality, removing or replacing noise, removing or replacing unwanted values, or increasing the contrast between targets.

Filtering the image in the spatial domain, or simply spatial filtering, involves operations with the neighborhood of each pixel under analysis. The group of neighboring pixels involved in the operation, including the pixel under analysis, is called a window. The window is, by definition, a subset of the image and can have various sizes, also referred to as a mask. The most common is to work with square windows, which have an equal number of rows and columns, such as 3 × 3 windows (3 rows by 3 columns) or 5 × 5 windows (5 rows by 5 columns) (Fig. 3.23). As there needs to be a central pixel, windows are defined with an odd number of rows and columns.

In the filtering process, the defined window is positioned, considering each pixel of the image as the central one each time the filter is applied. Sequentially, the window moves along the rows and columns of the image, justifying this being referred to as a moving window. Special treatment is necessary to apply the filters to pixels located on the edge or near the edge of the image. Setting up the window with its full size is impossible in these locations. One may apply filtering analogous to the standard approach, but only considering neighboring pixels within the window bounded by the edge. An alternative rule could be to skip filtering for edge pixels that don't have the complete set of neighbors defined by the filter window.

When applying a filter with the moving window, a new value for the central pixel is calculated according to the filter definition and the values in the window. This operation is repeated throughout the image, with one pixel taken as the center each time. As the neighboring pixels influence the result, the larger the window, the higher the filtering effect. This occurs because the relative importance of the central pixel,

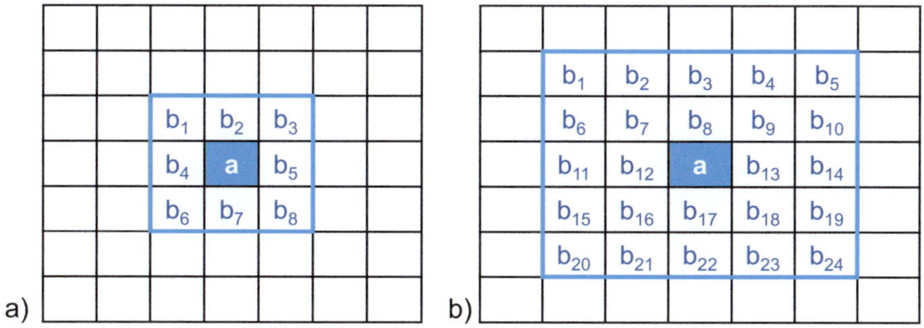

Fig. 3.23 Moving windows 3 × 3 (**a**) and 5 × 5 (**b**) for application of spatial filtering in a raster, where '*a*' represents the central pixel to be filtered in the function of the neighboring pixels 'b_i', with *i* varying from 1 to 8 in the 3 × 3 window and *i* varying from 1 to 24 in the 5 × 5 window

which holds the original value of the image, is reduced due to the larger number of neighbors included in the window. In the 3 × 3 window, there are eight neighboring pixels beside the central pixel, while in the 5 × 5 window, there are 24 neighboring pixels around the central pixel.

This type of spatial filtering is known as convolution filtering, with two main subtypes: smoothing filters and edge detection filters (Fig. 3.24). Edge detection filters aim primarily to highlight objects or regions that stand out from the surroundings of the image by increasing the contrast between them. It is beyond the scope of this book to delve into this type of filtering, as it is not usually applied to Digital Elevation Models.

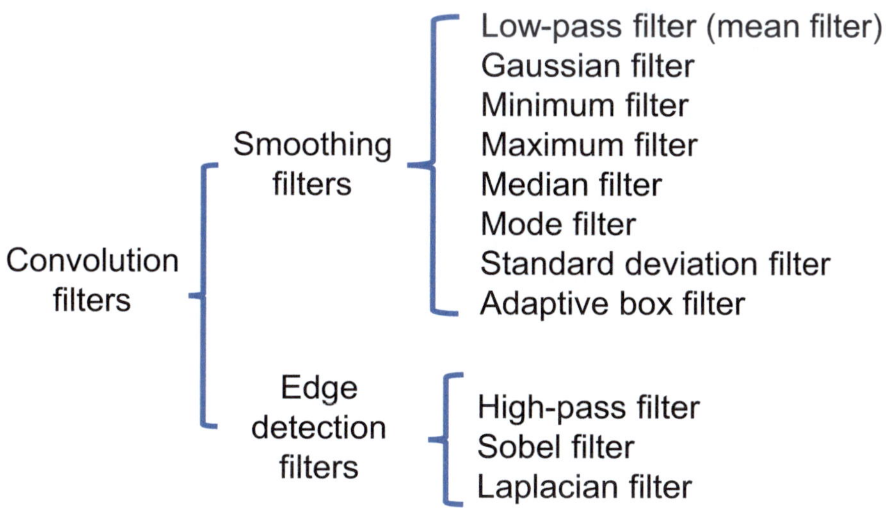

Fig. 3.24 Classification of the main types of spatial image filtering

3.13.2 Smoothing Filters

The application of smoothing filters refers mainly to situations where the interest is to dampen or reduce the occurrence of noise. The mean filter, also known as the low-pass filter, is the simplest method of smoothing filtering. The low-pass nomenclature refers to letting low-frequency information pass and filtering high-frequency information, which would be the noise in the image. With this filter, the central pixel of the moving window is replaced by the arithmetic mean of the pixels of the complete neighborhood, including itself.

In the standard low-pass filter, each pixel in the window has the same weight, and the sum of these weights equals 1. For example, in a 3 × 3 window, each pixel has a weight equal to 1/9; in the 5 × 5 window, this uniform weight is 1/25; in the 7 × 7 window, it is 1/49 (Fig. 3.25). Using weights in a filtering window with a sum equal to the unit value is called normalization. This procedure is recommended to ensure that the resulting image has the same range of values as the input image.

It is possible to work with different weights among the pixels of the neighborhood when applying an average filter and, at the same time, maintain the unit sum for the weights, as is the case with the weighted average. For example, the central pixel

1/9	1/9	1/9
1/9	1/9	1/9
1/9	1/9	1/9

a)

1/16	2/16	1/16
2/16	4/16	2/16
1/16	2/16	1/16

b)

1/25	1/25	1/25	1/25	1/25
1/25	1/25	1/25	1/25	1/25
1/25	1/25	1/25	1/25	1/25
1/25	1/25	1/25	1/25	1/25
1/25	1/25	1/25	1/25	1/25

c)

1/49	1/49	1/49	1/49	1/49	1/49	1/49
1/49	1/49	1/49	1/49	1/49	1/49	1/49
1/49	1/49	1/49	1/49	1/49	1/49	1/49
1/49	1/49	1/49	1/49	1/49	1/49	1/49
1/49	1/49	1/49	1/49	1/49	1/49	1/49
1/49	1/49	1/49	1/49	1/49	1/49	1/49
1/49	1/49	1/49	1/49	1/49	1/49	1/49

d)

Fig. 3.25 Weights used for the application of a low-pass filter in 3 × 3 (**a**, **b**), 5 × 5 (**c**), and 7 × 7 (**d**) windows; Windows **a**, **c**, **d** correspond to the standard low-pass filter, equivalent to the arithmetic average; Window **b** is an example of a weighted average low-pass filter; In all windows, the central pixel is highlighted in blue could have a higher weight than all the neighbors, and the neighbors orthogonal to the central pixel could have a higher weight than the diagonal neighbors (Fig. 3.25b).

The mode, minimum, maximum, standard deviation, and median smoothing filters replace the central pixel of the window with the values, respectively, of the mode, minimum, maximum, standard deviation, and median calculated within the neighborhood window.

To illustrate, Fig. 3.26 shows the filtering of a raster using mean, mode, and minimum filters. For better understanding, for four pixels chosen individually as the central pixel, the corresponding 3 × 3 window is indicated, along with the resulting value when applying such filters. For instance, for the pixel with a value of 1, highlighted in Fig. 3.26c, the mean filter replaces this value with 3.5. In contrast, the mode and minimum filters result in 1, considering only the four pixels within the window and the image as the neighborhood.

It should be noted that the results of the filters differ considerably. However, as a rule, applying any of them smoothed the image, bringing the values between neighboring pixels closer. For example, suppose that the value 9 in the image's 2nd row and 2nd column is noise. With the application of the filters, this value was replaced by values that were more coherent concerning the neighborhood. However, almost all the other pixels in the image were also replaced. Indeed, a striking feature of smoothing filters is the tendency to homogenize the image, which is seen as a disadvantage in some types of applications.

Gaussian filtering fits a Gaussian curve to the moving window, resulting in a higher weight to the central pixel and lower weights as it distances itself from it along the neighborhood. In the case of a 3 × 3 window, the weights indicated in Fig. 3.25b also correspond to a Gaussian filter, considering that the diagonal neighbors are more distant from the central pixel than the orthogonal neighbors. With this filter, the smoothing effect also occurs, but is slightly lower compared to the average, for example, due to the differentiation of weights and the higher weight attributed to the central pixel.

The adaptive box filter is an interesting smoothing filter that can remove noise in an image and soften the unwanted homogenization effect in other parts of the image without noise. The essence of this filter is to replace the pixel of the central window only if certain conditions are met. If properly adjusted, these conditions induce the filtering of unwanted values while preserving the other values of the image.

An example of conditions for checking whether a pixel should be replaced or not when applying an adaptive box filter is (Eliason & McEwen, 1990):

(i) The difference between the pixel and the average of the window's pixels is higher than a threshold 'L1';
(ii) The difference between the pixel and the average of the window's pixels is higher than a certain number 'L2' of times the window's standard deviation

The values 'L1' and 'L2' are parameters defined by the user. If any of these conditions are not met for a particular pixel, that pixel is not replaced, remaining in the resulting image with the original value of the input image. If both conditions

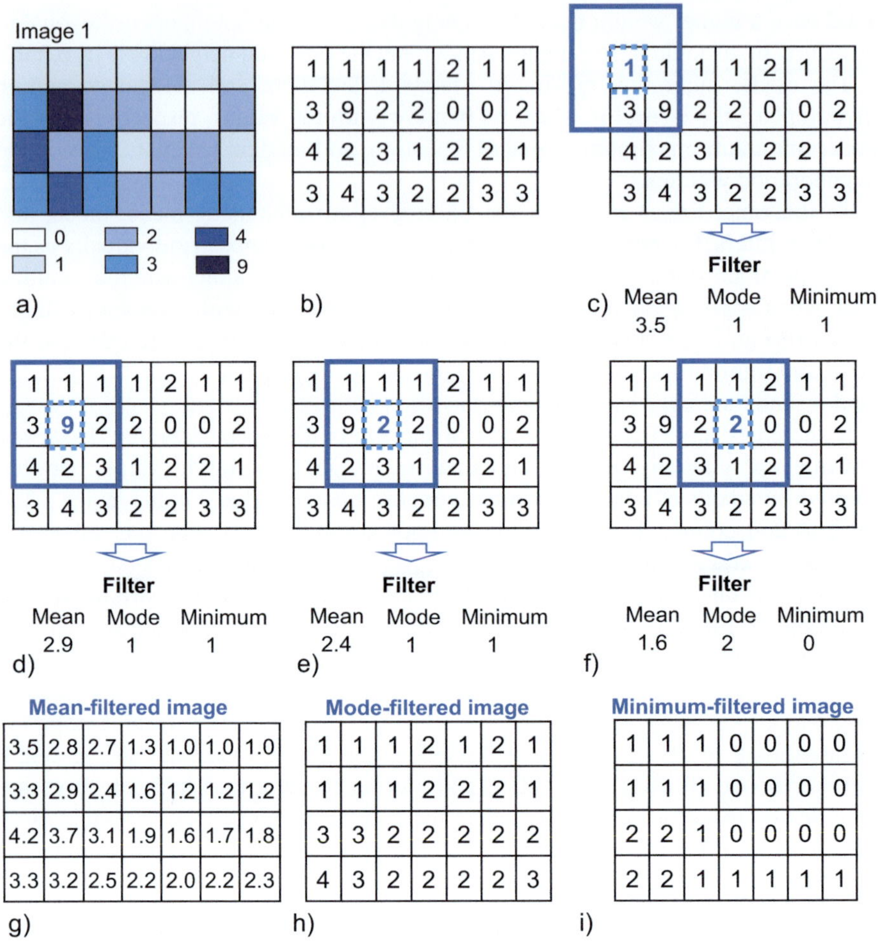

Fig. 3.26 Example of application of average, mode, and minimum filters to a raster: **a**, **b** Raster to be filtered; **c** Moving 3 × 3 window with restriction because it is centered on an edge pixel, and corresponding values obtained from filtering the central pixel; **d–f** Displacement of the moving 3 × 3 window along the image and corresponding values obtained with the filtering of the central pixel; **g–i** Result of image 1 filtered by the average, mode, and minimum, respectively

are met for a pixel, that pixel is replaced by the average of the window neighbors, excluding itself. Alternatively, value zero can be used in this replacement.

In Fig. 3.27, the adaptive box filter is applied to the same image worked on in Fig. 3.26. The supposed noisy pixel, with value 9, was identified as such by the filter by meeting the two imposed conditions and was replaced by the average of the eight neighbors around it in the 3 × 3 window. All other pixels in the image remained unchanged as they did not meet the necessary conditions to be replaced.

Refining the adaptive box filter application can define thresholds $P_{\max}$ and $P_{\min}$, representing the maximum and minimum acceptable values for pixels to be included

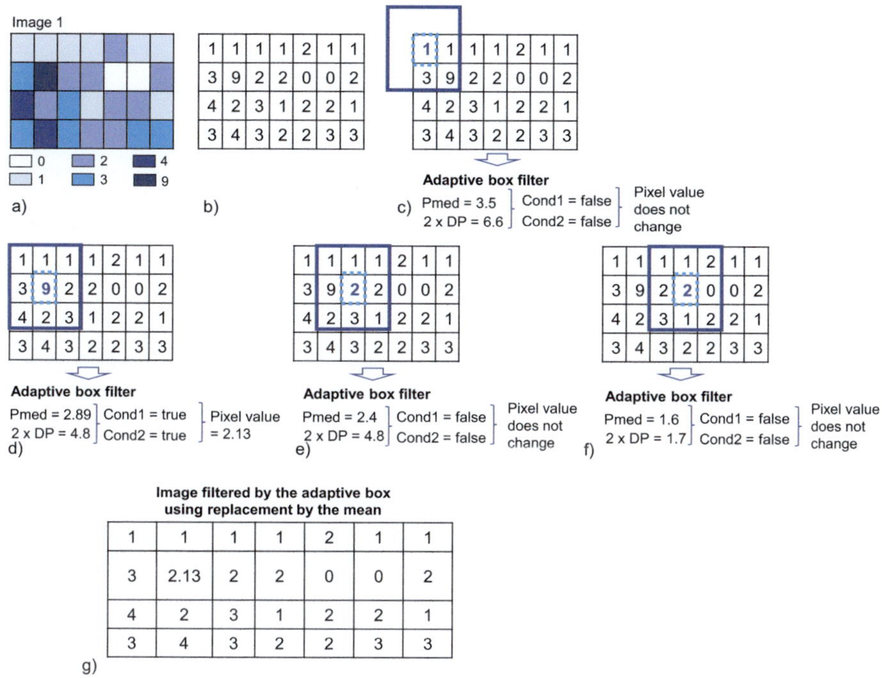

Fig. 3.27 Example of applying an adaptive box filter to a raster (image 1) with criteria to replace pixels if two conditions are met: the difference relative to the average (P_{med}) of the window is higher than 4 (Condition 1—Cond1), and this difference is higher than twice the standard deviation (2 × SD) of the window (Condition 2—Cond2); **a, b** Raster to be filtered; **c–f** Displacement of the 3 × 3 moving window along the image and corresponding values of P_{med} and 2 × SD, with indication of true or false for the two conditions; if both are true, pixel changes value to the average of the window neighbors, without including it; **g** Resulting filtered image

in the calculations as neighbors. During the filtering process, the mean and standard deviation are computed for each moving window using only the neighboring pixels whose values fall within the defined acceptable range. The idea is to work only with values considered more reliable or suitable for the phenomenon represented in the raster.

Reference

Eliason, E. M., & McEwen, A. S. (1990). Adaptive box filters for removal of random noise from digital images. *Photogrammetric Engineering*.

Chapter 4
DEM: Concepts, Data Sources, and Preparation

This chapter presents the main fundamentals of Digital Elevation Models (DEMs). It introduces concepts, describes the main characteristics of a DEM, and depicts basic operations involving this type of data, such as clipping, concatenation, spatial filtering, and change of spatial resolution.

This chapter also discusses the primary data sources for obtaining DEMs. Particular attention is given to the description of dozens of DEM data sources at a global level and free of charge, tracing the historical evolution of the availability of this type of information.

4.1 Approaches for Topography Digital Representation

The terrain's topography or the Earth's surface can usually be represented digitally in several ways, such as by a set of elevation points, contour lines, triangulated irregular networks, and raster format (Fig. 4.1).

The points with topographic information can be acquired using traditional ground-based instruments. It also includes the concept of a point cloud, which refers to data generated by laser surveys, as discussed in more detail later in this chapter. Contour lines achieve a topographic representation through lines or contours associated with given elevations. These lines can be obtained by aerophotogrammetry, topographic surveys with ground instruments, or other procedures.

The TIN (Triangulated Irregular Network) represents the topography by a set of triangular faces of varying sizes and inclinations. From sample points with topographic information, triangles are generated without overlap, covering the entire region of the points. The generation of the triangles follows a series of rules, which vary according to the method of generating the triangular mesh. The Delaunay method is the most widely used.

A. Rolim da Paz, *Digital Elevation Models for Environmental Studies*,
https://doi.org/10.1007/978-3-032-04523-2_4

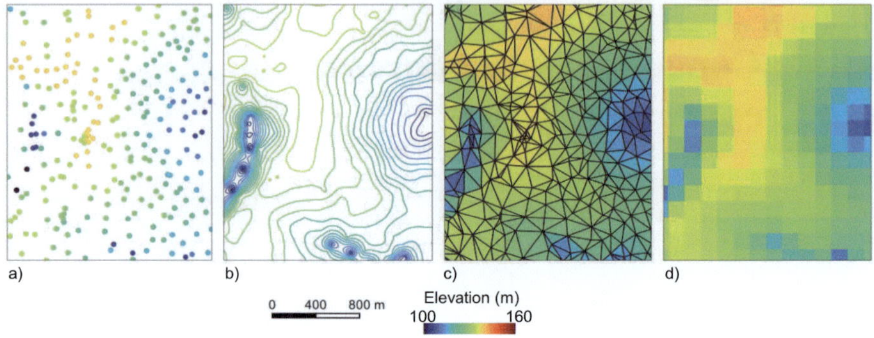

Fig. 4.1 Topographic representation of the same area, at the same scale and with symbology according to the same color ramp, in the formats of elevation points (**a**), contour lines (**b**), triangulated irregular network (**c**), and raster (**d**)

In areas with more rugged relief, i.e., higher topographic variation, the triangulated irregular network presents higher density, with elements of smaller dimensions. In flatter areas, there is less density of elements, which have larger dimensions. This approach of topographic representation facilitates the three-dimensional visualization of the relief. It is computationally efficient in terms of information storage by varying the density and dimensions of the elements according to the need for higher or lower detail in each portion of the area.

However, due to the irregularity of the triangles, the organization, storage, and application of this type of topographic representation are relatively more complicated than in the raster format, for example. In the TIN, it is necessary to store information about the planimetric position of the triangles and the topological relationship between neighboring triangles, in addition to the elevation of the points themselves. Overlaying and analyzing a TIN with other layers derived from remote sensing data, usually provided as rasters, is also relatively more complicated.

The other way to digitally represent topography is the raster format. This consists of a grid or matrix of elevation values representing the terrain or surface. Among all digital topographic representation methods, the raster format is currently the most widely used for three reasons. First, it is a straightforward and practical way to organize and work with topographic data, since a raster is essentially a numerical matrix. Second, the raster format greatly facilitates the development of analyses and overlays with other information layers, especially when those layers are also available in raster format—this is the primary format for data obtained through remote sensing and one of the most commonly used formats in GIS. The third reason is that global topographic datasets are predominantly made available in this format, as discussed in more detail later in this chapter.

The disadvantages of the raster format for representing elevation data include that this structure does not adequately handle abrupt elevation changes and that, in general, the pixel size influences the use of information and the computational efficiency of dealing with these data (Moore et al., 1991).

4.2 What Is a Digital Elevation Model?

A Digital Elevation Model (DEM) is an ordered set of numbers representing the spatial distribution of elevation measured relative to a vertical reference (Moore et al., 1991). In summary, a DEM is a digital representation of the topography of the terrain or the surface of the Earth. The most common format of a DEM is the raster. In this case, there is a raster whose attribute is the elevation of the terrain. The numerical value stored in each pixel represents the elevation relative to a given vertical reference.

However, the reader should remember that the pixel occupies a geographical area. Therefore, the pixel value represents a variable over the entire pixel area. When working with a raster, it is assumed that the defined variable is spatially uniform in the area occupied by each pixel. The variable is not generally uniform in this pixel area in the real world. There is, in fact, a variation, but it is ignored in the raster, as this variation is sub-pixel, smaller than the pixel size.

In the area occupied by each pixel of a raster that represents a DEM, the only information about the elevation of the terrain is the pixel attribute. For example, consider a DEM with a spatial resolution of 30 m, as exemplified in Fig. 4.2. In the 900 m^2 (=30 m × 30 m) occupied by each of the three highlighted pixels, the topography is uniform according to the DEM, with an elevation equal to each attribute. In the real world, there is topographic variation in the area covered by each pixel, but such variation is not represented in this DEM, given its spatial resolution.

Looking closely at the three highlighted pixels in Fig. 4.2, it becomes clear how the DEM represents the topography in steps. This is the limiting effect of spatial discretization when representing this field of information, which is topography,

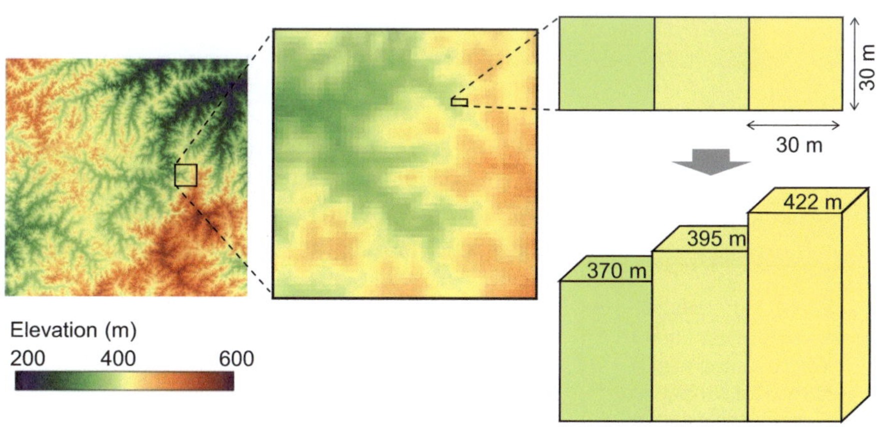

Fig. 4.2 Digital Elevation Model in raster format with a spatial resolution of 30 m, sufficiently detailing a region to visualize the pixels, with higher emphasis on a set of three pixels, whose 3D visualization is also presented

whose real variation is continuous. The higher the refinement of the spatial discretization (smaller pixel dimensions), the closer the discretized representation of the DEM becomes to the real continuous topographic variation. This is more pronounced in regions with more rugged terrain.

4.3 Digital Surface Model Versus Digital Terrain Model

The DEM is a broad definition that encompasses the concepts of DSM (Digital Surface Model) and DTM (Digital Terrain Model) (Fig. 4.3). Technically, DSM and DTM are not synonyms for each other, nor equivalent to DEM. The DSM is the digital representation of the topography of the Earth's surface, including artifacts above the ground, such as buildings and vegetation (Fig. 4.4). The DTM is the topographic representation of the terrain without considering these artifacts. It can be stated that both the DSM and the DTM are DEMs, but not the other way around.

Due to confusion of concepts and lack of standardization, one often finds the indistinct use between the terms DSM, DTM, and DEM in the literature or even the separation between them, but in a way that is mistaken about what they should be. Some authors, however, prefer to adopt DEM for the model that represents the elevations of the surface, including the artifacts on the terrain, and keep DTM as previously defined here, not using the term DSM.

Another term also of similar context to the previous ones is the Numerical Surface Model (NSM), whose words that compose it induce the mistaken understanding that NSM would be synonymous with DEM or even more so with DTM. The NSM, conceptually, refers to the representation of a numerical surface in geographic space. However, it does not necessarily refer to topography, although the term is most used

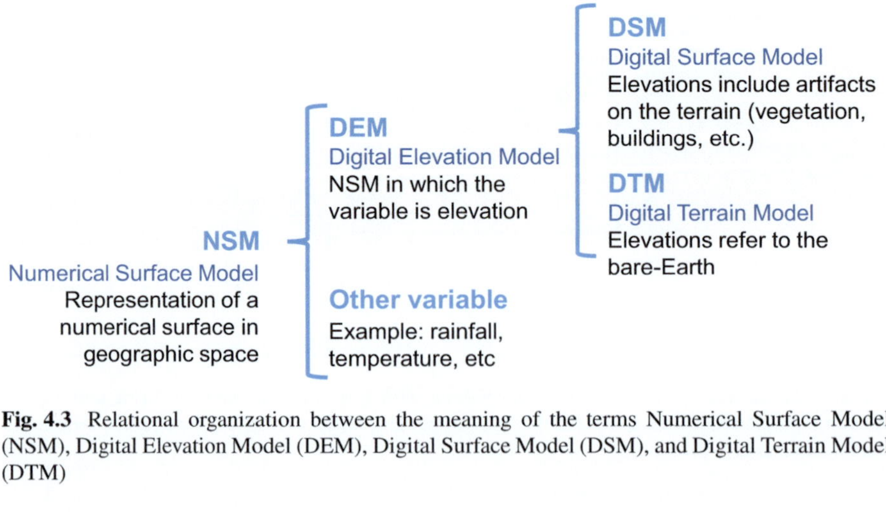

Fig. 4.3 Relational organization between the meaning of the terms Numerical Surface Model (NSM), Digital Elevation Model (DEM), Digital Surface Model (DSM), and Digital Terrain Model (DTM)

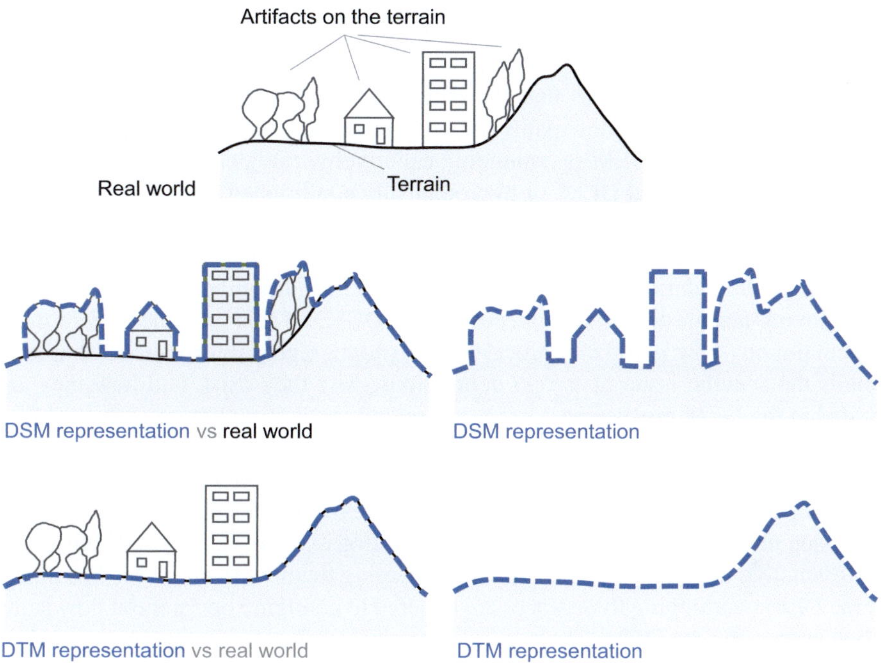

Fig. 4.4 Scheme illustrating what a Digital Surface Model (DSM) represents and what a Digital Terrain Model (DTM) represents

in this context. Other variables, such as precipitation, can be represented in an NSM. Thus, it can be considered that every DEM is an NTM but not the other way around.

4.4 Vertical Reference for the Elevation Data

A Digital Elevation Model digitally represents the elevations of the surface or terrain in a matrix format, but… elevations are measured relative to what? What is the vertical reference of these measurements? There are two main geodetic or vertical datum types: the geoid and the ellipsoid. And there are different geoids and different ellipsoids. The details of these concepts related to vertical datum are presented in item 2.5 of Chap. 2.

4.5 Hydrologically Conditioned Digital Elevation Model

A DEM can be modified so that the elevation values represent the topographic surface in a way that the flow paths derived from it align with the expected spatial patterns. This type of DEM is commonly called a hydrologically consistent DEM, hydrologically corrected DEM, or hydrologically conditioned DEM.

There is no standardized procedure or even an absolute definition for what establishes that a DEM is considered hydrologically conditioned. There are two main ideas. The first is the understanding that the hydrological conditioning of the DEM refers to removing terrain depressions present in the DEM, so it is possible to determine the continuous flow path from processing this topographic information. Chapter 6 details the specific issue of terrain depressions, why they exist, and how they are treated in the DEM processing.

The second primary belief is that a hydrologically conditioned DEM is obtained by removing terrain depressions and an additional step explicitly directed to make the flow paths derived from this DEM follow the expected traces for the rivers. This extra step mainly refers to procedures incorporating existing vectors and considered representative of the reality of river traces, merging them with the DEM. The pixels of the DEM overlapping these traces are lowered to configure preferential flow paths when processing the DEM. The general procedure performed in this sense is called stream burning, detailed in Chap. 7 (item 7.10).

In addition to these two main ideas, some authors also associate the concept of hydrologically corrected DEM with the procedure of removing above-ground artifacts represented in the DEM, such as vegetation. An example is the work of Ettritch et al. (2018).

Several researchers applied procedures to remove or at least reduce the effect of vegetation and even buildings on the elevations represented in a DEM, mainly in the case of global DEMs, as detailed later in this chapter. This type of DEM treatment seeks to modify it so that it is less characterized as a DSM and gets closer to a DTM. The better a DEM approximates a DTM, the higher the chance that the derived flow paths resemble the actual spatial patterns.

Many caveats should be noted regarding the search for a hydrologically conditioned DEM. First, it is essential to remember the purpose for which you want to perform this operation. What purpose will this DEM be used for? What level or type of hydrological conditioning is intended and why? If an existing vector river network is going to be used, is this network satisfactorily representative of reality to be used for this aim?

Finally, the hydrologically conditioned DEM may result in the best representation of flow paths but not necessarily a better topographic representation of the terrain surface. The elevations are artificially altered according to criteria and procedures established to better represent the flow paths. This does not mean these altered elevations get closer to the actual topographic representation at these points.

More considerations and details about the hydrological conditioning of a DEM are presented in Chaps. 6 and 7.

4.6 Spatial Resolution in Angular and Linear Units

As detailed later in this chapter, many DEMs are available, with data covering almost the globe. In these cases, the geographic coordinate system (latitude and longitude) is usually adopted, with units in degrees. Each DEM's pixel size and spatial resolution correspond to a specific number of degrees, depending on each case. One can explicitly work with this quantity in degrees or angular units such as arc minutes (or minutes of arc) and arc seconds (or seconds of arc). An arc-minute corresponds to 1/60 degrees, and an arc-second equals 1/60 arc-minutes. So, one degree has 60 arc-minutes, an arc-minute has 60 arc-seconds, and one degree has 3600 arc-seconds (Table 4.1). To make the equivalence between a measure in one of these units and the other, follow a simple rule of three with some of these correspondences. For example, a spatial resolution of 5 arc-minutes corresponds to 0.08333… degrees (Fig. 4.5).

These angular measures are necessary due to their representation in geographic coordinates, but they are difficult for most users to understand. The main reason is the lack of knowledge or difficulty in recognizing the meaning of the values measured in these units regarding distances. There is also a limitation in making the equivalence to planar units, such as meters or kilometers.

When the value is expressed in degrees, the spatial resolution almost always assumes decimal numbers much lower than one and of the recurring decimal type. In these cases, the understanding is not straightforward and reduces the ease of memorizing and presenting such a number. Generally, integer values are obtained when expressed in arc minutes or arc seconds. However, these units have much less understanding of correspondence with values in planar units than the degree.

For these reasons, it is much more common to find references to spatial resolutions to values converted to units such as meters or kilometers than to mention the original

Table 4.1 Equivalences between arc units: degree, arc-minute, and arc-second

Value	Equivalence
1 degree	60 arc-minutes
1 degree	3600 arc-seconds
1 arc-minute	60 arc-seconds
1 arc-minute	1/60 degrees
1 arc-second	1/60 arc-minutes
1 arc-second	1/3600 degrees

1 arc-minute ———— 1/60 degrees
5 arc-minutes ———— x

$$\Rightarrow x = 5 \cdot 1/60 \Rightarrow x = 0.08333\ldots \text{ degrees}$$

Fig. 4.5 Example of applying a simple rule of three to determine the value in degrees of a measure equal to 5 arc-minutes

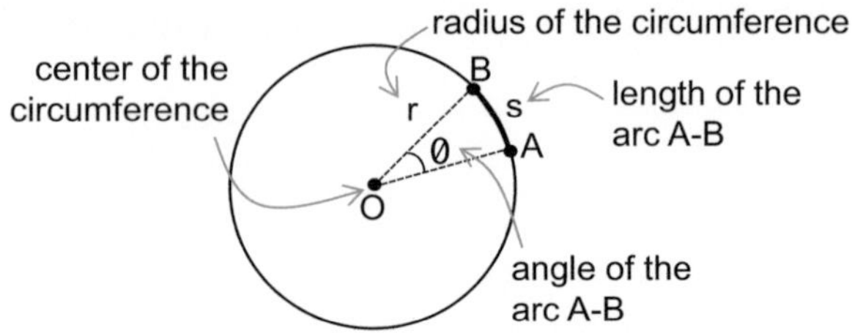

Fig. 4.6 Correspondence between the angle and length of an arc of a circumference

values in angular units. For example, as seen later in this chapter, the DEM of some sources of global topographic data has a spatial resolution of 1 arc-second. This value is equivalent to 0.0002777 degrees. However, the value that became famous when mentioning the spatial resolution of this DEM is 30 m. This is the approximate equivalence to 1 arc-second in meters.

But how is the correspondence between an angular measure and the distance in meters or kilometers determined? To introduce this idea, consider the circumference shown in Fig. 4.6. A geometric arc is defined as a segment of this circumference, such as the arc A-B identified in the figure. This arc corresponds to a central angle (θ) formed between the two lines that connect the center of the circumference to each point A and B. The length of the arc A–B, in units of distance such as meters, is directly proportional to the size of the radius of the circumference and the central angle measured in radians (Eq. 4.1).

$$s = r \cdot \theta \tag{4.1}$$

where s is the length of an arc in meters, r is the radius of the circumference in meters, and θ is the measure of the central angle in radians.

When working with angular measures in geographic coordinates, the radius of the reference circumference for the conversion between the angular measure and the length of the arc changes according to the latitude. To illustrate this concept, let's take an ellipsoid of revolution representing the Earth's surface (Fig. 4.7). As seen in Chap. 2, cross-sections to the axis of rotation, the minor axis (north–south orientation), result in circles. The parallels that make up the geographic coordinate system are circumferences. At the Equator, zero latitude, this circumference has a maximum radius.

As one moves toward the poles, a cross-section of the ellipsoid's axis of rotation results in ever-smaller circumferences. This occurs symmetrically for north and south latitudes. As shown in Fig. 4.7, a consequence is that, for a fixed central angle, the length of the corresponding arc decreases as the circumference radius reduces with increasing distance from the Equator.

For the representation of the Earth according to the WGS84 ellipsoid and considering the circumference radius at the equatorial level, a measure of 1° of longitude corresponds to an arc extension of approximately 111.32 km (Figs. 4.8 and 4.9). This

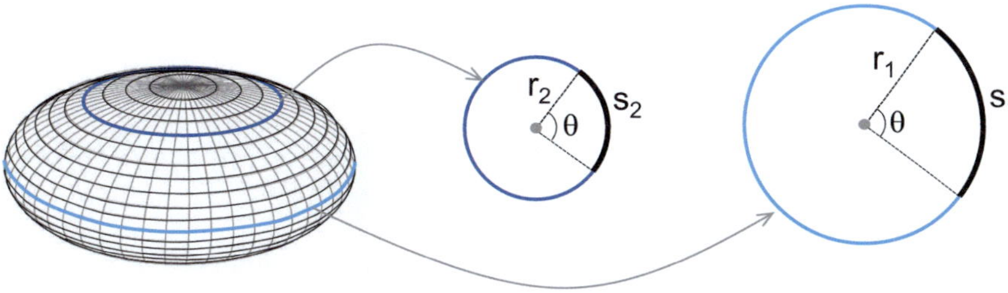

Fig. 4.7 Ellipsoid of revolution used to represent the earth's surface with emphasis on two parallels: the closer to the poles, the circumference that describes the parallel has a smaller radius ($r_2 < r_1$), causing that, for the same angular measure of the central angle θ, the equivalent arc length is reduced ($s_2 < s_1$)

At 0° of latitude (Equator):	At 30° of latitude (North or South):	At 60° of latitude (North or South):
r = 6,378,137 m θ = 1° ⇒ θ = π/180 = 0.017453 rad ⇒ s = r · θ = 6,378,137 · 0.017453 ⇒ s = 111,319.49 m	r = 5,528,257 m For θ = 1° (0.017453 rad) ⇒ s = 96,486.28 m	r = 3,197,105 m For θ = 1° (0.017453 rad) ⇒ s = 55,800.00 m
1° ≅ 111.32 km	1° ≅ 96.49 km	1° ≅ 55.80 km

Fig. 4.8 Example of determining the arc length (s) equivalent to an angular measure of 1° (θ) of longitude on the Earth's surface, at different latitudes, as a function of the radius of the Earth's circumference (r) based on the WGS84 ellipsoid of revolution

equivalence is independent of the longitude where the analysis is made (for example, from 22° to 23° of west longitude or from 59° to 60° of east longitude).

At 30° north or south latitude, 1° of longitude corresponds to an arc of about 96.49 km, regardless of its longitude. At 60° north or south latitude, the equivalence to 1° of longitude reduces to around 55.80 km, less than half the value at the equatorial level. The difference between the equivalences as a function of latitude results from the variation of the Earth's radius according to the circumference corresponding to each parallel.

The equivalence between degrees of latitude and arc length in kilometers differs from the correspondence between degrees of longitude and arc length. This is due to the ellipsoid of revolution used to represent the Earth's surface having been generated by rotating the original ellipse around the minor semi-axis, which results in meridians that configure elliptical figures, as explained in Chap. 2. As a result, 1° of latitude corresponds to an arc length of approximately 111 km, having a relatively slight variation of equivalence as one moves from the Equator to the poles (Fig. 4.9) and being independent of the longitude where it is located.

With the same reasoning, a spatial resolution of 1 arc-second (or 0.0002777... degrees) at the Equator corresponds to about 30.9 m. Also, for convenience and

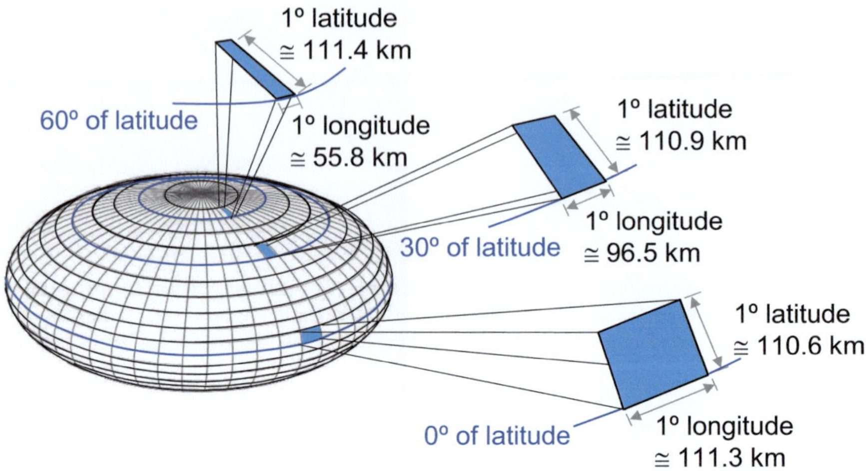

Fig. 4.9 Scheme of correspondence between 1° of longitude and 1° of latitude and distances measured in kilometers, as a function of latitude and considering the representation of the earth's surface by the WGS84 ellipsoid of revolution (drawing out of scale)

simplification, it is customary to report an equivalent spatial resolution value in meters or kilometers in a rounded way, not the exact equivalence value. The usual mention in the previous example is 30 m and not 30.9 m. For a coarser DEM, with a resolution of 30 arc seconds, for example, it is usually referred to as 1 km, although the equivalence is for approximately 927.7 m at the equatorial latitude.

4.7 Changing the DEM Spatial Resolution

In some cases, coarsening the spatial resolution of the DEM before using it for a study is necessary. Examples of this need are overlaying the DEM with other, more coarsened spatial resolution datasets or reducing the cost of processing the DEM to obtain products such as drainage networks and watershed delineation.

To illustrate the quality of the topographic representation as a function of the DEM's pixel size, Fig. 4.10 indicates a line referring to the actual topographic variation and the representation achieved with DEMs of incrementally coarsened spatial resolutions, going from a particular value up to 16 times this value. At each spatial resolution, the representation was drawn so that the actual topographic curve crosses the center of the surface of each pixel (i.e., its centroid).

The larger the pixel size, the less accurately one can represent the topographic variation. As spatial resolution coarsens, there are more pronounced steps between neighboring pixels. At the same time, there is a tendency for a smaller amplitude between maximum and minimum terrain values, i.e., the representation gets smoother.

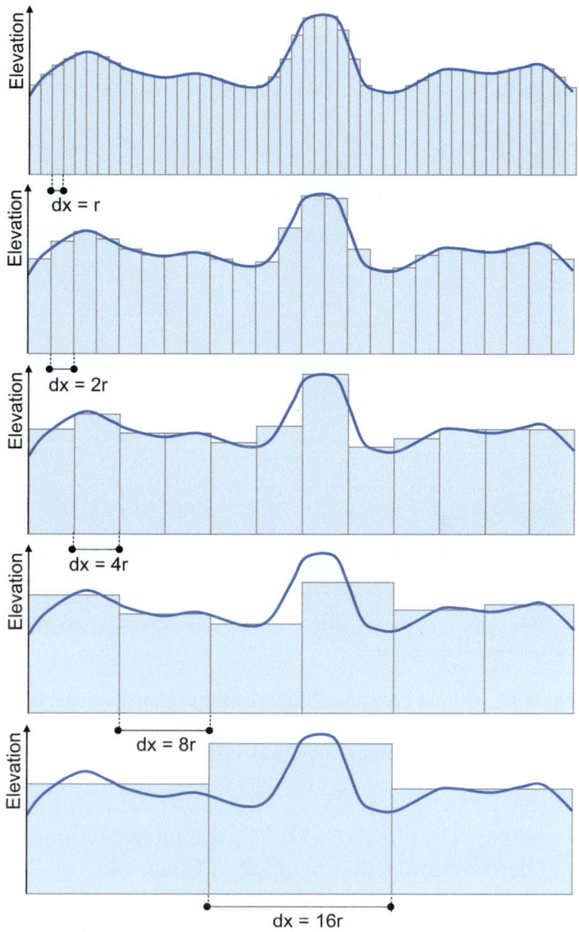

Fig. 4.10 Schematic illustration of the effect of a DEM's pixel size (dx, the spatial resolution) on the representation of topography compared to the indicative line of the actual terrain: variation from d$x = r$ to d$x = 16r$

The example in Fig. 4.11 illustrates the process of coarsening the spatial resolution of a DEM at different levels. At the finest resolution (30 m), there is elevation information for each area of 30 × 30 m. When moving to the 90 m resolution, there is now a single elevation value representing an area of 90 × 90 m, whereas in the previous DEM, there were nine elevations. At the 500 m resolution, each pixel of this DEM covers an area covered by more than 250 pixels of the 30 m DEM. In other words, at the 30 m resolution, there is a distinction of more than 250 elevation values in this area. In contrast, only a single elevation value represents this area at the 500 m resolution.

In Fig. 4.11, the spatial pattern of topographic variation is roughly maintained even when coarsening to resolutions more than 10 × coarser than the original. However, with the increase in pixel size, there is a progressive loss of representation of local

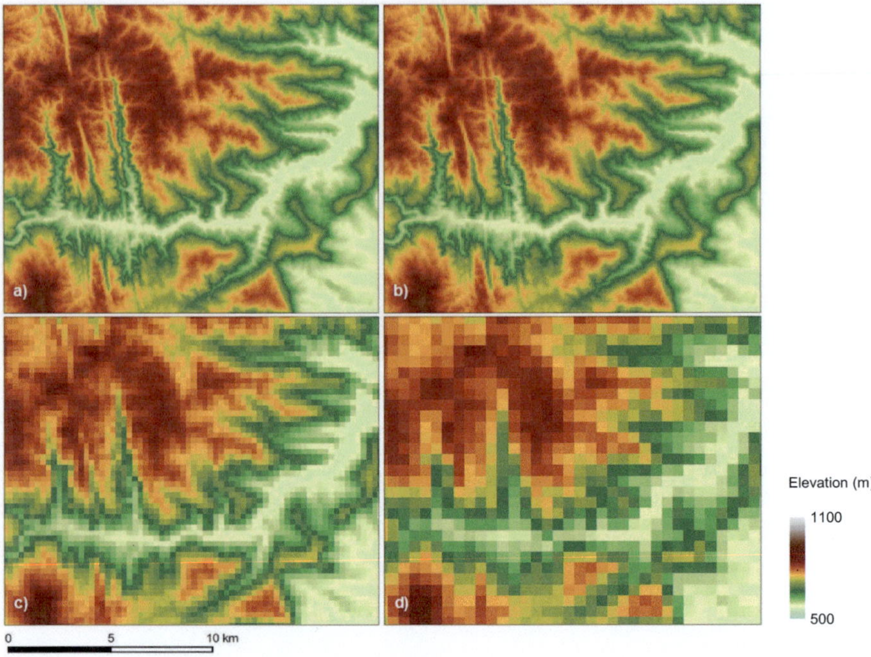

Fig. 4.11 Digital Elevation Model of the region upstream of the Três Marias hydroelectric power plant on the São Francisco river from SRTM data with an approximate resolution of 30 m (**a**) and resampled for resolutions of 90 m (**b**), 250 m (**c**), and 500 m (**d**)

variations. The effect of a DEM's spatial resolution on the topographic representation and derived products is detailed in Chap. 14.

The coarsening of the spatial resolution of a DEM is a procedure just like that performed for other rasters representing a quantitative attribute, as described in Chap. 3. The average is one of the most used methods of DEM spatial resolution coarsening. This procedure results in a coarsened DEM whose elevation of each pixel is given by the average of the elevations of the corresponding pixels in the geographical coverage area in the input DEM. Approaches via interpolation methods are also widely used.

As seen in Chap. 3, the average method and the interpolation methods result in a coarsened resolution DEM that is smoothed and whose pixel values do not necessarily exist in the original DEM. On the other hand, methods such as coarsening by the central pixel or the nearest neighbor method result in a coarsened DEM whose elevations correspond to values already existing in the original DEM.

There is no a priori definition of which of these methods is the best, with all the approaches mentioned here considered valid. However, the user must understand that each method affects the chosen information coarsening process. Eventually, some methods may not be recommended due to some particularity of the purpose of further using the coarsened DEM.

As also seen in Chap. 3, it is possible to refine the spatial resolution of any raster. Following this approach, a DEM can be processed to obtain a finer spatial resolution elevation dataset. However, this is not a recommended procedure. Interpolation cannot improve the topographic representation in the DEM or insert additional elevations (representative of reality) that had not been captured in the original data (Bielski et al., 2023). In summary, refining the spatial resolution of a DEM implies creating information where there is none. The resulting DEM from this process may appear to have a smoother and visually pleasing surface, but they are data that does not reproduce the actual topography.

Suppose the topographic representation of a DEM is shown in Fig. 4.12a, with a spatial resolution of $dx = r$. By an interpolation procedure, a DEM of spatial resolution $dx = r/4$ can be obtained (i.e., the area covered by each pixel of the input DEM is now occupied by 16 pixels of the refined DEM).

The topography resulting from two DEMs with resolution $dx = r/4$, obtained through different procedures from the DEM with resolution $dx = r$, is illustrated in Fig. 4.12b and c. Which of the two is more consistent with the actual topography? Would it be the DEM in Fig. 4.12b due to its smoother variations? Or would it be the DEM in Fig. 4.12c, which has variations of higher amplitude? This question can never be answered definitively, as the available information is limited to the resolution $dx = r$, and any representation more refined than that can be considered hypothetical.

However, a finer resolution DEM from an available DEM is sometimes required. For example, this may occur due to the need for a DEM with spatial resolution equal to other rasters for cross-tabulation analysis or input into some hydrological model. In these cases, it may be more interesting to adopt the rule of simply repeating the elevation of the pixel of dimension r for all pixels of dimension $r/4$ (Fig. 4.12d). The topographic representation remains the same as the DEM of resolution r, but now there are 16 times more pixels in the DEM of resolution $r/4$.

4.8 DEM Cropping

Depending on the data source and how the data is made available, the DEM can be obtained in the raster format with a much more extensive spatial coverage than the area of interest for the study. As seen in Chap. 3, the DEM can be cropped to a smaller region, encompassing the study area like any raster. This reduces the amount of data to be stored and processed, especially considering that various other raster products, such as slope maps, drainage networks, watershed boundaries, distances to watercourses, and so on, are typically generated from the DEM.

The DEM windowing can be performed as described in Chap. 3, based on the user's specification of the geographic coordinates of the vertices that define the area of interest. Suppose the DEM or its derived products will later be overlaid with another existing raster. In that case, it is recommended to use the coordinates of that raster's vertices as input information for clipping the DEM.

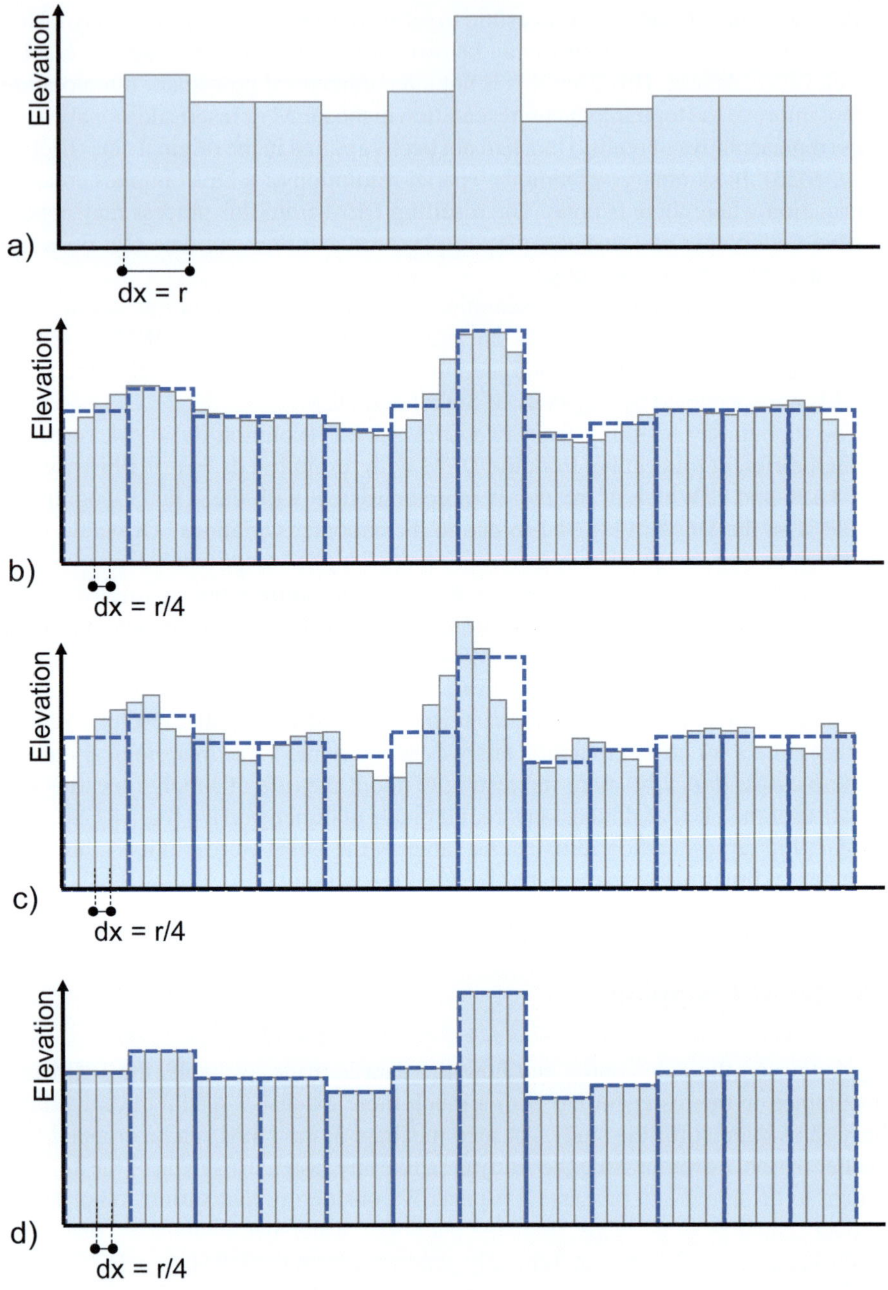

Fig. 4.12 Schematic illustration of the refinement of the spatial resolution d$x = r$ of a DEM (**a**) to generate three other DEMs with spatial resolution d$x = r/4$ (**b–d**)

One mistake that should be avoided is the exaggerated restriction of the spatial extent of the DEM. The image should contain a substantial margin around the study area. Lack of attention to this procedure can lead to a serious problem with one of the main processes to be performed with a DEM, which is the delineation of watersheds, as described in Chap. 8.

The exaggerated cropping of the DEM can hinder the correct basin delineation (Fig. 4.13a). In the left region of the image, in the part indicated by the arrow, the contour of the basin became rectilinear, visibly limited by the edge of the DEM. This means an area contributing to the basin is outside the DEM window, invalidating the delineation. Other DEM-derived data layers for this watershed, such as flow directions and drainage networks, are equally invalid. The correct delineation of the basin obtained by considering a more extensive spatial coverage for the DEM shows how an area to the west had been hindered from being included in the previous delineation due to the incorrect cropping of the DEM (Fig. 4.13b).

A similar type of operation, referred to by some as DEM clipping or DEM masking, uses a raster or polygon that defines a region of interest, such as a watershed or a municipality, to generate a new DEM raster. The output DEM has valid attributes exclusively for pixels in this area of interest (Fig. 4.13c).

This procedure can be done when this DEM is used to prepare a map or figure in some technical document, or for some particular analyses, with due attention. This is because using the masked DEM can invalidate some results. For example, for determining flow paths, river drainage networks, and watershed delineation, the DEM must have valid values outside the delineation of the same basin used to clip it. Each algorithm for determining flow directions (the basic information layer to derive the basin delineation) can achieve different results. Therefore, the basin delineation

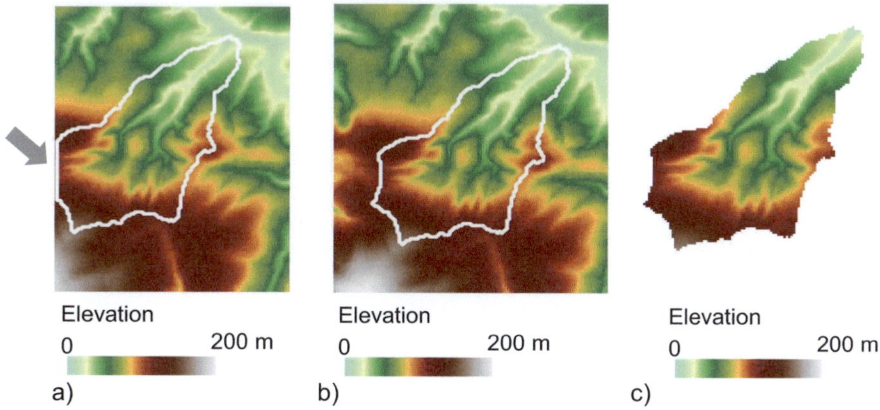

Fig. 4.13 **a** Incorrect delineation (the error in indicated by the arrow) of a watershed (white contour) due to the restriction imposed by the exaggerated cropping of the spatial coverage of the DEM (in the background); **b** Correct delineation of the same basin (white contour) obtained from the DEM with more extensive spatial coverage (in the background); **c** DEM cropped by the mask of the basin delineation, with invalid values outside this delineation

used previously to mask it may have included areas that would not be part of the delineation to be obtained in the processing of the DEM or, worst case, have omitted areas that should be part of the delineation.

Another situation is calculating the surface slopes pixel by pixel. The edge pixels of the DEM cropped by the mask of the area of interest have external neighbors to the mask with invalid attributes. The slopes obtained for these edge pixels may be excessively high and unreal if the software treats the pixels outside the mask with a zero value, or these slopes may be without numerical determination and have invalid values. In both cases, the raster of slopes cannot be used for further analysis of this watershed.

4.9 Concatenation of Images to Obtain the DEM

In the opposite sense of the procedure described in the previous item, joining more than one raster is sometimes necessary to obtain the DEM that covers the entire area of interest. The operation consists of concatenating the rasters, as described in Chap. 3, for any rasters. If there is an intersection of spatial coverage between the rasters, a decision rule involving the priority between the images and the prevalence of valid values from one image over invalid values in another should be implemented.

For example, the concatenation of 12 scenes of SRTM 90 data, each covering an area of 5° × 5°, was used to obtain a DEM necessary to represent the São Francisco River basin (Fig. 4.14).

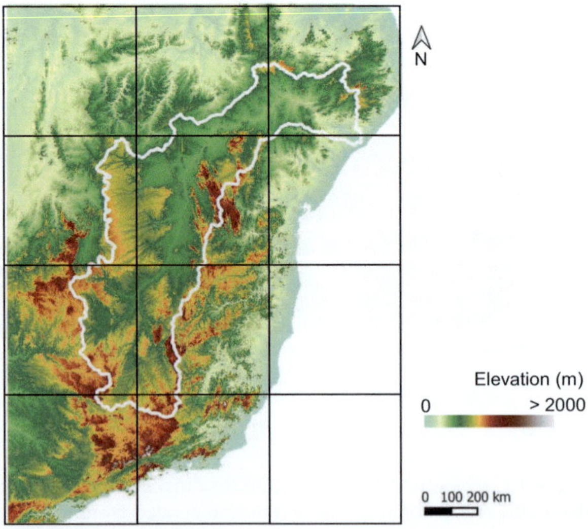

Fig. 4.14 Concatenation of 12 scenes of SRTM 90 with 5° x 5° of extension each (approximately 500 × 500 km) to represent the Digital Elevation Model of the São Francisco river basin (white outline)

4.10 Filtering of Noise or Invalid Data

A DEM may contain noise or invalid data, mainly when it originates from remote sensing surveys. This can occur due to several factors, such as data acquisition or processing issues. Data may be missing in global DEMs obtained by remote sensing, like some SRTM and ASTER data versions. Missing data can also occur in LiDAR-type surveys. It is beyond the scope of this text to delve into the discussion about the origins and causes of missing data in the generation of DEMs.

Generally, the data provider uses some marker ('flag') to indicate the pixels with missing data to the user. Depending on the software used to work with the DEM, the marker can be interpreted as 'no data' or 'null.' Some software interprets it as a numerical value, like −32,768. The value 32,768 equals 2^{15} and is considered the maximum value that an integer variable can assume in some software.

The occurrence of missing data in a DEM can be problematic, and it may be recommended that it be filled via image filtering. This depends on the type of analysis or processing done with the DEM and how the software deals with this missing data. For example, some software identifies it by a flag and automatically ignores the corresponding pixels for many types of processing. On the other hand, software that interprets missing data with numerical values tends to use this numerical value as an attribute of the corresponding pixels for the calculations that are part of the DEM analysis.

For example, to generate descriptive statistics of the topography from the DEM or obtain a hypsometric curve, pixels identified as missing data should not be part of the survey. If the software includes them, the analysis is not valid. In this case, the DEM must be treated previously to remove missing data. A usual alternative is the replacement by values estimated from the neighborhood. A procedure for this is spatial filtering, as discussed in Chap. 3. A recommended filter for this situation is the adaptive box, as it presents the versatility of identifying and replacing missing data without changing the other valid pixels of the image. That is, the tendency of smoothing caused in general by spatial filtering is avoided.

The conditions for applying an adaptive box filter to replace invalid or noise pixels in a DEM can be related to the difference between the value of the central pixel of the moving window of the filter and the average of the neighborhood. A maximum threshold is defined for this difference in absolute terms and relative to the number of times the standard deviation of the attributes of the neighborhood pixels. Pixels that meet both conditions are considered invalid. The replacement of these invalid values can be performed based on the arithmetic average of the valid pixels of the neighborhood in the moving window, excluding the own pixel, since it is considered invalid.

An example of spatial filtering of a DEM via adaptive box is shown in Fig. 4.15 for SRTM 30 data in an area of the municipality of Frei Martinho, in Paraíba, on the border with the state of Rio Grande do Norte. A group of 44 pixels with missing data is in the DEM. The applied filtering identified these 44 pixels and altered the DEM only in these elements. The pixels with missing data were replaced by the average of the valid neighboring pixels, taking a moving window of 5×5.

As illustrated by the topographic profile (Fig. 4.16), the elevation estimates generated by the filtering process for the invalid pixels appeared reasonably consistent with the surrounding topography. However, it is essential to remember that these are still estimates since there is no way to know the actual elevations that would best represent reality for these pixels, unless reliable topographic information from another source is available for this area.

The case of invalid data in a DEM refers to the most extreme case of noise, which is generally already labeled as such by the institution that processed and provided the data. However, other types of noise require a broader analysis to be detected in the DEM values, such as contamination of elevation information with artifacts on the

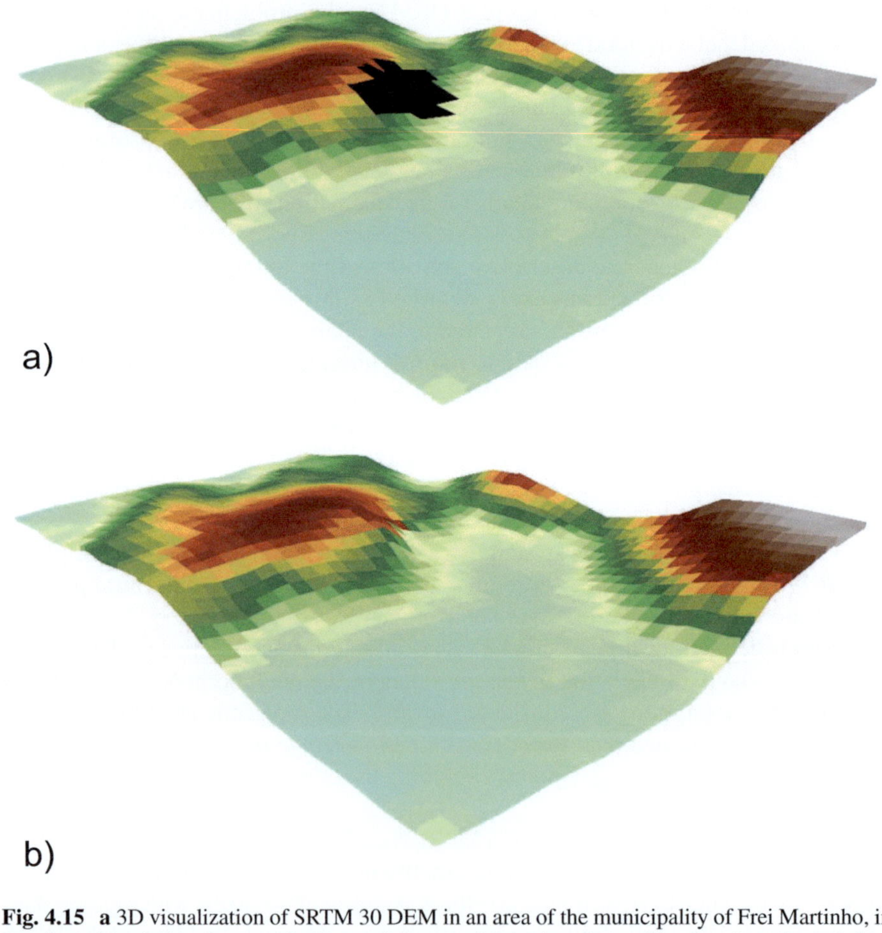

Fig. 4.15 **a** 3D visualization of SRTM 30 DEM in an area of the municipality of Frei Martinho, in Paraíba, with missing data, and **b** after applying an adaptive box filter with a 5×5 moving window for filling missing values

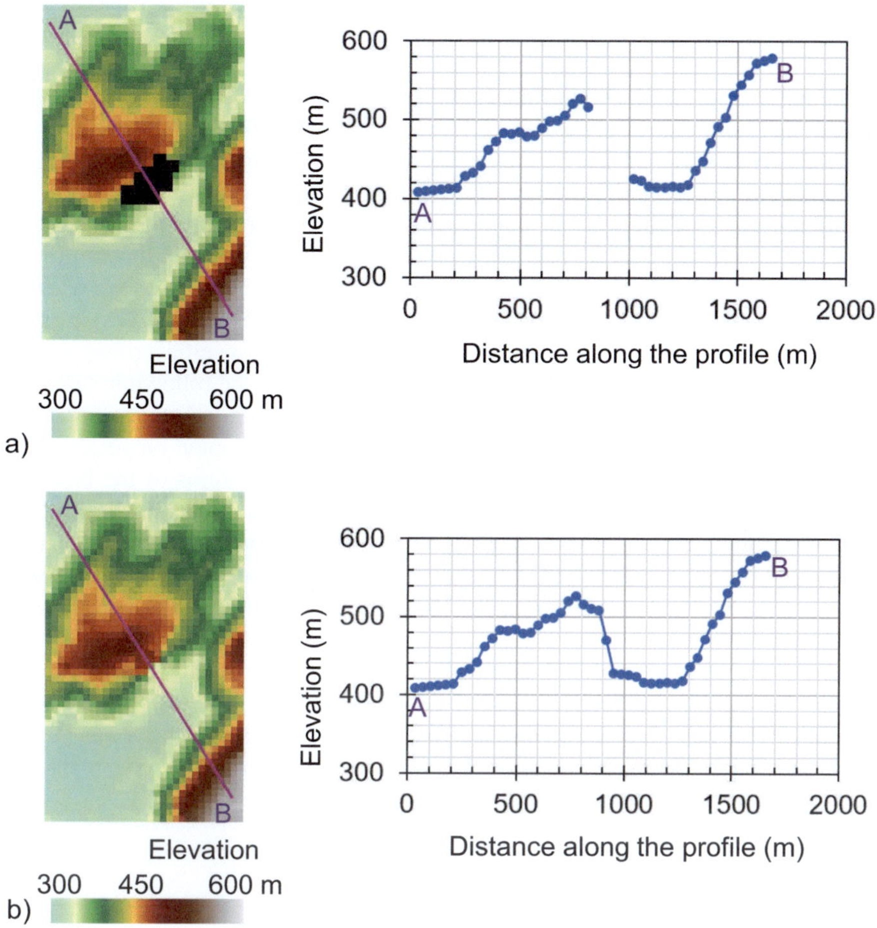

Fig. 4.16 Detailing of the SRTM 30 DEM with missing data presented in the previous figure: **a** Topographic profile of the original DEM along the line AB; **b** Topographic profile of the DEM after applying an adaptive box-type filter with a moving window of 5 × 5 for replacement of missing data, along the same line AB

ground (vegetation, transmission lines, buildings, etc.), atmospheric effects resulting from data acquisition, and signal scattering, among others. The treatment of this type of noise to improve the topographic information represented in the DEM requires a deeper understanding of each situation, which is beyond the purpose of this text.

4.11 Obtaining Topographic Data

The complete characterization of the Earth's relief concerns determining the topography of the Earth's surface, the bathymetry of oceans and lakes, and information on ice sheets (Hirt & Rexer, 2015). Obtaining data for this complete characterization requires multiple data sources from different survey types.

Different types of surveys can acquire topographic data from the Earth's surface. One of the most traditional ways is using ground equipment, such as a theodolite and a total station. There are also surveys with airborne equipment and sensors, whether conventional aircraft, such as airplanes, or uncrewed aerial vehicles (UAVs), a group in which drones stand out. Another type of survey is via specific orbital remote sensors aboard satellites or orbital spacecraft, as detailed in the following item.

The bathymetry of lakes and oceans can be surveyed using remote sensing techniques (such as altimetry), equipment like echo sounders mounted on vessels, or a combination of these methods. The information on the elevations of the bedrock, below the ice sheets, and the thickness of these layers is usually obtained by airborne orbital radar measurements. More details and examples of these bathymetry surveys of lakes and oceans can be found in Becker et al. (2009), while the work of Fretwell et al. (2012) provides more information about the characterization of bedrock and ice layers.

Several global or near-global databases represent the Earth's topography. Some of these databases specifically focus on ocean bathymetry (seafloor) or ice sheet elevations. This book focuses on datasets that describe the Earth's surface as Digital Elevation Models (DEMs), whether or not they include information on oceans, ice sheets, or bedrock.

4.12 DEM Data from Global Surveys

Over time, several digital elevation models of global (or near-global) scope have been developed and made available for free from different types of surveys. Some of these DEMs are restricted to topographic data of the Earth's surface, while others combine such information with data from the bathymetry of lakes and oceans and/or bedrock and ice sheets.

The interest of researchers, professionals, and students in topographic data available as global DEMs is enormous. The DEMs developed from data obtained by orbital remote sensing stand out and constitute the primary source of worldwide topographic information. This occurs for several reasons: the free availability, the global (or near) coverage, the data standardization, the ease of handling, the extensive description of the products, and the scientific literature with validation results, accuracy analyses, and other studies from these data.

According to the survey conducted for this book (Fig. 4.17), global topographic databases were made available as tables roughly resembling what would be a DEM

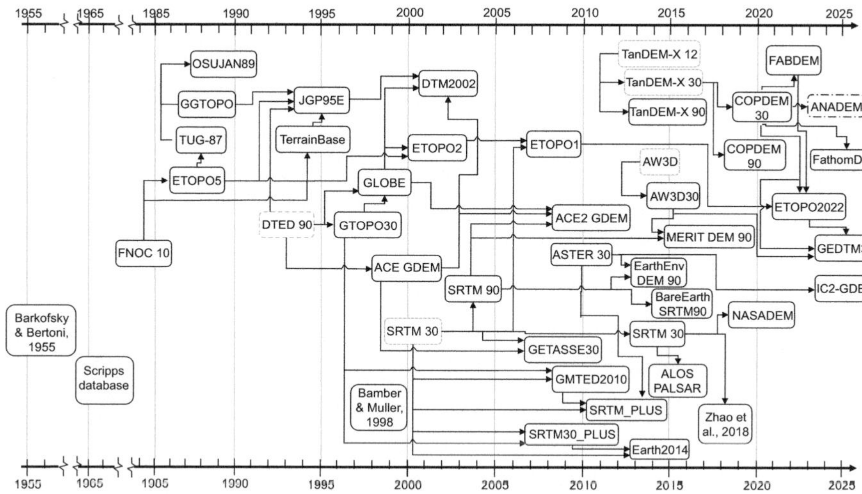

Fig. 4.17 Synthesis of the historical evolution of the availability of global or near-global DEMs (arrows indicate dependence in the elaboration of a DEM concerning others; DEMs with dashed outlines are presented for their relevance to serve as a basis for others, but were not made available at the time presented in the table)

numeric structure, since the 1950s–1960s, with the works of Berkofsky and Bertoni (1955) and Smith et al. (1966).

Rudimentary procedures like the one carried out by Berkofsky and Bertoni (1955) illustrate the difficulties of the time: transparent acetate plates with a $1° \times 1°$ grid (about 100×100 km) drawn were positioned over air navigation paper charts, and the average elevation values at each grid point were visually estimated.

In the following decades, 1970, 1980, and 1990, the first DEMs of global scope and made available for free were developed, such as the FNOC 10, ETOPO5, TUG-87, GGTOPO, OSUJAN89, TerrainBase, JGP95E, GTOPO30, GLOBE, and ACE GDEM models. Most of these models have a spatial resolution of approximately 10 km, except GTOPO30, GLOBE, and ACE GDEM, with 1 km, and GGTOPO, with 100 km. But all of them, without exception, result from combining multiple data sources, including regional databases, vector topographic databases, nautical charts, and air navigation charts. Some of these DEMs are based on previous DEMs, as indicated by the relationships established by the arrows in Fig. 4.17.

The difficulty of prospecting and compiling elevation data from the time is illustrated in the approach used by Wolf and Wingham (1992). From catalogs of more than 200 map producers around the world and a review in the scientific literature, these authors identified and contacted individuals and institutions considered potential sources of elevation data with public access; more than 350 letters questioning the existence of this type of data were sent, but in less than 10% of the cases were responses obtained with a description of the existing data.

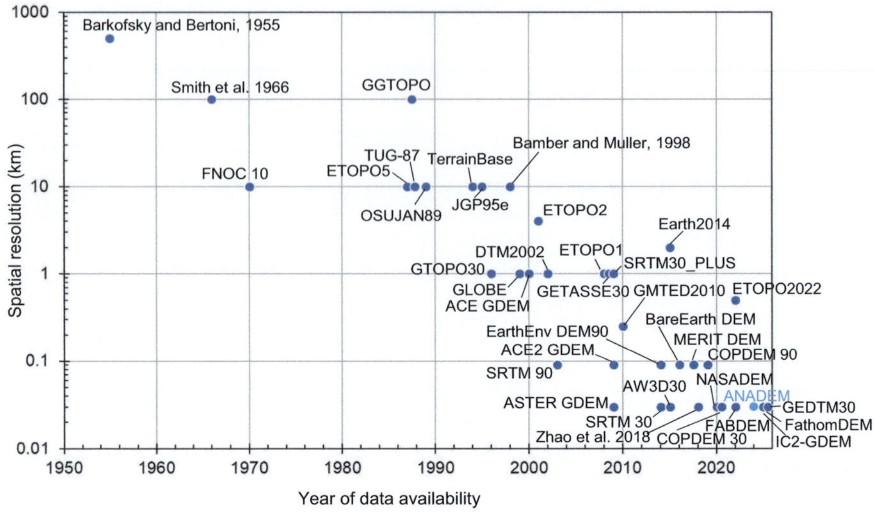

Fig. 4.18 Evolution over time of the spatial resolution of global or near-global DEMs according to the year of data availability (the ANADEM product, which is restricted to South America, was also included)

At the end of the 1990s and the beginning of the 2000s, the first DEMs of global or near-global scope, whose origin was entirely orbital remote sensing, were created. The DEM developed by Bamber and Muller (1998) is the first example of this type, with a resolution of about 10 km.

The DEM generated from the mission known as SRTM (Shuttle Radar Topography Mission), conducted in 2000, is a significant milestone in this regard, given the quality of the data produced and its spatial resolution on the order of tens of meters (initially, data were made available with 90 m resolution and later with 30 m resolution). Figure 4.18 schematically illustrates the evolution of the spatial resolution of the available DEMs globally over time.

Following the SRTM mission, several other global DEMs were (and continue to be) generated and made available for free with resolutions of around 90 and 30 m. Some of these DEMs correspond to new data acquisitions with different types of orbital surveys, such as ASTER GDEM, AW3D30, and TanDEM-X. Other DEMs correspond to reprocessing already available DEMs, seeking to benefit from the combination of more than one of them and/or to perform noise removal, such as the MERIT DEM, the NASADEM, and the FABDEM.

Among these DEMs available at a spatial resolution of 30 m, the ASTER GDEM is considered of lower quality (Bielski et al., 2023). The SRTM 30 and the NASADEM are very similar to each other and of higher quality than the ASTER GDEM, but they can also be considered outdated. According to the study of Bielski et al. (2023), DEMs AW3D30, COPDEM 30, and FABDEM stand out as having superior quality to their predecessors. When launched, the FABDEM was the only global DEM proposed as a DTM, an improvement from the COPDEM. Bielski et al. (2023) assess that there was indeed an improvement in some places in the representation of the FABDEM compared to the COPDEM 30. However, the opposite occurs in other areas, recommending the COPDEM 30 as of superior quality to the FABDEM in general. The continuous development in this scientific field is enormous. More recent global DEMs have been produced, such as the IC2-GDEM (Li et al., 2025), the FathomDEM, pointed out as of superior quality than FABDEM (Uhe et al., 2025), and GEDTM30, evaluated as outperforming the FathomDEM (Ho et al., 2025).

Specifically for South America, the ANADEM product has recently stood out as an improvement of the COPDEM 30 but is restricted to this continent (Laipelt et al., 2024), as described later.

Parallel to the search for refined DEMs, coarser spatial resolution global DEMs continued to be generated and made available, such as the GMTED2010, the Earth2014, and the ETOPO2022. There are two primary motivations for this: (i) to provide a model of relatively coarse resolution compared to 30 and 90 m data, since for many applications, mainly at the continental and global scale, there is no interest in data as refined as tens of meters; (ii) to provide a representation of the complete relief of the Earth, not just the topography of the terrestrial surface, but also elevations of ice sheets, information on bedrock, and bathymetry of lakes and oceans.

Figure 4.19 illustrates the topographic representation provided by different data sources, comparing cutouts of the DEMs of the GTOPO30 (resolution of 1 km), GMTED2010 (resolution of 250 m), SRTM (resolution of 90 m), and FABDEM (resolution of 30 m) for the same areas and scales. Visually, no distinction is perceived between the DEMs at the coarser scales, with all reproducing the same spatial patterns of topographic variability. As the scale increases, that is, when viewing the targets more closely, the differences become increasingly noticeable.

In the following sub-items, each global or near-global DEM with freely available data found in the survey conducted for this book is described. The level of detail in the description of each product varies according to the relevance of these data and the availability of information in official or consolidated sources.

4.12.1 Berkofsky and Bertoni (1955)

In 1955, the work of Berkofsky and Bertoni produced a database in tabular format with terrain elevations for the entire globe from aeronautical navigation charts, mainly the charts known as World Aeronautical Charts (scale 1:1,000,000), supplemented by data from the Aeronautical Planning Charts (scale 1:5,000,000). The elevations from the data sources were structured as average values on a 1° spacing grid by a visual procedure of the operators. Interpolation and extrapolation procedures were applied to estimate values in areas without data. Subsequently, the 1° data were converted to average values at intervals of $5° \times 5°$. This database was used, for example,

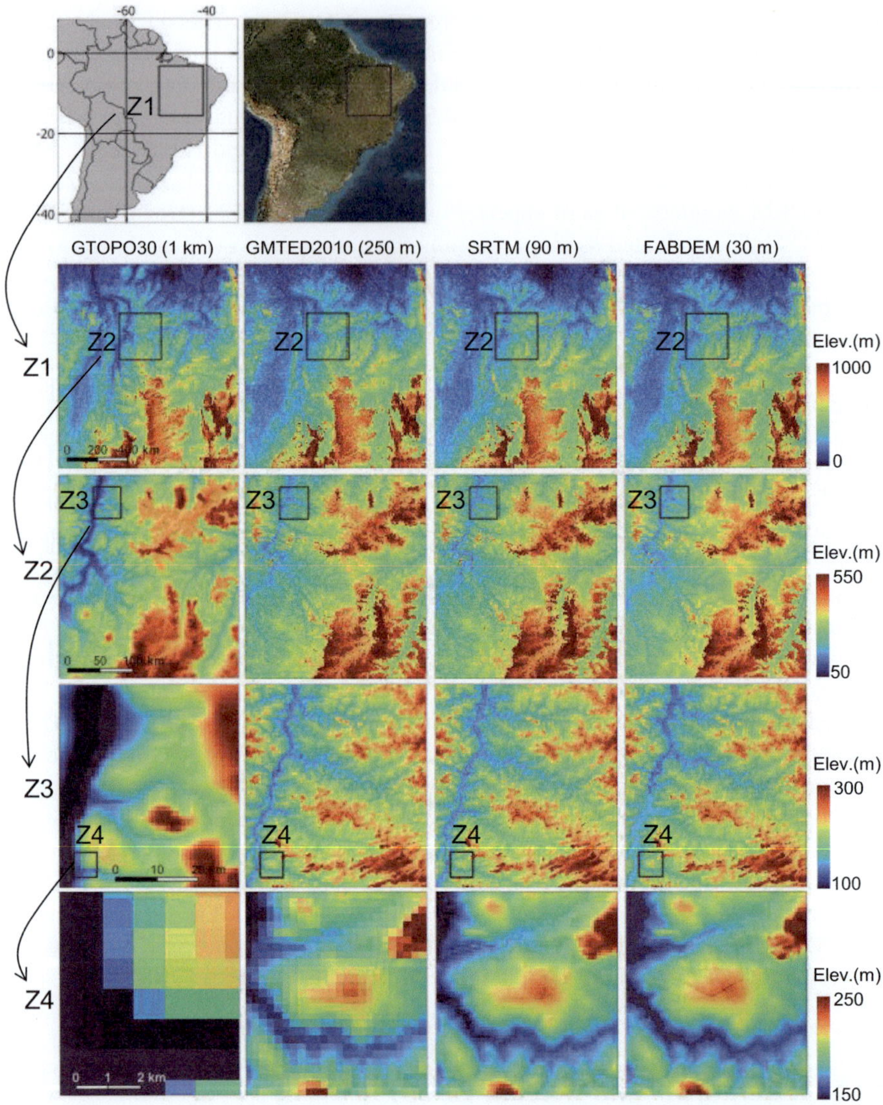

Fig. 4.19 Illustrative comparison of the topographic representations of a part of the Brazilian territory at different spatial scales, considering the DEMs GTOPO30 (spatial resolution of 1 km), GMTED2010 (250 m), SRTM (90 m), and FABDEM (30 m)

in a wind circulation study (Vianello, 1985) and a work with global hydrological modeling (Kavvas & Anderson, 2002).

The data generated by Berkofsky and Bertoni (1955) are presented explicitly in such a publication as a printed map with elevations positioned at each corresponding point of the 5° × 5° grid (or about 500 × 500 km at the Equator). These elevations are presented in the unit of hundreds of feet. This elevation dataset is not referred to by the authors nor reported in the literature as a DEM but as a topographic chart (Vianello, 1985). However, the regular nature of the discretization of the data, positioned according to a grid with geographic coordinates, can characterize this dataset at least as a rudimentary matrix format of a raster DEM.

4.12.2 Smith et al. (1966)

In 1966, Smith et al. presented a database of terrestrial surface elevation and sea bottom depths in a grid of approximately $1^\circ \times 1^\circ$, with units in meters. The primary motivation was to have a global database with more detail than what was generated in the work of Berkofsky and Bertoni (1955).

The average values of elevation or depth in each grid square were obtained by visual estimation from contour lines presented in topographic and bathymetric maps at scales of 1:1,000,000 and 1:25,000,000. As this information development occurred at the Scripps Institution of Oceanography, University of California (USA), this dataset is sometimes referred to as the Scripps database or simply Scripps Topography. The details of procedures and the listing of the topographic and bathymetric maps considered in the preparation of this database are presented in Smith et al. (1966).

In 1975, this Scripps database was still recognized as "the best available source of large-scale continental elevations," according to Gates and Nelson (1975). However, these authors consider the format of these data to be of minimal convenience, hindering and restricting most applications. The authors developed a different way of presenting these data to circumvent this difficulty and revised the information to treat inconsistencies and improve it.

Gates and Nelson (1975) presented the revised and updated Scripps database as tables with longitude coordinates in the columns and latitude coordinates in the rows, and both discretized every 5°. The elevation or depth values are in the unit of decameters ($\times 10$ m), with an asterisk to indicate when it is depth instead of elevation. This database is also not reported by the authors or the literature as a DEM, but it has a data structure in a format roughly similar to a raster DEM.

4.12.3 FNOC 10

The Fleet Numerical Meteorology and Oceanography Center (FNOC) of the US Navy developed a global DEM with a spatial resolution of 10 arc-minutes (0.16666... degrees or close to 20 km at the equatorial level). This DEM is usually referred to by the acronym FNOC 10 as a reference to the developing institution and the spatial resolution in arc minutes. The development of this DEM began in the mid-1960s and advanced until the beginning of the following decade (NCGIA, 2023), with updates continuing until 1984, the date of the last version.

The FNOC 10 was developed from various topographic data sources, with the primary one being the navigation charts of the US Navy (US Department of Defense Operational Navigational Charts) at a scale of 1:1,000,000 (NCGIA, 2023). Air navigation charts were selected from the Jet Navigation Charts and World Aeronautical Charts databases in areas not covered by such nautical charts.

In addition to the DEM itself, a dataset was developed that included, for each grid point, the minimum and maximum elevation, among other topographic characteristics (Joseph, 1982). An example of the FNOC 10 data application is the study of atmospheric modeling and simulation of CO_2 concentration developed by Fujita et al. (2003).

4.12.4 ETOPO

The ETOPO acronym, from Earth TOPOgraphy, refers to a group of global DEMs developed chronologically: ETOPO5, ETOPO2, ETOPO1, and ETOPO2022.

The global DEM called ETOPO5 (the number 5 comes from 5 arc-minutes of spatial resolution) was initially produced jointly by the National Geophysical Data Center of NOAA and the US Navy from databases of different sources and with varying degrees of detail. Some of these data sources are equally at the resolution of 5 arc-minutes (0.08333… degrees, or approximately 10 km at the Equator) as ETOPO5, while other databases are at coarser resolutions. In areas without the availability of different data, the elevations of FNOC 10 were used (NCGIA, 2023).

ETOPO5 has global coverage and represents both the elevations of continental masses, characterized by positive values, and the depths of the ocean floor, represented by negative values. With integer-type data, this DEM was made free in 1987 in the WGS 84 horizontal reference system and the mean sea level vertical reference.

The successor to ETOPO5, ETOPO2, was launched in 2001. This global DEM with a spatial resolution of 2 arc-minutes (0.0333… degrees or close to 4 km at the Equator) was updated in 2006. ETOPO2 differs from ETOPO5 in the topographic representation of the continental masses mainly by including GLOBE data (described later in this chapter), a global DEM that did not exist at the time of ETOPO5. ETOPO2 was produced with integer format data in the WGS84 horizontal reference system and with the mean sea level as the vertical reference.

ETOPO1 was produced in 2008, with a spatial resolution of 1 arc-minute (0.01666… degrees, or almost 2 km at the Equator). This DEM is available in two versions, which differ only in the elevations in the ice areas of Antarctica and Greenland (Amante & Eakins, 2009). In one of the versions, the elevations of the surface of the ice sheets of these regions are represented. In contrast, in the other version, the elevations refer to the rocky base where these ice sheets are settled.

ETOPO1 was developed from various sources of global and regional topographic data, such as SRTM and GLOBE, in addition to depth data from the sea floor, ice sheets, bedrock, and elevations of the coastal part of the continents. All data sources were compatible with the horizontal and vertical references, and other edits were made for adjustments and the coherence of the values, as described in Amante and Eakins (2009). ETOPO1 covers elevations of the entire globe and includes the sea floor's depths. This DEM was generated with integer data using the WGS84 horizontal reference system and has the average sea level as the vertical reference.

Continuing the family of ETOPO global DEMs, ETOPO2022 was generated in 2022. It is a DEM with a spatial resolution of 15 arc seconds (about 0.0041666... degrees or almost 500 m at the equator level), developed by NOAA (2022). This DEM has global coverage that includes the continental masses' elevations and the ocean floor's depths, just like its ETOPO predecessors.

The ETOPO2022 version was elaborated from several data sources, including airborne LiDAR sensors, orbital topographic data, and bathymetric data obtained by vessels. Globally, the elevation data of the continental masses of ETOPO2022 come from the global DEMs COPDEM and FABDEM, both at a resolution of 30 m and described later in specific items of this chapter. ETOPO2022 is free, with data in integer format, and refers to the WGS 84 horizontal reference system and the EGM 2008 geoid vertical datum.

4.12.5 GGTOPO

Trent University (Canada) developed the GGTOPO (Global Grid TOPOgraphy) with a spatial resolution of 1° (or about 100 km at the Equator). This DEM is mainly the result of the compilation of the Scripps Topography database, described in Gates and Nelson (1975) and commented on in a previous item of this book, but with extensive revision (Cogley, 1987, cited in Pavlis & Rapp, 1990).

4.12.6 TUG-87

The global DEM TUG-87 consists of a slight improvement of ETOPO5. Its name refers to the developing institution, the Technical University of Graz (TUG) in Austria, and the year of its elaboration (1987). As ETOPO5, the TUG-87 represents terrain elevations and sea floor depths at a spatial resolution of 5 arc-minutes (0.08333... degrees, or approximately 10 km at the Equator), with global coverage. The difference between TUG-87 and ETOPO5 is that additional bathymetric data were used to compose the ocean depths, while the elevations referring to the terrain are almost the same in these two models (Wieser, 1987, cited in Wolf & Wingham, 1992). The additional bathymetric data added to TUG-87 are mainly from the database known as DBDB5 (Digital Bathymetric Data Base-5 min) (Pavlis & Rapp, 1990). In addition, TUG-87 was made available in three spatial resolutions: 5 arc minutes, 30 arc minutes, and 1° (Lemoine et al., 1998).

4.12.7 OSUJAN89

The OSUJAN89 is a global DEM developed from the combination of two other global DEMs available at the time, the TUG-87 and the GGTOPO. It covers information from both terrain elevations and sea floor depths (Pavlis & Rapp, 1990). Comparisons between these two models (TUG-87 and GGTOPO) indicated considerable discrepancies between the elevations, leading to a preference for the TUG-87 data to elaborate the OSUJAN89. This was done for both the land surface and the ocean depth elevations. The exception was for areas classified as oceans with ice cover, for which GGTOPO data were used. This procedure was adopted because there is information in the GGTOPO to identify this region type. At the same time, it is not known at each point if the TUG-87 elevations refer to which type of surface (land, ice cover, etc.) (Pavlis & Rapp, 1990).

The name OSUJAN89 is an abbreviation for Ohio State University Joint Archive for Digital Elevation Models, combined with the year of development (1989). The OSUJAN89 was developed with a spatial resolution of 5 arc-minutes (0.08333... degrees, or approximately 10 km at the equator level).

4.12.8 TerrainBase

The DEM called TerrainBase is considered a prototype of the DEM later developed and named GLOBE and represents considerable advances relative to ETOPO5. A significant difference between TerrainBase and ETOPO5 is the quantity and quality of the data sources used in the DEM elaboration. For the TerrainBase, 15 data sources were used, 10 additional to the 5 used for the ETOPO5 (Arabelos, 2000). One of the pieces of information used for the ETOPO5 and the TerrainBase was the elevations of the FNOC 10 (Lemoine et al., 1998). The development of TerrainBase was intended to solve many of the problems related to errors and artifacts reported in the ETOPO5 data (NOAA, 2023).

In the elaboration of TerrainBase, all data sources were converted to a resolution of 5 arc-minutes and combined to obtain the global DEM, as described in NOAA (2023). In areas where two or more data sources overlapped, a hierarchical selection procedure was established first to get data from the most relevant source, fill gaps using data from the second-highest ranked source, and so on. Additional processing was also performed to smooth and ensure compatibility across transitions between areas covered by different datasets, primarily to reduce elevation discontinuities caused by combining different data sources.

The TerrainBase was produced in 1994 by the USA's National Geophysical Data Center (NGDC) in conjunction with the World Data Centers-A for Geophysics and Marine Geology (WDC-A). It has global coverage and a spatial resolution of 5 arc-minutes (approximately 10 km at the equator latitude), including elevations of the land surface (positive values) and depths of the oceanic areas (negative values).

4.12.9 JGP95E

The DEM named JGP95E was developed in 1995 from the combination of various regional and global data sources, such as the DTED (Digital Terrain Elevation Data), prepared by the National Geospatial-Intelligence Agency (NGA) of the USA with restricted use, and global DEMs available TerrainBase, ETOPO5 and GGTOPO, in addition to ice cover data obtained by altimetry (Lemoine et al., 1998). The nomenclature JGP95E refers to Joint Gravity Project (JGP) combined with the development year 1995, while the letter E refers to the final version, since throughout the development process, the versions JGP95A, JGP95B, JGP95C, and JGP95D were made.

The motivation for the creation of JGP95E is associated with the perceived need by the Defense Mapping Agency (DMA) of the USA and NASA to have a global DEM of better quality in the spatial resolution of 5 arc-minutes (about 10 km at the Equator), aiming at the elaboration of a global geopotential model (Arabelos, 2000). From JGP95E, the representation of the geoid known as EGM96 was subsequently developed (Lemoine et al., 1998).

4.12.10 GTOPO30

GTOPO30 is a global DEM with a spatial resolution of 30 arc seconds (0.008333… degrees, or approximately 1 km at the Equator) generated from different sources of topographic data in vector and raster format (EROS, 2018). This DEM was completed in 1996 under the coordination and leadership of the U.S. Geological Survey's Center for Earth Resources Observation and Science (USGS-EROS), with collaboration from other institutions in different countries.

GTOPO30 was developed to meet the general scientific requirements for topographic data established by the Topographic Science Working Group in 1988 (Gesch et al., 1999). The primary database was the DEM named DTED from the NGA of the USA. This DTED data has a spatial resolution of approximately 90 m and was resampled to a resolution of almost 1 km.

About 50% of GTOPO30 is a result of DTED. The second central database used to generate GTOPO30 (about 30% of the area), mainly to fill areas not covered or gaps in the DTED data, was the Digital Chart of the World (DCW), a vector base of topographic data at the 1:1,000,000 scale also produced by the NGA. The other datasets used for GTOPO30 have regional or local coverage and were used for areas without DTED and DCW data availability. The complete description of these data and details of the development of GTOPO30 can be found in Gesch et al. (1999) and EROS (2018).

GTOPO30's coverage is global, and it was designed to be freely available. It uses integer-type data in the WGS 84 horizontal reference system and EGM96 vertical reference (Podest & Crow, 2013).

4.12.11 Bamber and Muller (1998)

Bamber and Muller (1998) developed a global DEM from data from the radar altimetry of the ERS-1 mission (European Remote Sensing Satellite-1) during the geodetic data acquisition phase over a cycle of 336 days. As described in such a publication, various data processing procedures were carried out, including filtering and interpolation. The resulting DEM has a spatial resolution of 5 arc-minutes (or almost 10 km at the equatorial level).

4.12.12 GLOBE

The DEM named GLOBE, from the acronym Global Land One-km Base Elevation, was developed through the cooperation of various international experts on behalf of a task force called the GLOBE Task Team. GLOBE has global coverage and a spatial resolution of 30 arc seconds (0.008333… degrees), or almost 1 km at the Equator (Hastings et al., 2000).

Various sources of topographic data were used in the development of GLOBE, with emphasis on the NGA's DTED. About 58% of GLOBE in continental areas is based on DTED, while 23% is based on the vector base of topographic data, DCW, also produced by the NGA (Hastings & Dunbar, 1998).

GLOBE is, in addition to a DEM, the name of a broad project with various purposes far beyond the production of the DEM of the same name. Two other global DEMs were produced during the development of GLOBE with this exact spatial resolution (Hastings & Dunbar, 1998): a DEM developed by NASA for internal use and support of space missions and GTOPO30. GLOBE data has been free since 1999, with elevations relative to the WGS 84 horizontal reference system and EGM96 vertical reference.

4.12.13 ACE GDEM

The results of analyses that revealed issues and errors in the GTOPO30 and GLOBE datasets motivated the development of a new global Digital Elevation Model (DEM) with a spatial resolution of 30 arc-seconds (approximately 1 km at the Equator), based on altimetry data from the ERS-1 geodetic mission of the European Space Agency (ESA), supplemented by other global datasets such as DTED and DCW (Berry et al., 2020). This DEM was developed by researchers from De Montfort University (Leicester, United Kingdom) and named the ACE GDEM (Altimeter Correction Elevations, Global Digital Elevation Model).

The details of the ACE GDEM development are described in Johnson et al. (2000) and summarized here. A fundamental step was implementing a spatial interpolation procedure to adjust a surface to the altimetry data and estimate elevation values at points not sampled by altimetry but within the area covered by this survey. A second significant step was the fusion of the data from altimetry with the available global topographic data. It was assumed that these available data were incorrect and needed to be evaluated for use, prioritizing using altimetry data when available or adjusting the available global data based on the comparison with altimetry. When neither of these situations was possible, the information from the existing global data was used.

Around 29% of the ACE GDEM data for land surfaces is derived directly from altimetry data, and about 11% is related to global data corrected or adjusted by altimetry data. The remaining 60% of the data refers to other topographic bases without adjustment by altimetry. The ACE GDEM was developed in the horizontal reference system WGS 84 and vertical reference EGM96, with integer data and global coverage.

From the completion of the SRTM mission in 2000 (described in detail in an item later in this chapter) and the availability of the corresponding DEM, researchers involved in the development of the ACE GDEM saw an opportunity to benefit from the altimetry data of the ERS-1 and techniques used in the development of the ACE GDEM to generate an improved version of the SRTM 90 DEM. While the main limitation of SRTM elevations is related to areas with vegetation cover, as described later, the altimetry data of the ERS-1 have a higher capacity to represent the elevation of the terrain itself, given the higher penetration capacity of the altimetric sensor signal through the vegetation cover (Smith & Berry, 2010).

The ACE2 GDEM was developed by merging data from the SRTM DEM and about 70 million elevation data points obtained by altimetry from the ERS-1, complemented by other data, such as from GLOBE (Berry, 2009; Berry et al., 2010; Smith & Berry, 2010). In areas with a higher density of vegetation cover, the ERS-1 altimetry data prevailed. In some regions, SRTM data were used after being corrected based on comparisons with the ERS-1 altimetry. In other areas, the SRTM data remained due to the unavailability of ERS-1 data. In areas outside the coverage of the SRTM, the ERS-1 altimetry data were merged with other available databases, such as GLOBE.

This procedure resulted in a DEM with global coverage, a spatial resolution of 3 arc-seconds (approximately 90 m at the Equator), integer data, a horizontal reference system WGS 84, and a vertical reference EGM 96. The development of this DEM was coordinated by the European Space Research Institute (ESRIN) of the European Space Agency and made available in 2009.

4.12.14 DTM2002

The global model called DTM2002 (an acronym for Digital Terrain Model 2002, year of development) was developed by Saleh and Pavlis (2002). This model represents the elevations of the land surface and depths of the sea floor, with global coverage and spatial resolution of 30 arc-seconds (0.008333... degrees or close to 1 km at the Equator), with versions also available at resolutions of 2 arc-minutes (0.0333… degrees or almost 4 km) and 5 arc-minutes (0.08333… degrees or about 10 km).

The DTM2002 was developed by combining several data sources described in Saleh and Pavlis (2002). For the coastal parts, the primary information was the database of Wessel and Smith (1996). The land surface elevations were mainly based on GLOBE, using data from the ACE DEM model and other sources. Ocean bathymetry data were compiled from Smith and Sandwell (1997), whose work is based on altimetry information and data collected by vessels, complemented by data from the JGP95E model. Specific databases were also included to compose information on ice layer thickness and depths of large lakes for the 2 and 5-arc-minute versions of DTM2002.

4.12.15 SRTM DEM

Even today, the most famous and used data regarding global DEM are those from the SRTM mission (Shuttle Radar Topography Mission; Farr et al., 2007). An orbital spacecraft collected these data in 2000 via interferometry with synthetic aperture radar sensors operating with wavelengths in the spectral band C (wavelength 5.6 cm) and spectral band X (wavelength 3.1 cm) (Walker et al., 2007).

The data from the SRTM mission marked a milestone, making a paradigm shift in global digital topography due to their coverage, quality, and free availability (Bielski et al., 2023; Grohmann, 2018).

The SRTM DEM was widely available initially with a spatial resolution of approximately 90 m (3 arc-seconds or 0.000833… degrees), commonly referred to as SRTM 90 and referring to the data obtained in the C band. Later, data from the SRTM mission were also made available at a spatial resolution of about 30 m (1 arc-second or 0.000277… degrees), which refers to the data obtained in the X band. This 30 m DEM is usually referred to as SRTM 30. The elevation data from SRTM are in the horizontal reference system WGS 84 and have the geoid EGM 96 as a vertical datum, available in integer data format. The SRTM DEM covers approximately from 60° N to 56° S.

The first version of SRTM 90 was made available in 2003, and the second version was made available in 2005, with improvements related to noise removal and filling in missing data. In addition to the versions of the SRTM DEM made available by the institutions responsible for acquiring and processing the data, NASA and USGS, relevant versions of these data were made available by CGIAR (Consultative Group on International Agricultural Research). CGIAR released versions 3 and 4 of the SRTM 90 in 2006 and 2008 (Jarvis et al., 2008), using other procedures for improving coastal areas, noise removal, gap filling, and assistance from different data sources.

Version 3, developed by NASA, is also known as SRTM_PLUS. This dataset was elaborated from version 2, with gap-filling mainly based on ASTER GDEM2 and GMTED2010 (both described later in this chapter; Buckley et al., 2020).

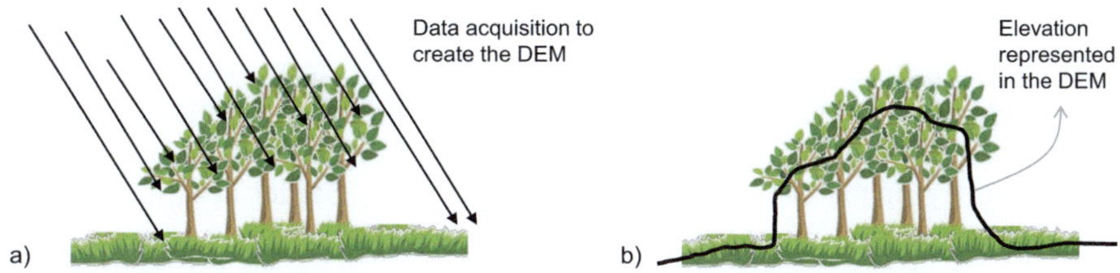

Fig. 4.20 Illustrative scheme of data acquisition (**a**) and the resulting SRTM DEM representation (**b**), showing the influence of vegetation cover in limiting the complete penetration of electromagnetic waves to the true ground surface

The SRTM 30 DEM for the United States was released in 2003 and has been available worldwide since 2014.

Due to the data acquisition type, the SRTM elevations are contaminated with vegetation, especially when it is large in size and density (Fig. 4.20). The electromagnetic wave of the radar used in this imaging cannot fully penetrate the vegetation canopy and reach the terrain properly (bare ground). The main reason for this impediment has to do with the occurrence of the scattering of this electromagnetic wave within the vegetation canopy due to the wavelength used being of similar size to the dimension of the leaves, which act as a scattering element of the signal, as well as branches and trunks also act in this sense. For a better understanding of this effect, it is recommended that one read the works of Baugh et al. (2013) and Walker et al. (2007).

In areas with dense vegetation, the elevations of the SRTM DEM do not refer to the bare ground, but they do not necessarily represent the top of the vegetation. Instead, they are in an intermediate position between the top of the canopy and the ground. Various factors such as the structure of the vegetation in terms of size and quantity of leaves, height and density of vegetation, branching angle, and even the humidity of the trunks and branches and the soil moisture interfere with making variable how much the electromagnetic wave of the sensors used in the SRTM mission penetrated the vegetation canopy (Brown et al., 2010; Kellndorfer et al., 2004). Theoretically, assuming constant vegetation cover conditions and similar characteristics such as incidence angle and polarization, the electromagnetic wave in the C-band can penetrate the canopy more than the X-band wave, due to its longer wavelength, albeit with a relatively small difference (Kellndorfer et al., 2004).

Due to the influence of vegetation on elevation measurements, the SRTM DEM is referred to by several authors as a Digital Surface Model. The prior recognition of this vegetation effect on the data even influenced data acquisition planning during the winter in the Northern Hemisphere (Smith & Berry, 2010), when temperate forests shed their leaves.

Furthermore, in urban areas, SRTM elevations also correspond to the tops of buildings. However, due to the DEM's pixel size, the information is typically mixed, representing a mix of buildings, bare ground, and even vegetation.

In 2009, SRTM30_PLUS, a global DEM with a spatial resolution of 30 arc seconds (0.008333… degrees or about 1 km at the latitude of the Equator), was developed. This should not be confused with version 3 of the SRTM, also called SRTM30_PLUS. The SRTM30_PLUS covers the representation of the terrestrial surface and the bathymetry of the oceans, as described in detail in Becker et al. (2009) and summarized here.

For the topographic representation of the terrestrial surface, the SRTM30_PLUS combines data from the SRTM 30 (1 arc-second or ~ 30 m) resampled to 30 arc-seconds, complemented by data from GTOPO30 in the Arctic and elevations derived from ICESat in Antarctica. The bathymetry of the oceans represented in the SRTM30_PLUS was elaborated from multiple data sources, including surveys by orbital remote sensing and databases of bathymetric soundings by vessels. This model also includes the bathymetry of the large lakes of the terrestrial surface (Lakes Superior, Michigan, Huron, Erie, Ontario, and Baikal) and the Caspian Sea (Hirt & Rexer, 2015).

4.12.16 GETASSE30

The DEM called GETASSE30 (Global Earth Topography And Sea Surface Elevation at 30 arc second resolution) was elaborated with spatial resolution of 30 arc-seconds (0.08333… degrees or almost 1 km at the Equator) from existing databases: the SRTM 30 DEM, the ACE GDEM and MSS (Mean Sea Surface), the latter with data referring to the sea surface. EGM96 geoid data were also used to produce GETASSE30 elevations referenced to the WGS84 ellipsoid.

The GETASSE30 was elaborated in 2008 by the European Space Research and Technology Centre (ESTEC) of the European Space Agency (ESA). It is composed of data in integer format, with horizontal and vertical references WGS84 (SNAP/ESA, 2023).

4.12.17 GMTED2010

The Global Multi-resolution Terrain Elevation Data 2010, or simply GMTED2010, is a global DEM developed in 2010 in collaboration between the USGS and the NGA, aiming to replace the GTOPO30 for continental and global applications (Danielson & Gesch, 2011). The GMTED2010 was generated in three distinct spatial resolutions: 30 arc-seconds (about 1 km at the Equator), 15 arc-seconds (about 500 m), and 7.5 arc-seconds (approximately 250 m). The coverage of GMTED2010 is almost global, between 84° N and 56° S for most products and between 84° N and 90° S for some products. (Danielson & Gesch, 2011). Areas such as Antarctica and Greenland only have data available for the 30-arc-second version due to the unavailability of more refined data for elaborating the DEM.

GMTED2010 is produced using 11 elevation data sources. The primary data source is the SRTM elevations, representing almost 70% of the data from continental areas. This is the main reason for the better quality and better detailing of the topographic representation achieved with GMTED2010 compared to the GTOPO30, primarily based on elevations from coarser databases, such as the DTED and the DCW.

In areas outside the SRTM's coverage or with missing data from this mission, several other sources were used, such as the DTED (about 14% of the continental areas) and other regional databases, like the Canadian Digital Elevation Data (CDED), Satellite Pour l'Observation de la Terre (SPOT 5) Reference 3D, and National Elevation Dataset (NED), among others. Data from the GTOPO30 itself was also used, but almost as a last resort.

The elevations of the GMTED2010 are in integer format, and all products are related to the WGS84 horizontal reference system, with most of these data referring to the EGM96 geoid. For areas with data from sources other than SRTM and SPOT 5, the vertical datum in the GMTED2010 is according to the source of the data (Danielson & Gesch, 2011). In addition to the conventional DEM product, representing the elevations at each point of the grid, seven other derived products were also generated and made available to users, such as maximum, minimum, average, median elevations, and standard deviation.

4.12.18 ASTER GDEM

The DEM obtained from ASTER data (Advanced Spaceborne Thermal Emission and Reflection Radiometer) is called ASTER GDEM (ASTER Global Digital Elevation Model). This DEM was developed from data acquired by orbital photogrammetry using sensors on board the Terra satellite, which was launched in 1999 (Tachikawa et al., 2011). ASTER is one of the five instruments on board this satellite. This sensor was built in Japan, but the project, calibration, and data validation are the joint responsibility of Japanese and American teams, mainly involving the institutions NASA, NGA, and USGS in the USA, and METI (Ministry of Economy, Trade and Industry), ERSDAC (Earth Remote Sensing Data Analysis Center) and the University of Tokyo, on the Japanese side (EarthData, 2023).

ASTER is composed of three sensor subsystems operating in infrared spectral bands: visible-near-infrared (VNIR), shortwave-infrared (SWIR), and thermal-infrared (TIR) (Abrams et al., 2020). Each subsystem has specific spectral and spatial resolution characteristics, and together, they compose data collected in 14 distinct spectral bands. Among the information acquired, reflectance, emissivity, surface temperature, and elevation data stand out.

The elevation data from ASTER are the result of the processing of stereographic optical images, as detailed in Taschikawa et al. (2011). Due to this data acquisition, the elevations obtained refer to the top of the surface coverage, whether due to vegetation or buildings. Therefore, several authors also consider the ASTER DEM a DSM. These data are also sensitive to the occurrence of clouds (Grohmann, 2018).

Up to the writing of this book, three versions of the DEM obtained from ASTER data were released in 2009, 2011, and 2019 and named ASTER GDEM-1, ASTER GDEM-2, and ASTER GDEM-3, respectively. All versions have a spatial resolution of approximately 30 m, and each version benefits from the continued data acquisition since the previous version and improvement processes. The GDEM-1 results from data collected in about 1.264 million scenes from March 1, 2000, to November 30, 2007, while the GDEM-2 had an additional 260,000 scenes (Abrams et al., 2020). The WGS84 is considered the horizontal reference system, and the EGM96 is the vertical datum.

The GDEM-3 results from processing 2.3 million scenes (EarthData, 2023). The acquisition of ASTER GDEM data covers almost the entire Earth's surface (from 83°N to 83°S), and the third version of the data presents improvements that reduce areas without data and anomalies of water-covered areas (EarthData, 2023).

4.12.19 ALOS AW3D

From 2011 to 2015, the PRISM sensor (Panchromatic Remote-Sensing Instrument for Stereo Mapping) acquired data on board the ALOS satellite (Advanced Land Observing Satellite) with a spatial resolution of 2.5 m. PRISM is an optical type sensor comprising three panchromatic radiometers, collecting data by orbital photogrammetry. These data covered a total of 6.5 million scenes. They were used by the Japanese space agency (JAXA) to develop a global DEM with a spatial resolution of 5 m, which is called ALOS World 3D or simply AW3D (Tadono et al., 2015, 2016).

The DEM with a 5 m spatial resolution resulting from the ALOS PRISM data considers the WGS84 horizontal reference system and the EGM96 vertical datum and is commercially available. From this product, a resampling was done to obtain a DEM with a spatial resolution of 30 m, referred to as ALOS World 3D 30 mesh or AW3D30, covering from 82° N to 82° S. JAXA makes the AW3D30 available for free.

The first version of AW3D30 (version 1.0) was made available in May 2015, and scenes from the most recent version up to the writing of this book (version 3.2) have been released since March 2021 (JAXA/EORC, 2021). New versions are created from data improvements, such as noise removal, gap filling, and adopting and updating masks with delineating continental water bodies and ocean coastlines (JAXA/EORC, 2021).

Due to the data acquisition type, the elevations of AW3D30 refer to the top of the vegetation cover and buildings, turning this product into a DSM. Additionally, the AW3D30 is subject to cloud interference during data collection. However, according to an evaluation by Grohmann (2018), as the AW3D30 is a result of the AW3D data, elaborated with a resolution of 5 m and with a higher capacity to detect open areas between trees, the elevations of the AW3D30 represent average values influenced by the 5 m elevations in these open areas, resulting in values lower than the average height of the vegetation cover. A study by Bielski et al. (2023) points out AW3D30 as a high-quality and one of the best free products globally.

4.12.20 Earth2014

The model called Earth2014 was developed by Hirt and Rexer (2015) to obtain a complete representation of the Earth's relief, including the terrestrial surface, ice sheets, and bedrock. Four databases were combined to obtain Earth2014. The SRTM 90 DEM version 4.1, developed by CGIAR (Jarvis et al., 2008), with an approximate spatial resolution of 250 m, was used to represent the surface topography between 60° N and 60° S. Data from SRTM30_PLUS, developed by Becker et al. (2009), were used in the representation of ocean bathymetry and also of continental topography, in the case of areas above 60° N. A third dataset was the Bedmap2 (Fretwell et al., 2012), with data from the bedrock, bathymetry, and thickness of ice sheets in Antarctica. The fourth database was the Greenland Bedrock Topography v3 (Bamber et al., 2013), which is about the bedrock of Greenland.

The SRTM30_PLUS data were considered the primary layer of information from which the other databases were combined. A series of procedures was applied to make this combination regarding the different spatial resolutions involved and the treatment of inconsistencies between the data. As a result of these procedures, Earth2014 is composed of five different representations of the Earth's relief (Hirt & Rexer, 2015): (i) representation of the terrestrial surface layer (considers zero depth of the oceans and elevations of the surface of lakes according to SRTM v4.1); (ii) representation of the bedrock layer (the planet without water and ice); (iii) representation that includes topography, bedrock, and ice (disregards liquid water); (iv) representation of ice and water masses converted into equivalent layers of rock; (v) exclusive representation of the thicknesses of the ice sheets.

Earth2014 was made available in 2015 with products at a spatial resolution of 1 arc-minute (0.01666… degrees or almost 2 km at the Equator), with the GRS80 ellipsoid as the horizontal reference, which in practice has a negligible difference relative to WGS84. The vertical reference is the mean sea level (Hirt & Rexer, 2015). Earth2014 is considered an improvement over its predecessor, called Earth2012, whose spatial resolution is 5 arc-minutes (0.08333… degrees or about 10 km at the Equator) and which was based on older data, such as elevations from ETOPO1 (Hirt & Rexer, 2015).

4.12.21 TanDEM-X

The mission called TanDEM-X (TerraSAR-X add-on for Digital Elevation Measurement) was developed in a public–private partnership between the German Aerospace Center (DLR) and the company Airbus to produce a global DEM with a spatial resolution of 12 m (Krieger et al., 2007). The primary data for generating this DEM were acquired from 2010 to 2015 by interferometry with synthetic aperture radars operating in the X spectral band (wavelength 3.1 cm).

These data were acquired on board two satellites, TerraSAR-X and TandDEM-X, almost identical to each other and positioned in close formation, around 120–500 m apart during flight (DLR, 2023). These data acquisition characteristics of the TanDEM-X mission are an essential aspect of innovation regarding the technique of radar interferometry (Krieger et al., 2013; Rizzoli et al., 2017).

The TanDEM-X mission generated different products with global coverage (Airbus, 2023a, 2023b, 2023c; DLR, 2023; Gdulová et al., 2020). The main product is the DEM called TanDEM-X DEM, produced by DLR and commercially available with an approximate spatial resolution of 12 m. A version with a coarser spatial resolution (30 m) of the TanDEM-X DEM is available for specific scientific research purposes. In contrast, a third version with a spatial resolution of 90 m is openly available.

As the X-band radar has limited signal penetration in the vegetation cover, the TandDEM-X mission's DEMs are considered DSMs, having interference from vegetation and buildings on the elevations. However, studies like that of Grohmann (2018) indicate that the 30 m DEM of TanDEM-X would have higher penetration in the vegetation cover than the SRTM 30 DEM due to the TanDEM-X 30 m being derived from the TanDEM-X of 12 m, whose more spatially refined data would have been able to capture open vegetation areas between trees. These TanDEM-X DEMs have the WGS84 horizontal reference system and the WGS84 vertical datum.

The partner company of DLR, Airbus, also produced the global DEM called WorldDEM for commercial purposes at a spatial resolution of 12 m. A data treatment was carried out to remove the effect of vegetation cover and buildings on the elevations. The company offers both the DSM, whose elevations are impacted by vegetation and buildings, and the DTM, with such effect removed, and a spatial resolution of 24 m (a product called WorldDEM $DTM_{lite;}$ Airbus, 2023b). Subsequently, from the 12 m WorldDEM and with more recent data acquired from 2017 to 2021, Airbus also developed the WorldDEM Neo, a global DEM with 5 m spatial resolution, also commercially available (Airbus, 2023c).

4.12.22 ALOS PALSAR

On board the ALOS satellite, the PRISM sensor was previously mentioned regarding data acquisition for generating the global DEM called AW3D30. Another ALOS sensor is the PALSAR (Phased Array type L-band Synthetic Aperture Radar). It is a synthetic aperture radar operating in the L spectral band (wavelength 23.5 cm) and with repeat-pass interferometry. It obtains information from the Earth's surface regardless of natural light (day and night) and weather conditions.

The data acquired by the ALOS PALSAR have various applications, such as analysis and classification of land use and land cover, deforestation monitoring, mapping of wetland areas, disaster monitoring, and delineating areas with wet snow cover (Small, 2011). However, the robustness of these analyses requires accurate estimates of the backscatter of the electromagnetic energy emitted by the radar (Das et al., 2015). The side-looking imaging of a synthetic aperture radar, like the PALSAR, can cause overlap effects, shadowing, ramp shortening, relief inversion, and radiometric variations with the slope, which results in geometric and radiometric distortions in the collected data (Shimada, 2010). These effects are more intense in terrains with higher undulation, like mountainous regions, due to the considerable variation in local incidence angles of the radar signal (Das et al., 2015).

An alternative to improving the radar signal's backscatter estimates is terrain radiometric correction, which seeks to compensate for the influence of topographic variations on the radar backscatter. The key information for this procedure is a DEM, whose data acquisition source can be the radar itself or an external source, which is more common (Das et al., 2015).

For the data acquired by PALSAR, radiometric corrections and adjustments of terrain geometric distortions were made using as topographic information the SRTM 30 DEM at a global level and regional DEMs for the territories of the USA, Canada, and Mexico. These different DEMs were combined and concatenated into a single DEM, for which geoidal correction was performed (conversion of elevations from the geoidal reference to the ellipsoidal reference) and resampling to a spatial resolution of 12.5 m. This resolution was adopted to match the PALSAR data. The resulting DEM from this procedure has global coverage, is freely available, and is referred to as the ALOS PALSAR DEM.

However, what is available as a global DEM from ALOS PALSAR is not a product of new data acquisition but a reprocessing of the SRTM 30 DEM for most of the globe and regional DEMs for areas of the three mentioned countries. This DEM associated with ALOS PALSAR was explicitly developed to serve as a reference for interpreting the results of the terrain radiometric correction process of the PALSAR data. Using this DEM for other purposes is not recommended, as warned on the developer and maintainer's site (Alaska Satellite Facility, 2023). The so-called ALOS PALSAR DEM should also not be confused with a DEM that can be developed for a specific region from data collected by interferometry by PALSAR itself, as done in the work of Huang et al. (2015), for example.

4.12.23 EarthEnv-DEM90

Robinson et al. (2014) developed a global DEM (covering approximately 91% of the globe) called EarthEnv-DEM 90 from the fusion of SRTM data with a spatial resolution of 90 m (version 4.1; Jarvis et al., 2008) and the ASTER GDEM 2 (spatial resolution of 30 m, version 2), relying on other auxiliary data. The proposal was to take advantage of the data from these existing DEMs to produce a new DEM with less occurrence of missing data and noise than those observed in the DEMs used as input in the processing. Interpolation, smoothing, and other techniques were used to merge the data and obtain the final DEM, as described in the referenced publication.

The EarthEnv-DEM 90 is freely available and has a spatial resolution of 90 m. The data refer to the WGS84 horizontal reference system and EGM96 vertical reference.

4.12.24 Bare Earth SRTM DEM

The Bare Earth SRTM DEM is almost global (approximately from 60° N to 54° S) and was developed by O'Loughlin et al. (2016) from the combination of remote sensing data from different sources. The primary database used is the SRTM DEM with a spatial resolution of 90 m (CGIAR version 4). As this DEM has elevations influenced by vegetation cover, the purpose of O'Loughlin et al. (2016) was to use data from other sources to remove this effect to obtain a DEM based on the SRTM but representative of the terrain itself (hence the term bare earth in the product name). However, the impact of buildings on the elevations of the SRTM 90 was not treated.

One of the main types of data used in conjunction with the SRTM DEM was a set of terrain point elevations obtained by laser altimetry by the GLAS sensor (Geoscience Laser Altimeter System) on board the ICESat satellite (Ice, Cloud, Land, and Elevation Satellite), developed by NASA and operated between 2003 and 2009.

Another primary data used was a map of vegetation height estimates produced by Simard et al. (2011) with a spatial resolution of 1 km, also obtained based on GLAS-ICESat data from 2005 and auxiliary data, such as elevations from the SRTM DEM itself and data on the percentage of vegetation cover from MODIS (MoDerate-resolution Imaging Spectroradiometer, MOD44 product for the year 2005), in addition to precipitation and temperature data.

The third primary dataset used in conjunction with the SRTM DEM to obtain the Bare Earth SRTM DEM refers to the data on the percentage of vegetation cover that is part of the MODIS Vegetation Continuous Field (VCF) product, with a spatial resolution of 250 m and corresponding to the year 2000.

The complete methodology used to correct errors in the SRTM 90 DEM due to vegetation's effect and obtain the Bare Earth SRTM DEM is described in O'Loughlin et al. (2016). The synthesis of this approach was to use auxiliary data to estimate actual terrain elevations at some points, estimate vegetation height, and define the vegetation height to be removed from the SRTM DEM. In other words, in the end, the elevations represented in the Bare Earth SRTM DEM are those of the SRTM DEM lowered by a certain vegetation height, estimated for each point.

The Bare Earth SRTM DEM is freely available. It has a spatial resolution of 90 m, a WGS84 horizontal reference system, and an EGM96 vertical reference.

4.12.25 MERIT DEM

The MERIT DEM is a global DEM developed by Yamazaki et al. (2017) from the SRTM 90 DEM and the AW3D30 (spatial resolution of 30 m) using other auxiliary data. Among the auxiliary data, the Viewfinder Panorama DEM stands out, which is the result of an effort by researcher Jonathan de Ferranti to fill gaps in SRTM data and is freely available (http://www.viewfinderpanoramas.org/). Other relevant auxiliary data were terrain elevations captured by laser altimetry by ICESat and vegetation height estimates made by Simard et al. (2011).

Yamazaki et al.'s (2017) main proposal was to apply methods to treat and remove the four main types of errors in the elevation captured by DEMs derived from orbital remote sensing data: systematic noises associated with acquisition geometry (stripe noise), point noises associated with signal scattering (speckle noise), absolute deviations, and the effect of vegetation cover.

The authors of the MERIT DEM developed a methodology to treat the four components of errors following four main steps, applied to both the SRTM and the AW3D30, as detailed in Yamazaki et al. (2017) and summarized here. Initially, a two-dimensional Fourier transform filter was applied to remove systematic noises associated with acquisition geometry (stripe noise) from detecting unreal regular terrain undulations. This type of noise is apparent as stripes in orbital images, hence the term stripe. Then, the absolute deviation of the elevations was determined from the differences between the DEM elevations and those captured by ICESat, which are presumably related to the terrain. The third significant step was to estimate and treat the error due to vegetation as a function of vegetation density and height, since the SRTM and AW3D30 elevations are contaminated by vegetation. The fourth step was the application of an adaptive smoothing filter to separate and remove the random and point noise associated with signal scattering (speckle noise). This type of noise gives parts of the orbital image a granulated aspect, called the salt-and-pepper effect, due to random extreme values.

Because of these processing characteristics to obtain the MERIT DEM, mainly regarding the removal of the vegetation effect, this DEM can be considered closer to a DTM at a global level than the other global DEMs freely available and released before it (Hawker et al., 2022).

The MERIT DEM has a spatial resolution of 90 m and adopts a WGS84 horizontal reference system and EGM96 vertical reference. This DEM covers the entire globe except Antarctica (coverage between 90° N and 60° S; Yamazaki et al., 2019) and is freely available. Using the MERIT DEM along with water body delineation datasets, flow direction rasters, and a global river network named MERIT Hydro were generated and made freely accessible to the public (Yamazaki et al., 2019).

4.12.26 Zhao et al. (2018)

Zhao et al. (2018) used the SRTM 30 DEM to produce a new DEM proposed as a product of the SRTM with corrections due to the effects of vegetation. LiDAR survey data covering about 150,000 km^2 of areas around the world and data from the Ice, Cloud, and Land Elevation Satellite (ICESat)/Geoscience Laser Altimeter System (GLAS) were used to correct the systematic bias of the SRTM elevations in vegetated areas.

The method of correcting the elevations was proposed by Su et al. (2015) and consists of the following main steps: (i) determination of the errors of the SRTM DEM by subtracting this DEM and the elevations of the LiDAR/ICESat-GLAS data; (ii) 2/3 of these data were randomly selected for the development of regression models based on estimates of tree height, vegetation cover, and terrain slope; iii) these regression models were used to estimate the error distribution of the SRTM DEM and correct the elevations.

The LiDAR data were filtered and interpolated to generate 30 m resolution grids, as described in Zhao et al. (2018). The ICESat/GLAS data were complementary to the LiDAR data, referring to the acquisition period from February to November 2003. Checking and adjustment procedures of the ICESat/GLAS data were performed to avoid inconsistencies.

For the regression models, the slopes were estimated from the SRTM 30 itself. The information related to vegetation was obtained from previous studies: global estimates of tree height from the work of Simard et al. (2011), with 1 km spatial resolution; data of vegetation cover generated by Hansen et al. (2013) for the year 2000; information on vegetation type derived from land cover data from the MODIS MCD12Q1 product, with a spatial resolution of 500 m.

The DEM generated by Zhao et al. (2018) with the corrections of the effect of vegetation on the SRTM 30 data also has a spatial resolution of 30 m and was made available for free.

4.12.27 NASADEM

The global DEM called NASADEM is a reprocessing of the data from SRTM with new algorithms and using data that did not exist at the time of the SRTM mission, such as the laser altimetry data from ICESat-GLAS and the elevations of ASTER GDEM-2, ASTER GDEM-3, and AW3D30 (Buckley et al., 2020).

A simplified summary of the processing steps to produce NASADEM is that ICESat-GLAS data were initially used to improve and generate a new version of the

SRTM dataset, with support from other secondary data sources. Subsequently, the filling of remaining voids and refinement of this new SRTM version were performed using ASTER GDEM-2, GDEM-3, and AW3D30, along with additional auxiliary data. A detailed description of these procedures for generating NASADEM is provided by Buckley et al. (2020).

The NASADEM has been free since 2020, has a spatial resolution of 30 m, and uses integer-type data, the WGS84 horizontal reference system, and the EGM96 vertical reference system. NASA considers this DEM to be the successor to previous versions of DEMs developed with data from the SRTM mission.

4.12.28 Copernicus DEM

The Copernicus DEM product, nicknamed COPDEM, was developed from the TanDEM-X mission data acquired from 2011 to 2015. Among the products generated by this mission is the global DEM called WorldDEM, which is the primary information used to derive the COPDEM. Improvements mainly occurred in representing water bodies and river tracing, editing of coastlines, and special features such as airports and unacceptable terrain structures (Fahrland et al., 2022).

As seen in a previous item, the DSM and the DTM were generated for the WorldDEM, but the DSM was used to derive the COPDEM. Therefore, the COPDEM is also considered a DSM, with vegetation cover and buildings influencing elevations. In addition to the WorldDEM DSM, data from other global DEMs such as ASTER GDEM, SRTM 90, SRTM 30, GMTED2010, TanDEM-X DEM, and AW3D30, and regional data such as for Norway and Spain, assisted in the generation of the COPDEM (Fahrland et al., 2022).

The Copernicus DEM comprises three data instances: EEA-10, GLO-30, and GLO-90. The EEA-10 instance covers only the territory of the countries of the European Economic Area (EEA) plus six cooperating countries. This instance, restricted to such geographical coverage, represents a DEM with a spatial resolution of 10 m.

The GLO-30 and GLO-90 instances represent DEMs with global coverage and spatial resolutions of 30 and 90 m, respectively, called COPDEM30 and COPDEM90. These DEMs are free with data referring to the WGS84 horizontal and EGM2008 vertical reference systems. These data are available in floating-point and integer formats. According to Guth and Geoffroy (2021), constraining elevation values to integer meters in a 30 m resolution DEM should be avoided, as this practice tends to degrade the quality of spatial derivatives by introducing additional noise. Therefore, these authors recommend avoiding the version of the COPDEM with integer data, highlighting as a positive differential the COPDEM with elevation data as decimal numbers compared to the global DEMs made available before it (all with integer numbers at this 30 m spatial resolution).

The first version of COPDEM 90 was made available in 2019, while the first version of COPDEM 30 was released for free in 2020, with improved versions since then (ESA, 2023). According to the evaluation of Guth and Geoffroy (2021), the

COPDEM 30 presents superior quality to the global and free DEMs that precede it at the exact resolution (SRTM, ASTER GDEM, AW3D30, and NASADEM).

4.12.29 FABDEM

The Forest And Buildings removed copernicus DEM (FABDEM) is a global DEM developed by Hawker et al. (2022) from the COPDEM. It aims to remove the effect of vegetation cover and buildings present in the elevations of this DEM. A methodology based on machine learning techniques with random forest regression models was adopted. The removal of the effects of the vegetation and buildings was carried out separately, with specific data considered as predictor variables in each case.

The primary database for the vegetation effect treatment was constituted from the vegetation height estimates of the Global Forest Canopy Height mapping from 2019. Such estimates were elaborated by Potapov et al. (2021) from LiDAR data from the Global Ecosystem Dynamics Investigation (GEDI), acquired by the International Space Station, integrated with the Landsat dataset called Landsat Analysis Ready Data (ARD) from the Global Land Analysis and Discovery (GLAD) team. Other auxiliary data were also considered to treat the effect of the vegetation present in the COPDEM, as described in Hawker et al. (2022).

For urban areas, multiple datasets were considered predictors of the removal of the effect of buildings on the elevations of the COPDEM. These data covered characteristics related to the height of the buildings themselves, but also population data and other information related to the degree of urbanization, such as mapping of night lights, built area per capita, and travel time data, in addition to socioeconomic indicators such as Gross Domestic Product (GDP).

After removing the effects of vegetation from the COPDEM and removing the influence of buildings, separately and in parallel, a post-processing stage was carried out to join these two resulting DEMs, identify and fill gaps, and apply smoothing filters, finally obtaining the FABDEM. This DEM is considered the first free global DEM that seeks to represent the terrain properly without the effects of vegetation and buildings, although such effects have not been eliminated (Bielski et al., 2023; Dandabathula et al., 2022; Hawker et al., 2022).

The availability of the FABDEM is free, and the data cover from 80° N to 60° S. This DEM has a spatial resolution of 30 m, and the data are in the horizontal reference system WGS 84 and vertical reference EGM 2008, similar to the COPDEM from which it is derived.

4.12.30 IC2-GDEM

The global digital elevation model named IC2-GDEM was produced by Li et al. (2025) by correcting the elevation data from the ASTER GDEM (version 3) based on ICESat-2 satellite laser altimeter data and other ancillary data, such as global land cover and vegetation index products. This global DEM is available at 1 arc-second spatial resolution (~30 m), considering WGS84 as the horizontal reference system, and EGM96 as the vertical datum. Several source errors in ASTER GDEM elevations were treated, such as platform positioning, atmospheric noise, terrain undulation, land cover, and data processing related to data filling to produce the IC2-GDEM. The evaluation presented by Li et al. (2025) demonstrates the improvements in IC2-GDEM relative to the ASTER GDEM product. However, these authors recognize the error correction limitation in areas of steeper slope and high vegetation cover and the shortcoming of combining datasets with large temporal acquisition date intervals (ASTER GDEM versus ICESat-2).

4.12.31 FathomDEM

The Fathom Digital Elevation Model (FathomDEM) was proposed by Uhe et al. (2025) as an improvement of the Copernicus DEM with 30 m spatial resolution. The authors aimed at removing all height biases, including those from trees and buildings, to get a product that closely matches a global DTM. The methodology used to derive the FathomDEM combines convolutional neural networks and a vision transformer model, a neural network architecture adapted to computer vision tasks. As described in Uhe et al. (2025), elevations from this product outperformed COPDEM and FABDEM as well as global DEMs restricted to coastal areas, such as DeltaDTM. FathomDEM data is available at 1 arc-second spatial resolution (approximately 30 m at the Equator) in $1^\circ \times 1^\circ$ tiles, with WGS84 as the horizontal reference system and EGM08 as the vertical datum.

4.12.32 GEDTM30

The Global Ensemble Digital Terrain Model 30 m (GEDTM30) was proposed by Ho et al. (2025) by combining Copernicus DEM, ALOS World3D, and object height models in a data fusion approach. A global-to-local transfer learning method was adopted based on globally distributed lidar datasets from ICESat and GEDI. The GEDTM30 is available on a 1 arc-second spatial resolution (approximately 30 m), considering WGS84 as the horizontal reference system and EGM08 as the vertical datum. Results presented by Ho et al. (2025) depict improvements in GEDTM30 elevations compared to COPDEM, FABDEM, and FathomDEM, justifying that these authors consider this dataset to be a globally consistent DTM.

4.13 DEM Data for Coastal Areas

There are also ongoing advances in the search for the topographically specific representation of the coastal regions. The rise in sea levels due to global climate change is the primary concern that motivates such a search. The detailing of DEMs explicitly conceived with the concern of representing coastal areas is beyond the scope of this book. However, four of the most prominent recent coastal DEMs are presented here as representative examples of global datasets:

i. CoastalDEM, proposed by Kulp and Strauss (2018);
ii. Global LiDAR Lowland DTM version 1 (GLL_DTM_v1) and version 2 (GLL_DTM_v2) described in Vernimmen et al. (2020) and Vernimmen and Hooijer (2023), respectively;
iii. DiluviumDEM, developed by Dusseau et al. (2023);
iv. DeltaDTM, presented by Pronk et al. (2024).

4.14 Regional DEM for South America: ANADEM

The ANADEM (Brazilian National Water Agency Digital Elevation Model) is a DEM restricted to South America generated by Laipelt et al. (2024) from the global DEM COPDEM30. Procedures for removing noise due to vegetation in the COPDEM30 were applied using machine learning techniques, elevation data from the GEDI mission, and multispectral remote sensing products.

The comparative analyses in such a study indicate that the ANADEM is the best quality DEM product for the South American continent. This product has a spatial resolution of 30 m and is available for free.

4.15 DEM Data Obtained by LiDAR Survey

In addition to DEMs derived from orbital remote sensing, airborne LiDAR (Light Detection and Ranging) surveys have gained prominence for topographic data acquisition. This active remote sensing system acquires horizontal and vertical surface coordinates by detecting the reflections of pulsed laser beams. Elevation is measured based on the time interval between the emission and return of the laser signal (Dong & Chen, 2018). The laser returns are recorded as millions or even billions of individual points, forming what is known as a point cloud.

A feature that makes this type of topographic data survey compelling is that the point cloud is formed by points referring to different information acquired by the same laser pulse, known as multiple returns. These signal returns occur when the laser pulse encounters reflective surfaces. The first return can be from the vegetation canopy, a building roof, or the terrain itself (ground) when there are no artifacts or

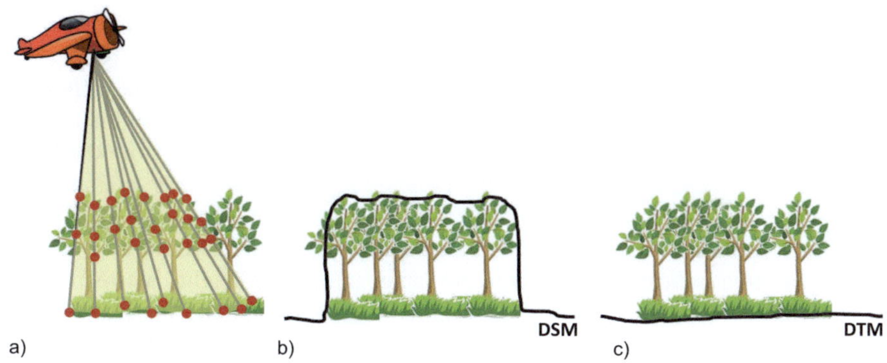

Fig. 4.21 **a** Data acquisition via airborne LiDAR survey indicating the points referring to the multiple returns of the laser pulses; **b**, **c** Indication of what represents (black line) the Digital Surface Model (DSM) and the Digital Terrain Model (DTM) derived from this LiDAR survey

vegetation on it. The second and third or more intermediate returns can occur as the pulse encounters other surfaces for reflection in the sequence of its trajectory. The last return, in general, is from the terrain itself, although it may occur earlier if the surfaces above the terrain have reflected the entire pulse. An illustration of the laser pulse with the point cloud generated by a LiDAR-type survey is shown in Fig. 4.21a.

Typically, the laser penetrates the vegetation cover and reaches the bare ground. This survey provides insightful information about the vegetation, such as estimating the tree heights, the plant biomass volume, and tree mortality rates. Interesting studies on these topics are the works of Dalagnol et al. (2021), Van Leeuwen and Nieuwenhuis (2010), and Wang and Glenn (2008).

The multiple returns from each laser pulse of a LiDAR survey allow for a three-dimensional representation of other objects and artifacts on the surface, such as bridges, houses, buildings, and other structures, in addition to the terrain itself. It is then possible to produce the DSM and the DTM from the same point cloud of the LiDAR survey. Different techniques exist to process the LiDAR point cloud to remove the artifacts and achieve the bare ground representation in the DTM. Detailing these methods is beyond the scope of this book.

Both the DSM and the DTM from LiDAR are usually obtained with a very refined spatial resolution, such as 1 m or less, on the order of centimeters. The DSM's elevations represent the surface, considering the vegetation, objects, and identified artifacts. At the same time, the DTM characterizes the bare ground elevations, as illustrated in Fig. 4.4 and discussed in item 4.3.

The difference between the DSM and DTM obtained by a real-world LiDAR survey for an area of the municipality of Flores (PE), with data from the project Pernambuco Tridimensional (PE3D; Cirilo et al., 2014, 2015), corroborates the difference between these two models (Fig. 4.22). The orthophoto of the same region allows identifying the type of surface and the targets on the surface, such as houses, trees, and bridges. In the DSM, building features are identifiable due to their elevated heights relative to the surroundings and their geometric shapes. In contrast, these features are not visible in the DTM, as they represent the bare ground surface.

In this example from the municipality of Flores, the raster of the difference between DSM and DTM shows the height of the artifacts estimated by the LiDAR in each pixel. It highlights how much the DTM distinguishes itself from the DSM by eliminating these artifacts in the topographic representation. For example, the warehouse identified at point A has a difference between DSM and DTM of about 10 m. In comparison, on the bridge indicated by the letter C, this difference is around 8 m. In the area indicated by D, there are larger trees with heights between 7 and 11 m. In region E, there is smaller vegetation, around 2–4 m, with individual trees

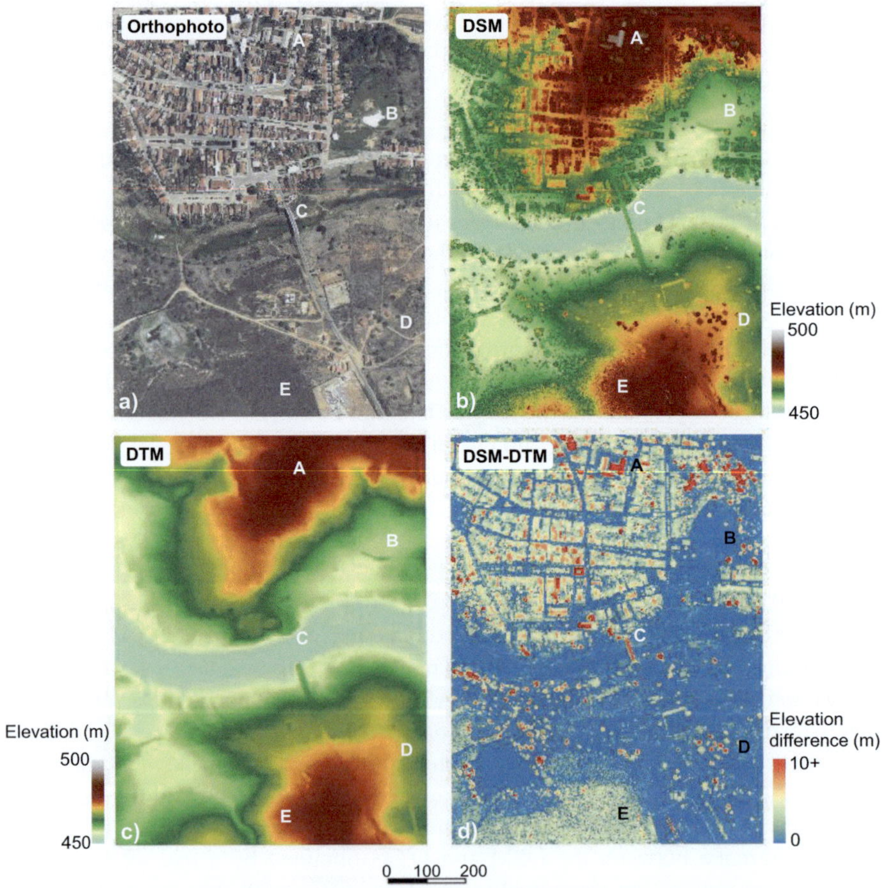

Fig. 4.22 Data from the Pernambuco Tridimensional (PE3D) LiDAR survey project in the region of the municipality of Flores (Pernambuco, Brazil): **a** Orthophoto; **b** Digital Surface Model (DSM); **c** Digital Terrain Model (DTM); **d** Raster of the differences between DSM and DTM; A, B, C, D, E indicate notable points for comparison between the images

reaching 7 m. In B, an area of exposed soil or low grass is next to a lake, resulting in no differences between DSM and DTM.

With the high spatial detail of the data and the generation of the Digital Terrain Model, LiDAR topographic surveys have enabled the characterization of terrain in an unprecedented and previously unimaginable way, even just a few decades ago (Murphy et al., 2008; Persendt and Gomez, 2016; Roelens et al., 2018). The disadvantage is the challenge of the computational cost associated with the data from LiDAR surveys.

For a study area with 1000 km^2, a DTM with 1 m spatial resolution has 1 billion pixels (Table 4.2). When processing this DTM, several other rasters with an equal number of pixels are generated. For comparison purposes, the same 1000 km^2 area requires just over 1 million pixels to be represented by a global DEM with 30 m spatial resolution (such as SRTM 30), and only slightly more than 120,000 pixels when using a DEM with 90 m resolution (such as SRTM 90).

The cost of storing and processing data derived from LiDAR can prohibit its use in some applications. An alternative is to take advantage of methods and solutions adjusted to deal with such a large amount of data, adapting more computationally efficient processing algorithms, cloud processing, or parallel processing, and allocating more space for data storage. These issues are discussed in greater depth in Chaps. 6 and 7, specifically regarding the computational cost of DTM processing for terrain depression removal, flow direction assignment, and accumulated drainage area calculation.

In some cases, the DTM can be changed to a coarser spatial resolution, like 5 m or 10 m, to make manipulating such a large amount of data feasible. For an area of 1000 km^2, the DTM would be composed of 40 million pixels at a spatial resolution of 5 m, a reduction of 25 times compared to the 1 billion pixels at a resolution of 1 m (Table 4.2). If the DTM of this area is coarsened to a spatial resolution of 10 m, it would be a reduction of 100 times in the amount of resulting data compared to the resolution of 1 m.

The increase in pixel size substantially reduces the need for data processing and storage. Depending on the study's purpose and the area's characteristics, this coarsening may not represent a significant loss of information quality. However, this

Table 4.2 Dimensions of a DTM necessary to represent a study area of 1000 km^2, considering different spatial resolutions (for simplification, the area is taken with square geometry)

Study area (km^2)	DTM spatial resolution (m)	Number of lines	Number of columns	Number of pixels
1000	1	31,622	31,622	1,000,000,000
	2	15,811	15,811	250,000,000
	5	6325	6325	40,000,000
	10	3162	3162	10,000,000
	30	1054	1054	1,111,111
	90	351	351	123,457

should be evaluated in each case, depending on the study's purpose and the area's characteristics, among other factors.

References

Abrams, M., Crippen, R., & Fujisada, H. (2020). ASTER Global Digital Elevation Model (GDEM) and ASTER Global Water Body Dataset (ASTWBD). *Remote Sensing, 12*, 1156. https://doi.org/10.3390/rs12071156

Airbus. (2023a). *Airbus reference layers—The right level of information in 2D & 3D for every project need.* Airbus Defence and Space Intelligence

Airbus. (2023b). *WorldDEM DTMlite technical product specification version 1.1.* Airbus—Airbus Defence and Space Intelligence.

Airbus. (2023c). *WorldDEMTM Neo—The next level of global elevation models.* Airbus Defence and Space Intelligence.

Alaska Satellite Facility (ASF). (2023). P*ALSAR RTC DEM information.* University of Alaska Fairbanks. https://asf.alaska.edu/

Amante, C., & Eakins, B. W. (2009). *ETOPO1 1 arc-minute global relief model: Procedures, data sources and analysis (NOAA Technical Memorandum NESDIS NGDC-24).* National Oceanic and Atmospheric Administration, Boulder (EUA).

Arabelos, D. (2000). Intercomparisons of the global DTMs ETOPOS, TerrainBase and JGP95E. *Physics and Chemistry of the Earth (A), 25*, 89–93.

Bamber, J. L., Griggs, J. A., Hurkmans, R. T. W. L., Dowdeswell, J. A., Gogineni, S. P., Howat, I., Mouginot, J., Paden, J., Palmer, S., Rignot, E., & Steinhage, D. (2013). A new bed elevation dataset for Greenland. *The Cryosphere, 7*, 499–510. https://doi.org/10.5194/tc-7-499-2013

Bamber, J. L., & Muller, J.-P. (1998). Derivation of a global land elevation data set from satellite radar altimeter data for topographic mapping. *Journal of Geophysical Research, 103*, 32159–32168. https://doi.org/10.1029/98JD01154

Baugh, C. A., Bates, P. D., Schumann, G., & Trigg, M. A. (2013). SRTM vegetation removal and hydrodynamic modeling accuracy: SRTM vegetation error removal. *Water Resources Research, 49*, 5276–5289. https://doi.org/10.1002/wrcr.20412

Becker, J. J., Sandwell, D. T., Smith, W. H. F., Braud, J., Binder, B., Depner, J., Fabre, D., Factor, J., Ingalls, S., Kim, S.-H., Ladner, R., Marks, K., Nelson, S., Pharaoh, A., Trimmer, R., Von Rosenberg, J., Wallace, G., & Weatherall, P. (2009). Global bathymetry and elevation data at 30 arc seconds resolution: SRTM30_PLUS. *Marine Geodesy, 32*, 355–371. https://doi.org/10.1080/01490410903297766

Berkofsky, L., & Bertoni, E. A. (1955). Mean topographic charts for the entire Earth. *Bulletin of the American Meteorological Society, 36*, 350–354.

Berry, P. (2009). International Digital Elevation Model Service (IDEMS). Report of the International Association of Geodesy 2007–2009.

Berry, P. A. M., Smith, R. G., & Benveniste, J. (2010). ACE2: The new global digital elevation model. In S. P. Mertikas (Ed.), *Gravity, Geoid and Earth Observation.* International Association of Geodesy Symposia (pp. 231–237). Springer Berlin Heidelberg. https://doi.org/10.1007/978-3-642-10634-7_30

Berry, P. A. M., Pinnock, R. A., Hilton, R. D., Johnson, C. P. D. (2020). ACE: a new global digital elevation model incorporating satellite altimeter derived heights. Presented at the European Space Agency, ERS-ENVISAT Symposium, Gotemburgo (Suécia), p. 9.

Bielski, C., López-Vázquez, C., Grohmann, C. H., Guth, P. L., & Group, T. T. D. W. (2023). DEMIX wine contest method ranks ALOS AW3D30, COPDEM, and FABDEM as Top 1" Global DEMs.

Brown, C. G., Sarabandi, K., & Pierce, L. E. (2010). Model-based estimation of forest canopy height in Red and Austrian pine stands using shuttle radar topography mission and ancillary data: A proof-of-concept study. *IEEE Transactions on Geoscience and Remote Sensing, 48*, 1105–1118. https://doi.org/10.1109/TGRS.2009.2031635

Buckley, S., Agram, P., Belz, E., Crippen, R., Gurrola, E., Hensley, S., Kobrick, M., Lavalle, M., Martin, J., Neumann, M., Nguyen, Q., Rosen, P., Shimada, J., Simard, M., & Tung, W. (2020). *NASADEM user guide (No. versão 1)*. National Aeronautics and Space Administration, Jet Propulsion Laboratory, Pasadena (EUA).

Cirilo, J. A., Alves, F. H. B., Silva, B. M., & Campos, P. H. A. L. C. (2015). Pernambuco tridimensional: base de dados espaciais para planejamento urbano e gestão territorial. Presented at the 12° Simpósio de Hidráulica e Recursos Hídricos dos Países de Expressão Portuguesa, Brasília.

Cirilo, J. A., Alves, F. H. B., da Silva, L. A. C., & Campos, J. H. A. L. (2014). Suporte de Informações Georreferenciadas de Alta Resolução para Implantação de Infraestrutura e Planejamento Territorial. *Revista Brasileira De Geografia Física, 7*, 755–763.

Dalagnol, R., Wagner, F. H., Galvão, L. S., Streher, A. S., Phillips, O. L., Gloor, E., Pugh, T. A. M., Ometto, J. P. H. B., & Aragão, L. E. O. C. (2021). Large-scale variations in the dynamics of Amazon forest canopy gaps from airborne lidar data and opportunities for tree mortality estimates. *Scientific Report, 11*, 1388. https://doi.org/10.1038/s41598-020-80809-w

Dandabathula, G., Hari, R., Ghosh, K., Bera, A. K., & Srivastav, S. K. (2022). Accuracy assessment of digital bare-earth model using ICESat-2 photons: Analysis of the FABDEM. *Modeling Earth Systems and Environment*. https://doi.org/10.1007/s40808-022-01648-4

Danielson, J. J., & Gesch, D. B. (2011). *Global multi-resolution terrain elevation data 2010 (GMTED2010)* (U.S. Geological Survey Open-File Report 2011-1073). USGS.

Das, A., Nagendra, H., Anand, M., & Bunyan, M. (2015). Topographic and bioclimatic determinants of the occurrence of forest and grassland in tropical Montane Forest-Grassland mosaics of the Western Ghats, India. *PLoS ONE, 10*, Article e0130566. https://doi.org/10.1371/journal.pone.0130566

DLR. (2023). TanDEM-X—Digital Elevation Model (DEM)—Global, 90m. German Aerospace Center, Earth Observation Center. https://geoservice.dlr.de/data-assets/ju28hc7pui09.html. Accessed 23 Feb.

Dong, P., & Chen, Q. (2018). *LiDAR remote sensing and applications*. Taylor & Francis.

Dusseau, D., Zobel, Z., & Schwalm, C. R. (2023). DiluviumDEM: Enhanced accuracy in global coastal digital elevation models. *Remote Sensing of Environment, 298*, Article 113812. https://doi.org/10.1016/j.rse.2023.113812

EarthData. (2023). New Version of the ASTER GDEM. National Aeronautics and Space Administration, Earth Observational Data. https://www.earthdata.nasa.gov/news/new-aster-gdem. Accessed 4 February 2023.

EROS. (2018). *GTOPO30 Documentation ReadMe*. Earth Resources Observation and Science Center.

ESA. (2023). Copernicus DEM—Global and European digital elevation model (COP-DEM). European Space Agency. https://spacedata.copernicus.eu/collections/copernicus-digital-elevation-model. Accessed 4 February 2023.

Ettritch, G., Hardy, A., Bojang, L., Cross, D., Bunting, P., & Brewer, P. (2018). Enhancing digital elevation models for hydraulic modelling using flood frequency detection. *Remote Sensing of Environment, 217*, 506–522. https://doi.org/10.1016/j.rse.2018.08.029

Fahrland, E., Paschko, H., Jacob, P., & Kahabka, H. (2022). *Copernicus digital elevation model—Product handbook*. Airbus Defense and Space Agency.

Farr, T. G., Rosen, P. A., Caro, E., Crippen, R., Duren, R., Hensley, S., Kobrick, M., Paller, M., Rodriguez, E., Roth, L., Seal, D., Shaffer, S., Shimada, J., Umland, J., Werner, M., Oskin, M., Burbank, D., & Alsdorf, D. (2007). The shuttle radar topography mission. *Reviews of Geophysics, 45*, 2004. https://doi.org/10.1029/2005RG000183

Fretwell, P., Pritchard, H. D., Vaughan, D. G., Bamber, J. L., Barrand, N. E., Bell, R., Bianchi, C., Bingham, R. G., Blankenship, D. D., Casassa, G., Catania, G., Callens, D., Conway, H., Cook, A. J., Corr, H. F. J., Damaske, D., et al. (2012). Bedmap2: Improved ice bed, surface and thickness datasets for Antarctica (preprint). *Antarctic*. https://doi.org/10.5194/tcd-6-4305-2012

Fujita, D., Ishizawa, M., Maksyutov, S., Thornton, P. E., Saeki, T., & Nakazawa, T. (2003). Inter-annual variability of the atmospheric carbon dioxide concentrations as simulated with global terrestrial biosphere models and an atmospheric transport model. *Tellus B: Chemical and Physical Meteorology, 55*, 530. https://doi.org/10.3402/tellusb.v55i2.16721

Gates, W. L., & Nelson, A. B. (1975). A new(revised) tabulation of the Scripps Topography on a 1o global grid. In *Part I: Terrain heights for Defense Advanced Research Projects Agency*, Santa Monica (EUA).

Gdulová, K., Marešová, J., & Moudrý, V. (2020). Accuracy assessment of the global TanDEM-X digital elevation model in a mountain environment. *Remote Sensing of Environment, 241*, Article 111724. https://doi.org/10.1016/j.rse.2020.111724

Gesch, D. B., Verdin, K. L., & Greenlee, S. K. (1999). New land surface digital elevation model covers the Earth. *Eos Transactions, 80*, 69. https://doi.org/10.1029/99EO00050

Grohmann, C. H. (2018). Evaluation of TanDEM-X DEMs on selected Brazilian sites: Comparison with SRTM, ASTER GDEM and ALOS AW3D30. *Remote Sensing of Environment, 212*, 121–133. https://doi.org/10.1016/j.rse.2018.04.043

Guth, P. L., & Geoffroy, T. M. (2021). LiDAR point cloud and ICESat-2 evaluation of 1 second global digital elevation models: Copernicus wins. *Transactions in GIS, 25*, 2245–2261. https://doi.org/10.1111/tgis.12825

Hansen, M. C., Potapov, P. V., Moore, R., Hancher, M., Turubanova, S. A., Tyukavina, A., Thau, D., Stehman, S. V., Goetz, S. J., Loveland, T. R., Kommareddy, A., Egorov, A., Chini, L., Justice, C. O., & Townshend, J. R. G. (2013). High-resolution global maps of 21st-century forest cover change. *Science, 342*, 850–853. https://doi.org/10.1126/science.1244693

Hastings, D. A., Dunbar, P. K., & Hittelman, A. M. (2000). Assessing the global land one-km base elevation DEM. In *Geodesy beyond 2000—The challenges of the first decade* (pp. 101–106). International Association of Geodesy Symposia. Springer.

Hastings, D. A., & Dunbar, P. K. (1998). Development & assessment of the global land one-km base elevation digital elevation model (GLOBE). *International Society of Photogrammetry and Remote Sensing Archives, 32*, 218–221.

Hawker, L., Uhe, P., Paulo, L., Sosa, J., Savage, J., Sampson, C., & Neal, J. (2022). A 30 m global map of elevation with forests and buildings removed. *Environmental Research Letters, 17*, Article 024016. https://doi.org/10.1088/1748-9326/ac4d4f

Hirt, C., & Rexer, M. (2015). Earth 2014: 1 arc-min shape, topography, bedrock and ice-sheet models—Available as gridded data and degree-10,800 spherical harmonics. *International Journal of Applied Earth Observation and Geoinformation, 39*, 103–112. https://doi.org/10.1016/j.jag.2015.03.001

Ho, Y.-H., Grohmann, C. H., Lindsay, J., Reuter, H. I., Parente, L., Witjes, M., & Hengl, T. (2025). Global Ensemble Digital Terrain modeling and parametrization at 30 m resolution (GEDTM30): A data fusion approach based on ICESat-2, GEDI and multisource data (preprint).

Huang, Y., Yu, M., Xu, Q., Sawada, K., Moriguchi, S., Yashima, A., Liu, C., & Xue, L. (2015). InSAR-derived digital elevation models for terrain change analysis of earthquake-triggered flow-like landslides based on ALOS/PALSAR imagery. *Environmental Earth Sciences, 73*, 7661–7668. https://doi.org/10.1007/s12665-014-3939-5

Jarvis, A., Reuter, H. I., Nelson, A., & Guevara, E. (2008). *Hole-filled seamless SRTM data V4*. International Centre for Tropical Agriculture (CIAT).

JAXA/EORC. (2021). ALOS Global Digital Surface Model (DSM) ALOS World 3D-30m (AW3D30) Version 3.2/3.1 Product Description Edition 1.2. Japan Aerospace Exploration Agency, Earth Observation Research Center.

Johnson, C. P. D., Berry, P. A. M., & Hilton, R. D. (2000). *Report on ACE Version1 generation.* Geomatics Unit, De Montfort University, Leicester (Reino Unido).

Joseph, D. (1982, updated 1984). Data format for global 10 minute elevation data from the U.S. Navy. Boulder, CO: NCAR. Data Support Section.

Kavvas, M. L., & Anderson, M. (2002). A global hydrologic model. In *Mathematical models of large watershed hydrology*. Water Resource Publications, Highlands Ranch.

Kellndorfer, J. M., Walker, W. S., Dobson, M. C., Vona, J., & Clutter, M. (2004). Vegetation height derivation from shuttle radar topography mission data in Southeast Georgia, USA. In *IEEE International Geoscience and Remote Sensing Symposium, 2004* (pp. 4512–4515). IGARSS '04. Anchorage, AK, USA. https://doi.org/10.1109/IGARSS.2004.1370157

Krieger, G., Moreira, A., Fiedler, H., Hajnsek, I., Werner, M., Younis, M., & Zink, M. (2007). TanDEM-X: A satellite formation for high-resolution SAR interferometry. *IEEE Transactions on Geosciences and Remote Sensing, 45*, 3317–3341. https://doi.org/10.1109/TGRS.2007.900693

Krieger, G., Zink, M., Bachmann, M., Bräutigam, B., Schulze, D., Martone, M., Rizzoli, P., Steinbrecher, U., Walter Antony, J., De Zan, F., Hajnsek, I., Papathanassiou, K., Kugler, F., Rodriguez Cassola, M., Younis, M., Baumgartner, S., López-Dekker, P., Prats, P., & Moreira, A. (2013). TanDEM-X: A radar interferometer with two formation-flying satellites. *Acta Astronautica, 89*, 83–98. https://doi.org/10.1016/j.actaastro.2013.03.008

Kulp, S. A., & Strauss, B. H. (2018). CoastalDEM: A global coastal digital elevation model improved from SRTM using a neural network. *Remote Sensing of Environment, 206*, 231–239. https://doi.org/10.1016/j.rse.2017.12.026

Laipelt, L., Comini De Andrade, B., Collischonn, W., Amorim, A., Paiva, R. C. D. D., & Ruhoff, A. (2024). Anadem: A digital terrain model for South America. *Remote Sensing, 16*(13), 2321. https://doi.org/10.20944/preprints202404.1305.v1

Lemoine, F. G., Kenyon, S. C., Factor, J. K., Trimmer, R. G., Pavlis, N. K., Chinn, D. S., Cox, C. M., Klosko, S. M., Luthcke, S. B., Torrence, M. H., Wang, Y. M., Williamson, R. G., Pavlis, E. C., Rapp, R. H., & Olson, T. R. (1998). The development of the joint NASA GSFC and the National Imagery and Mapping Agency (NIMA) Geopotential Model EGM96.

Li, B., Xie, H., Liu, S., Ye, Z., Hong, Z., Weng, Q., Sun, Y., Xu, Q., & Tong, X. (2025). Global digital elevation model (GDEM) product generation by correcting ASTER GDEM elevation with ICESat-2 altimeter data. *Earth System Science Data, 17*, 205–220. https://doi.org/10.5194/essd-17-205-2025

Moore, I. D., Grayson, R. B., & Ladson, A. R. (1991). Digital terrain modelling: A review of hydrological, geomorphological, and biological applications. *Hydrological Processes, 5*, 3–30. https://doi.org/10.1002/hyp.3360050103

Murphy, P. N. C., Ogilvie, J., Meng, F.-R., & Arp, P. (2008). Stream network modelling using lidar and photogrammetric digital elevation models: A comparison and field verification. *Hydrological Processes, 22*(12), 1747–1754. https://doi.org/10.1002/hyp.6770

NCGIA. (2023). *Locating terrain data.* National Center for Geographic Information & Analysis. www.ncgia.ucsb.edu/cctp/units/unit06/elev2.html. Accessed 3 June 2023.

NOAA. (2022). *ETOPO 2022 user guide.* National Oceanic and Atmospheric Administration.

NOAA. (2023). TerrainBase Global Terrain Model—Summary documentation. National Oceanic and Atmospheric Administration. https://www.ngdc.noaa.gov/mgg/gravity/1999/document/html/tbase.html. Accessed 3 July 2023

O'Loughlin, F. E., Paiva, R. C. D., Durand, M., Alsdorf, D. E., & Bates, P. D. (2016). A multi-sensor approach towards a global vegetation corrected SRTM DEM product. *Remote Sensing of Environment, 182*, 49–59. https://doi.org/10.1016/j.rse.2016.04.018

Pavlis, N. K., & Rapp, R. H. (1990). The development of an isostatic gravitational model to degree 360 and its use in global gravity modelling. *Geophysical Journal International, 100*, 369–378. https://doi.org/10.1111/j.1365-246X.1990.tb00691.x

Podest, E., & Crow, W. (2013). Preliminary, v.1 SMAP Science Document no. 043.

Potapov, P., Li, X., Hernandez-Serna, A., Tyukavina, A., Hansen, M. C., Kommareddy, A., Pickens, A., Turubanova, S., Tang, H., Silva, C. E., Armston, J., Dubayah, R., Blair, J. B., & Hofton, M. (2021). Mapping global forest canopy height through integration of GEDI and Landsat data. *Remote Sensing of Environment, 253*, Article 112165. https://doi.org/10.1016/j.rse.2020.112165

Pronk, M., Hooijer, A., Eilander, D., Haa, A., Jong, T., Vousdoukas, M., Vernimmen, R., Ledoux, H., & Eleveld, M. (2024). DeltaDTM: A global coastal digital terrain model. *Scientific Data, 11*, 273. https://doi.org/10.1038/s41597-024-03091-9

Persendt, F. C., & Gomez, C. (2016). Assessment of drainage network extractions in a low-relief area of the Cuvelai Basin (Namibia) from multiple sources: LiDAR, topographic maps, and digital aerial orthophotographs. *Geomorphology, 260*, 32–50. https://doi.org/10.1016/j.geomorph.2015.06.047

Rizzoli, P., Martone, M., Gonzalez, C., Wecklich, C., Borla Tridon, D., Bräutigam, B., Bachmann, M., Schulze, D., Fritz, T., Huber, M., Wessel, B., Krieger, G., Zink, M., & Moreira, A. (2017). Generation and performance assessment of the global TanDEM-X digital elevation model. *ISPRS Journal of Photogrammetry and Remote Sensing, 132*, 119–139. https://doi.org/10.1016/j.isprsjprs.2017.08.008

Robinson, N., Regetz, J., & Guralnick, R. P. (2014). EarthEnv-DEM90: A nearly-global, void-free, multi-scale smoothed, 90m digital elevation model from fused ASTER and SRTM data. *ISPRS Journal of Photogrammetry and Remote Sensing, 87*, 57–67. https://doi.org/10.1016/j.isprsjprs.2013.11.002

Roelens, J., Rosier, I., Dondeyne, S., Van Orshoven, J., & Diels, J. (2018). Extracting drainage networks and their connectivity using LiDAR data. *Hydrological Processes, 32*(8), 1026–1037. https://doi.org/10.1002/hyp.11472

Saleh, J., Pavlis, N.K., 2002. The Development and Evaluation of the Global Digital Terrain Model DTM2002. 3rd Meeting of the International Gravity and Geoid Commission, Thessaloniki, Greece.

Shimada, M. (2010). Ortho-rectification and slope correction of SAR data using DEM and its accuracy evaluation. *IEEE Journal of Selected Topics in Applied Earth Observations and Remote Sensing, 3*, 657–671. https://doi.org/10.1109/JSTARS.2010.2072984

Simard, M., Pinto, N., Fisher, J. B., & Baccini, A. (2011). Mapping forest canopy height globally with spaceborne lidar. *Journal of Geophysical Research, 116*, G04021. https://doi.org/10.1029/2011JG001708

Small, D. (2011). Flattening gamma: Radiometric terrain correction for SAR imagery. *IEEE Transactions on Geoscience and Remote Sensing, 49*, 3081–3093. https://doi.org/10.1109/TGRS.2011.2120616

Smith, R. G., & Berry, P. A. M. (2010). *ACE2: Global Digital Elevation Model—User guide*. De Montfort University, Leicester (Reino Unido).

Smith, S. M., Menard, H. W., & Sharman, G. (1966). *World-wide ocean depths and continental elevations averaged for areas approximating one degree squares of latitude and longitude*. University of California.

Smith, W. H. F., & Sandwell, D. T. (1997). Global sea floor topography from satellite altimetry and ship depth soundings. *Science, 277*, 1956–1962.

SNAP/ESA. (2023). GETASSE30 Elevation Model. Sentinel Application Platform, European Space Agency. https://seadas.gsfc.nasa.gov/help-8.2.0/desktop/GETASSE30ElevationModel.html. Accessed 3 February 2023.

Su, C., Yu, W., Feng, C., Yu, C., Huang, Z., & Zhang, X. (2015). An efficient algorithm for calculating drainage accumulation in digital elevation models based on the basin tree index. *IEEE Geoscience and Remote Sensing Letters, 12*, 424–428. https://doi.org/10.1109/LGRS.2014.2345561

Tachikawa, T., Iwasaki, A., Gesch, D. B., & Danielson, J. J. (2011). *ASTER Global Digital Elevation Model Version 2—Summary of validation results*. ASTER GDEM Validation Team.

Tadono, T., Takaku, J., Tsutsui, K., Oda, F., & Nagai, H. (2015). Status of "ALOS World 3D (AW3D)" global DSM generation. In *2015 IEEE International Geoscience and Remote Sensing Symposium (IGARSS)*, Milan, Italy (pp. 3822–3825). https://doi.org/10.1109/IGARSS.2015.7326657

Tadono, T., Nagai, H., Ishida, H., Oda, F., Naito, S., Minakawa, K., & Iwamoto, H. (2016). Generation of the 30 m-mesh global digital surface model by ALOS PRISM. In *The International Archives of the Photogrammetry, Remote Sensing and Spatial Information Sciences* XLI-B4 (pp. 157–162). https://doi.org/10.5194/isprsarchives-XLI-B4-157-2016

Uhe, P., Lucas, C., Hawker, L., Brine, M., Wilkinson, H., Cooper, A., Saoulis, A., Savage, J., & Sampson, C. (2025). FathomDEM: An improved global terrain map using a hybrid vision transformer model. *Environmental Research Letters, 20*(3), 034002. https://doi.org/10.1088/1748-9326/ada972

Van Leeuwen, M., & Nieuwenhuis, M. (2010). Retrieval of forest structural parameters using LiDAR remote sensing. *European Journal of Forest Research, 129*, 749–770. https://doi.org/10.1007/s10342-010-0381-4

Vernimmen, R., & Hooijer, A. (2023). New LiDAR-based elevation model shows greatest increase in global coastal exposure to flooding to be caused by early-stage sea-level rise. *Earth's Future, 11*, e2022EF002880. https://doi.org/10.1029/2022EF002880

Vernimmen, R., Hooijer, A., & Pronk, M. (2020). New ICESat-2 satellite LiDAR data allow first global lowland DTM suitable for accurate coastal flood risk assessment. *Remote Sensing, 12*, 2827. https://doi.org/10.3390/rs12172827

Vianello, R. L. (1985). Vertical velocity forced by topography in the southern hemisphere. *Journal of Climatology, 5*, 213–219. https://doi.org/10.1002/joc.3370050207

Walker, W. S., Kellndorfer, J. M., & Pierce, L. E. (2007). Quality assessment of SRTM C- and X-band interferometric data: Implications for the retrieval of vegetation canopy height. *Remote Sensing of Environment, 106*, 428–448. https://doi.org/10.1016/j.rse.2006.09.007

Wang, C., & Glenn, N. F. (2008). A linear regression method for tree canopy height estimation using airborne lidar data. *Canadian Journal of Remote Sensing, 34*, S217–S227. https://doi.org/10.5589/m08-043

Wessel, P., & Smith, W. H. F. (1996). A global, self-consistent, hierarchical, high-resolution shoreline database. *Journal of Geophysical Research, 101*, 8741–8743. https://doi.org/10.1029/96JB00104

Wolf, M., & Wingham, D. (1992). The status of the world's public-domain digital topography of the land and ice. *Geophysical Research Letters, 19*, 2325–2328. https://doi.org/10.1029/92GL01886

Yamazaki, D., Ikeshima, D., Sosa, J., Bates, P. D., Allen, G. H., & Pavelsky, T. M. (2019). MERIT hydro: A high-resolution global hydrography map based on latest topography dataset. *Water Resources Research, 55*, 5053–5073. https://doi.org/10.1029/2019WR024873

Yamazaki, D., Ikeshima, D., Tawatari, R., Yamaguchi, T., O'Loughlin, F., Neal, J. C., Sampson, C. C., Kanae, S., & Bates, P. D. (2017). A high-accuracy map of global terrain elevations: Accurate global terrain elevation map. *Geophysical Research Letters, 44*, 5844–5853. https://doi.org/10.1002/2017GL072874

Zhao, X., Su, Y., Hu, T., Chen, L., Gao, S., Wang, R., Jin, S., & Guo, Q. (2018). A global corrected SRTM DEM product for vegetated areas. *Remote Sensing Letters, 9*, 393–402. https://doi.org/10.1080/2150704X.2018.1425560

Chapter 5
Basic Terrain Characterization

This chapter presents procedures and methods to characterize the terrain of a region from the interpretation and processing of topographic data in the format of a DEM. This preliminary analysis is essential for various environmental studies, preceding more in-depth procedures such as deriving drainage networks or specific topographic indices.

Initially, we discuss the descriptive statistical type analysis of the DEM, the generation of contour lines and hypsometric classes, the evaluation of the elevation frequency histogram, and the elaboration of the hypsometric curve for a watershed.

Then, we address the preparation and interpretation of cross-sectional topographic profiles, particularly cross-sections of rivers and floodplains. In the last part of the chapter, we present the approaches for obtaining and analyzing attributes derived directly from the DEM: surface slopes, aspect, shading, and curvature.

For each item, conceptual aspects are presented, methods for generating the analyses are discussed, and results and examples of how to interpret them are presented.

5.1 Descriptive Statistics

Generating descriptive statistics is one of the most basic terrain analyses from processing a Digital Elevation Model. The procedure is straightforward, similar to that for any other raster, and follows the steps outlined in Chap. 3. As the DEM represents the region's topography, the statistical analysis of this raster means a statistical analysis of the area's topography.

Measures of central tendency are the most commonly used for statistically analyzing a DEM, emphasizing obtaining the maximum, minimum, and average elevation values. We also have the range from the maximum and minimum, referred

A. Rolim da Paz, *Digital Elevation Models for Environmental Studies*,
https://doi.org/10.1007/978-3-032-04523-2_5

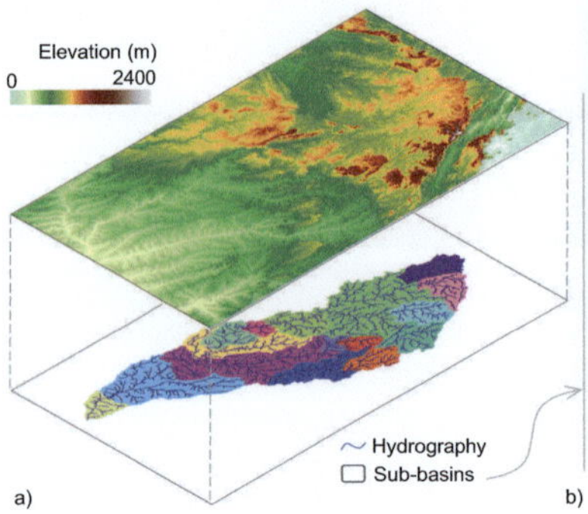

Sub-basin	Area (km²)	Min. elev. (m)	Max. elev. (m)	Mean elev. (m)	Elevation range (m)	St. dev. (m)
Jusante	5,486	301	552	420.6	251	48.1
Água Vermelha	21,484	352	768	500.0	416	59.3
Porto Colômbia	11,510	406	1,273	602.2	867	112.4
Marimbondo	19,689	448	1,284	733.6	836	164.3
Volta Grande	5,743	476	1,179	705.4	703	156.6
P. Guatapara	9,494	505	1,667	690.1	1,162	134.8
Jaguara	2,148	519	1,283	886.5	764	175.5
Pádua Sales	4,965	578	1,669	945.3	1,091	253.4
Peixoto	7,566	630	1,470	901.7	840	177.6
Limoeiro	1,875	569	1,556	945.7	987	173.4
Caconde	2,568	855	1,791	1,182.6	936	151.1
Furnas	25,819	686	1,768	918.4	1,082	119.0
Caruaçu	7,376	800	2,292	1,108.3	1,492	272.3
P. dos Buenos	6,331	785	2,724	1,059.0	1,939	239.6
Ibituruna	6,054	846	1,370	1,042.3	524	81.9
Camargos	6,322	870	2,622	1,128.9	1,752	210.5

b)

Fig. 5.1 **a** Digital Elevation Model derived from SRTM 90 data coarsened to a spatial resolution of 250 m for the region of the Rio Grande watershed (between Minas Gerais and São Paulo) and subdivision of this basin into 16 sub-basins; **b** Listing of sub-basins with parameters of descriptive statistics extracted from the DEM

to as the altimetric range in the case of a watershed's DEM. In other words, this parameter represents the elevation difference between the basin's highest and lowest points. In hydrology, altimetric elevation is relevant because it influences the maximum potential energy of surface water within a watershed, affects flow velocity, and plays a role in erosion processes (Collischonn & Dornelles, 2013).

In Fig. 5.1, central tendency measures are obtained for 16 sub-basins of the Rio Grande watershed, located on the border between Minas Gerais and São Paulo, from SRTM 90 data coarsened to a spatial resolution of 250 m. The statistics for each basin are generated by considering only the DEM pixels in each subdivision.

In general, the sub-basins in the lower part of the Rio Grande basin have a lower altimetric range (example: Água Vermelha, with 416 m) and, therefore, are subject to lower surface flow speeds and lower erosion rates. The headwater sub-basins have higher altimetric ranges (for example, Camargos, with 1752 m for this variable).

5.2 Contour Lines

The topographic representation in contour lines constitutes a simple and traditional concept in hydrological studies, which is the representation by isolines of equal terrain elevation. It is a practical way to understand the topographic variation of the region and is very useful, particularly for irrigation projects, hydrosedimentological studies, terracing, and civil engineering works. It is possible to generate contour lines from a DEM using an operation that is available in most GIS software.

Usually, contour lines are generated for equally spaced elevation values (equidistant from each other by a given value Δz). For example, curves every 5 m or every 10 m are used for relatively small areas, or every 100 m or 200 m for areas of more extensive spatial coverage and larger topographic range.

The contour line representation can be visually interpreted to identify steeper or gentler terrain areas. As the elevation difference is constant between consecutive curves, the higher spacing in the plane between the tracing of the curves indicates areas of lesser topographic variation. This occurs because the same height difference Δz is achieved with higher horizontal distances, as in the part to the right of the region illustrated in Fig. 5.2. Areas of steeper terrain, like the left part of the region shown in the referred figure, present a closer proximity between consecutive curves. In this case, the height difference Δz of 5 m is overcome in much shorter horizontal distances than those of the right part of the region.

Figure 5.3 exemplifies a DEM of part of the middle São Francisco river basin, in the north of Minas Gerais, and the contour lines obtained from it, in this case, with vertical spacing every 50 m. Each contour line indicates the contiguity of the terrain with the elevation indicated by the value represented in the symbology. Curves refer to elevations 450–800 m, every 50 m.

The visual comparison between DEM and contour lines allows a better understanding of the representativeness of these curves. Areas with contour lines close together on the plane and regions with wider gaps can be observed due to the higher distance between the contour lines. In the first case, the proximity between curves indicates a higher topographic gradient, as with shorter distances, one moves from one curve to the next. Suppose there is a higher separation between curves. In that

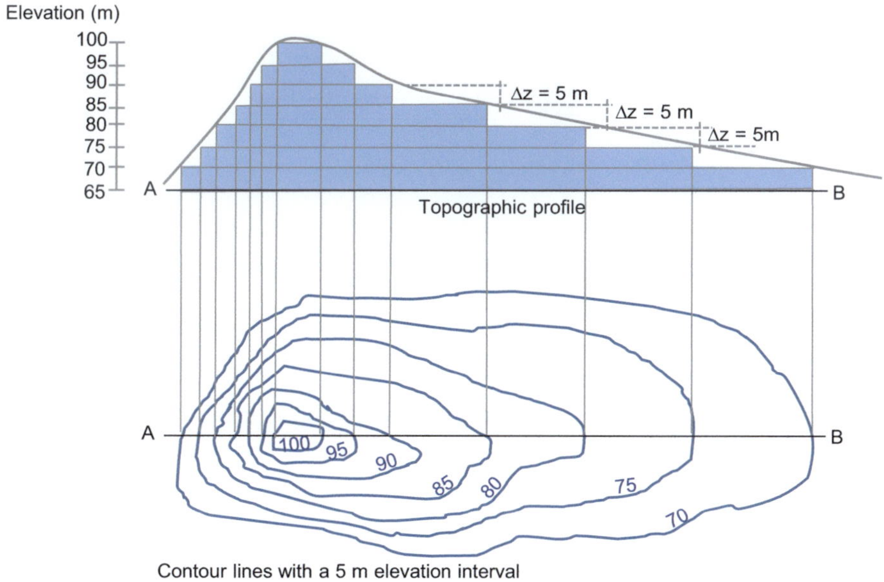

Fig. 5.2 Contour lines with equal elevation difference between them ($\Delta z = 5$ m) for a hypothetical region, whose left face has a much steeper terrain than the part to the right, as illustrated in the topographic profile along the transect A–B

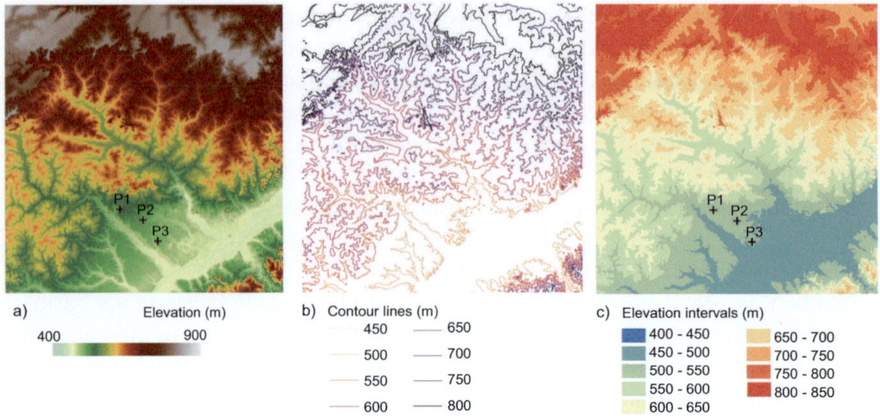

Fig. 5.3 **a** Digital Elevation Model of the area of the middle São Francisco river basin, in the north of Minas Gerais, from SRTM 90 m data; **b** Corresponding contour lines derived from the DEM, spaced every 50 m; **c** Classification in elevation intervals of 50 m; points P1, P2 and P3 have elevations of 542, 518 and 503 m, respectively, all belonging to the range of 500–550 m

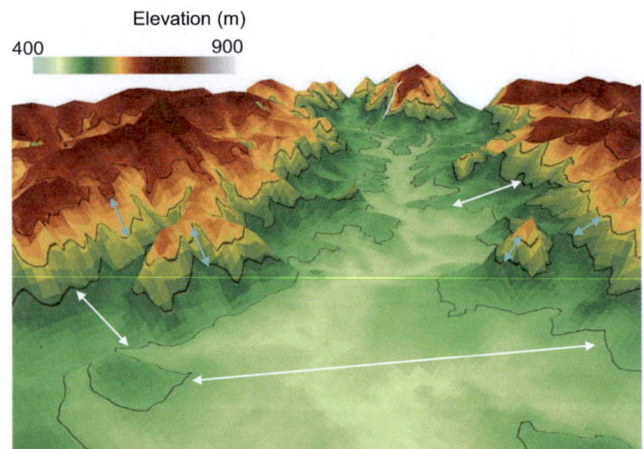

Fig. 5.4 3D visualization of a DEM of the middle part of the São Francisco River basin (north of Minas Gerais), with an overlay of contour lines (contours in black color) spaced 50 m: white arrows indicate areas with flatter relief, where the distance on the plane between contour lines is higher; blue arrows indicate areas of steeper relief, with contour lines closer to each other

case, this indicates flatter areas with relatively slight topographic variation, as illustrated in Fig. 5.4, which shows the 3D view of part of the DEM from Fig. 5.3 overlaid with contour lines.

5.3 Terrain Classification

The reclassification procedure described in Chap. 3, applicable to any raster, is also widely used to analyze and characterize the topography of a region from the DEM. The DEM is segmented, considering rules that establish the elevation classes intended by the user. Adopting classes at regular intervals, such as every 50 or 100 m, is common. The raster resulting from this procedure is usually called a hypsometric map.

For convenience, some GIS software allows automating this regular slicing of the DEM, with the user only needing to inform the desired spacing between classes or the number of intervals, in addition to the lower and upper limits for defining the classes. Imagine, for example, reclassifying in 50 m intervals a DEM whose topographic variation is from 0 to 2000 m. This implies 40 classes, making it advantageous to automate the definition of the interval limits to reduce the effort and chances of error associated with manually typing each one.

It is important to note that the result of reclassifying a DEM into elevation classes is not another DEM. The result is a raster whose attribute is no longer the elevation of the terrain but rather the number of the elevation class, which is a qualitative attribute. For example, pixels with an elevation between 300 and 350 m became class 5 in the previous example. In the DEM, the attribute is the elevation itself, such as a pixel with an elevation of 338 m. In the reclassified DEM raster, this pixel has attribute 5, which is not the elevation but a label indicating which elevation class it belongs to.

It is recommended that the class labels be established from the lowest to the highest elevation. The class number remains a qualitative attribute since class 4 does not necessarily represent twice the elevation defined by class 2, for example. However, this class ordering according to the elevation brings two benefits: it facilitates the reclassification operation itself, avoiding omission or intersection errors caused by the reclassification rules, and it allows a better visual interpretation of the spatial patterns of topography since a color gradient symbology can be used to represent this reclassified DEM raster.

In Fig. 5.3, already discussed earlier, a DEM of part of the middle reach of the São Francisco river basin is presented, with the corresponding reclassification in intervals of 50 m elevation. As expected, the raster obtained through reclassification reproduces the main spatial patterns of topographic variation, but the level of detail present in the DEM is reduced. This occurs because the elevation information, continuous in the DEM, has been discretized into 50-m intervals in the reclassified raster.

For example, all the pixels of the DEM that have an elevation between 500 and 550 m, like points P1 (542 m), P2 (518 m), and P3 (502 m) indicated in Fig. 5.3c, are relatively close to each other in terms of attribute and the symbology used for the DEM. However, they are distinct, and the variation is visually perceived in the DEM. With the reclassification, these pixels in the resulting raster have equal attributes, class number 3 (range of 500–550 m). Nevertheless, it is in this way, in elevation classes, that topographic characterization is needed for some studies.

One of the advantages of reclassifying a DEM into elevation classes is quantifying the areas occupied by each one via a procedure like the one described in Chap. 3 (quantifying raster areas). This analysis is the basis for generating a frequency histogram and elaborating a hypsometric curve from the DEM, as detailed in the following item. For the DEM itself, it is not recommended to calculate the area occupied by each attribute since there is a continuous variation of values. If the attribute has decimal values, this calculation does not make sense.

5.4 Frequency Histogram of Elevations

A frequency histogram of a DEM can be created by establishing elevation classes and quantifying the occurrence of pixels in each class. This procedure is analogous to obtaining a histogram for any raster, as described in Chap. 3. Figure 5.5a presents the elevation frequency histogram for the São Francisco River basin (Brazil). Based on this data, it can be observed that approximately 22% of the basin area falls within the 400–450 m elevation range. In contrast, only about 5% of the area lies between 900 and 1000 m.

In addition to the frequency histogram, one can work with the accumulated frequency of occurrence, as illustrated in Fig. 5.5a. In the accumulated frequency curve, each point indicates the percentage of the basin area that has an elevation up to the corresponding value. For example, about 30% of the São Francisco river basin has elevations up to 500 m.

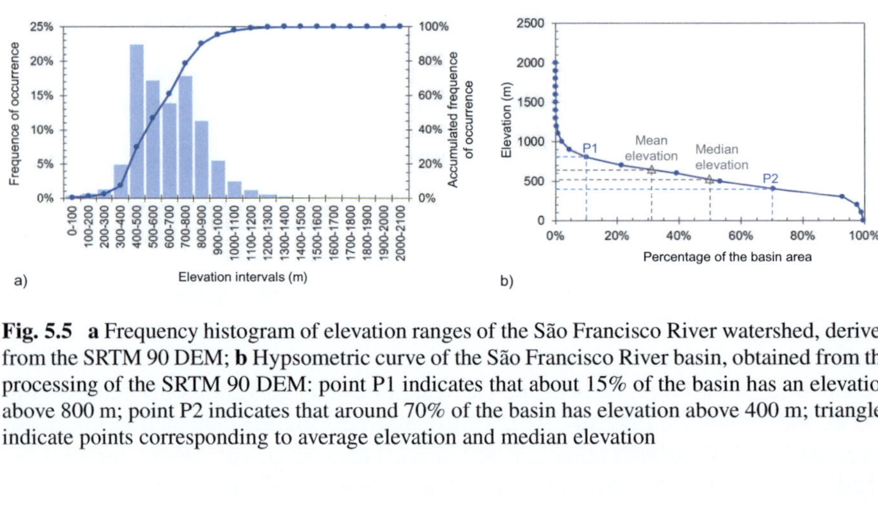

Fig. 5.5 **a** Frequency histogram of elevation ranges of the São Francisco River watershed, derived from the SRTM 90 DEM; **b** Hypsometric curve of the São Francisco River basin, obtained from the processing of the SRTM 90 DEM: point P1 indicates that about 15% of the basin has an elevation above 800 m; point P2 indicates that around 70% of the basin has elevation above 400 m; triangles indicate points corresponding to average elevation and median elevation

5.5 Hypsometric Curve

5.5.1 Concept and Interpretation

The most common way to describe topography through the analysis of the distribution of areas of a region according to elevations is to construct a hypsometric curve. Essentially, the hypsometric curve is a graph that shows the proportion of the region's area above (or below) different elevation levels.

In the curve of Fig. 5.5-b, for example, point P1 indicates that around 10% of the São Francisco river basin has an elevation above 800 m. In contrast, point P2 illustrates that about 70% of the basin is at an elevation above 400 m. The DEM of this basin and the mapping of the areas that correspond to the 10% above 800 m (point P1 of the hypsometric curve) and the 70% above 400 m (point P2) are highlighted in Fig. 5.6. According to the legend, this mapping was obtained by reclassifying the DEM, but only two classes were considered.

On the hypsometric curve, one can directly identify the point corresponding to the median elevation, indicating 50% of the area with elevation above and 50% below this value. For the example in Fig. 5.5, the median elevation is around 520 m. Considering the same DEM of the example, the average elevation obtained for the São Francisco river basin is 640 m, corresponding to a percentage of about 31%—a point indicated on the hypsometric curve.

The shape of the hypsometric curve is related to the stage of the watershed's geomorphological development (Strahler, 1957). Suppose the curve is predominantly convex. In that case, the region is considered to be of more recent development and

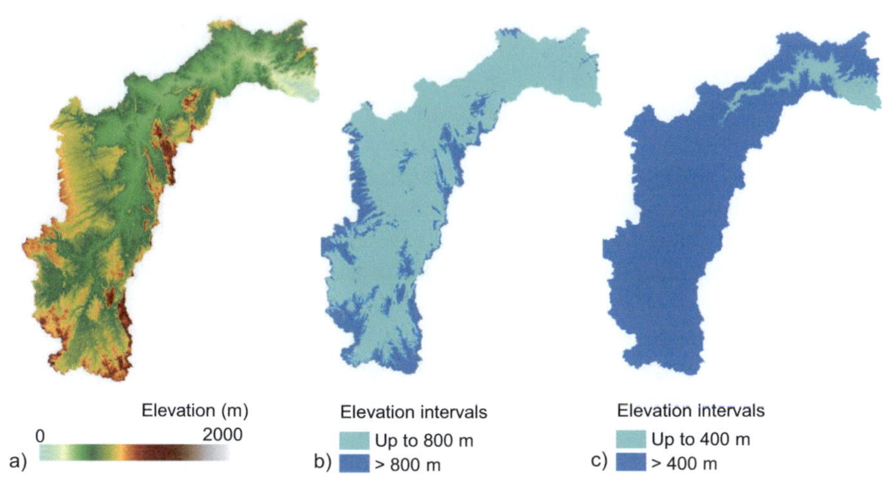

Fig. 5.6 **a** Digital Elevation Model of the São Francisco River watershed, from SRTM 90 data; **b** Reclassification of the DEM for elevation classes of up to 800 m and above 800 m; **c** Reclassification of the DEM for elevation classes of up to 400 m and above 400 m

with higher erosive potential than the region whose curve is concave and considered older. A curve in an S-like shape indicates a geological stage of maturity for the area intermediate between the other two stages.

5.5.2 Deriving from the DEM

A sequence of reclassification procedures and area quantification can derive the hypsometric curve from a DEM:

i. For each elevation value, reclassification is done to delimit the portion of the DEM above this value (areas with elevation above receive attribute 1, while other regions remain with attribute zero);
ii. Then, the area of this region with attribute 1 is quantified and transformed into a relative percentage value concerning the total area;
iii. Operations i and ii are repeated sequentially for equally spaced elevations (for example, every 50 m).
iv. The set of points formed by the pairs of elevation (x-axis) and corresponding area percentage (y-axis) arranged in a graph constitutes the hypsometric curve.

Alternatively, the hypsometric curve can be elaborated as an additional step from the data already obtained to elaborate the DEM histogram. As seen in previous items of this chapter, from the histogram data of a raster, one can get the accumulated frequency and the percentage of the area along the classes. For the hypsometric curve, the percentage of the area with elevation above each histogram class can be deduced as the complementary percentage of the accumulated value.

For example, Table 5.1 presents the histogram generated for the DEM of the São Francisco River basin, using 100-m elevation classes. For class number 3, with elevations from 200 to 300 m, 1.3% of the basin falls within this range (column 5), and 2.4% of the basin is included cumulatively from the first class up to this one (column 7). The corresponding point on the hypsometric curve is 97.6% (column 8), which is the complement of 2.4% to reach 100%. This means that 97.6% of the basin area lies above this third class, above 300 m. In practice, to construct the hypsometric curve, it is common to adopt the mean value of each class (the average between the lower and upper limits; column 9) for the x-axis, which makes the graph more straightforward to interpret.

In addition to these approaches to combine multiple operations and construct the hypsometric curve, some software provides specific commands to automatically generate such a curve, given a DEM as input. Whether using a specialized command or via the described approaches, attention should be paid to differentiate whether the desired hypsometric curve is for the entire DEM (all of the raster) or if it corresponds to a specific region, such as a watershed, a municipality, or a farm, for example.

In the first case, when you want to construct the hypsometric curve for the entire image, the reclassification and area quantification operations are typically performed

Table 5.1 Frequency histogram of elevations and hypsometric curve for the São Francisco River basin, derived from the SRTM 90 DEM

Elevation interval			Number of basin pixels within this elevation interval	Fraction of the basin within this elevation interval (%)	Cumulative number of pixels up to this elevation interval	Cumulative fraction of the basin up to this elevation interval (%)	Fraction of the basin above this elevation interval (%)	Mean value (m)	Relative dimensionless elevation (–)
Order	Lower limit (m)	Upper limit (m)							
1	0	100	289,180	0.4	289,180	0.4	99.6	50	0.02
2	100	200	561,433	0.7	850,613	1.1	98.9	150	0.07
3	200	300	1,013,177	1.3	1,863,790	2.4	97.6	250	0.12
4	300	400	3,780,023	4.9	5,643,813	7.3	92.7	350	0.17
5	400	500	17,240,999	22.4	22,884,812	29.7	70.3	450	0.21
6	500	600	13,258,677	17.2	36,143,489	46.9	53.1	550	0.26
7	600	700	10,712,868	13.9	46,856,357	60.8	39.2	650	0.31
8	700	800	13,761,398	17.9	60,617,755	78.6	21.4	750	0.36
9	800	900	8,707,886	11.3	69,325,641	89.9	10.1	850	0.40
10	900	1000	4,251,696	5.5	73,577,337	95.5	4.5	950	0.45
11	1000	1100	1,905,772	2.5	75,483,109	97.9	2.1	1050	0.50
12	1100	1200	935,994	1.2	76,419,103	99.1	0.9	1150	0.55
13	1200	1300	413,188	0.5	76,832,291	99.7	0.3	1250	0.60
14	1300	1400	177,488	0.2	77,009,779	99.9	0.1	1350	0.64
15	1400	1500	52,695	0.1	77,062,474	100.0	0.0	1450	0.69
16	1500	1600	13,919	0.0	77,076,393	100.0	0.0	1550	0.74
17	1600	1700	4398	0.0	77,080,791	100.0	0.0	1650	0.79
18	1700	1800	2233	0.0	77,083,024	100.0	0.0	1750	0.83
19	1800	1900	509	0.0	77,083,533	100.0	0.0	1850	0.88
20	1900	2000	105	0.0	77,083,638	100.0	0.0	1950	0.93
21	2000	2100	11	0.0	77,083,649	100.0	0.0	2050	0.98

without any additional condition, and the percentage calculations are relative to the total size of the raster.

In the second case, however, care should be taken not to include in the area quantification the regions outside the area of interest (outside the watershed, for example), and the percentages are calculated relative to this area of interest (watershed drainage area, for example).

When using a specific command to determine the hypsometric curve in software, it is generally possible to provide as additional input data a vector or raster file that defines the area of interest for constructing the hypsometric curve. It may vary between software, but the most common is that, in the case of a raster, this file should consist of pixels with a value of zero or invalid outside the area of interest and pixels with attribute 1 inside this area, for which the hypsometric curve is constructed.

5.5.3 Dimensionless Hypsometric Curve

The hypsometric curve can also be worked in a dimensionless structure, with relative values ranging from 0 to 1. For this, the elevations of the classes (h) are converted to dimensionless relative values (h_{rel}) as a function of the minimum (h_{min}) and maximum (h_{max}) elevations of the area of interest (Eq. 5.1). On the y-axis, you can keep the area fractions in percentages or express them as dimensionless fractions (0–1).

$$h_{rel} = \frac{h - h_{min}}{h_{max} - h_{min}} \tag{5.1}$$

The advantage of the dimensionless relative hypsometric curve is the possibility of a more coherent and direct comparison between curves of different regions, which is hindered when working with absolute values. To illustrate this calculation, the relative elevations for the São Francisco River basin are determined in Table 5.1 (last column), considering a h_{min} of 0 and a h_{max} of 2100 m.

As an example of comparison, in Fig. 5.7a, we have the hypsometric curves with absolute values referring to the watersheds of the São Francisco River and the Grande River, located between São Paulo and Minas Gerais. According to these curves, 30% of the São Francisco River basin has elevations above 640 m (point P1), while 30% of the Grande River basin has elevations above 820 m (point P2).

This type of analysis determines comparatively who has parts of the basin at higher or lower elevations in absolute terms. However, another interesting comparison is obtained with the dimensionless relative hypsometric curves of these basins shown in Fig. 5.7b. It is noted that, in this case, the curve of the São Francisco River basin is above the curve of the Grande River basin, inversely to their position when absolute values were considered (Fig. 5.7a).

On the one hand, by analyzing the absolute hypsometric curves, the Grande River basin has higher absolute elevations than the São Francisco River basin for the same fractions of the basin. On the other hand, the dimensionless curves indicate that,

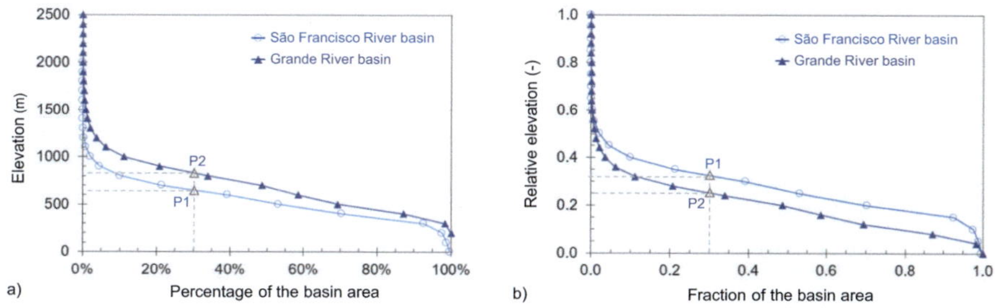

Fig. 5.7 **a** Hypsometric curves of the São Francisco River and the Grande River basins, obtained from the processing of the SRTM90 DEM; **b** Hypsometric curves of the same basins converted to the relative dimensionless structure

regarding relative elevations, the São Francisco River basin presents higher values than the Grande River basin for most fractions of the area.

As an example, a fraction of 0.3 (30%) of the São Francisco River basin has elevations above approximately 1/3 (0.33) of the maximum elevation (point P1 in Fig. 5.7b). In contrast, the elevations for the Grande River basin are above 1/4 (0.25) of the maximum elevation for the same relative fraction (30%) of the basin (point P2 in Fig. 5.7b).

Various dimensionless attributes can be derived from the dimensionless hypsometric curve and used to deepen the comparison between watersheds. Examples of attributes are (Strahler, 1957): the integral of the curve (area below the hypsometric curve), the slope of the curve at the inflection point, and the degree of sinuosity of the curve.

5.6 Cross-Sectional Topographic Profiles

The cross-sectional topographic profile analyzes the topographic variation along a transect. As seen in Chap. 3, transverse profiles can be developed along the space for a raster representative of any phenomenon or variable. But when the raster is a DEM, this profile is called a topographic profile. For example, this is a standard analysis for transects aligned transversely to watercourses, aiming to observe the conformation of the river's main channel and adjacent floodplain. The operational procedure is the same whether from the DEM or any other raster, as described in Chap. 3.

For example, two cross-sectional topographic profiles of the São Francisco river basin are shown in Fig. 5.8, developed from the SRTM 90 DEM. Notable points are indicated in the DEM and the profiles to facilitate understanding. The P–T transect has a length of about 42 km. It cuts transversely to the São Francisco River, which flows in a south-north direction in that figure. This transect is downstream of the Três Marias hydroelectric power plant, whose dam axis is near point X. In this P–T profile, the main river channel is located near point R, about 26 km from the beginning of the

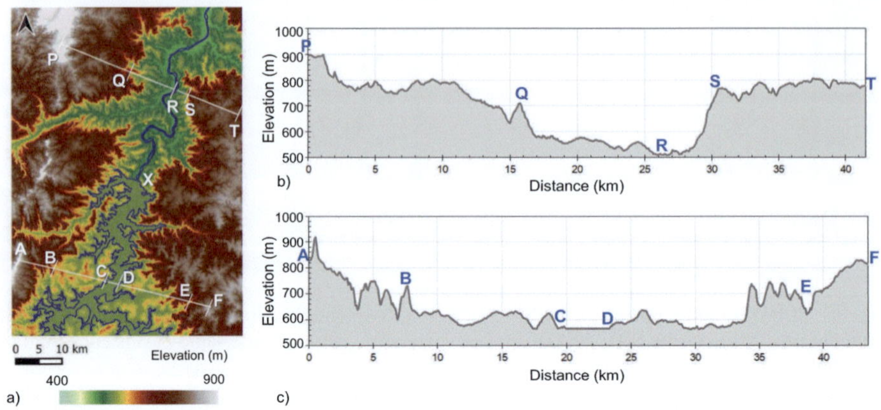

Fig. 5.8 **a** Digital Elevation Model of part of the upper São Francisco river basin, in Minas Gerais, developed with SRTM 90 data, partially showing the lake of the Três Marias reservoir (dam axis near point X); reservoir and river banks downstream of the dam are indicated in blue outline, digitized from Landsat 7 ETM + images acquired on 07 august 2001; **b**, **c** Topographic profiles along the P–T and A–F transects, respectively, with indication of intermediate reference points

profile (point P). At point Q, 16 km from P, there is a peak with a relatively sudden increase in elevation. In contrast, at point S, to the right of the main channel, there is a region of more intense topographic variation, followed by an area like a plateau up to point T.

The A–F transect crosses the lake of the hydroelectric power plant, whose water-covered area in the profile occupies the reach from point C to D, with a length of more than 3.3 km, and is flat. The lake area detected in the DEM data refers to the condition of the reservoir at the time of the sensor's data acquisition from the SRTM mission. In the A–F topographic profile, the letters B and E indicate other points highlighted by the rapid relative variation in elevation.

Another example of a topographic profile is presented in Fig. 5.9. For an area of the municipality of Santa Maria da Boa Vista, in Pernambuco, the orthophoto, the DTM, and the DSM generated from the LiDAR survey within the scope of the PE3D project (Cirilo et al., 2014, 2015) are shown. The topographic variation along the A-F transect in the topographic profile clearly illustrates the difference between DSM and DTM and the quality of the data from this project. For example, large trees near point B are over 15 m tall. While the DSM topographic profile represents the canopy of these trees, the DTM profile represents the bare ground beneath them. In the area indicated by C, the ground was almost without vegetation at the time of the LiDAR data acquisition. Therefore, the topographic profiles of the DSM and DTM are virtually identical. A building nearly 4 m high, where the letter E is indicated, is also clearly represented in the DSM, while this artifact does not exist in the DTM.

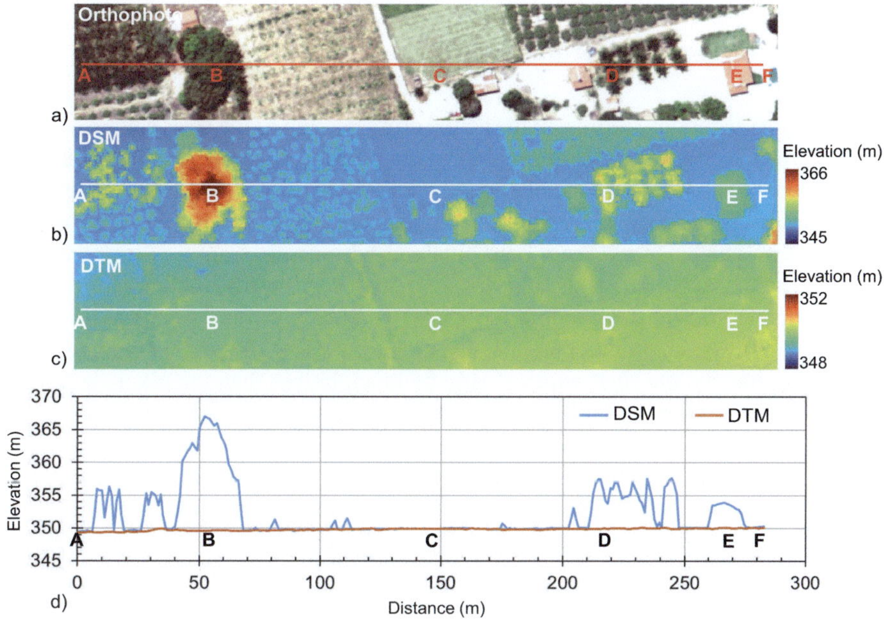

Fig. 5.9 **a** PE3D project orthophoto of a part of the municipality of Santa Maria da Boa Vista; **b**, **c** DSM and DTM obtained from LiDAR data from the PE3D project; d) Cross-sectional topographic profile along the A–F transect considering DSM and DTM

5.7 Cross-Sections of Rivers with Floodplains

The flow routing simulation along a river is a necessary step for various studies, such as flood risk estimation, sediment transport estimation, design and operation of streamflow forecasting and alert systems, quantification of aquatic habitats, estimation of the effect of reservoirs on the flow regime, assessment of water availability and impact of climate changes. One primary data required for this hydraulic simulation is the cross-sections along the river. These sections should cover the river's main channel (where the flow occupies most of the time, mainly during dry periods) and the floodplain areas.

It is possible to extract cross-sections of a river channel and the adjacent floodplain from a DEM. The general operational procedure is the same for extracting cross-sectional topographic profiles, especially since a cross-section is a transverse topographic profile. Given a line-type vector that indicates the alignment of the cross-section to be developed, this line is traversed, quantifying the distance traveled and the elevations obtained from the DEM.

5.7.1 *Difficulties and Limitations for Extracting Cross-Sections from a DEM*

The ability to accurately represent the actual cross-section of the river's main channel using a cross-sectional profile derived from the DEM is generally minimal for three main reasons. The first is the channel width and DEM spatial resolution relationship. For example, rivers with relatively narrow channels, like 20 m, cannot have their cross-section characterized from data obtained with pixels from a 30 m DEM. As a result, the extracted cross-section does not represent the main channel, only the main characteristics of the floodplain (Fig. 5.10). Ideally, several points of the DEM grid with elevation information are needed along the width of the river channel. This means that, ideally, the river width should be much higher than the pixel size.

The second reason is that the DEM usually comprises data not representing the river bathymetry, i.e., the topographic variation of the river bottom below the water line. As a rule, DEMs obtained by orbital remote sensing present this limitation, as seen in this chapter. If the DEM used to extract cross-sections is of this type, each section will appear as indicated in Fig. 5.11a. In the river's main channel, the elevations refer approximately to the water level at the data acquisition time instead of representing the topographic variation below this level (Fig. 5.11b). If it is an intermittent river and the bed was dry during DEM data acquisition, the cross-section derived from this DEM may represent the bottom of the dry river channel.

The third main reason that makes it difficult to characterize a river cross-section from DEM data adequately is the effect of vegetation cover on DEM elevations. As seen in Chap. 4, global DEMs obtained by orbital remote sensing usually present this influence. This occurs due to the inability of the electromagnetic waves used in data acquisition to penetrate through the vegetation canopy and effectively result in elevations of the terrain itself. Consequently, these DEMs are considered a Digital Surface Model, not a Digital Terrain Model.

Along riverbanks, there is a high likelihood of dense and relatively tall vegetation, which causes a cross-sectional profile extracted from a DSM-type DEM to appear as illustrated in Fig. 5.12. In this schematic situation, riparian vegetation prevents the

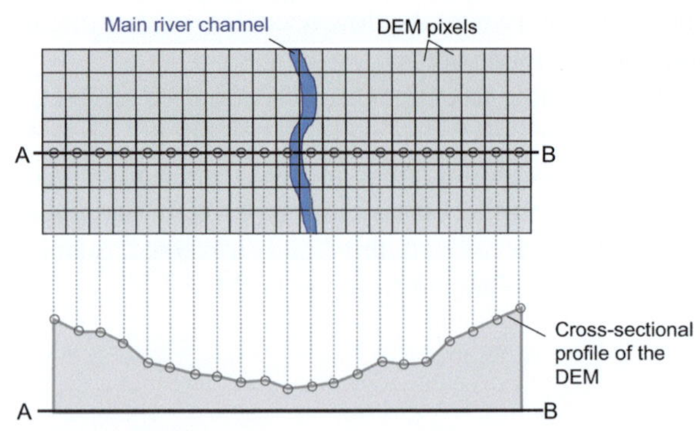

Fig. 5.10 Schematic DEM-derived cross-sectional topographic profile, characterizing the floodplain cross-section, but not being able to represent the cross-section of the very narrow river channel relative to the size of the DEM pixels

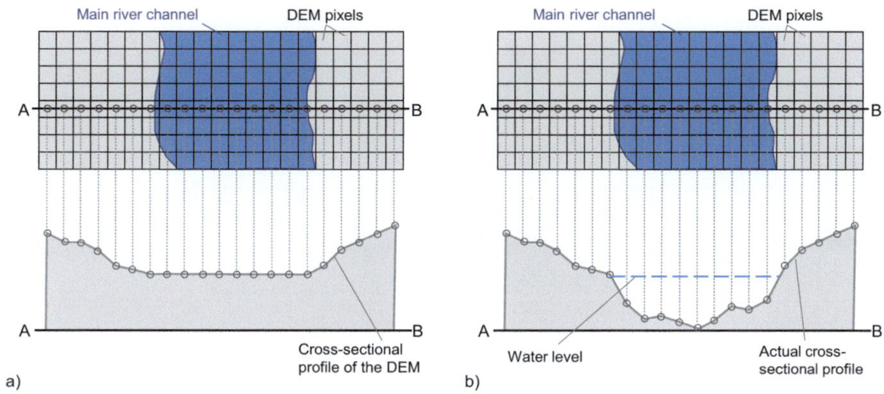

Fig. 5.11 **a** Cross-sectional profile developed from a DEM to represent the cross-section of a river, but with elevations along the main channel that refer approximately to the water level at the time of DEM data collection instead of representing the topography below the water level as in (**b**)

cross-sectional profile extracted from the DEM from displaying even a single point of lower elevation that would indicate the location of the main river channel (marked as point C). This limitation is more likely to occur in rivers relatively narrower than the DEM pixel size (Fig. 5.10). In contrast, larger rivers tend to be represented in the DEM by pixels whose elevation corresponds to the water surface level (Fig. 5.11; Paiva et al., 2011).

The effect of vegetation also makes it challenging to characterize the topography of the floodplain. An area of canopy opening (point D of Fig. 5.12) can even be mistakenly interpreted as corresponding to a river channel. This effect was perceived, for example, in analyses carried out for a small watershed in the Amazon (Valeriano et al., 2006). Other studies that characterize the effect of vegetation on cross-sections derived from global DEMs, with examples for treatment and attempts to minimize this effect, are the works of Baugh et al. (2013) and Modi et al. (2022).

5.7.2 Combination of Field Data and DEM Data

Suppose topographic/bathymetric data from field surveys for the main channel part of the river are available. In that case, they can be combined with topographic data from the DEM referring to the floodplain. The result is a cross-sectional profile that includes the detailed cross-section of the river channel and roughly the floodplain over a long distance.

Figure 5.13 schematically exemplifies the combination of topographic and bathymetric data. On the left side of the figure, the cross-sectional profile comprises data obtained solely from the DEM. Each point of the profile corresponds to a pixel found

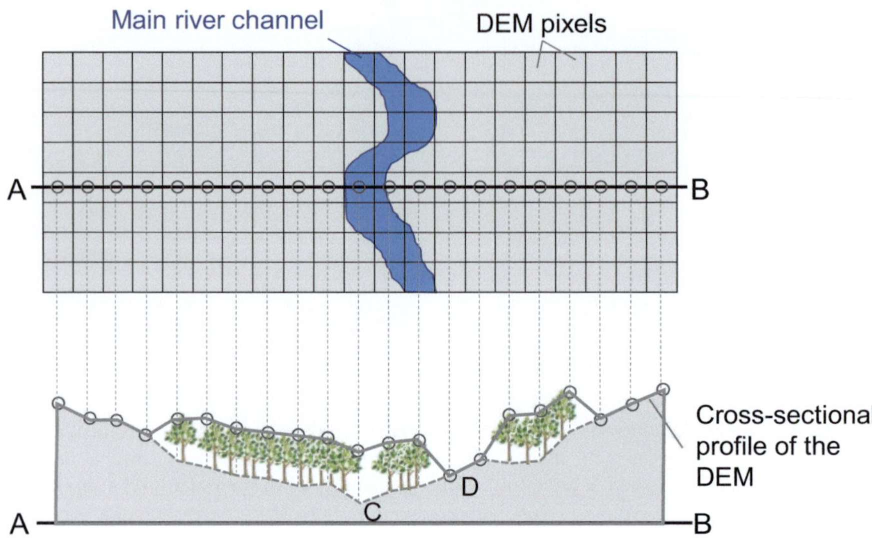

Fig. 5.12 Cross-sectional profile developed from a DSM-type DEM, indicating the effect of vegetation cover on the topographic representation: the main channel crossing point is hidden (C); in contrast, an artificial appearance of crossing a river channel at point D is created

along the transect; clearly, there is no adequate representation of the section of the river channel.

On the right side of the figure, the point of the profile derived from the DEM corresponding to the river channel was replaced by a set of 10 points obtained by a field survey. In this situation, the section composed of the two datasets configures

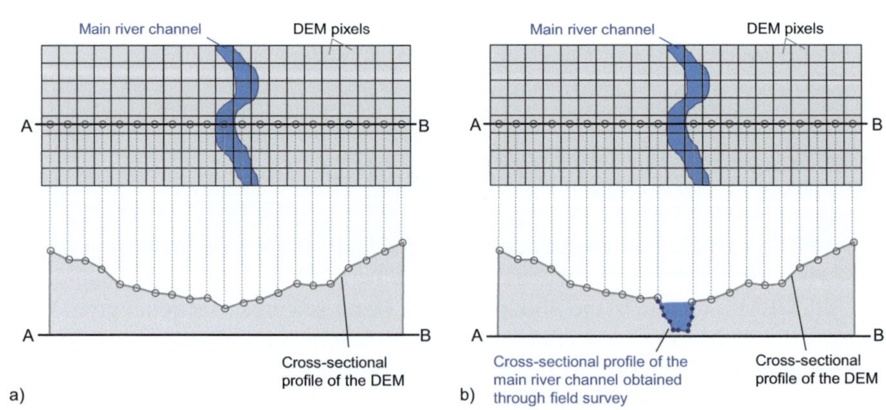

Fig. 5.13 **a** Cross-sectional profile obtained from DEM that does not characterize the main channel of the river; **b** Cross-sectional profile developed by the combination of field survey data for the main channel of the river and DEM data for the floodplain

a detailed cross-sectional profile that explicitly represents the river channel's cross-section. For instance, this valuable and quality information can be entered into a hydrodynamic model for flow routing simulation along this river.

In the example of Fig. 5.13, field data were used to replace only one point of the profile obtained from the DEM. However, each situation should be analyzed to decide whether to replace the necessary points. The combination of data sources is interesting because it takes advantage of the details of the field data, which are nonexistent in the DEM, with the large extent of the DEM coverage in the floodplain. Field surveys are generally limited to the main channel or to just a few additional meters along the banks due to high costs, time constraints, and access difficulties caused by vegetation, rugged terrain, or private property boundaries.

5.7.3 Difficulties in Combining Field Data and DEM Data

When combining topographic data from a field survey and a DEM to compose cross-sections, one must ensure that both information have the same vertical reference. If they are in different vertical references, one should adapt one of these data sources to the reference of the other, as discussed in Chap. 4.

Additionally, the coherence of combining floodplain and main channel data should be analyzed for each developed cross-section (Fig. 5.14a). For example, artificial elevation discontinuities may occur at the transition between floodplain elevations (derived from the DEM) and main channel elevations (obtained from field data).

One possible situation is when the edge points of the floodplain that connect to the channel have a lower elevation than the elevations of the adjacent points of the main channel (Fig. 5.14b). If data are used this way for input into a hydrodynamic model, for example, it is as if there were artificial dikes on the banks of the main channel. An opposite situation can also occur when the river's main channel is artificially lowered relative to the connection with the floodplain (Fig. 5.14c). In this case, when using such information as input for hydrodynamic modeling, the discontinuities will function as if the main channel had a higher depth than it does.

Both situations can impact the results during a flood wave routing simulation, as illustrated in Fig. 5.14 for the passage of three flows (Q_0, Q_1, and Q_2). While the flow is restricted to the main channel (Q_0), the three representations of the cross-section do not imply differences (Fig. 5.14d–f). When the flow increases to Q_1, there is overflow from the main channel in the actual situation (Fig. 5.14g). This overflow also occurs in the section with the artificial dike (Fig. 5.14h). However, this overflow may not happen in the case of the erroneously lowered section relative to the floodplain (Fig. 5.14i) since the main channel section has enlarged and has a higher flow capacity. In other words, there is an underestimation of floods in this case.

After this flood passes and the flow is reduced to a level limited to the main channel (from Q_1 to Q_2; Fig. 5.14j), the artificial dike prevents the return of part of the flooded

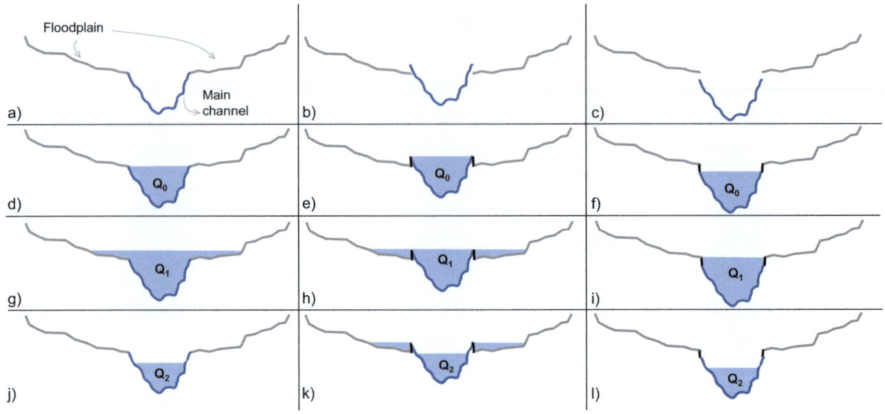

Fig. 5.14 Composite cross-section with a transition between floodplain and main channel data **a** properly connected, **b** forming an artificial levee, and **c** with an artificial elevation drop from the floodplain to the channel; **d–f** Behavior of the three cross-sections under the flow of a constant discharge Q_0, which does not exceed the capacity of the main channel in case (**a**); **g–i** Behavior of the three cross-sections under a constant discharge Q_1, which causes overbank flow in case (**a**); **j–l** Behavior of the three cross-sections after the passage of discharge Q_1 and its reduction to a discharge Q_2, which is confined to the main channel in case (**a**)

volume to the floodplain (Fig. 5.14k). In this case, the simulated floods tend to have higher depths and longer durations, indicating a tendency to overestimate the floods.

A real-world example of a combination of detailed data from more than 300 cross-sections obtained by topobathymetric survey and data extracted from the SRTM DEM for the floodplain is presented in Paz et al. (2010) for the flood wave routing along a reach of more than 1200 km of the Paraguay River, crossing the Pantanal, one of the most extensive wetlands on the planet. Adjustments for vertical reference adequacy, transition coherence analysis between channel and floodplain, and interpolation procedures to estimate main channel data in sections without the availability of field information were necessary for the simulation of spatiotemporal flood patterns in the Pantanal.

There are alternative procedures proposed in the literature to take advantage of DEM information to characterize cross-sections without having the availability of complementary field data and, even so, maintain coherence and reduce the effect of vegetation on the elevations. For example, Paiva et al. (2011) elaborated cross-sections for the Purus River in the Amazon from the SRTM 90 DEM with an approach in two primary steps. The first was estimating the bottom elevation of the cross-sections from the longitudinal profile extracted from the SRTM DEM but corrected to remove the effect of the water level, vegetation, and random noise. The second step was to adopt a simplified rectangular shape for the cross-sections and estimate height and width from geomorphological relations with the drainage area at each one.

5.8 Surface Slope

5.8.1 Definition and Determination Methods

The surface slope can be calculated pixel by pixel from a DEM, generating a raster with this attribute. It is a data layer with practical applications in various areas. For example, it can assist in determining the land suitability for crops, considering both the soil's intrinsic characteristics and its mechanization conditions. Additionally, it can be used to classify the landforms or identify suitable areas for installing sanitary landfills.

Given a DEM, determining surface slopes is a standard procedure in GIS software and is a simple operation. This slope represents the rate of change of the elevations in the DEM pixels; therefore, it is mathematically referred to as the first derivative of a DEM. It is assumed that the terrain surface represented in the DEM is mathematically continuous. Therefore, it follows the rule that, theoretically, mathematical derivatives can be calculated at any point (Burrough & McDonnell, 1998). However, since the surface is discretized in the DEM, derivatives are approximated by computing differences using a moving window filter—typically 3×3 or larger—or by fitting polynomial curves to the data within a defined neighborhood.

The slope can be defined as a local function calculated for a pixel based on the elevations of its four orthogonal neighbors: neighbors to the right, left, top, and bottom (Fig. 5.15).

The vertical slope (S_v) can be determined as the difference in elevations between the top and bottom neighbors (ΔZ_v), divided by the distance between these pixels (ΔL_v), which is equal to twice the spatial resolution (Eq. 5.2). Multiply this result by 100% to obtain slopes as percentages (Eq. 5.3).

$$S_v = \frac{\Delta Z_v}{\Delta L_v} = \frac{Z_b - Z_c}{2 \cdot dx} \tag{5.2}$$

$$S_v(\%) = S_v \cdot 100\% \tag{5.3}$$

where dx is the spatial resolution and Z_b and Z_c are the elevations of the top and bottom pixels, respectively.

Similarly, the horizontal slope (S_h) can be obtained by the difference in elevations between the neighboring pixels to the left and right (ΔZ_h), divided by the distance between such pixels (ΔL_h), also equal to 2dx (Eq. 5.4), with the multiplication by 100% to obtain the result in percentage (Eq. 5.5).

$$S_h = \frac{\Delta Z_h}{\Delta L_h} = \frac{Z_d - Z_e}{2 \cdot dx} \tag{5.4}$$

$$S_h(\%) = S_h \cdot 100\% \tag{5.5}$$

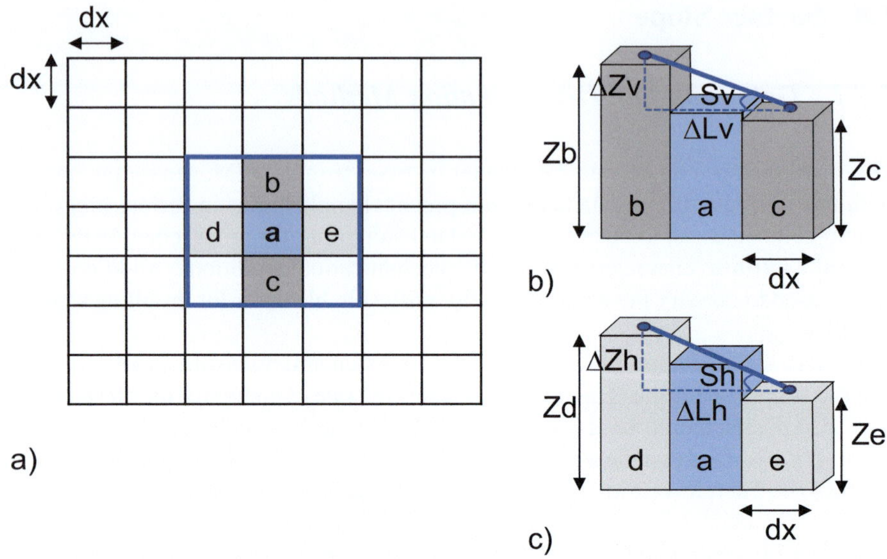

Fig. 5.15 Schematic indication of the determination of the slope of the surface from a DEM with spatial resolution equal to dx: **a** DEM highlighting a central pixel 'a' and its four orthogonal neighbors 'b', 'c', 'd,' 'e' in a 3 × 3 window; **b** Representation of the slope of the central pixel on the vertical axis of the 3 × 3 window (S_v), as a function of the elevations Z_b and Z_c of the neighboring pixels vertically and the distance between them (ΔL_v); **c** Representation of the slope of the central pixel in horizontal (S_h), as a function of the elevations Z_d and Z_e of the neighboring pixels horizontally and the distance between them (ΔL_h)

where Z_d and Z_e are the elevations of the pixels to the left and right, respectively.

Other approaches exist to determine the slopes in the horizontal and vertical directions, and several methods exist to obtain the slope of the pixel itself. One of the most used and implemented methods in GIS software is calculating the pixel slope based on the vertical and horizontal slopes indicated above. By this method, the slope of the pixel (S_p) is independent of the elevation of the pixel itself and a function exclusively of the elevations of the four orthogonal neighbors (Eq. 5.6; Zevenbergen & Thorne, 1987). The moving 3 × 3 window is shifted along the image, replicating the slope calculation for each pixel taken as the center of the window (Fig. 5.15).

To obtain the pixel slope in percentage, multiply the result of Eq. 5.6 by 100% (Eq. 5.7), while the arc tangent function should be applied to obtain the slope in degrees (Eq. 5.8). The percentage slope varies from 0 to infinity (it is not constrained to 100%). In degrees, the slope varies from 0° to 90°. For example, if $S_p = 1$, then S_p (%) = 100% and S_p (°) = 45°.

$$S_p = \sqrt{S_v^2 + S_h^2} = \sqrt{\left(\frac{Z_b - Z_c}{2 \cdot dx}\right)^2 + \left(\frac{Z_d - Z_e}{2 \cdot dx}\right)^2} \tag{5.6}$$

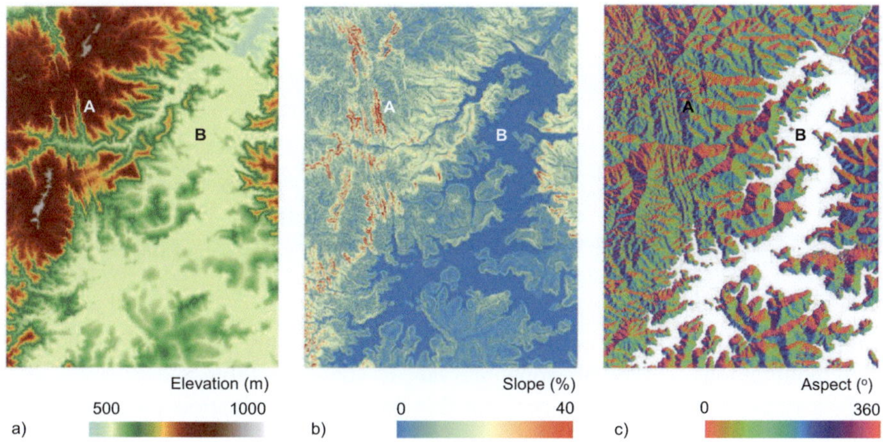

Fig. 5.16 Digital Elevation Model (**a**) of part of the upper São Francisco river basin around the Três Marias hydroelectric power plant, from SRTM 90 data, and derived products: surface slope raster (**b**); aspect raster (**c**)

$$S_p(\%) = S_p \cdot 100\% \tag{5.7}$$

$$S_p(°) = \arctan(S_p) \tag{5.8}$$

Figure 5.16 presents a real-world example of determining slopes from the SRTM 90 DEM for an area around the Três Marias hydroelectric power plant in the upper course of the São Francisco River (Minas Gerais). The slopes are presented in percentage, with raster color symbology saturated to the maximum value of 40%, although the actual maximum slope was 294%.

In general, the spatial patterns of the slope raster in the given example reflect the patterns identified in the DEM, as conceptually expected. Areas with steeper slopes correspond to regions where elevation variations in the DEM are more pronounced, such as to the left of point A. In the hydroelectric reservoir area, slopes are zero (e.g., point B) since the dam was already in place and with a given water-covered area when the SRTM DEM data were acquired. Another DEM-derived product, the aspect, is also presented in Fig. 5.16. This product aids in describing the region's topography, as explained in the following section.

5.8.2 *Relief Classification According to Slope*

Slope values can be used to classify the relief of a given region. In Brazil, for example, EMBRAPA (the Brazilian Agricultural Research Corporation) defines six

Table 5.2 Relief classes according to slope following EMBRAPA (1999)

Slope interval (%)	Relief class
0–3	Flat
3–8	Gently undulating
8–20	Undulating
20–45	Strongly undulating
45 – 75	Hilly
> 75	Steep mountainous

relief classes—flat, gently undulating, undulating, strongly undulating, hilly, and steep mountainous—based on slope intervals, as shown in Table 5.2.

Two steps are necessary to perform this type of classification for a particular region:

(i) First, the slope raster is obtained directly from the processing of the DEM.
(ii) Then, a classification operation of the slope raster is performed to get a new raster whose attributes indicate the relief classes and no longer the value of the slope itself. The reclassification procedure follows as described in Chap. 3, adopting the ranges of values shown in Table 5.2 (or another classification according to interest).

As an example of application to a real-world case, we have the DEM and the corresponding slopes obtained, in percentage, for an area in the north and west of Paraná (Fig. 5.17). The color scale of the slopes was saturated at 30% for better visualization, as only a few pixels have values higher than this threshold. The raster of relief classes was obtained according to the intervals defined in Table 5.2 through reclassification of the slope raster. Noticeably, there is a significant predominance of flat, gently undulating, or undulating relief, with localized areas of strongly undulating relief and few occurrences of higher slopes. Analyses such as the quantification of regions occupied by each class and the frequency histogram could complement understanding of the slope behavior in the area, following the procedures described in Chap. 3.

5.8.3 *Analysis of Agricultural Suitability Depending on Slopes*

The classification of a region according to slope ranges can also be carried out with other focuses, such as the suitability of the area for a specific agricultural crop. Various factors relate slope to agricultural suitability, such as mechanization limitations, vegetation structure, soil erosion, and water drainage (Akpoti et al., 2019). The slope is related to soil characteristics, such as texture, porosity, and stoniness. However, it is also related to soil productivity due to the influences of the slope on the conditions

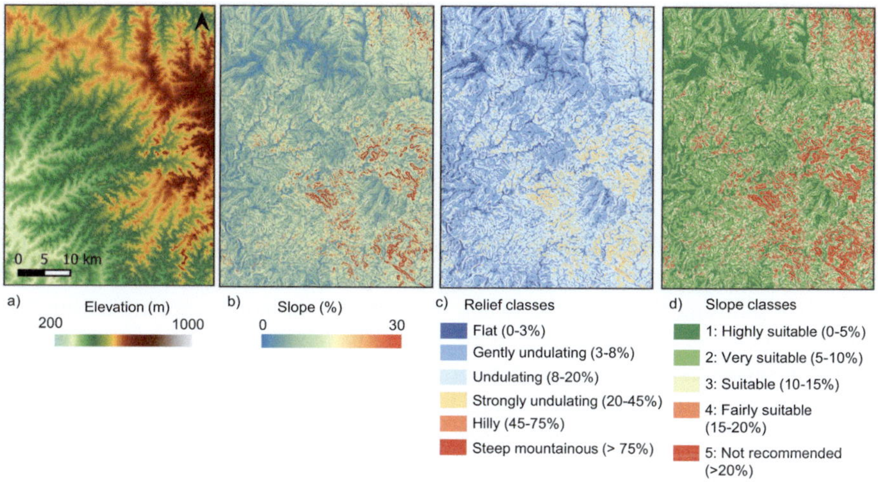

Fig. 5.17 **a** Digital Elevation Model of part of the north and west of Paraná, from SRTM 90 data; **b** Raster of surface slopes determined from the DEM; **c** Raster of slope classes according to relief types; **d** Raster of slope classes established according to the suitability for coffee cultivation

of humidity and surface drainage, the accumulation of organic matter, and exposure to solar radiation.

Generally, the slope is a limiting factor for the mechanization of an agricultural crop. This is related to vehicle operation since it influences movement speed, fuel consumption, and stability. The mechanization of sugarcane cultivation in Brazil is an example, as it requires areas with a slope of up to 12%.

The slope classification adopted by Hofig and Araújo-Júnior (2015) illustrates an example of this type of analysis for coffee cultivation in Paraná, in southern Brazil. The slopes are grouped into five ranges, and there is a reduction in the potential for mechanization as the slopes increase (Table 5.3).

For the same region as the previous example, slope classification was carried out to categorize the area according to the degrees of suitability for the mechanization of coffee cultivation (Fig. 5.17d), following the classes specified in Table 5.3. The classification procedure is analogous to that described for the previous case: starting from the slope raster derived from the DEM, the classification operation is performed

Table 5.3 Classification of potential for mechanization of coffee cultivation based on slope, adopted by Hofig and Araújo-Júnior (2015)

Slope interval (%)	Classe de potencial à mecanização
0–5	Highly suitable
5–10	Very suitable
10–15	Suitable
15–20	Fairly suitable
> 20	Not recommended

as described in Chap. 3, using each class's lower and upper limits. In this example, extremely suitable areas are predominant for the mechanization of coffee cultivation.

5.9 Aspect

The aspect is an attribute derived from the DEM representing the direction of the maximum slope in each pixel. It is measured in degrees, varying clockwise, from zero when pointing north to 360° when it returns to the north, making a complete turn. Flat areas without a defined slope generally receive attribute −1 or another value that serves as a flag for this indetermination condition.

Figure 5.16c illustrates the aspect raster for each pixel in the Três Marias Hydropower Plant region, in the upper São Francisco River basin. The reservoir area appears white (e.g., point B), indicating no defined aspect value. For better understanding, a portion of the same aspect raster is displayed in three dimensions based on the DEM elevations in Fig. 5.18. This figure clearly shows how the aspect varies according to the orientation of pixel faces relative to north. Pixels in orange or reddish tones tend to face north, while those transitioning from green to blue face the opposite direction.

Another way to analyze the aspect raster is to reclassify this attribute, which has a continuous value from 0° to 360°, into groups around the four cardinal directions (north, south, east, and west) and the four intercardinal directions (northeast, southeast, northwest, and southwest). Pixels whose aspect is 0° to 22.5° or between 337.5° and 360° become a single class relative to the north. Pixels between 22.5° and 67.5° become the northeast category, and so on for the remaining cardinal and intercardinal directions (Fig. 5.19).

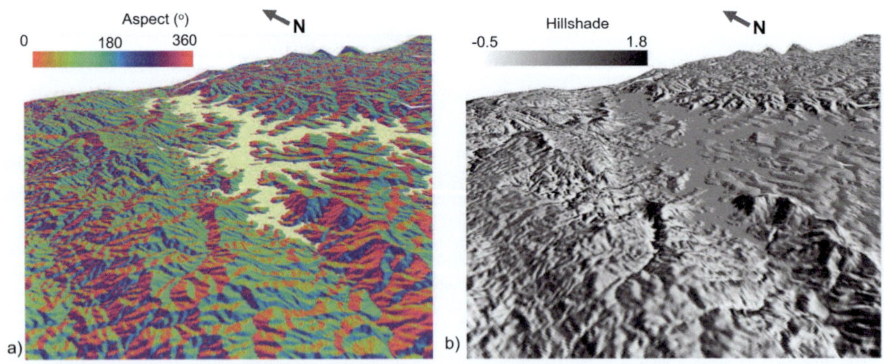

Fig. 5.18 3D view of the DEM of part of the upper São Francisco river basin around the Três Marias hydroelectric power plant, from SRTM 90 data, overlaid to the aspect raster (**a**) and the hillshade raster (**b**)

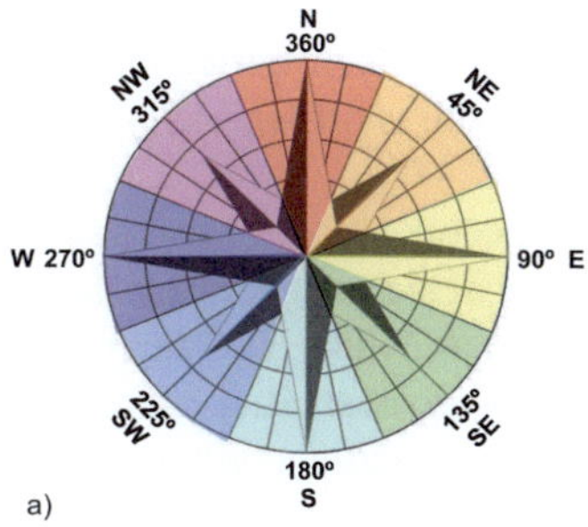

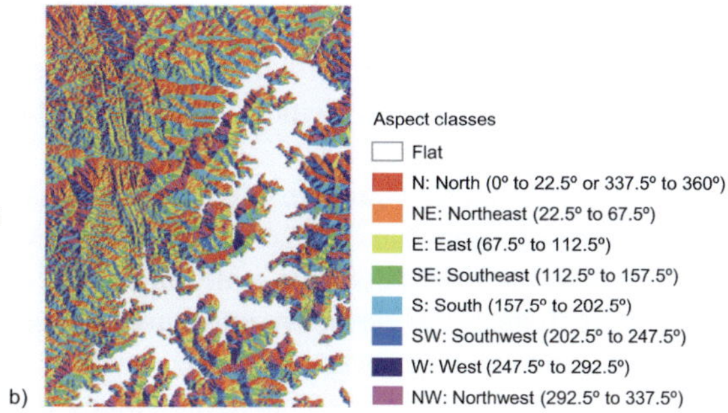

Fig. 5.19 **a** Compass rose with indication of the cardinal and intercardinal directions and the classes of aspects indicated by the colors; **b** Aspect raster of part of the upper São Francisco river basin surrounding the Três Marias hydroelectric power plant classified to group the aspects around the cardinal and intercardinal directions

5.10 Hillshade

The shading of the DEM is a technique used to give a 3D effect in visualizing the topography represented in this DEM. The idea is to simulate a particular position of the Sun and generate the corresponding shaded or illuminated areas. For this operation, two fundamental parameters are related to the position of the light source, the Sun (Fig. 5.20).

The first parameter is the horizontal angle of the Sun (α_1), in this case, the azimuth, which is the angle of the Sun's projection on the plane relative to the north, measured clockwise. This angle varies from 0° when the Sun is in the north position and increases clockwise to 360° when it returns to the north.

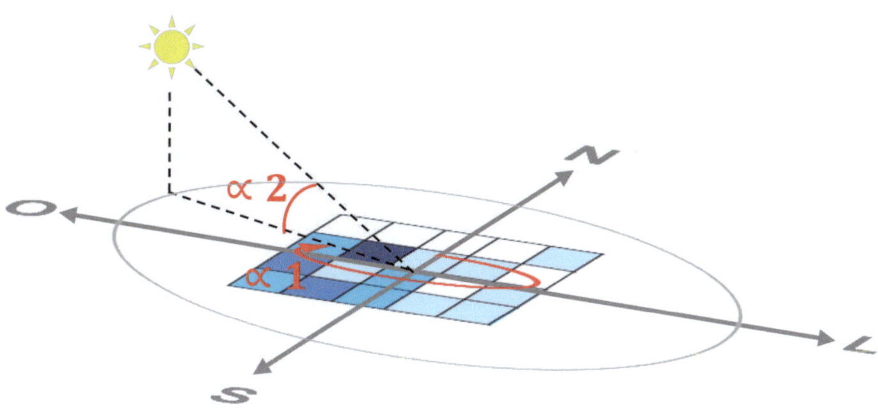

Fig. 5.20 Indication of the horizontal (azimuth, α_1) and vertical (α_2) angles for applying the shading operation of a DEM

The second parameter is the vertical angle of the Sun (α_2), which refers to the elevation of this light source relative to the plane. This parameter varies from 0°, when the Sun is aligned on the horizon, to 90°, when the Sun is orthogonally above the image.

The choice of these parameters defines the areas that become more or less shaded or illuminated by the Sun. For some software, a vertical scale exaggeration factor relative to the horizontal distance scale can also be defined as a third parameter. This parameter enhances the elevation amplitude to create a more substantial 3D effect.

Usually, the shading raster is visualized with a grayscale gradient, where darker tones represent shaded areas and lighter tones indicate illuminated areas. This is exemplified by the shading generated from a DEM of part of the Lower São Francisco River basin (Fig. 5.21). The effect of the change in the Sun's horizontal angle is also illustrated in the figure.

The reader can observe that when the horizontal angle is 300°, the Sun is positioned to the left and above (near northwest), making the faces of the relief facing this direction more illuminated. In contrast, those facing the opposite side are more shaded. When the horizontal angle of the Sun was defined as 60°, the approximate position of the northeast, the shading raster is quite distinct from the previous one.

One of the applications of hillshading is to enhance the visual representation of terrain in a DEM. To achieve this, a common approach is to set the hillshade layer to a certain level of transparency and overlay it on top of the DEM. In the example shown in Fig. 5.21, an 80% transparency (or 20% opacity) was applied. The DEM visualization combined with the hillshade overlay allows for a more precise identification of topographic patterns than the DEM alone.

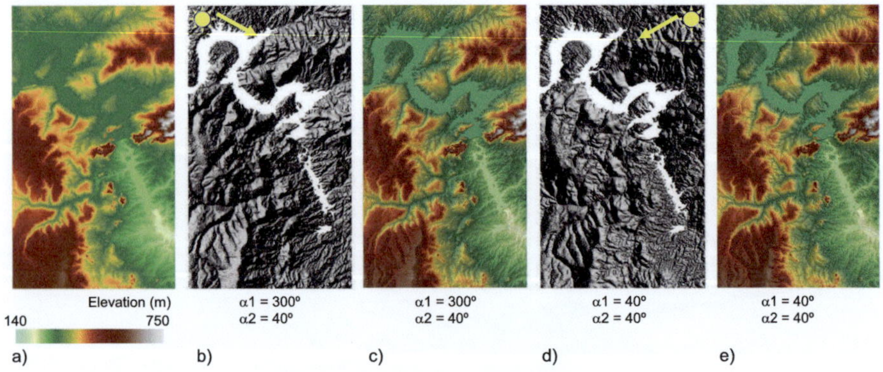

Fig. 5.21 **a** DEM obtained from SRTM 90 data for part of the lower São Francisco river basin, between Pernambuco and Bahia (surrounding the Paulo Afonso hydroelectric power plant); **b**, **d** Shading generated from this DEM, considering a vertical angle of the Sun (α_2) of 40° and a horizontal angle (α_1) of 300° in the first case and 40° in the second case (the yellow arrow indicates the solar radiation direction); **c**, **e** Enhancement of the DEM obtained by overlaying it to the shading (with opacity reduced to 20%) obtained in (**b**) and (**d**), respectively

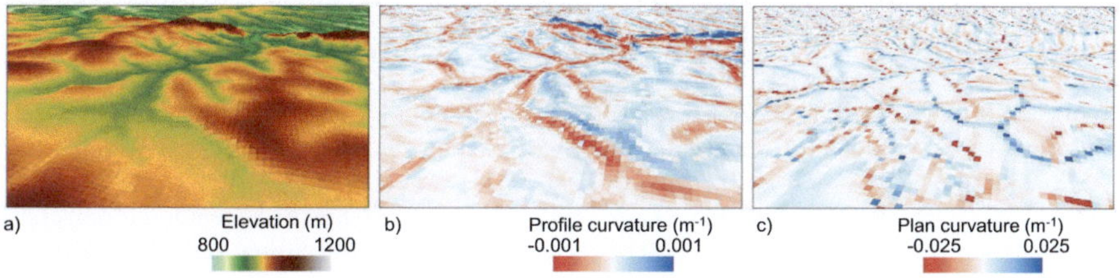

Fig. 5.22 **a** 3D view of the DEM of part of the municipality of Patrocínio (Minas Gerais) obtained from SRTM 90 data; **b**, **c** 3D view of the curvature in profile and curvature in the plane

5.11 Curvature

The slope is mathematically the first derivative of the elevation, representing the elevation change rate. The second derivative is the curvature, which represents the rate of change of the slope. There are several types of curvature, the three main ones being the profile curvature, the plan curvature, and the mean curvature, given by the average of the profile and plan curvatures.

The profile curvature, or vertical curvature, is the slope change rate in the direction of the aspect, which is the direction of the maximum slope in which the action of gravity is maximized. The profile curvature largely influences the speed of water flow, erosion, and sediment deposition processes. Positive values for this magnitude indicate the relative acceleration of the flow, while negative values indicate the relative deceleration of the flow. The surface can be classified as concave, convex, or rectilinear depending on the profile curvature. The usual unit of curvature in the profile is m^{-1}.

The plan curvature, planimetric curvature, or even horizontal curvature, as it is also referred to, represents the rate of change of the aspect. In other words, it means the rate of change of the slope in the direction orthogonal to the aspect in which the action of gravity is minimized. Therefore, this magnitude is related to the ease of accumulating water on the surface, being positive when there is convergence of lateral flow and negative when there is divergence of lateral flow.

As a real-world example of curvature in profile and curvature in the plane, Fig. 5.22 shows results obtained from the SRTM 90 DEM for an area of the municipality of Patrocínio in Minas Gerais.

References

Akpoti, K., Kabo-bah, A. T., & Zwart, S. J. (2019). Agricultural land suitability analysis: State-of-the-art and outlooks for integration of climate change analysis. *Agricultural Systems, 173*, 172–208. https://doi.org/10.1016/j.agsy.2019.02.013

Baugh, C. A., Bates, P. D., Schumann, G., & Trigg, M. A. (2013). SRTM vegetation removal and hydrodynamic modeling accuracy: SRTM vegetation error removal. *Water Resources Research, 49*, 5276–5289. https://doi.org/10.1002/wrcr.20412

Burrough, P. A., & McDonnell, R. A. (1998). *Principles of geographical information systems.* Oxford University Press.

Cirilo, J. A., Alves, F. H. B., da Silva, L. A. C., Campos, J. H. A. L. (2014). Suporte de Informações Georreferenciadas de Alta Resolução para Implantação de Infraestrutura e Planejamento Territorial. *Revista Brasileira de Geografia Física, 7*, 755–763.

Cirilo, J. A., Alves, F. H. B., Silva, B. M., & Campos, P. H. A. L. C. (2015). Pernambuco tridimensional: base de dados espaciais para planejamento urbano e gestão territorial. In *Presented at the 12o Simpósio de Hidráulica e Recursos Hídricos dos Países de Expressão Portuguesa, Brasília.*

Collischonn, W., & Dornelles, F. (2013). *Hidrologia para Engenharia e Ciências Ambientais.* Associação Brasileira de Recursos Hídricos (ABRH).

EMBRAPA. (1999). *Sistema brasileiro de classificação de solos.* Empresa Brasileira de Pesquisa Agropecuária, Centro Nacional de Pesquisa de Solos, Brasília.

Hofig, P., & Araújo-Júnior, C. F. (2015). Classes de declividade do terreno e potencial para mecanização no estado do Paraná. *Coffee Science, 10*, 195–203.

Modi, P., Revel, M., & Yamazaki, D. (2022). Multivariable integrated evaluation of hydrodynamic modeling: A comparison of performance considering different baseline topography data. *Water Resources Research.* https://doi.org/10.1029/2021WR031819

Paiva, R. C. D., Collischonn, W., & Tucci, C. E. M. (2011). Large scale hydrologic and hydrodynamic modeling using limited data and a GIS based approach. *Journal of Hydrology, 406*, 170–181. https://doi.org/10.1016/j.jhydrol.2011.06.007

Paz, A. R., Bravo, J. M., Allasia, D., Collischonn, W., & Tucci, C. E. M. (2010). Large-scale hydrodynamic modeling of a complex river network and floodplains. *Journal of Hydrologic Engineering, 15*, 152–165. https://doi.org/10.1061/(ASCE)HE.1943-5584.0000162

Strahler, A. N. (1957). Quantitative analysis of watershed geomorphology. *Transactions AGU, 38*, 913. https://doi.org/10.1029/TR038i006p00913

Valeriano, M. M., Kuplich, T. M., Storino, M., Amaral, B. D., Mendes, J. N., & Lima, D. J. (2006). Modeling small watersheds in Brazilian Amazonia with shuttle radar topographic mission-90m data. *Computers and Geosciences, 32*, 1169–1181. https://doi.org/10.1016/j.cageo.2005.10.019

Zevenbergen, L. W., & Thorne, C. R. (1987). Quantitative analysis of land surface topography. *Earth Surface Processes and. Landforms, 12*, 47–56. https://doi.org/10.1002/esp.3290120107

Chapter 6
Flow Directions

When working with Digital Elevation Models for environmental studies, one of the primary purposes is extracting the drainage network and delineating watersheds. The decisive step behind these processes is the determination of the flow directions. This involves defining the hydrological connectivity between DEM pixels using specialized algorithms based on the interpretation of topographic variability and the resulting flow paths. Various other DEM-derived products also depend on establishing flow directions.

Flow direction determination is the core operation behind a series of DEM-based processes and represents the most complex stage. This step is often embedded into broader GIS workflows, with users giving minimal attention to its inherent complexity. For this reason, this chapter is specifically dedicated to flow directions. It addresses the main concepts, methods of determination, the dependency on terrain depression treatment, and the challenges that have led to the development of dozens of algorithms in specialized literature, among other aspects.

6.1 Definition and Types of Approach

When visualizing a DEM in 3D, the preferential flow paths are roughly perceived due to the topographic variation, considering the effect of gravity on the flow. This is visually perceived due to the spatial patterns present in the DEM. You can even roughly trace these preferential paths (Fig. 6.1) manually as an approximation. For example, points A and B are located relatively close together. However, by visual interpretation of the DEM, the flow paths downstream of these points follow entirely different paths, draining to points P and Q, respectively.

Through direct interpretation of the DEM, it can also be inferred that point C has a flow that converges toward the downstream flow path of B, continuing to point Q. On the other hand, point D, near C, drains through a different path to point R.

A. Rolim da Paz, *Digital Elevation Models for Environmental Studies*,
https://doi.org/10.1007/978-3-032-04523-2_6

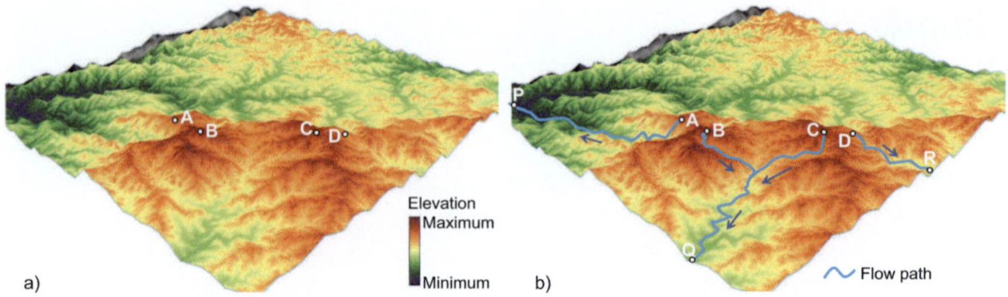

Fig. 6.1 **a** 3D visualization of a DEM; **b** Approximate indication of the downstream flow paths of points A, B, C, and D from the visual interpretation of the DEM

This type of interpretive analysis of the DEM, whether in standard 2D visualization or 3D structure, can give a general idea of which areas contribute to the flow to a certain point in the image. Visually interpreting the main spatial patterns of the image is helpful for a preliminary understanding of hydrological processes and other types of studies.

However, the visual interpretation of the DEM is subjective and quite limited, as it does not allow identifying the flow paths with more details, pixel by pixel, or precisely identifying which area contributes to flows to a certain point in the image. Making these determinations accurately is fundamental to properly taking advantage of the information in the DEM and generating the necessary data for studies in various areas. Defining the downstream flow path of each pixel is required, for example, to trace drainage networks, quantify flow distances, delimit watersheds, and develop hydrological, hydrosedimentological, and water quality studies, among many others.

The key issue in determining the flow paths from the DEM topographic representation is defining the flow directions, i.e., where each image pixel's flow goes in the immediate vicinity. The following items describe three main approaches to determining the flow directions.

6.2 Single Flow Direction Approaches (D8 and D4)

The most widely used approach considers that each pixel of the DEM can have the flow directed to one of its eight immediate neighbors in a 3 × 3 window (window formed by 3 rows and 3 columns of the raster). This approach is known as D8 (Deterministic eight neighbors) and allows each pixel to have one of eight possibilities of flow directions (Fig. 6.2a). The four orthogonal neighbors, located to the right, left, above, and below the pixel, and the four diagonal neighbors are considered.

The D4 approach is a simplification of the D8 procedure. In the D4 approach, each pixel's flow can only go to one of its four neighbors located orthogonally in a 3 × 3 window. A pixel has a flow direction of only one of the four options: north,

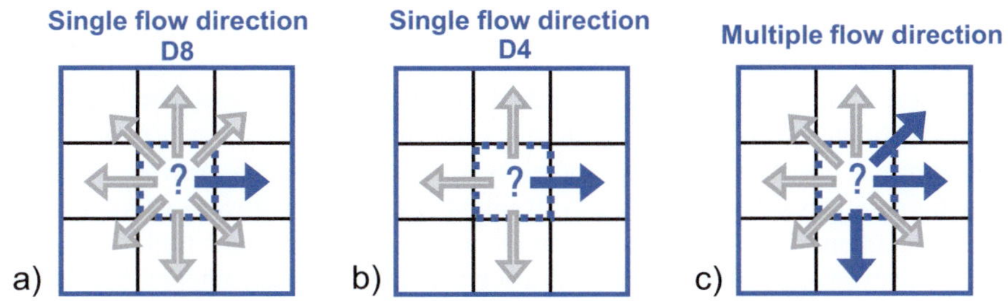

Fig. 6.2 Types of approach for determining flow directions: **a** Approach with single flow direction selected from the four orthogonal and four diagonal (D8); **b** Approach with single flow direction selected from only the four orthogonal (D4); **c** Approach with multiple flow directions, selected from 8 possibilities (4 orthogonal and four diagonal)

south, east, and west (Fig. 6.2b), and no longer the directions related to the diagonal neighboring pixels, as in the D8 approach.

The D4 and D8 approaches consider single flow directions. Each pixel of the DEM has a single flow direction and thus drains to only one of its neighbors. In this book, for simplification and the author's preference, the single flow directions of the D8 approach are adopted in the DEM processing described in the other items of this chapter and in the different chapters, except when explicitly mentioned otherwise.

6.3 Multiple Flow Directions Approach

The multiple flow directions approach distinguishes itself from the D8 and D4 methods by allowing a pixel to have more than one flow direction (Fig. 6.2c). The idea is to partition the flow generated or passing through a pixel to more than one neighboring pixel in the 3 × 3 window. Therefore, this third approach is called multiple flow directions, with versions initially developed by Freeman (1991) and Quinn et al. (1991).

The proposal to consider multiple flow directions is fundamentally motivated to overcome the main simplification of the single flow direction approach, which assumes that the whole surface of a pixel drains entirely and exclusively to one of its eight neighbors. It is not expected, as a rule, that the regular geometric area defined by the pixel boundaries has a topographic conformation draining entirely to the same point or neighboring pixel. The natural expected situation is that different portions of the pixel drain to distinct neighboring pixels.

However, what would be advantageous at first glance can become a problem. The main disadvantage of considering multiple flow directions is the numerical dispersion of the contributing area when passing the flow from one pixel to the downstream neighbors (Orlandini et al., 2003). This dispersion can be considered inconsistent with the physical definition of a contributing area, which does not occur when working with single flow directions. Some users think this characteristic is a reason to avoid working with multiple flow directions.

In some multiple flow directions methods, the flow that reaches a particular pixel can be distributed to up to eight neighboring pixels in the 3×3 window. The approach of Quinn et al. (1991) is an example of this type. In this method, the area accumulated up to the central pixel is distributed to all neighboring pixels with a descending topographic gradient. This flow distribution among neighboring pixels follows proportionally to two factors. The first factor is a geometrically defined contour length based on the relative position between each neighboring pixel and the central pixel. The second factor is the topographic gradient generated by each neighboring pixel, estimated as the elevation difference between the central pixel and the neighbor, divided by the distance between them. Neighboring pixels that produce a steeper topographic gradient receive a higher fraction of the central pixel's drainage.

The method proposed by Freeman (1991) is generally similar to that of Quinn et al. (1991), as it allows the distribution of a pixel's accumulated area among all neighboring cells with lower elevation. A distinguishing feature of Freeman's method is that the weighting is based on the topographic gradient raised to a power. This power is treated as a procedure parameter, with a recommended value of 1.1.

Tarboton's (1997) proposal allows partitioning the accumulated flow to only two neighboring pixels, unlike the two previous multiple flow directions methods. This method initially consists of determining a single flow direction in the direction of an angle θ, which can assume any value continuously in the interval from 0 to 360 degrees based on the neighboring elevations. There is an infinity of values that the angle θ can assume, justifying it to be called D-infinity ($D\infty$), with the letter D referring to the term Deterministic of the D4 and D8 approaches.

As the second primary step of the $D\infty$ method, the flow direction according to the angle θ is used to distribute the cell's flow to up to two of the neighboring pixels in the 3×3 window. In the example of Fig. 6.3, the neighbors identified by labels 2 and 3 receive part of the flow initially indicated by the direction according to the angle θ. The flow partition between these neighboring pixels is proportional to the angles formed between the initial single flow direction and the direction aligned with the centroid of each neighboring pixel. If θ coincides with this alignment in the direction of the centroid of a neighboring pixel, then the central pixel drains exclusively to that neighboring pixel.

For the example of Fig. 6.3, the flow to the neighboring pixel 2 is proportional to the relationship between the angle $\alpha 3$ and the sum $\alpha 2 + \alpha 3$. Similarly, the determination of the flow directed to the neighboring pixel 3 is based on the relationship between $\alpha 2$ and the sum $\alpha 2 + \alpha 3$. The values of these flow partition proportions could indicate, for example, that 63% of the central pixel's flow drains to the neighboring pixel 3 and 37% drains to the neighboring pixel 2.

Several other methods consider the assignment of multiple flow directions. Interesting studies that evaluate and compare some of these methods are the works of Wilson et al. (2007) and Erskine et al. (2006).

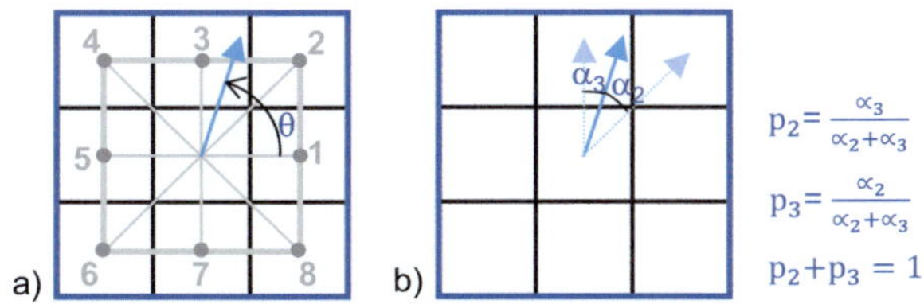

Fig. 6.3 General idea of the D-infinity approach for determining flow directions: **a** Definition of flow direction according to an angle that assumes any value between 0 and 360° and an indication of the eight faces of flow sharing between neighboring pixels in the 3 × 3 window; **b** Partition of the flow to two of the neighboring pixels proportionally to the angles between the calculated flow directions and the directions aligned to the centroids of these neighbors

6.4 Numerical Representations

Flow directions need to be stored as available information for other geoprocessing methods and even as input for hydrological models. When explaining flow directions in figures from books, academic works, and scientific papers, it is common to use arrows for easier understanding, as seen in several examples in this book. However, this is merely a didactic and schematic representation. Flow direction information is, by definition, a numerical attribute and is stored as a raster.

There are different numerical representations of flow directions. Each software or author has proposed distinct numerical codes associated with the eight possible flow directions (Fig. 6.4). The oldest one dates back to the first methods of determining flow directions proposed in the 1970s and 1980s. These early methods used numbers representative of sequential powers of 2 for the flow directions: 1, 2, 4, 8, 16, 32, 64, and 128, which are the result of the number 2 raised to the powers of 0, 1, 2, 3, 4, 5, 6, and 7, respectively.

The primary motivation for choosing the numerical representation of flow directions as powers of 2 at that time was the data storage cost reduction during the DEM processing. One reason is that, with this encoding, each flow direction can be stored by only one digit, which is the exponent of the power of 2. However, the most critical issue is regarding ties in the choice between two or more flow directions. As seen in the next sub-items of this chapter, there are situations where the resolution of the flow direction is undetermined after the first processing stages due to a tie of conditions between two or more possible flow directions. In this situation, storing information is necessary to treat it later and resolve the tie in another stage of the calculation algorithm. This information includes which neighboring pixels represent the ties or which flow directions are in a tie regarding the highest slope.

With the encoding based on powers of 2, it is enough to add the numerical codes of the flow directions involved in the tie and store this sum. The sum of any two or more numbers from the set 1, 2, 4, 8, 16, 32, 64, and 128 always results in a number distinct from the other sums. Storing only the sum is sufficient to, in a later step of the flow direction definition algorithm, know which directions provide that tie and make

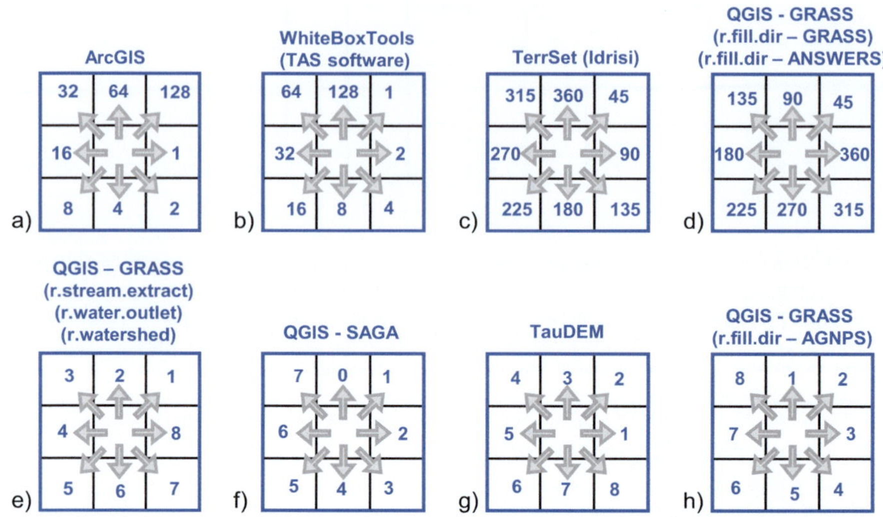

Fig. 6.4 Examples of single flow direction numerical codes according to the D8 approach adopted in different software: **a** ArcGIS (ESRI); **b** WhiteBoxTools (formerly terrain analysis system software); **c** TerrSet (formerly IDRISI); **d** GRASS via direct use or QGIS plugin, using r.fill.dir command with output choice for other GRASS modules or the ANSWERS hydrological model; **e** GRASS via direct use or QGIS plugin, using r.stream.extract, r.water.outlet or r.watershed commands; **f** SAGA GIS via direct use or QGIS plugin; **g** Terrain analysis using digital elevation models (TauDEM); **h** GRASS via direct use or QGIS plugin, using r.fill.dir command with output choice for the AGNPS hydrological model

the treatment and solution of the tie. For example, if the tie was between directions 2, 4, and 32, their sum equals 38. Knowing that there is a tie and the sum is 38, it is undoubtedly recognized that the directions involved in this tie are 2, 4, and 32. There is no other possible combination of directions whose sum result is 38. This type of encoding in powers of 2 is adopted in ArcGIS and WhiteBoxTools software, but with values positioned differently in the 3×3 window (Fig. 6.4a and b).

Another type of flow direction encoding is the adoption of numbers representative of the angle of the direction relative to a reference. For example, the TerrSet software encodes each direction's angle relative to the north (upper orthogonal neighbor), going from 45° to 360° clockwise (Fig. 6.4c). The encoding by angles is also adopted in the GRASS software, combined with QGIS when using the r.fill.dir command (output options in the GRASS or ANSWERS style), but the reference of the angle is the east position (right orthogonal neighbor), and the count follows the counterclockwise direction (Fig. 6.4d).

Similarly to encoding associated with angles, some software adopts sequential numbers for the flow directions, such as 1 to 8, as the GRASS software combined with QGIS when using the commands r.stream.extract, r.water.outlet, and r.watershed (Fig. 6.4e), or from 0 to 7, as in the case of the SAGA software combined with QGIS (Fig. 6.4f). The numbers themselves do not indicate the angles, but if multiplied by

45, they result in the angles of each direction. In the case of GRASS-QGIS, these angles are relative to the east position and consider counterclockwise direction—for example, direction 2 indicates 90° (2 × 45) relative to the east. In the case of QGIS-SAGA, this angle relates to the north position and considers the clockwise direction.

The TauDEM and GRASS software, combined with QGIS (via the r.fill.dir command in the output option of results in the format of the AGNPS hydrological model), also adopt numbers varying from 1 to 8 for the flow directions. However, in these cases, no apparent reason exists for association with angles represented by the directions rather than an ordered sequence of numbers (Fig. 6.4g and h).

Other software not listed here may adopt similar or distinct encodings from the exemplified ones. In each case, it is recommended that the user be aware of the software and algorithm documentation to avoid misinterpretation.

Any numerical representation of flow directions can be converted to another via the usual procedure of reclassifying raster data, as detailed in Chap. 3. This is necessary when exchanging information from one software to another or using the flow directions calculated by some software as input into a hydrological model that requires another encoding.

Regardless of the encoding, the flow direction raster usually uses a color symbology that follows a color gradient for the directions in the sequence of numbering, but with relatively distinct shades from each other, as exemplified in Fig. 6.5. In this figure, different zoom levels (Fig. 6.5a–d) are sequentially used until the individual pixels are visualized. To facilitate the understanding of the flow directions, the corresponding arrows are schematically drawn (Fig. 6.5e and f).

By interpreting the flow directions in Fig. 6.5, one can easily follow the flow paths that cross the indicated region. For example, starting from the pixel highlighted by a

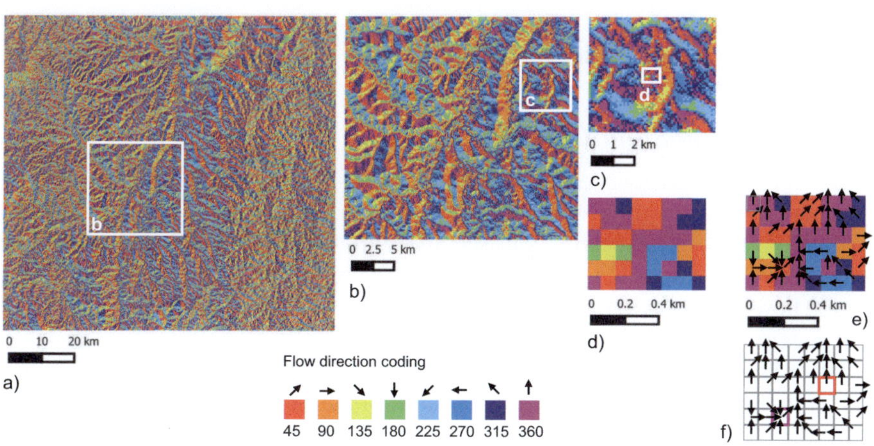

Fig. 6.5 **a–d** Flow directions determined from the DEM processing at different zoom levels as indicated by the white rectangles and the corresponding letters; **e** Arrow representation overlaid to the flow direction raster; **f** Representation of the flow directions only by arrows

red border in Fig. 6.5f, the flow follows in the vertical direction to the upper neighbor and from this again in the vertical direction, leaving the image. Starting from any chosen pixel, one can follow the flow direction iteratively until reaching the edge of the image.

It is also noticed in this figure that there are source pixels, which are pixels for which no flow arrives, only downstream drainage, like the pixel indicated by the red outline. In contrast, the pixel indicated by the pink outline receives the flow from five neighbors in the 3×3 window around it.

6.5 Edge Pixel Handling

As seen in the following items, there are several rules and procedures to determine the flow directions from the DEM. However, pixels at the edges of the image are most commonly handled in a specific way, distinct from the other pixels. The most common rule is to impose on the edge pixels flow directions in the direction out of the image so that these pixels represent exit points of the image flow to external regions. This occurs in the algorithms present in the TerrSet (by default) and ArcGIS (optionally) software, for example.

Other algorithms calculate the flow directions of edge pixels based on the elevations of valid neighbors in the 3×3 window. In this case, some software signals are different when edge pixels drain to neighboring pixels of the image (positive flow directions) than when edge pixels drain out of the image (negative flow directions). This consideration is adopted, for example, in the algorithm of the r.watershed command of the GRASS software integrated with QGIS.

6.6 The Steepest Descent Slope Rule

6.6.1 Definition and Overview

A common issue among various algorithms for obtaining flow directions is adopting the steepest descent slope rule and specific rules unique to each algorithm. This steepest descent slope rule is part of the earliest algorithms that addressed DEM-derived flow directions and proposed the first D8 approaches in the 1970s and 1980s, such as Jenson and Domingue (1988).

According to this rule, the slope of the pixel under analysis is calculated relative to each of its eight neighbors in the 3×3 window. In this case, the slope is determined by the difference in elevations between the central pixel and its neighbor divided by the distance between them (Fig. 6.6). When the pixel is a neighbor in an orthogonal orientation to the central pixel, the distance between them equals the spatial resolution (Δx), as in the case of pixel '*d*' relative to the central pixel '*p*' in the figure. The

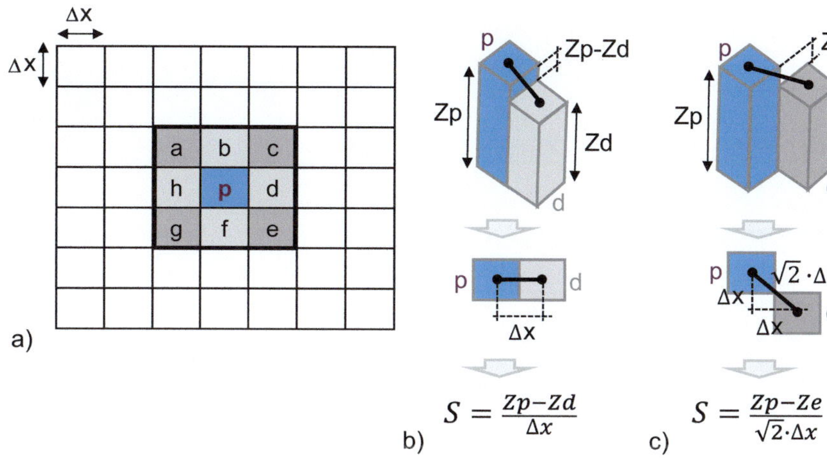

Fig. 6.6 **a** Scheme of a DEM of spatial resolution dx, highlighting the pixels in a 3 × 3 window centered on pixel *p*, with four orthogonal neighbors (b, d, f, h) and four diagonal neighbors (a, c, e, g); **b** Calculation of the slope of the central pixel '*p*' relative to the orthogonal neighbor '*d*', based on the difference in elevation and the distance between them; **c** Calculation of the slope of the central pixel '*p*' relative to the diagonal neighbor '*d*', based on the difference in elevation and the distance between them

distance between the central pixel and a diagonal neighbor pixel equals $\sqrt{2} \cdot \Delta x$, as in the case of pixel '*e*' relative to the central pixel '*d*'. The flow direction is assigned as the direction of the neighboring pixel that provides the steepest slope.

The application of the rule of the steepest descent slope is exemplified numerically in the DEM shown in Fig. 6.7. For pixel A in this example, adopting this rule and the D8 approach leads to deciding the direction in the lower right diagonal direction due to the steepest slope provided by the neighboring pixel corresponding to this direction (slope of 0.06 m/m; Fig. 6.8a). If a D4-type approach were adopted, the diagonal pixels would be disregarded and, therefore, the flow direction of pixel A would be horizontally to the right side, as it provides the steepest descent slope among the four orthogonal neighbors (0.02 m/m; Fig. 6.8b).

When considering the multiple flow direction approach with a procedure like the one proposed by Quinn et al. (1991) or by Freeman (1991), three flow directions would be assigned to pixel A from the previous example, as three neighboring pixels provide positive slopes, that is, it has a lower elevation than it (Fig. 6.8c). However, there are variations among the multiple flow direction methods proposed in the literature. For example, the procedure suggested by Tarboton (1997) allows a maximum of two multiple flow directions for each pixel.

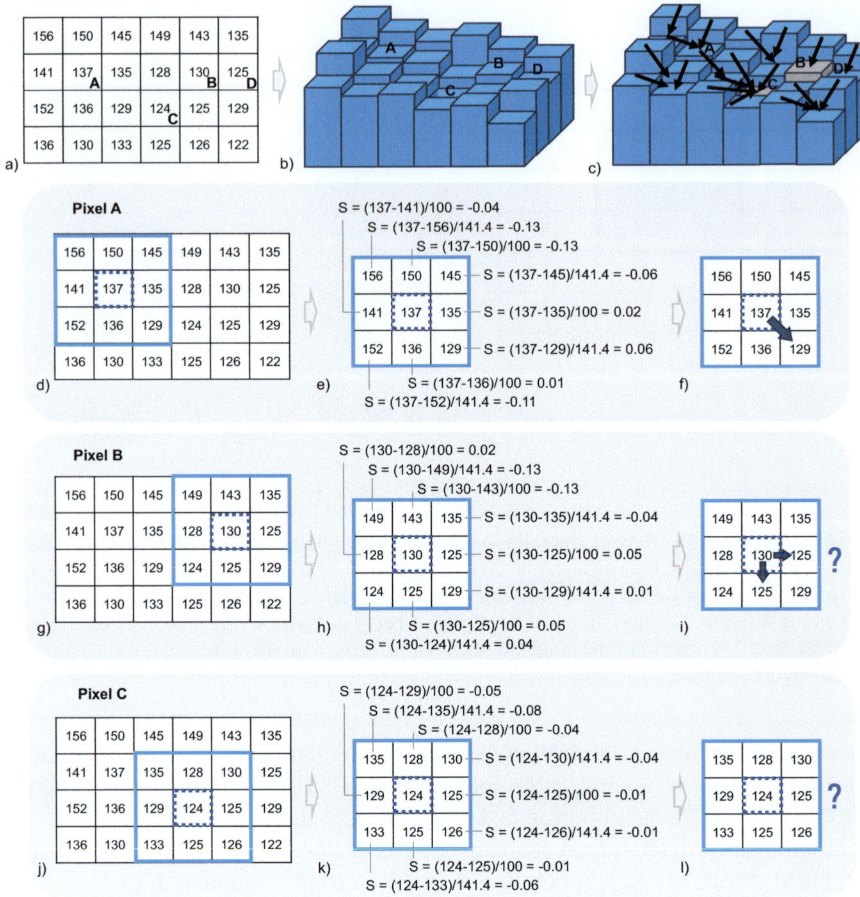

Fig. 6.7 **a** DEM with an indication of elevation in meters and highlight for three pixels A, B, C; **b** Three-dimensional visualization of the DEM; **c** Flow directions derived from the DEM solely from the steepest descent slope rule; **d, e, f** Detailing of flow direction determination for pixel A, with the calculation of the slope S between this pixel and each of its eight neighbors in the 3 × 3 window; **g, h, i** Analogously for pixel B; **j, k, l** Analogously for pixel C

6.6.2 Situations Not Solved by the Steepest Descent Slope Rule

Considering the D8 approach, most pixels in a DEM already have the flow direction resolved just by applying the steepest descent slope rule. In the case of the example in Fig. 6.7, 21 of the 24 pixels were resolved with this rule, including pixel A. In this example, it was assumed that the edge pixels of the image had their slopes analyzed concerning the existing neighboring pixels, and the flow direction was decided based

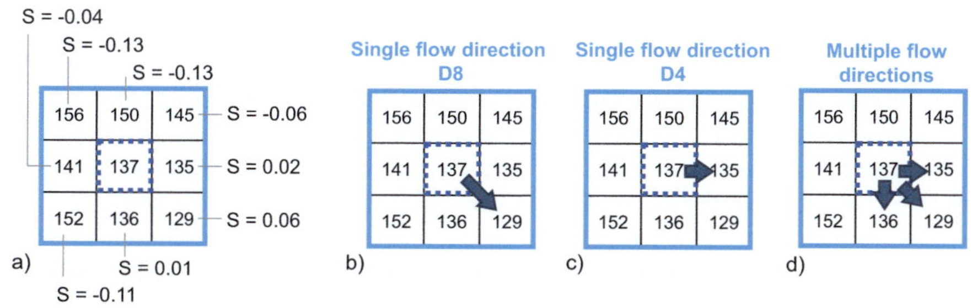

Fig. 6.8 Example of assigning flow direction to DEM pixel (central dashed pixel) based on the calculated slopes relative to the neighboring pixels of the 3 × 3 window **a**, considering the approach of single flow directions D4 **b**, D8 **c**, and of multiple flow directions **d**

solely on this. Only the pixels indicated by the letters B, C, and D did not have their flow directions directly resolved by the steepest descent slope rule.

Two pixels from the example in Fig. 6.7 that are not located at the edge of the image (pixel B and pixel C) presented an undefined flow direction, in addition to the edge pixel D. This occurs because two conditions must be simultaneously met for the steepest descent rule to determine the flow direction of a pixel: (1) at least one neighboring pixel must produce a positive slope, meaning it has a lower elevation than the pixel being analyzed; and (2) there must be no tie among the steepest slopes provided by neighboring pixels.

Condition 1 is not met when two situations occur. The first situation is when the pixel is lower than all its neighbors, representing a lower terrain area relative to the neighborhood. This pixel is usually called a spurious depression or pit. This is the case of pixel C in Fig. 6.7, whose slope is negative relative to any neighboring pixel. A more complex case is when a set of pixels is surrounded by a neighborhood of higher elevation, configuring an entire region of depression and not a single pixel.

Whether each depression (at the pixel level or a larger region) exists in the actual topography or is caused by artifacts in the DEM, a specific treatment is needed to obtain continuous flow paths.

The second situation that violates condition 1 is when there is no neighboring pixel with an elevation lower than the central one, and there are one or more pixels with an elevation equal to the central one. This configures a flat area (null slope) that needs additional rules to resolve the flow direction. This is the case of pixel D in Fig. 6.7, which has an elevation of 125 m, equal to its lowest neighbor, resulting in a null slope. However, as pixel D is on the edge of the image, it could be treated differently, imposing a flow direction out of the image. Analogous to the case of depressions, an entire region may be formed by a set of pixels configured as a flat area.

Condition 2 is not met when two or more pixels provide the same highest positive slope, causing a tie in the definition of the steepest descent slope. Pixel B in Fig. 6.7 is an example of this situation. Two of its neighbors provided a slope of 0.05 m/m, steeper than all the other neighbors. Additional rules need to be adopted to resolve this tie, generally to decide which of the two neighboring pixels involved in the tie the flow direction goes to.

Applying the steepest descent rule may result in four distinct situations (Table 6.1). The first involves simple depressions, characterized by pixels whose neighbors all have higher elevations—thus, all calculated slopes are negative. Another situation involves flat areas, where the maximum slope toward any neighbor is zero.

The third situation is when there are positive slopes, but the maximum value is tied between two or more neighbors. The fourth situation is when there is a positive slope with a maximum value distinct from the other slopes. Only this last situation represents a case that directly defines the flow direction. All three previous situations require treatment and the adoption of additional steps to determine the flow direction according to rules, criteria, and methods that vary among the various existing algorithms.

Figure 6.9 shows a DEM derived from the SRTM 30 data for a region in southern Brazil, primarily located in the watershed of the Ijuí River. This figure also shows the identification of the pixels according to the type of result obtained by applying the steepest descent slope rule. This DEM has 2215 rows and 2839 columns, totaling around 6.3 million pixels. According to the analysis of the slopes relative to the eight neighbors of the 3 × 3 window, about 88.1% of the pixels have a neighboring pixel that provides a maximum positive slope and is distinct from the other neighbors. Therefore, for 88.1% of the pixels, it is possible to directly define the flow direction to one of the eight neighbors with the steepest descent slope rule.

In this example, other analysis and decision rules are necessary to assign the flow directions for the remaining 11.9% pixels. This percentage is distributed in three categories. One category is formed by about 6.8% of the pixels having neighbors that offer a positive slope, but there is a tie between two or more neighbors in terms of maximum slope. Another category refers to 4.8% of the pixels whose maximum slope relative to the eight neighbors is null. These pixels configure flat areas. This category tends to be more concentrated in valley areas, particularly along the width

Table. 6.1 Description of the possible situations when analyzing the slopes of a pixel concerning its eight neighbors, according to the application of the steepest descent slope rule for deriving flow directions from a DEM

Condition relative to the eight neighbors	Interpretation	Procedure for assigning flow direction
All slopes are negative	Pixel is a single depression	Application of additional rules and procedures, varying according to different existing algorithms
The maximum slope is zero (for one or more neighbors)	Pixel represents a flat area with this/these neighbor(s)	
The maximum slope is positive, but it occurs for two or more neighbors	There is a tie among the steepest positive slopes	
The maximum slope is positive and occurs uniquely for one neighbor	Clear identification of the steepest positive slope	Flow direction is assigned toward the steepest positive slope

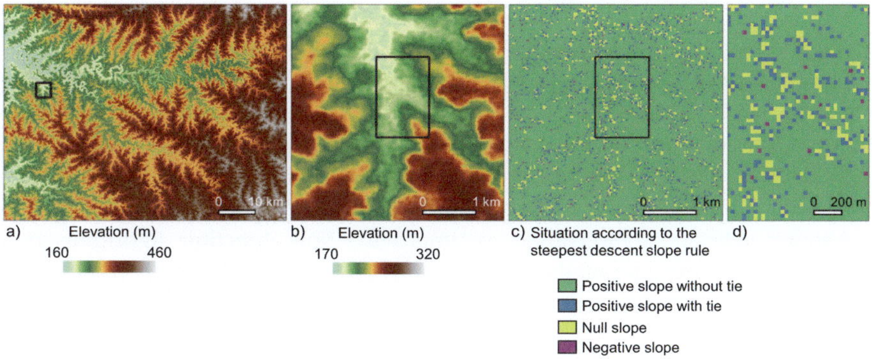

Fig. 6.9 **a** SRTM 30 DEM for an area in the southern Brazil; **b** Detail of the DEM for the area indicated in the rectangle in **a**; **c** Classification of pixels according to the situation in terms of maximum slope relative to the eight neighbors in the 3×3 window (in the area highlighted by the rectangle in **a**); **d** Detail of the pixel classification for the area indicated by the rectangle in **c**

of rivers, especially when the rivers have dimensions larger than the pixel size. The remaining category concerns the pixels representing simple depressions (i.e., pixels whose slopes relative to the eight neighbors are negative). This category covers approximately 0.3% of the DEM pixels in this example.

6.6.3 Handling Specific Cases

The analysis of the example in Fig. 6.9 shows a real-world example of how the occurrence of pixels that do not have flow directions resolved directly by the steepest descent slope rule is a minority, in general. However, this minority is the one that demands the most computational and creative effort to propose solutions, especially the group of least occurrence, which is the depression area.

The percentages to which each specific case shown in Table 6.1 may occur within a DEM vary according to the study area characteristics and the topographic data source. However, a general pattern like the one shown in Fig. 6.9 is expected in most cases. Moreover, depending on the algorithm applied, resolving ties, flat areas, or depressions may require redefining the flow directions of pixels that had already been assigned directions based on the steepest descent rule.

A classic example is when a more in-depth analysis indicates that several pixels should be grouped and treated as a large depression area. Many of these pixels that become part of this depression may have presented a maximum positive slope relative to the neighbors and received the flow direction assignment directly from the steepest descent slope rule. In a broader contextual analysis, the entire surrounding area is identified as a depression and must be addressed as an integrated unit. This process may involve re-evaluating and potentially modifying the flow directions of pixels initially defined based on the standard steepest descent slope criteria.

The solution to these situations that were not resolved by the steepest slope rule, especially terrain depressions and flat areas, has been challenging. Additionally, addressing larger areas considered depressions adds to this complexity. This scenario justifies the existence of dozens of distinct algorithms proposed since the 1970s and 1980s, which remain a research subject today.

Another substantial motivation for improving flow direction algorithms is to propose more computationally efficient methods for dealing with more pixels regarding storage and data processing speed. The following items address these topics.

6.7 Terrain Depressions

6.7.1 Occurrence and Types of Terrain Depressions

When representing a continuous surface like the terrain by values arranged in a gridded dataset such as the DEM, it can be considered inevitable that some of these grid elements will be surrounded by neighbors of higher value (Burrough & McDonnell, 1998). Depressions can be realistic representations of the surface, i.e., areas may be physically lowered relative to the neighborhood. Depressions can also be artifacts from acquiring, processing, and representing topographic data as a raster-DEM.

For example, areas of narrow valleys in which the width of the valley bottom is smaller than the pixel size tend to cause artifact-type depressions in the DEM (Burrough & McDonnell, 1998). On the other hand, natural depressions can be typical of specific regions. An example is the landform called dolines, resulting from limestone rocks and characterized as natural depressions of circular or elliptical shape, very present in the region of the Pelotas river basin and other areas in the south of Brazil (Maciel Filho et al., 2013).

When there is only one pixel integrating the terrain depression, it is considered a simple or spurious depression (a pit). When the depression is constituted by a grouping of pixels surrounded by neighboring pixels of higher elevation, it is configured as a multiple depression.

6.7.2 Approaches for Removing Terrain Depressions

The treatment of terrain depressions is one of the biggest concerns when proposing algorithms to determine flow directions, regardless of whether the depressions are of natural origin or DEM artifacts. The presence of depressions, whether simple, formed by a single pixel surrounded by neighbors of higher elevation, or formed

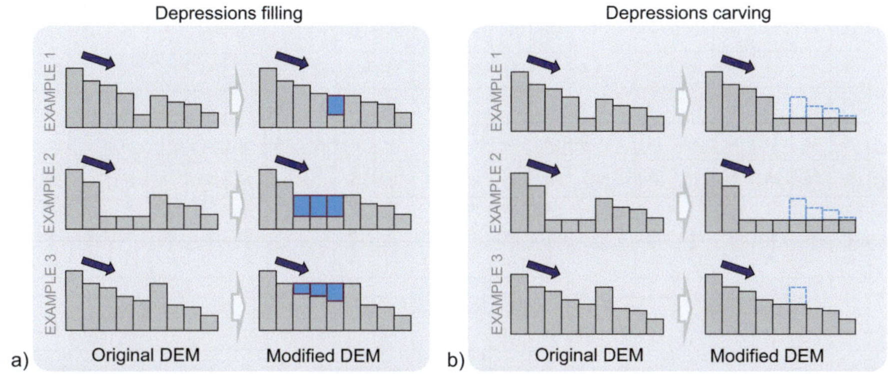

Fig. 6.10 Simplified scheme of the functioning of the filling **a** and carving **b** procedures adopted by depression removal algorithms to determine flow directions from the DEM

by a grouping of pixels, prevents the continuity of the water flow derived from the DEM. There is an indeterminacy about where the flow that arrives or originates in the depression goes. It is beyond the scope of this book to detail specific algorithms that deal with the issue, but only to illustrate the problem and comment in general terms on how the algorithms seek to solve it.

Two procedures are commonly used to remove terrain depressions in algorithms that adopt the steepest descent slope rule. One is the filling of depressions, usually called pit filling. This approach consists of artificially increasing the elevation of the pixels that are part of the depression to a height equivalent to the lowest elevation of the neighboring pixels (Fig. 6.10a). This modification leads to a flat area, and rules are applied to determine the flow directions of these modified pixels.

The other primary procedure for treating terrain depressions is to lower pixels in the neighborhood to find an outlet for the flow from this region (Fig. 6.10b), whose preferred English term is *carving*.

According to the simplified examples illustrated in Fig. 6.10, it is intuitively perceived that there are situations where filling the depressions seems to be the most recommended option, as it requires little intervention and modification in the DEM, as in the case of example 1. However, depression filling may have a more significant impact on other situations, as in examples 2 and 3. The proposal to carve out an outlet seems much more interesting in the latter.

The examples shown in this figure are a simplified, one-dimensional schematic representation of the actual situations involving configurations of terrain depressions in DEMs. These situations reach different levels of complexity and have motivated the diversity and improvement of algorithms to address this issue. Some perform a combination of filling and carving, with varied rules and criteria for applying these procedures. In general, it is recommended to use algorithms that seek to minimize modifications of the DEM when removing terrain depressions, that is, algorithms that can solve the different situations with the least possible impact on the original elevations present in the DEM.

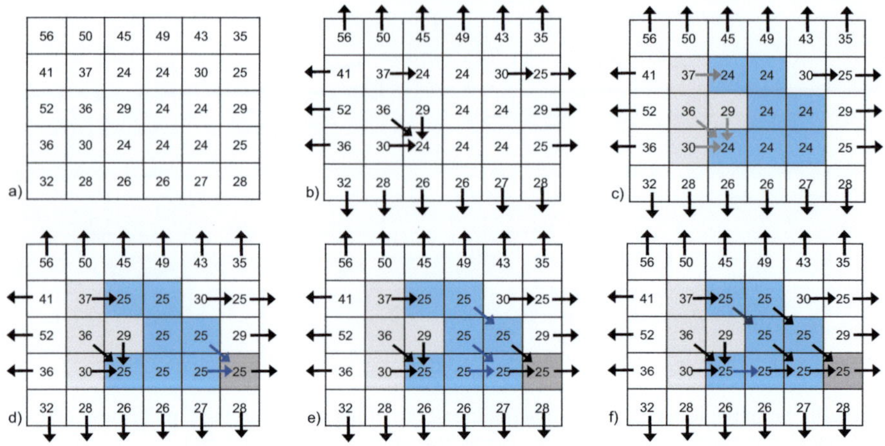

Fig. 6.11 **a** Digital elevation model (in meters); **b** Flow directions determined directly by the steepest descent slope rule, with edge pixels draining orthogonally out of the image; **c** Indication of pixels that are part of a depression (in blue) and neighboring pixels whose directions drain into the depression (in light gray); **d** Identification of the depression's neighboring pixel with the lowest elevation (in dark gray; the depression's outlet pixel) and assignment of the flow direction from the depression's pixels directly to this outlet pixel; **e**, **f** Sequential assignment of flow directions for depression pixels neighboring those of the depression with an already-defined direction

From the DEM modified by the treatment of terrain depressions, other decision rules to assign the flow directions of the region are established in the different algorithms. The detailed description of various existing algorithms in the literature is beyond the purpose of this book, and the reader is recommended to see works such as Buarque et al. (2009), Siqueira et al. (2016), and the ones listed in item 6.9 further in this chapter. However, to give a more concrete idea of how to deal with the treatment of depression until the final resolution of flow directions, some procedures that go back to the simplest algorithms proposed by the pioneers in the 1980s are exemplified in Fig. 6.11.

For the DEM of Fig. 6.11a, composed of 30 pixels, the flow directions of the 18 edge pixels were assigned orthogonally out of the image, and the steepest descent slope was applied to the other 12 pixels (Fig. 6.11b). Among these 12 pixels, 5 had their flow direction resolved, and 7 are characterized as a region of flat depression (pixels in blue in Fig. 6.11c), surrounded by neighbors of higher elevation.

When a depression is identified, the flow directions of pixels draining into this region are placed under review (indicated by gray arrows). Depending on each algorithm's specifics, these flow directions may be re-evaluated and modified. This is because the resolution of depression pixels may render the original flow directions of some neighboring pixels inappropriate, potentially creating dead-end paths or 'short-circuits' in the drainage network.

To solve the pixels inside the depression, a simple rule involves identifying the neighborhood pixel of the depression with the lowest elevation as a first step. In the example, the neighbor with the lowest elevation is the pixel with an elevation of 25 m (highlighted in dark gray in Fig. 6.11d). This pixel is the depression outlet through which the depression flows.

According to a classic filling approach of solving this type of depression, all the pixels of the depression are then raised to this same value (25 m). Next, among these pixels modified to the elevation of 25 m, those that are immediate neighbors to the one highlighted in dark gray are identified, which is the external pixel to the neighborhood with the lowest elevation (outlet pixel of the depression).

The pixels identified in the previous step are assigned a flow direction toward the pour point pixel (blue arrows in Fig. 6.11d). Next, the remaining depression pixels adjacent to those just being assigned directions also have their flow direction resolved, pointing toward their already-defined neighbors. This process is repeated progressively until all pixels within the depression are included (blue arrows shown in Fig. 6.11e and f). As a result, flow paths are established that drain the depression area toward its pour point.

In this example, the flow directions of the pixels neighboring the depression and draining to the depression, which were initially put into suspicion, do not need to be changed, as they create acceptable flow paths after entering the depression.

6.7.3 Depressionless DEM

The result of the depression treatment is the so-called depressionless DEM, also referred to as 'pit-free DEM'. As mentioned in Chap. 4, some authors consider this depressionless DEM to be a hydrologically conditioned DEM. Other authors assume that a hydrologically conditioned DEM refers to a more profound treatment of the DEM, incorporating vector river traces by the stream burning procedure described in Chap. 7.

Regardless of the algorithm used to treat and remove terrain depressions, the user cannot identify differences between the original DEM and the depressionless DEM by visually comparing them. This occurs for two main reasons. First, as seen in the previous item, the number of pixels identified as depressions is generally a tiny fraction of the total number of DEM pixels. The second reason is that the modifications made to the elevations are mostly slight enough not to appear visually highlighted, considering a specific gradual color symbology associated with the DEM. In other words, the general spatial patterns of topographic variability are not altered by removing depressions.

The DEM modified by the depression removal process should not be used as general and unrestricted topographic data unless one wishes explicitly to work with this DEM. This caveat is independent of which algorithm has been used for depression removal. The modifications made to the DEM are artificial changes that aim to determine the flow directions to constitute continuous flow paths. Which pixels are modified and how much the elevations are changed depend on each algorithm. The topographic representation of the region continues to be the DEM without these modifications, although it presents imperfections, uncertainties, and associated errors.

In some specific situations, having the depressionless DEM as input data is explicitly necessary or recommended. A classic example in this sense is the determination of flow directions, as already mentioned. Depressionless DEM is also recommended in other applications, such as estimating runoff travel times from the DEM. In this case, depressions imply negative slopes and the consequent indefiniteness of incremental travel times of runoff between these pixels that constitute the depression. An opposite example is the DEM estimate of the area-volume-elevation curve of a lake formed by reservoir construction. In this application, the recommendation is to use the original DEM and not the DEM artificially altered by the treatment of terrain depressions, since the former is the topographic representation of the flooded terrain.

A real-world example of a DEM processed to remove terrain depressions is shown in Fig. 6.12 for the region of the Icamaquã river basin, a tributary of the Uruguay River, in southern Brazil. Figure 6.12a shows the SRTM 90 DEM with the watershed delineation, while Fig. 6.12b and c show a detail of part of the depressionless DEM obtained by two different procedures: the algorithms available in the r.fill.dir command of QGIS software and the algorithm available in TerrSET software, respectively. The first method works only with depression filling. In contrast, the second uses an approach that modifies the DEM to remove depressions according to a minimum cost path technique, combining actions of increasing and decreasing pixel elevations.

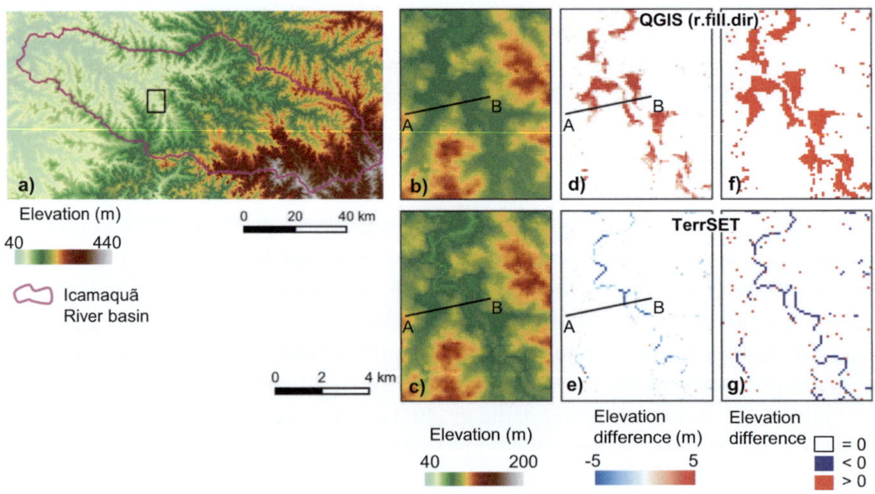

Fig. 6.12 **a** DEM of the Icamaquã river basin region (southern Brazil) from SRTM 90 data; **b**, **c** Depressionless DEM obtained by QGIS-r.fill.dir and TerrSET, respectively, for the region indicated by the rectangle in **a**; **d**, **e** Raster of differences between the depressionless DEM (obtained by QGIS-r.fill.dir and TerrSET, respectively) and the DEM; **f**, **g** Classification of the raster of differences between the DEM and each of the depressionless DEMs

In Fig. 6.12d and e, the raster of differences between the depressionless DEM obtained by the two algorithms and the original DEM is presented. In Fig. 6.12f and g, these rasters of differences are classified to highlight whether the modification increased or decreased elevations.

In this example, all differences are positive and reach 9 m throughout the basin following the QGIS-r.fill.dir method. This means there were depressions for which it was necessary to artificially increase the pixel elevations by up to 9 m to eliminate them. With the TerrSet method, there are positive differences (increases of up to 3 m in elevations) and negative differences (reductions in elevations of up to almost 13 m). However, there is a more significant predominance of areas with negative differences. The cutbacks in elevations caused by the TerrSET algorithm occur mainly in pixels aligned with the main paths of the drainage network. Due to the tendency to seek this path with a positive topographic gradient, elevations are altered more intensely regarding value reduction.

Considering both algorithms, the comparison between the original DEM and the modified DEM is well-illustrated with the tracing of cross-sections (Fig. 6.13), referring to the A-B transect indicated in Fig. 6.12. It is noticeable how, in the case of the QGIS-r.fill.dir algorithm, there were only modifications to increase the elevations, as in the sections indicated by the letters P, Q, R, and S. In the processing carried out in TerrSET, elevation reductions predominantly occurred, as in sections R and S. However, elevation increases also occurred (section P). Section Q was not altered by the TerrSET algorithm, possibly because it found a flow output for the pixels of this section via pixels in the surrounding space not aligned by the transect—the analysis by the transect is restricted to the pixels located along this alignment and, therefore, offers a limited view of the result of the treatment of terrain depressions.

This example from the Icamaquã river basin also illustrates how the modifications of the DEM due to the presence of depressions reach a small fraction of the total area. In this case, the total number of pixels modified by the depression removal process represents 6.9% of the Icamaquã river basin when using the algorithm present in QGIS-r.fill.dir, all with modification of the filling type. When TerrSet was used, 3.0%

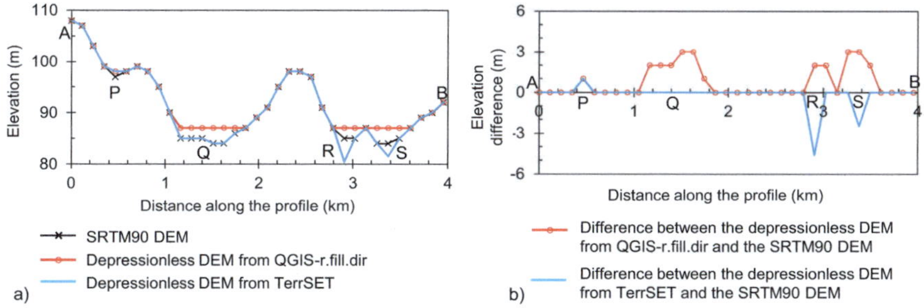

Fig. 6.13 **a** Profile of elevations along the A-B transect indicated in Fig. 6.12, considering the original DEM and the depressionless DEM obtained by two different algorithms; **b** Profile of the differences in elevation between the depressionless DEM and the original DEM along the same transect

of the basin's pixels were altered, with 2.1% of them with a reduction in elevation and 0.9% with an increase in elevation.

6.8 Flat Areas and Areas with Ties Between Maximum Slopes

6.8.1 Origin of Flat Areas in a DEM

Flat areas can be defined as those whose local topographic gradient is null and can occur in a DEM as a result of the characteristic of the region's topography or be the result of uncertainties, errors, and other technical issues associated with data acquisition and DEM elaboration (Barnes et al., 2014).

The previous item shows that treating terrain depressions by filling can result in flat area formation and/or an increase. Take the example of the SRTM 30 DEM of part of the Ijuí river watershed in southern Brazil, already worked on in Fig. 6.9. This DEM has 2215 rows and 2839 columns, totaling around 6.3 million pixels. As seen previously, by analyzing the relative slopes to the eight neighboring pixels of the 3 × 3 window, 4.8% of the pixels indicated an indefiniteness of the flow direction by presenting a maximum null slope relative to the neighbors. These pixels constitute isolated flat areas. When treating the removal of depressions by the filling procedure, the modified DEM presents 10.6% of the pixels as isolated flat areas.

The filling of the depressions more than doubled the number of pixels that integrate flat areas. Visualizing the location of these flat areas evidences their high concentration along the drainage network's main course and the floodplain areas (Fig. 6.14). A negative slope no longer occurs in the DEM modified by filling since all depressions have been removed.

6.8.2 Ties Between Flow Direction Candidates

In the individual analysis of a pixel relative to its eight neighbors of the 3 × 3 window, there is a flat area if the maximum slope is null. It is a situation of indefiniteness of the flow direction. Based solely on the elevations of the neighboring pixels and the application of the steepest descent slope rule, the flow direction cannot be defined.

A similar situation occurs even when a flat area is not configured. When applying the steepest descent slope rule to determine the flow directions, another situation of indefiniteness is the occurrence of two or more neighboring pixels that provide the same maximum positive slope. In this case, there is a tie between the directions of potential candidates.

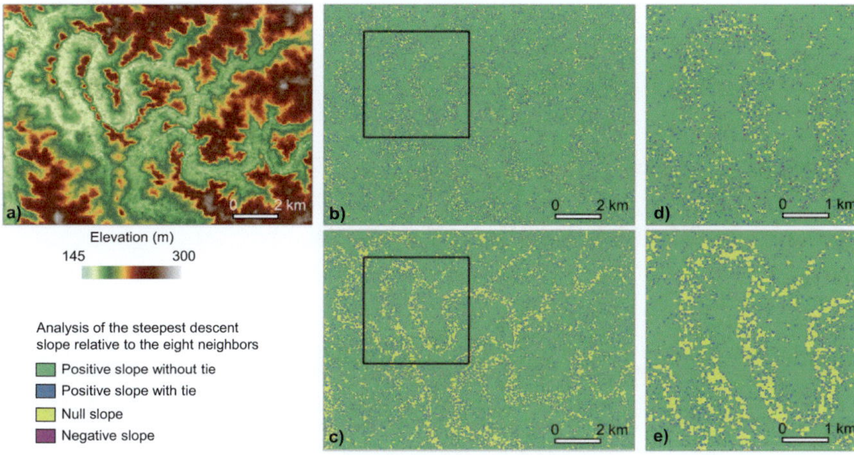

Fig. 6.14 **a** Part of the SRTM DEM of the Ijuí River watershed shown in Fig. 6.9; **b**, **c** Classification of pixels according to the analysis of the maximum slope relative to the eight neighbors applied, respectively, to the original DEM and the DEM modified by the depression filling approach; **d**, **e** Detail of the classification displayed in **b** and **c**, respectively, as indicated by the black rectangle

In both cases, the analysis of the elevations by the steepest descent slope rule is insufficient to establish the flow directions to obtain continuity of the flow paths along the image. Additional procedures and criteria are needed to resolve this indefiniteness reasonably and consistently.

6.8.3 *Approaches to Solve Flat Areas and Ties Between Candidate Flow Directions*

Similar to terrain depressions, numerous algorithms resolve flat areas and ties between potential candidate directions in a distinct way. It is beyond the scope of this book to detail various existing algorithms.

The earliest D8-type algorithms proposed in the literature adopt a simplified approach to address flow direction indeterminacies in flat areas or where ties occur between potential candidate directions. The core idea behind these algorithms is to iteratively test possible flow directions by assigning one of the candidate directions to the pixel and evaluating the outcome. For instance, the choice is considered invalid if the downstream pixel assigned to receive the flow drains in the opposite direction. Another invalid scenario occurs if the resulting flow path forms a short circuit by looping back to the original pixel. In such cases, another candidate direction is tested. This process continues until a direction is found whose resulting flow behavior is not invalidated upon further analysis.

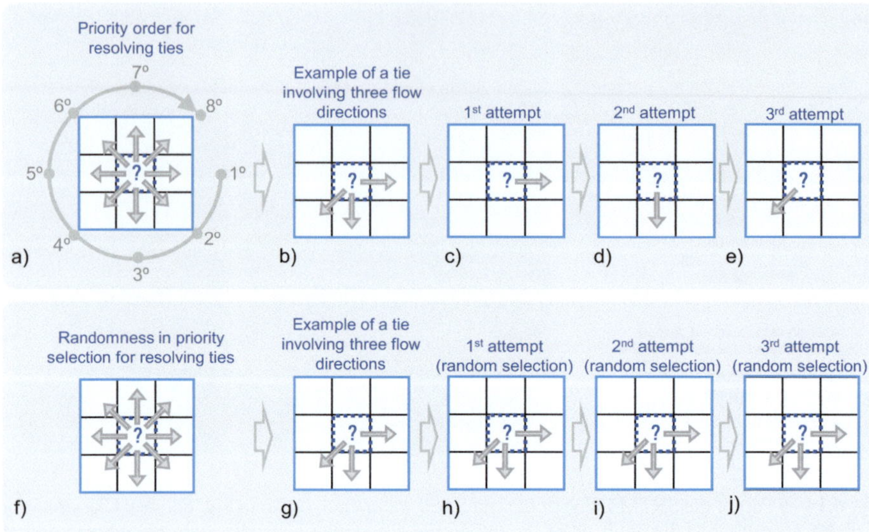

Fig. 6.15 Indication of sequential priority **a** and random **f** orders adopted by some algorithms for testing flow directions when there are ties in terms of the highest slope; **b**, **g** Example of a tie between three flow directions; **c**, **d**, **e** Attempts in sequential order to solve the tie example; **h**, **i**, **j** Attempts in random order to solve the tie example

The big problem with this procedure is that the original algorithms propose the order of priority for testing among the flow direction candidates according to a pre-established order. This order follows the increasing numbering of the flow direction coding adopted in these algorithms (Fig. 6.15a), which goes from 1 in the east position to 128 in the northeast position (Fig. 6.4a). Thus, if there is a tie, for example, between the directions coded by 1, 4, and 8 (Fig. 6.15b), the first test is always to see if direction 1 satisfies (Fig. 6.15c). If it does not solve, it starts testing direction 4 (Fig. 6.15d), and if it also does not meet, it moves to direction 8 (Fig. 6.15e).

In the case of flat areas—defined as a group of pixels with identical elevation values—the situation is analogous to a set of ties between multiple candidate flow directions. When using a systematic, predefined priority rule to resolve these ties, there is a clear tendency for parallel flow paths to emerge. This occurs because a flow direction initially tested and accepted as valid (i.e., leading to a downstream drainage path) is likely also suitable for neighboring pixels, resulting in parallel flows. In the hypothetical example shown in Fig. 6.11, flow directions within the flat area were assigned following such a procedure, and two parallel flow paths toward the southeast are observed, even though the flat area itself is pretty small.

A simple way to avoid unwanted parallelism in the flows is to change this type of algorithm so that the tie between the directions of flow candidates is done randomly (Fig. 6.15f), as tested by Paz and Collischonn (2008), inspired by the work of Fairfield and Leymarie (1991). For the case of the tie between the flow directions 1, 4, and 8 of the previous example (Fig. 6.15g), a random draw defines which of them is first tested (Fig. 6.15h), which is tested in the second attempt (Fig. 6.15i), and which is last (Fig. 6.15j). For example, the random choice could indicate first testing direction 4, then 8, and finally, direction 1. This random draw for the definition of choices occurs every time each tie situation is to be treated, independently of the other draws and choices.

In the case of rivers wider than the DEM's pixel size or the existence of floodplains, these areas are usually represented in the DEM as large, flat areas. Because of this, the D8-type algorithms that solve tie cases or flat areas with the systematic priority for testing possible directions show a strong tendency to generate parallel and unreal flow paths. This can cause considerable errors when using these data layers for other studies.

As a real-world example of this type of situation, there is a reach of the lower course of the Uruguay River in Rio Grande do Sul (Fig. 6.16a). The flow paths generated by the algorithm that treats the resolution of ties in a systematic way of priority result in an unwanted parallel flow (Fig. 6.16b) when there should be a single main path representing the reach of the Uruguay River. For comparison purposes, one can analyze the same reach resulting from processing the same DEM with a different algorithm (Fig. 6.16c) based on minimum cost paths, as implemented in the TerrSET software. The problem of the unwanted parallel flow path does not occur.

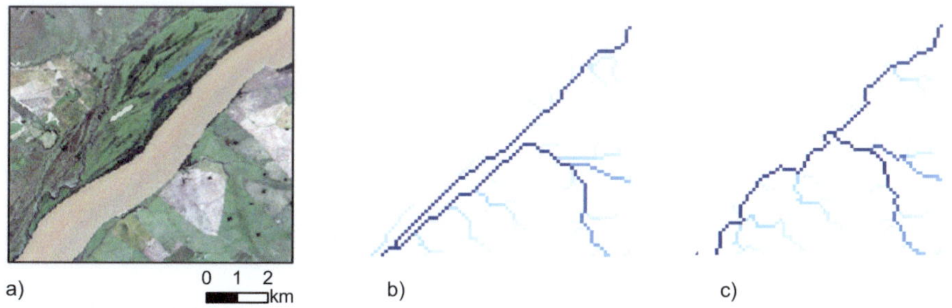

Fig. 6.16 **a** RGB color composition of Landsat images of a reach of the lower course of the Uruguay River in Rio Grande do Sul; **b**, **c** Flow paths (pixels in shades of blue) derived from the SRTM 90 DEM for the same area obtained by, respectively, an algorithm that uses the rule of the steepest descent slope and treats ties with choice of candidate flow directions in a pre-determined sequential way and an algorithm that uses an approach of minimum cost paths to resolve the flow directions

6.9 Efficient Algorithms for Depression Removal and/or Flow Direction Assignment

6.9.1 *Why Can the Computational Cost Be High?*

The DEM processing for removing depressions and assigning flow directions can usually be considered a computationally intensive operation. As illustrated in the descriptions and examples in this chapter, the analysis and solution of each pixel involves initially evaluating its immediate neighborhood in a 3×3 window. Depending on the type of situation found and the solution developed to resolve it, the analysis evolves to embrace a larger area, mainly when flat areas and terrain depressions exist. Different conditions can occur, with various configurations and varied levels of complexity distributed spatially along the DEM. Alternatives of flow paths can be traced and analyzed to make the best choice. The process is recursive in some solutions, with already analyzed pixels being reviewed.

Due to all these characteristics, treating terrain depressions to determine flow directions requires more intense processing than usual local or neighborhood operations involving rasters.

The general rule is that the larger the number of pixels in the DEM, the higher the computational cost for depression removal and flow direction assignment. Topographic characteristics can also play an essential role in augmenting the computational cost, such as the occurrence of large flat areas. The computational cost can be measured in two ways: the amount of disk memory for data storage and the computational time for processing.

6.9.2 *Situations Demanding Massive Data Processing*

The need to process a large number of pixels has occurred for two main reasons: (i) when the size of the study area is increased, maintaining a given source of topographic data and a given DEM spatial resolution; (ii) when working with a more refined spatial resolution, maintaining the same study area.

The first situation has emerged with the growing demand for regional, continental, and even global-scale studies, such as hydrological assessments of flood extent, hydroclimatological forecasts, scenario analyses related to land use and land cover change, and climate change. In such cases, using DEMs with 30 m resolution, such as SRTM 30, across areas spanning hundreds or even millions of square kilometers can result in high computational costs.

The Amazon basin, for example, which is the largest river basin in the world and has an area of almost 5.1 million km^2 (Lakshmi et al., 2018; Fig. 6.17), requires 5.7 billion pixels to be represented by a 30 m DEM (Table 6.2). Considering a DEM with a spatial resolution of 90 m, like the SRTM 90 or the MERIT DEM, the number of pixels is reduced to 0.63 billion. Working with this basin at even coarser resolutions considerably reduces the computational cost, involving only 82 million pixels when considering a resolution of 250 m and 5.1 million if considering a resolution of 1 km.

The second situation has clearly occurred with spatially refined DEMs of the order of 1 m spatial resolution, as is typical, for example, of surveys carried out by airborne LiDAR. In these cases, a few hundred or thousand km^2 basins may represent a computational cost comparable to the Amazon basin with 30 m pixels. For example, the Capibaribe River basin in the Pernambuco state has an area of around 5900 km^2 (Fig. 6.17). For this basin, the DEM representation derived from LiDAR surveying within the PE3D project (Cirilo et al., 2014, 2015) with a spatial resolution of 1 m consists of 5.88 billion pixels. This amount is higher than the Amazon Basin's representation using SRTM 30 data (Table 6.3). At a 30 m spatial resolution, the DEM of the Capibaribe river basin would contain only around 6.53 million pixels.

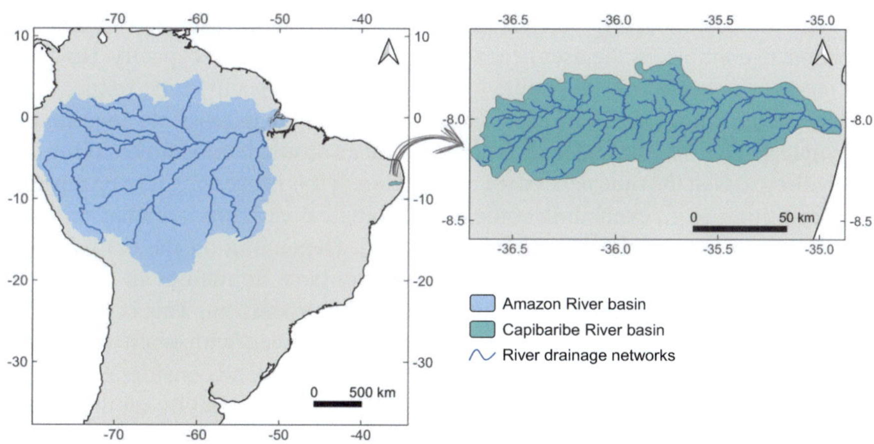

Fig. 6.17 Location of the Amazon and Capibaribe river basins

Table. 6.2 Number of DEM pixels needed to represent the Amazon River basin at four different spatial resolutions

Basin	Drainage area (km^2)	Spatial resolution			
		30 m	90 m	250 m	1000 m
Amazon river basin	5.1 million	5.67 billion	0.63 billion	81.6 million	5.1 million

Table. 6.3 Number of DEM pixels needed to represent the Capibaribe river basin at four different spatial resolutions

Basin	Drainage area (km^2)	Spatial resolution			
		1 m	2 m	10 m	30 m
Capibaribe river basin	5900	5.88 billion	1.47 billion	58.8 million	6.5 million

6.9.3 Issues that Raise the Computational Cost

In summary, when there is a massive amount of pixels in the DEM, processing can become unattractive, prohibitive, or even impeding for the user. This occurs for three main reasons related to disk storage, processing time, and physical memory:

- **Disk storage**. The DEM processing generates other rasters of equal dimensions, such as the depressionless DEM and the flow direction raster, in addition to other raster and vector products that can be derived in the continuity of the analyses, such as accumulated drainage areas, drainage networks, and watershed delineation. The sum of all these data requires considerable disk storage space when derived from a large dimension DEM, which may not be available to the user. This problem is technically simple, as the user needs storage disks with higher capacity. However, it may not be economically viable for some users in a specific time frame.
- **Processing time**. The DEM processing may require an excessively long time to complete. This may be undesirable, but it is not an absolute impediment if the user is willing to wait the time needed for completion. If this processing is to be repeated several times (e.g., evaluating some methodological variation), the feasibility of waiting for excessively long times is reduced. Depending on the software and the computational resources used, the user may have limitations in developing other analyses and parallel work during the DEM processing. This is unpleasant and unproductive. The increase in computational power, with improvements in computer hardware, can substantially improve this issue and considerably reduce processing time. However, this is not a rule and depends heavily on the type of algorithm used for DEM processing. Algorithms proposed to be computationally efficient can take much more advantage of the increased computational power. A typical example is an algorithm that works with parallelization, dividing tasks among the multiple processors available in the user's personal computer.
- **Physical memory**. During the processing of a massive DEM, there are temporary storage requirements for these data and various other information related to processing (arrays and variables related to the calculation algorithm). This storage occurs in the so-called RAM (Random Access Memory), which allows more agility and speed in the operation of a computational system since access (writing and reading) to data stored on disk is much slower. If the computer the user uses does not have enough RAM for processing, it impedes the execution of the calculation. Increasing the machine's RAM is the most direct measure of solving the problem. However, this may not be economically viable and may even be useless.

There are software/algorithms that internally limit the amount of RAM to be used for each operation, regardless of the machine's hardware. Some algorithms are computationally high-performance regarding processing time, but are not efficient in RAM usage, which often becomes a restriction for the user to process large amounts of data. On the other hand, there are software/algorithms designed to optimize memory management in favor of efficient use of this resource to enable processing DEMs of exaggerated dimensions in standard personal computers with a reduced available RAM.

6.9.4 Consequences of High Computational Cost

Due to the high computational cost of processing a DEM to remove terrain depressions and determine flow directions, four primary situations can occur. The worst scenario is the unavailability of the necessary computational resources for the task, which can prevent the processing.

The second worst scenario is having the resources available, but there is a decision of unfeasibility in front of other priorities of a project or study. A milder scenario is when you can perform the processing under a very high demand for time and resource allocation during the elaboration of a project, which can even impact the schedule and resources of other consecutive phases. The ideal scenario is to perform this processing without compromising time and resources for further project stages.

6.9.5 Computationally Efficient Algorithms

The literature proposes several computationally efficient algorithms for dealing with terrain depressions and assigning flow directions. These algorithms employ different approaches and strategies, which can distinguish them not only by computational performance but also by the quality of the results obtained with the solution adopted in removing depressions and assigning flow directions.

Table 6.4 lists some of the papers published in scientific journals that present the proposition of such algorithms. These papers can be consulted for knowledge of algorithms' functioning, application examples, and computational performance analysis.

Table. 6.4 Efficient algorithms for treating terrain depressions proposed in the literature

Reference	Article title
(Arge et al., 2003)	Flow computation on massive grid terrains
(Zhu et al., 2006)	An efficient depression processing algorithm for hydrologic analysis
(Gong & Xie, 2009)	Extraction of drainage networks from large terrain datasets using high throughput computing
(Metz et al., 2011)	Efficient extraction of drainage networks from massive, radar-based elevation models with least cost path search
(Wang & Liu, 2006)	An efficient method for identifying and filling surface depressions in digital elevation models for hydrologic analysis and modeling
(Barnes et al., 2014)	An efficient assignment of drainage direction over flat surfaces in raster digital elevation models
(Bai et al., 2015)	An efficient and comprehensive method for drainage network extraction from DEM with billions of pixels using a size-balanced binary search tree
(Yıldırım et al., 2015)	A virtual tile approach to raster-based calculations of large digital elevation models in a shared-memory system
(Zhou et al., 2016)	An efficient variant of the priority-flood algorithm for filling depressions in raster digital elevation models
(Lindsay, 2016)	Efficient hybrid breaching-filling sink removal methods for flow path enforcement in digital elevation models
(Barnes, 2016)	Parallel priority-flood depression filling for trillion cell digital elevation models on desktops or clusters
(Du et al., 2017)	Drainage network extraction from a high-resolution DEM using parallel programming in the .NET framework
(Wei et al., 2019)	Efficient priority-flood depression filling in raster digital elevation models
(Zhou et al., 2022)	Parallel assignment of flow directions over flat surfaces in massive digital elevation models

References

Arge, L., Chase, J. S., Halpin, P., Toma, L., Vitter, J. S., Urban, D., & Wickremesinghe, R. (2003). Efficient flow computation on massive grid terrain datasets. *GeoInformatica, 7*, 283–313.

Bai, R., Li, T., Huang, Y., Li, J., & Wang, G. (2015). An efficient and comprehensive method for drainage network extraction from DEM with billions of pixels using a size-balanced binary search tree. *Geomorphology, 238*, 56–67. https://doi.org/10.1016/j.geomorph.2015.02.028

Barnes, R. (2016). Parallel priority-flood depression filling for trillion cell digital elevation models on desktops or clusters. *Computers and Geosciences, 96*, 56–68. https://doi.org/10.1016/j.cageo.2016.07.001

Barnes, R., Lehman, C., & Mulla, D. (2014). An efficient assignment of drainage direction over flat surfaces in raster digital elevation models. *Computers and Geosciences, 62*, 128–135. https://doi.org/10.1016/j.cageo.2013.01.009

Buarque, D. C., Fan, F., Paz, A. R., & Collischonn, W. (2009). Comparação de métodos para definir direções de escoamento a partir de Modelos Digitais de Elevação. *Revista Brasileira de Recursos Hídricos, 14*, 91–103. https://doi.org/10.21168/rbrh.v14n2.p91-103

Burrough, P. A., & McDonnell, R. A. (1998). *Principles of geographical information systems.* Oxford University Press.

Cirilo, J. A., Alves, F. H. B., Silva, B. M., & Campos, P. H. A. L. C. (2015). Pernambuco tridimensional: base de dados espaciais para planejamento urbano e gestão territorial. Presented at the 12o Simpósio de Hidráulica e Recursos Hídricos dos Países de Expressão Portuguesa.

Cirilo, J. A., Alves, F. H. B., da Silva, L. A. C., & Campos, J. H. A. L. (2014). Suporte de Informações Georreferenciadas de Alta Resolução para Implantação de Infraestrutura e Planejamento Territorial. *Revista Brasileira De Geografia Física, 7*, 755–763.

Du, C., Ye, A., Gan, Y., You, J., Duan, Q., Ma, F., & Hou, J. (2017). Drainage network extraction from a high-resolution DEM using parallel programming in the .NET Framework. *Journal of Hydrology, 555*, 506–517. https://doi.org/10.1016/j.jhydrol.2017.10.034

Erskine, R. H., Green, T. R., Ramirez, J. A., & MacDonald, L. H. (2006). Comparison of grid-based algorithms forcomputing upslope contributing area. *Water Resources Research, 42*, W09416. https://doi.org/10.1029/2005WR004648

Fairfield, J., & Leymarie, P. (1991). Drainage networks from grid digital elevation models. *Water Resources Research, 27*, 709–717.

Freeman, T. G. (1991). Calculating catchment area with divergent flow based on a regular grid. *Computers and Geosciences, 17*, 413–422.

Gong, J., & Xie, J. (2009). Extraction of drainage networks from large terrain datasets using high throughput computing. *Computers and Geosciences, 35*, 337–346. https://doi.org/10.1016/j.cageo.2008.09.002

Jenson, S. K., & Domingue, J. O. (1988). Extracting topographic structure from digital elevation data for geographic information system analysis. *Photogrammetric Engineering, 54*(11), 1593–1600.

Lakshmi, V., Fayne, J., & Bolten, J. (2018). A comparative study of available water in the major river basins of the world. *Journal of Hydrology, 567*, 510–532. https://doi.org/10.1016/j.jhydrol.2018.10.038

Lindsay, J. B. (2016). Efficient hybrid breaching-filling sink removal methods for flow path enforcement in digital elevation models: efficient hybrid sink removal methods for flow path enforcement. *Hydrological Processes, 30*, 846–857. https://doi.org/10.1002/hyp.10648

Maciel Filho, C. L., Nummer, A. V., Da Silva, J. L. S., Sangoi, D. S., & Filho, L. D. (2013). Feições de carste sobre a formação Serra Geral na Região de Vacaria, RS, e Lages, SC. *Revista Caminhos de Geografia*, 14. https://doi.org/10.14393/RCG144821692

Metz, M., Mitasova, H., & Harmon, R. S. (2011). Efficient extraction of drainage networks from massive, radar-based elevation models with least cost path search. *Hydrology and Earth Systems Science, 15*, 667–678. https://doi.org/10.5194/hess-15-667-2011

Orlandini, S., Moretti, G., Franchini, M., Aldighieri, B., & Testa, B., (2003). Path-based methods for the determination of nondispersive drainage directions in grid-based digital elevation models: Technical note. *Water Resources Research*, 39. https://doi.org/10.1029/2002WR001639

Paz, A. R., & Collischonn, W. (2008). Derivação de rede de drenagem a partir de dados do SRTM. *Revista Geográfica Acadêmica, 2*, 84–95.

Quinn, P., Beven, K., Chevallier, P., & Planchon, O. (1991). The prediction of hillslope flow paths for distributed hydrological modelling using digital terrain models. *Hydrological Processes, 5*, 59–79. https://doi.org/10.1002/hyp.3360050106

Siqueira, V., Fleischmann, A., Jardim, P., Fan, F., & Collischonn, W. (2016). IPH-hydro tools: A GIS coupled tool for watershed topology acquisition in an open-source environment. *Revista Brasileira de Recursos Hídricos*, 21, 274–287. https://doi.org/10.21168/rbrh.v21n1.p274-287

Tarboton, D. G. (1997). A new method for the determination of flow directions and upslope areas in grid digital elevation models. *Water Resources Research, 33*, 309–319. https://doi.org/10.1029/96WR03137

Wang, L., & Liu, H. (2006). An efficient method for identifying and filling surface depressions in digital elevation models for hydrologic analysis and modelling. *International Journal of Geographical Information Science, 20*, 193–213. https://doi.org/10.1080/13658810500433453

Wei, H., Zhou, G., & Fu, S. (2019). Efficient priority-flood depression filling in raster digital elevation models. *International Journal of Digital Earth, 12*, 415–427. https://doi.org/10.1080/17538947.2018.1429503

Wilson, M. F. J., O'Connell, B., Brown, C., Guinan, J. C., & Grehan, A. J. (2007). Multiscale terrain analysis of multibeam bathymetry data for habitat mapping on the continental slope. *Marine Geodesy, 30*, 3–35. https://doi.org/10.1080/01490410701295962

Yıldırım, A. A., Watson, D., Tarboton, D., & Wallace, R. M. (2015). A virtual tile approach to raster-based calculations of large digital elevation models in a shared-memory system. *Computers and Geosciences, 82*, 78–88. https://doi.org/10.1016/j.cageo.2015.05.014

Zhou, G., Song, L., & Liu, Y. (2022). Parallel assignment of flow directions over flat surfaces in massive digital elevation models. *Computers and Geosciences, 159*, Article 105015. https://doi.org/10.1016/j.cageo.2021.105015

Zhou, G., Sun, Z., & Fu, S. (2016). An efficient variant of the Priority-Flood algorithm for filling depressions in raster digital elevation models. *Computers and Geosciences, 90*, 87–96. https://doi.org/10.1016/j.cageo.2016.02.021

Zhu, Q., Tian, Y., & Zhao, J. (2006). An efficient depression processing algorithm for hydrologic analysis. *Computers and Geosciences, 32*, 615–623. https://doi.org/10.1016/j.cageo.2005.09.001

Chapter 7
River Drainage Networks

River flow paths in digital format can be obtained from various sources using different data and procedures. For example, they can be obtained from aero photogrammetric surveys, via manual digitization of satellite images, or even from topographic maps.

One approach to obtain the river drainage networks in digital format is from DEM processing. After removing the terrain depressions and determining the flow directions, one can calculate the accumulated drainage areas and derive the drainage network. This chapter addresses these last two steps, considering that the previous steps were described in Chap. 6. In addition to obtaining the river drainage network, this chapter presents how to characterize the river reaches by extracting features such as length, slope, and sinuosity. The elaboration of the topographic longitudinal profile of a river is also discussed.

Two advanced items close the chapter. The first is the procedure to preprocess the DEM to incorporate an existing vector drainage network, called stream burning. The last item concerns the change of scale in the drainage network, presenting two approaches: one based on coarsening the spatial resolution of the DEM and the other on methods of flow direction upscaling.

7.1 Accumulated Drainage Areas

The accumulated drainage areas represent a data layer derived directly from the flow directions and are another product resulting from the DEM processing. It is a raster in which each pixel has as an attribute the number of pixels or the amount of area whose runoff drains to that pixel (for example, in m^2 or km^2).

Consider the raster of flow directions indicated in Fig. 7.1a, with a symbolic representation of directions by arrows for easier understanding. Initially, one can identify all the pixels that do not receive flows from any pixel. These pixels drain only themselves, representing headwaters of flow paths, and therefore have an accumulated

A. Rolim da Paz, *Digital Elevation Models for Environmental Studies*,
https://doi.org/10.1007/978-3-032-04523-2_7

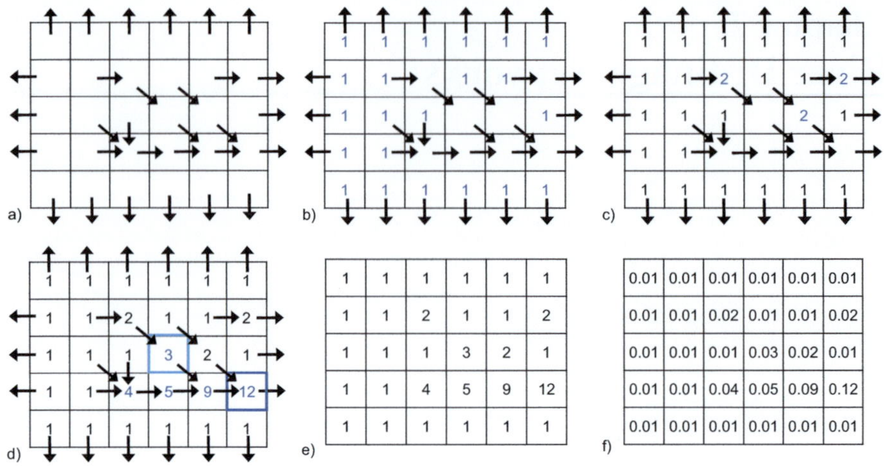

Fig. 7.1 **a** Flow directions, obtained from a DEM with a spatial resolution of 100 m; **b** Indication of headwater pixels, which drain only themselves; **c** Indication of the pixels that receive contribution from an upstream pixel, in addition to draining themselves; **d** Following all flow paths and accumulating the number of pixels that drain into each pixel of the image; **e** Raster of accumulated drainage areas, in number of pixels; **f** Raster of accumulated drainage areas, in km^2

drainage area equal to one pixel (Fig. 7.1b). In addition to draining themselves, some pixels in the area of this example receive the flow from one upstream pixel, having, in this case, an accumulated area of two pixels (Fig. 7.1c).

Following this logic, one can interpret and follow the flow directions to record the number of pixels they drain into each pixel. For example, the pixel highlighted by the light blue border in Fig. 7.1d drains three pixels: itself and two upstream pixels, while the pixel underscored by the dark blue border drains 12 pixels. The information that 12 pixels drain into a given pixel does not allow us to know which pixels these are. The analysis of the flow directions enables this association to be made.

The raster of accumulated drainage areas is the matrix of values indicating how many pixels flow into each pixel (Fig. 7.1e). In this case, the raster attribute is expressed as the number of pixels. This approach is not the most recommended due to the difficulty of directly interpreting the physical meaning of this value. Considering the pixel size in conjunction with this information is always necessary. For example, stating that a pixel has a drainage area of 1200 pixels is useless unless the pixel size is also provided.

Working with the raster of accumulated drainage areas with values in area units, such as m^2 or km^2, is more recommended and convenient. Assuming that the pixel size is 30 m, i.e., that this is the spatial resolution of the DEM used to derive the accumulated drainage areas, a pixel that has an accumulated area of 1200 pixels corresponds to a contribution area of 1.08 km^2 (=1200 × 30 × 30 = 1,080,000 m^2).

In the example of Fig. 7.1, considering the spatial resolution is 100 m, each pixel has an area of 0.01 km^2. To convert the raster of accumulated drainage areas expressed

0.02	0.01	0.01	0.01	0.01	0.01	0.01	0.01	0.01	0.01	0.01	0.02	0.02	0.01	0.01	0.01	0.01	0.01	0.02	0.04
0.02	0.02	0.02	0.02	0.02	0.01	0.03	0.02	0.02	0.02	0.01	0.32	0.29	0.03	0.01	0.01	0.01	0.01	0.01	0.02
2.91	0.10	0.07	0.03	0.03	0.06	0.03	0.01	0.06	0.02	0.33	0.01	0.01	0.21	0.19	0.03	0.01	0.01	0.01	0.01
0.87	1.92	1.91	1.89	1.87	1.80	1.71	1.55	1.47	0.36	0.02	0.02	0.02	0.01	0.01	0.13	0.11	0.04	0.02	0.01
0.09	0.73	0.01	0.01	0.02	0.02	0.03	0.08	0.01	1.07	0.07	0.03	0.01	0.02	0.01	0.01	0.01	0.04	0.02	0.01
0.06	0.72	0.02	0.01	0.01	0.01	0.02	0.02	0.04	0.04	0.95	0.06	0.03	0.01	0.02	0.01	0.01	0.01	0.01	0.01
0.03	0.07	0.59	0.04	0.03	0.02	0.01	0.01	0.03	0.03	0.07	0.82	0.71	0.03	0.02	0.02	0.02	0.01	0.01	0.01
0.03	0.02	0.51	0.02	0.01	0.01	0.01	0.01	0.01	0.02	0.06	0.04	0.06	0.65	0.51	0.06	0.02	0.02	0.01	0.01
0.01	0.02	0.07	0.41	0.03	0.01	0.01	0.01	0.01	0.02	0.02	0.03	0.04	0.07	0.07	0.40	0.25	0.07	0.04	0.02
0.01	0.02	0.03	0.04	0.32	0.27	0.03	0.01	0.01	0.01	0.01	0.02	0.03	0.06	0.06	0.03	0.10	0.15	0.12	0.09
0.01	0.01	0.01	0.03	0.02	0.02	0.21	0.11	0.09	0.06	0.02	0.01	0.02	0.04	0.04	0.02	0.03	0.06	0.02	0.02
0.01	0.01	0.01	0.01	0.01	0.01	0.03	0.06	0.02	0.02	0.02	0.01	0.01	0.03	0.03	0.01	0.02	0.03	0.01	0.01
0.02	0.01	0.01	0.02	0.01	0.01	0.01	0.01	0.01	0.01	0.01	0.01	0.01	0.01	0.01	0.01	0.01	0.01	0.01	0.01

Fig. 7.2 Schematic example of a raster of accumulated drainage areas, with an indication of the attribute in km^2, and symbology that associates a gradient of colors from white to dark blue in the direction from the smallest to the largest drainage area

in the number of pixels so that the attribute is drainage areas in km^2, multiply this raster by the scalar 0.01—the arithmetic operation between a raster and a scalar is explained in Chap. 3. The result of this operation is the raster of accumulated drainage areas in km^2 (Fig. 7.1f).

The raster of accumulated drainage areas has a quantitative attribute, which increases in value proportionally to the amount of areas drained by each pixel. Therefore, a good practice is associating this raster with a continuous color ramp symbology, with a gradient varying from white through shades of blue to dark blue (Fig. 7.2). With this symbology, the interpretation becomes more straightforward and intuitive, with the direct association that the bluer it is, the higher the accumulation of drainage area. It is even possible to perceive the primary flow paths, which follow in the direction where the blue shade increases. The image's main flows in the example go from right to left.

Figure 7.3 shows a real-world example of a raster of accumulated drainage areas for a section of the Ijuí river basin, a tributary of the Uruguay River, in southern Brazil. In parts 'a' and 'b' of this figure, the accumulated areas were generated without the depressions of the DEM being removed, while in parts 'c' and 'd', the accumulated areas follow the recommendation to be calculated from the depressionless DEM.

In this second case, the image allows identification of the continuous flow paths, the convergence between them, the formation of paths of higher relevance regarding the accumulation of contribution area, and the perception in a relatively approximate way of the areas that drain into each point of the image. In the first case, the terrain depressions in the DEM prevent the continuity of the flow so that the accumulation of areas follows a particular short flow path until it runs into a pixel that constitutes a depression, like the pixels pointed out by the arrows in Fig. 7.3b. Each depression acts as a sink to the contributing flow. Downstream of each depression, a new flow accumulation begins, making the main path from the upstream of the depression discontinuous.

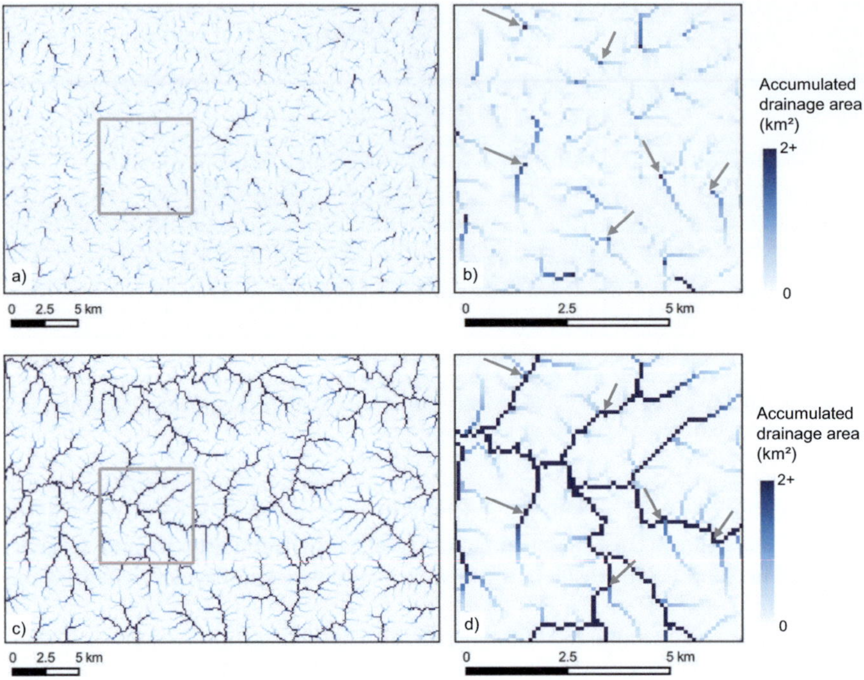

Fig. 7.3 **a**, **c** Raster of accumulated drainage areas for a section of the Ijuí river basin, a tributary of the Uruguay River (southern Brazil), obtained from the SRTM 90 DEM, taking the DEM with depressions and without depressions, respectively; **b**, **d** Detail of the rasters of accumulated drainage areas shown in (**b**) and (**c**), respectively, for the area indicated by the rectangle

The reader should note that, in Fig. 7.3's example, the color scale of the symbology is saturated at the value of 2 km^2. This indicates no distinction between shades of blue used from this value onward—all pixels with an accumulated drainage area higher than 2 km^2 have the same color shade in the raster. On the other hand, this allowed for better distinguishing and visualizing flow paths with smaller accumulation areas.

7.2 Algorithms for Calculating Accumulated Drainage Areas

Determining the accumulated drainage areas after defining flow directions is conceptually objective and straightforward. Regardless of how an algorithm is designed to carry out this procedure, the result should be the same as that obtained by other algorithms, provided that the input flow direction raster is identical. Ultimately, the task boils down to counting how many pixels—and the total area of those pixels—contribute to each pixel in the image. The critical aspect of the process has already been addressed: the definition of flow directions.

One factor that can cause an exception and generate a slight variation in results among algorithms for determining accumulated drainage areas is the conversion of angular units to planar units. When working with a reference system in geographic coordinates, with spatial resolution in angular units (degree, arc-minute, arc-second, as described in Chap. 4), a conversion is necessary to obtain accumulated areas in terms of planar units (such as m^2 or km^2). This conversion can be performed differently among the algorithms, causing variations in results between them. This variation tends to be higher the further from the equator the study area is located, due to the more significant effect of variation between degrees of latitude and the corresponding distance on the plane.

Even though it is a simple operation by definition and without room to propose variations in methodology that improve the final result, several scientific studies propose specific algorithms for determining accumulated drainage areas. The primary motivation for this is to minimize the excessive computational cost that can occur when working with DEMs with large amounts of data. Table 7.1 lists some works published in scientific journals proposing algorithms for efficiently calculating accumulated drainage areas.

Table 7.1 Computationally efficient algorithms for determining accumulated drainage areas

References	Article title
Arge et al. (2003)	Flow computation on massive grid terrains
Do et al. (2011)	Parallel computing flow accumulation in large digital elevation models
Qin and Zhan (2012)	Parallelizing flow accumulation calculations on graphics processing units—from iterative DEM preprocessing algorithm to recursive multiple-flow direction algorithm
Yao and Shi (2015)	Alternating scanning orders and combining algorithms to improve the efficiency of flow accumulation calculation
Su et al. (2015)	An efficient algorithm for calculating drainage accumulation in digital elevation models based on the basin tree index
Sten et al. (2016)	Parallel flow accumulation algorithms for graphical processing units with application to RUSLE model
Barnes (2017)	Parallel non-divergent flow accumulation for trillion cell digital elevation models on desktops or clusters
Zhou et al. (2019)	A fast and simple algorithm for calculating flow accumulation matrices from raster digital elevation
Kotyra et al. (2021)	High-performance parallel implementations of flow accumulation algorithms for multicore architectures
De Jong et al. (2022)	Scalability and composability of flow accumulation algorithms based on asynchronous many-tasks

7.3 Deriving a River Drainage Network from a DEM

7.3.1 *Flow Paths Versus River Drainage Network*

DEM processing concepts and methods have been addressed in the previous chapter and at the beginning of this chapter to obtain products such as depressionless DEM, flow directions, and accumulated drainage areas. Obtaining these three data layers is closely related to identifying and automatically tracing flow paths from the topography represented in the DEM. However, none of these products refers to the river drainage network itself.

Flow paths should not be confused with the river drainage network, which is the network of watercourses in the watershed. For every pixel in the image, the flow direction raster establishes the direction where the flow generated or reaching that pixel goes. Following these flow directions is a pixel-by-pixel path where the flow goes, called a flow path. Some sections of these flow paths may or may not effectively integrate part of the river drainage network, representing the watercourses. In summary, every section of the river drainage network represents a flow path, but not every flow path corresponds to a section of the river drainage network.

7.3.2 *Raster Representation of a River Drainage Network*

Usually, the pixels belonging to the drainage network receive attribute 1, and the others do not belong and receive attribute zero. This is the typical format of a river drainage network raster derived from the DEM. A variant of this format is when each river reach is identified and numbered sequentially, making each pixel belonging to the river drainage network have a non-null attribute of value equal to the river reach to which it belongs. The pixels that are not part of the network have a null attribute. This approach refers to the coding and discretization of river reaches, as commented on later in this chapter.

7.3.3 *Where Do Rivers Begin?*

The distinction between flow paths and river drainage networks goes through the key question: where do rivers begin? The best situation to resolve this question is when there is information about the beginning of the river drainage network, that is, the geographical location of the headwaters or the beginning of the most upstream river reaches. This information may be available directly as coordinates of these drainage start points or indirectly as a vector network of river reaches. In the second case, it is possible to identify the coordinates of the beginnings of the upstream reaches of this vector network.

When existing and available, the information about the location of the beginning of the river reaches can be used to distinguish between flow paths and river drainage networks derived from the DEM, based on the raster of accumulated drainage areas. The operation consists of identifying the pixels located at the drainage start points and assuming that, from then downstream, along the flow downstream, all the pixels of the path are part of the river drainage network (Fig. 7.4a, b). All other pixels in the image are not part of the network.

For most cases, information about the location of the beginning of the various headwaters of rivers is not available. In this situation, some criterion must be used to impose the condition of the beginning of rivers to obtain the river drainage network as a product of the DEM. There are several approaches in the literature, as described in the following items.

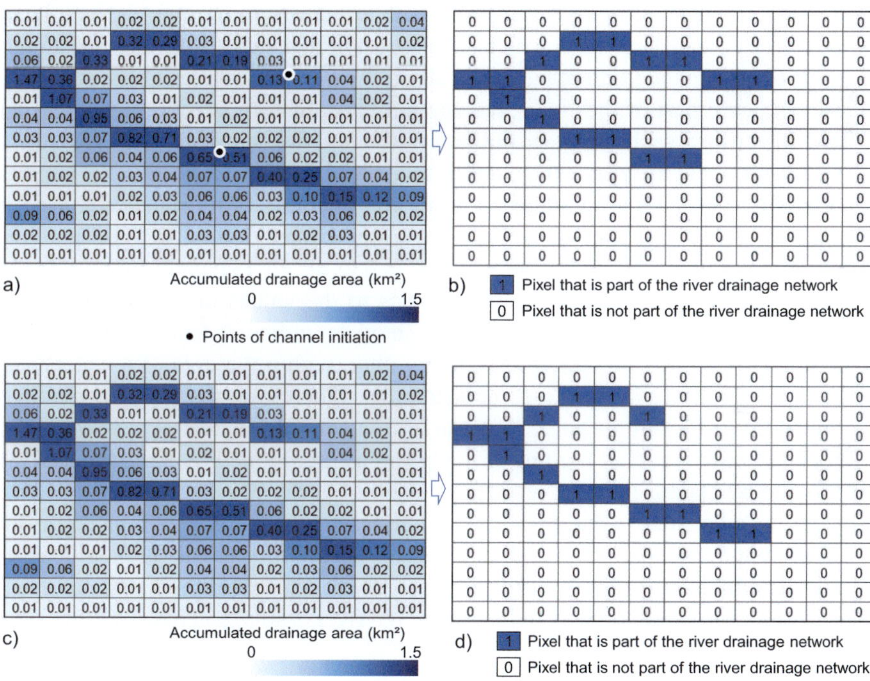

Fig. 7.4 **a** Raster of accumulated drainage areas indicating known starting points of river sections; **b** River drainage network obtained based on the indicated starting points; **c** Raster of accumulated drainage areas; **d** River drainage network obtained considering a minimum threshold (A_{min}) of 0.2 km^2

Table 7.2 Rules for classifying a raster of accumulated drainage areas and defining the river drainage network by considering a minimum area (A_{min}) necessary to river channel initiation

Classification rule	New value	Interpretation
Accumulated drainage area of the pixel < A_{min}	0	Pixel is **not** part of the river drainage network
Accumulated drainage area of the pixel $\geq A_{min}$	1	Pixel is part of the river drainage network

7.3.4 Beginning of Rivers from a Minimum Accumulated Drainage Area

The approach most widely used to define the beginning of rivers when processing a DEM is the simplest: assume the hypothesis that there is a minimum accumulated drainage area value (A_{min}) from which rivers form in that study area. Going through the raster of accumulated drainage areas, all pixels whose attribute is higher than or equal to A_{min} are part of the river drainage network. In comparison, those with an attribute lower than such value are not part of the drainage network (Fig. 7.4c, d).

The question that arises is how to define the value of Amin. Conceptually, various characteristics of the region influence the amount of contributing area necessary for forming a watercourse, such as soil type, geology, vegetation, slope, and precipitation regime, among others. In the work of Fan et al. (2013), an effort was made to relate some of these regions' characteristics to the values of Amin, which can serve as the basis for defining this parameter in other studies. A factor that increases the complexity of the problem is recognizing that, within the area of a watershed, there is spatial variability of these characteristics. Therefore, there would not be a recommendation of a single value of Amin for the entire basin area. There would be different values for distinct regions of the basin, as highlighted by Fan et al. (2013) and adopted by Dantas and Paz (2021).

Adopting a constant value of Amin to define the river drainage network from the analysis of the accumulated drainage areas is trivial. It is a raster reclassification operation available in virtually all GIS software. Two reclassification rules distinguish the categories belonging and not belonging to the river drainage network (Table 7.2).

7.3.5 Beginning of Rivers Considering Minimum Drainage Area and Minimum River Length

A refinement of the method of adopting the A_{min} parameter for establishing the river drainage network from the accumulated drainage areas is to impose a second criterion related to a minimum length threshold (L_{min}) for the most upstream river reaches (those that do not receive contributions from other reaches). This operation is available, for example, in QGIS-GRASS.

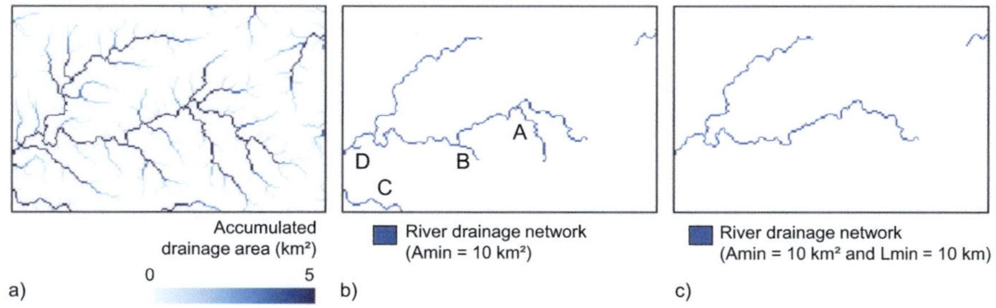

Fig. 7.5 **a** Raster of accumulated drainage areas; **b** River drainage network obtained by the criterion of minimum drainage area threshold (A_{min}) with a value of 10 km^2; **c** River drainage network obtained simultaneously by the criteria of A_{min} = 10 km^2 and minimum river reach length (L_{min}) of 10 km, in which the river reaches A, B, C and D that are included in the previous network no longer appear

The effect of adopting the second criterion for defining the river drainage network is exemplified in Fig. 7.5. The river reaches identified by A, B, C, and D appear in the river drainage network determined solely by the minimum area of contribution criterion, but do not appear in the network generated by adding the criterion of a minimum length for the most upstream reaches. The length of the river reaches A, B, C, and D, measured from the upstream start to the confluence with a larger river, is less than the adopted threshold for L_{min}. The purpose of this additional criterion is to exclude small river reaches that are tributary to other main reaches of the river drainage network. The practical difficulty is defining this parameter value (analogously to the Amin definition challenge).

7.3.6 Definition of the Parameters A_{min} and L_{min}

In practice, what is most commonly observed is the arbitrary adoption of a value for A_{min} and L_{min} (when the second criterion for defining the beginning of the river drainage network is additionally used). In some cases, users even uncritically accept the default value suggested by the software used for DEM processing without considering the implications of that choice.

The choice of A_{min} (and L_{min}) can drastically influence the subsequent results of studies with this river drainage network as input data. Both parameters firmly control the density of the river drainage network, as illustrated in the case of the Icamaquã river basin, a tributary of the Uruguay River (Fig. 7.6). The higher the value of A_{min}, the less extensive and dense the obtained river drainage network is; the higher the value of L_{min}, the fewer small river reaches in the headwaters.

For example, consider point P1 in Fig. 7.6, which is not part of the river drainage networks. Suppose the simulation of erosive processes resulting from precipitation and surface runoff originated at this point. From P1, the flow follows the terrain as

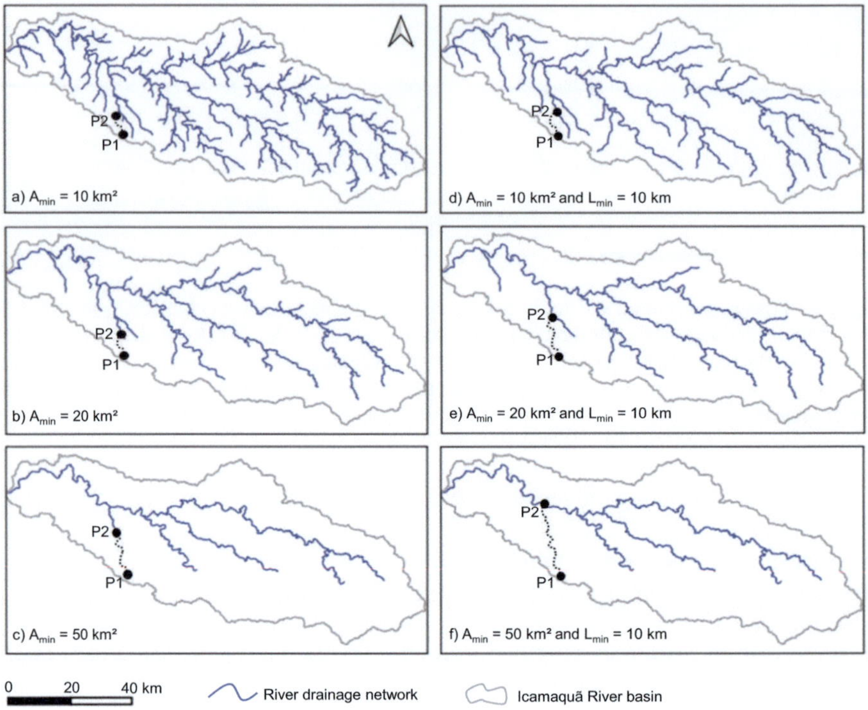

Fig. 7.6 River drainage networks of the Icamaquã river basin (tributary of the Uruguay River, southern Brazil), obtained by processing the SRTM 90 DEM, considering only the criterion of minimum drainage area (A_{min}) with different values (**a–c**) and additionally considering the criterion of minimum length of river reaches (L_{min}) (**d–f**)—with indication of flow path from point P1 until it meets the river drainage network at point P2

surface runoff until it reaches the river drainage network at the point indicated as P2. In each configuration of the river drainage network, point P2 is in a different spatial position.

Moving downstream from P1 up to the river drainage network at point P2, the sediment transport occurs as surface runoff (over the terrain itself). Downstream from P2, the sediments are conveyed by channel flow along the rivers. From P1 to P2, the sediment movement has much higher difficulty due to reduced water depth and high surface friction. In contrast, there is much higher ease of particle conveyance when the transport by the river drainage network begins (from P2 downstream).

Depending on the choice of A_{min} and L_{min} parameters, the river drainage networks obtained differ quite a bit, making the P1–P2 reach very distinct. Therefore, this can tremendously impact the results of a hydrosedimentological modeling study.

The recommendation is that the user at least perform a sensitivity analysis with the Amin parameter (and L_{min}, if they also adopt it as a criterion for obtaining the river drainage network). It is possible to get several river drainage networks and check, as far as possible, which has a higher degree of coherence with reality. This inference can be made by consulting information available for the study area, like maps from previous works, satellite image analyses, or information from technicians who know the region.

7.3.7 Beginning of the River Drainage Network from the Relationship Between the Drainage Area and Slope

A distinct approach to defining the beginning of the river drainage network is from geomorphology theory, based on analyzing the relationship between accumulated drainage areas and the slope of flow paths. Some works in the 1980s and 1990s stood out in terms of evaluation of the development of mathematical relationships between drainage area and slope of river reaches (Montgomery & Dietrich, 1988, 1992; Montgomery & Foufoula-Georgiou, 1993; Tarboton et al., 1991, 1992).

These works identified patterns of the relationship between drainage area and slope that would be useful for configuring the river drainage initiation. For example, a minimum threshold value C_{min} can be adopted for the product $A_s \cdot S^2$, where A_s is the specific drainage area and S is the slope, so that the river drainage network is identified as formed by all pixels whose product $A_s \cdot S^2$ has a value higher than such a threshold.

7.3.8 Other Approaches for Identifying the Beginning of the River Drainage Network

The problem of automatically identifying the beginning of the river drainage network continues to demand scientific research and has motivated the development of other approaches. To illustrate the range of methods being developed, we mention the study by Barbedo et al. (2022), based on the topographic position index in combination with the analysis of accumulated drainage area, and the works by Clubb et al. (2014), Pelletier (2013) and Passalacqua et al. (2010). These last three studies consider the analysis of the tangential curvature derived from the DEM (curvature of the surface in the vertical plane perpendicular to the flow gradient) but with additional analyses, such as the transformation of longitudinal topographic profiles in the case of the first work and the analysis of accumulated drainage areas, in the case of the third work.

7.4 Vector River Drainage Network

From the raster river drainage network, the vector river drainage network (a line-type vector) can be determined by the usual operation of converting the raster to a vector. As seen in Chap. 2, this is a standard operation in geoprocessing involving the conversion of data formats. With the raster river drainage network characterized by network pixels with attribute 1 and the remaining pixels of the image having attribute zero, in general, the conversion of this raster to the vector format of lines results in a set of lines with the same attribute 1 (Fig. 7.7).

In some software, the line vector resulting from the conversion of the raster river drainage network may have a variable attribute for each feature. Typically, each river reach between two junctions or each headwater river reach (from the beginning of the network to the downstream junction) is represented by a feature—a line object with

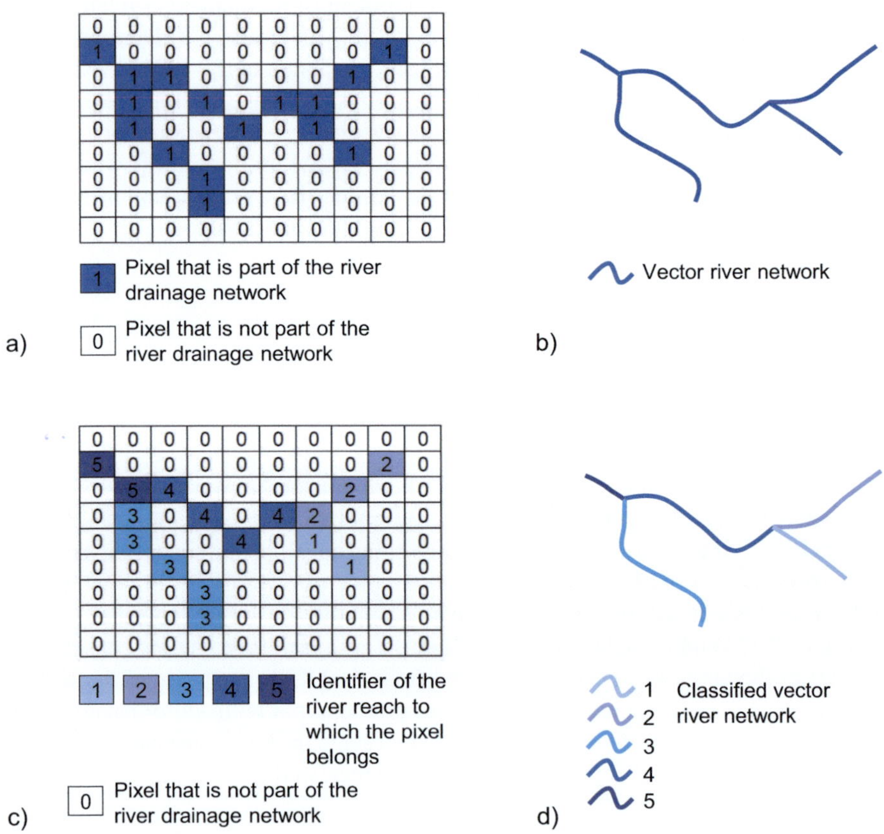

Fig. 7.7 **a** Raster river drainage network derived from a DEM; **b** River drainage network converted to vector format; **c** Raster river drainage network with classification of river reaches; **d** Classified river drainage network converted into vector format

its identifier corresponding to a specific record in the attribute table. The software assigns this identifier according to its rules during the raster-vector transformation process (for example, assuming the features' identifier in an increasing order of creation).

In the example of Fig. 7.7a, there is a raster river drainage network whose corresponding vector network is shown in Fig. 7.7b. This vector network has five features, corresponding to three headwater reaches and two downstream reaches. The result of converting the raster drainage network into vector format can vary according to the different algorithms used by the software, as already discussed in Chap. 2 more broadly.

7.5 Discretization of the River Drainage Network

In some cases, such as when applying hydrological models, discretizing the raster river drainage network into sections may be desirable. The most common method is a division according to the same logic used to convert the raster drainage to a vector. Sets of network pixels that configure continuous reaches between river junctions or from the headwaters to the downstream junctions are identified.

Each identified set of pixels receives a numerical code as an attribute that is the unique identifier of that river segment. In the example of Fig. 7.7c, the raster drainage network was discretized into five segments, with the pixels corresponding to each segment having equal attribute, which is the segment identifier.

The conversion of the classified raster river drainage network into a vector format also results in a discretized vector drainage network. In this format, the line features carry an attribute representing the identifier of each segment (Fig. 7.7d). There are other approaches to discretizing river drainage networks, whether raster or vector, some of which are associated with watershed and sub-watershed division rules. One such example is the segmentation into Ottobasins, as detailed in Chap. 8.

7.6 Length of River Reaches

Subsequent processing can be done to measure the lengths of river segments using the raster river drainage network derived from the DEM. Given upstream and downstream points defining the beginning and end of the river segment, the reach extension is quantified following the flow along the raster established by the flow directions and accounting for the distance traveled. By default, each orthogonal step between two drainage pixels equals a distance equal to the spatial resolution of the image, and a diagonal step equals the spatial resolution multiplied by the square root of two $\left(\sqrt{2}\right)$.

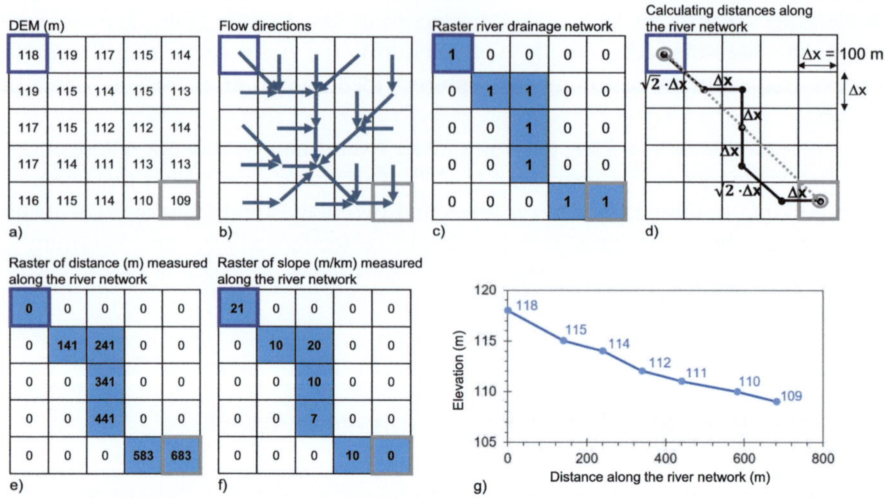

Fig. 7.8 **a** Digital elevation model; **b** Flow directions; **c** Raster river drainage network; **d** Distance calculation at each step along the river drainage network; **e** Raster of distances along the river drainage network, calculated from the pixel highlighted in dark blue to the pixel highlighted in gray; **f** Raster of slopes at each step along the river drainage network; **g** Topographic longitudinal profile along the river drainage network

In the example shown in Fig. 7.8, the river drainage segment from the blue to the gray pixel consists of four orthogonal steps and two diagonal steps, following the indicated flow directions. Considering a spatial resolution of 100 m, the total length of this segment is calculated as $4 \cdot 100 + 2 \cdot \sqrt{2} \cdot 100 = 682.8$ meters. This procedure can be applied to a single segment representing the main river in the study area or separately for each river reach based on an adopted discretization scheme.

7.7 Slope of River Reaches

Just as the lengths of the segments of the river drainage network can be obtained from DEM-derived products, it is also possible to get other relevant data for various hydrological studies, such as the slope of these segments.

An approach to generate this information is to account for the slope at each step when following the pixel-by-pixel path of the river segment. This results in an estimate of the slope distributed along the segment. In the example of Fig. 7.8, a raster of river drainage network slopes was obtained, and each pixel has the slope of the local sub-segment as an attribute. The slope at each pixel is calculated by the difference in elevations between the pixel and its downstream neighbor, divided by the distance of the local sub-segment. This distance is Δx or $\sqrt{2} \cdot \Delta x$, if the pixels are orthogonal or diagonal neighbors, respectively, as exemplified in Fig. 6.6. In the example of Fig. 7.8f, the slope of the pixel of the river drainage network located farthest downstream in the image (pixel highlighted with a gray border) cannot be determined, as there is no information downstream of it in this raster.

Suppose the interest is to have a single representative slope value for the entire river reach instead of values distributed per pixel. In that case, one can perform the arithmetic or harmonic average of the slope values of the pixels along the reach (Eqs. 7.1 and 7.2, respectively). A simpler alternative is to calculate the slope of the reach directly and solely based on the upstream and downstream pixels of the reach (Eq. 7.3). This latter alternative tends to overestimate slopes with increasing river reach length, as the river longitudinal profile gets farther from a straight line joining the upstream and downstream pixels.

$$S_{\text{reach}} = \sum_{i=1}^{n_p} S_i/n \tag{7.1}$$

$$S_{\text{reach}} = \left(\frac{L_{\text{reach}}}{\sum_{i=1}^{n_p} \frac{L_i}{\sqrt{S_i}}} \right)^2 \tag{7.2}$$

$$S_{\text{reach}} = \frac{Z_{\text{up}} - Z_{\text{down}}}{L_{\text{reach}}} \tag{7.3}$$

where S_{reach} and L_{reach} are the slope and length of the river reach, respectively; S_i is the slope of a pixel of the river drainage network relative to the downstream one, n_p is the number of pixels of the reach, L_i is the length of the river drainage network from one pixel to the downstream one, Z_{down} and Z_{up} are the elevations of the downstream and upstream pixels, respectively, of a river reach.

In the example of Fig. 7.8, the slope of the river reach in terms of arithmetic average would be 13 m/km or 11.5 m/km if the harmonic average is calculated, or even 13.2 m/km if the calculation is made directly based on the initial and final points of the reach.

7.8 Sinuosity of River Reaches

The sinuosity of a watercourse is defined as the relationship between the length of the river reach, measured along its course, and the length measured in a straight line from the upstream point to the downstream point (Eq. 7.4). The higher the difference in the length of the reach along its course relative to the distance obtained by a straight-line tracing, the higher the sinuosity.

$$\text{Si}_{\text{reach}} = \frac{L_{\text{reach}}}{L_{\text{straight}}} \tag{7.4}$$

where Si_{reach} is the sinuosity of a river reach, and $L_{straight}$ is the distance measured in a straight line from the upstream to the downstream point.

Conceptually, the smallest sinuosity value for a watercourse is 1 when its tracing is straight. As this value increases, the watercourse changes from a straight shape to a curvilinear or tortuous one.

From the DEM, it was seen how to determine the length of the river reach in the previous item. This information, in conjunction with the additional determination of the length measured in a straight line from upstream to downstream (dotted line in the example of Fig. 7.8d), allows the determination of the sinuosity of the reach. For the hypothetical river reach in the example of Fig. 7.8, the straight line length is 565.7 m, and the sinuosity is approximately $1.21(= 682.8/565.7)$.

7.9 Longitudinal Topographic Profile of the River

The longitudinal topographic profile of a watercourse represents how the topography varies along the course of the river (Fig. 7.9). It is a relevant characteristic of the watercourse for studies of fluvial geomorphology, the evolution of the river drainage network, landscape analysis, and hydrological and sediment transport studies, among others. With this type of profile, the variation of slopes along a watercourse is perceived, with higher slopes in the upper reach and lower slopes in the lower reach near the mouth.

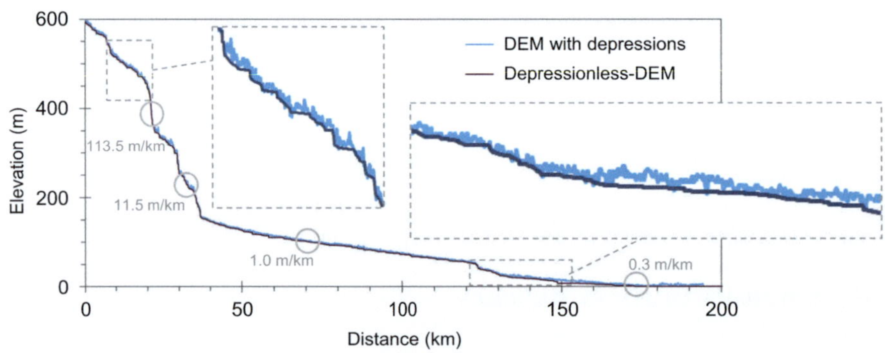

Fig. 7.9 Longitudinal topographic profile of the Mamanguape River (coast of Paraiba) elaborated from the SRTM30 DEM, considering both the original DEM with depressions and the DEM modified by the depression removal processing according to the TerrSET software algorithm

7.9.1 Approaches to Determine the Longitudinal Topographic Profile from the DEM

From a DEM, the longitudinal topographic profile of a watercourse can be elaborated following two primary approaches. The first is analogous to any profile traced in space, as explained in the case of cross-sectional profiles (Chap. 5). Given the tracing of the watercourse in the line vector format, overlaying this vector on the DEM allows the extraction of the elevation values along this line. The distance along the line from the origin is accounted for, and distance and elevation points are formed. The origin is usually set in the river's headwater, and this set of points represents the longitudinal topographic profile.

Following this methodology, the vector tracing of the river may or may not be derived from the DEM. In the case of the tracing originating from a data source independent of the DEM, such as from aero photogrammetric surveys or via the digitization of satellite images, this course of the river is not coincident with the course obtained by the DEM processing. If the study to be carried out incorporates products from the DEM, such as flow directions, accumulated drainage areas, or the river drainage network itself, a disadvantage of this approach to elaborate the longitudinal profile is that it does not necessarily translate how the topography varies along the river according to the DEM. Because the DEM processing results in a watercourse path that may differ from the actual one, with some reaches potentially coinciding and others not.

Another disadvantage of this approach is that it depends on the availability of the river's vector tracing and the quality of this tracing in terms of representing the actual river course. The advantages of this approach are the ease of use and convenience of the operation, which are generally available in geoprocessing software, and the possibility of repeating the profile elaboration, keeping the same river tracing for distinct DEM data sources.

The second approach to obtain a longitudinal topographic profile of the river from the DEM processing is based on the flow directions and the raster river drainage network derived from this DEM, without involving another external data source. Once the upstream and downstream points of the river reach of interest are chosen, one can walk from the former to the latter, following the flow directions. Along this path, at each pixel found, there is the incremental distance of this step (Δx or $\sqrt{2} \cdot \Delta x$, if the pixels are orthogonal or diagonally neighbors, respectively, as exemplified in Fig. 6.6). The accumulated distance from the trace's beginning and the pixel's elevation reached at each step are also accounted for. The pairs of points formed by the accumulated distance from upstream and the elevation of the pixel constitute the longitudinal topographic profile (Fig. 7.8). From a strictly technical perspective, this procedure generates a more coherent profile between elevations and river layout since both information come from the DEM.

Another advantage of this second approach is that it allows tracing the profile, considering the DEM modified by the depression removal process, obtaining a more homogeneous profile in the sense of always positive or null slope sections

(Fig. 7.9). Suppose the first approach to generate a longitudinal profile is used, with a vector layout of the river independent of the DEM. In that case, it is not necessarily guaranteed that the sections always have positive or null slopes when using the depressionless DEM.

The main disadvantages of this second approach are that it involves more processing steps, and the operation is usually unavailable in geoprocessing software, albeit it is relatively easy to develop a computational routine for this aim. A hybrid alternative between the two methods is to obtain the river layout from the DEM processing, convert this layout to vector format, and use this vector for a procedure similar to the one described in the first approach. It combines the convenience of using an existing tool in most GIS software to profile a raster given a vector of interest with the consistency of considering river trace and elevations from the same topographic source.

7.9.2 Longitudinal Topographic Profile with Distances Measured from Downstream

A longitudinal topographic profile of a river can also be elaborated upon by considering distances calculated relative to the mouth or the chosen downstream point. In this case, distances are accounted for from downstream to upstream. A convenience of elaborating the profile in this way of accounting distances is to facilitate the joint arrangement, in the same graph and in a comprehensive way, of the topographic profiles of several or all the watercourses of the basin.

For example, in Fig. 7.10, there are the longitudinal profiles of a set of 31 watercourses of the Mamanguape river basin on the coast of Paraíba simultaneously. The numbers on the basin map and the profile graph identify the head of each watercourse. For any point on the graph, there is a distance from this point to the basin outlet, but according to specific paths for each river. For example, points P1 and P2 are 150 km from the basin outlet but follow different paths, referring to the rivers whose heads have numbers 4 and 8, respectively. At the same distance from the outlet, point P3 is located in another tributary in a more southern part of the basin (number 10) and in a lower-altitude area, whose path is even more distinct from the previous ones.

7.9.3 Influence of Lakes and Dams on DEM-Derived Longitudinal Profiles

In the longitudinal topographic profiles extracted from DEMs, the conformation referring to existing dam lakes may appear, depending on the time and characteristics of the data acquisition and the condition of the dam's water level.

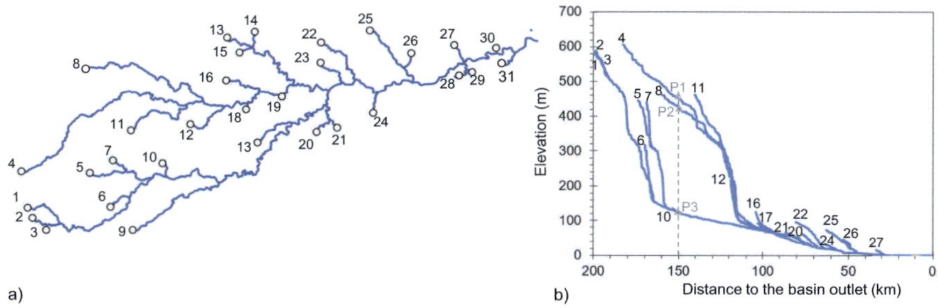

Fig. 7.10 **a** River drainage network of the Mamanguape river basin (PB), derived from the SRTM 30 DEM, with numbering that identifies each head point of a watercourse; **b** Longitudinal topographic profile of all the watercourses of the Mamanguape river basin, with the identification of the corresponding heads (some heads of smaller rivers are not identified in the profile for higher legibility of the figure; points P1, P2 and P3 are exemplified in the text)

In general, DEMs do not have bathymetric information, and due to the characteristics of data acquisition, elevation information refers to the water surface at the time of data collection. Dam lakes are water bodies of large dimensions relative to the size of the DEM pixel, so they usually appear in the topographic representation of the DEM as relatively flat areas.

In these cases, the longitudinal topographic profile elaborated from the DEM is expected to present a kind of plateau followed by a sharp drop, like a step, where the dam lake is located (Fig. 7.11). If the reservoir was dry at the DEM data's acquisition date, the topographic information represented refers to the bottom of the lake.

The spatial resolution of the DEM, the dimensions of the lake and the dam, the water level at the time of data acquisition, and the characteristics of the type of survey, data acquisition, and processing influence the shape and detail with which each lake appears in a longitudinal topographic profile derived from the DEM. The effects

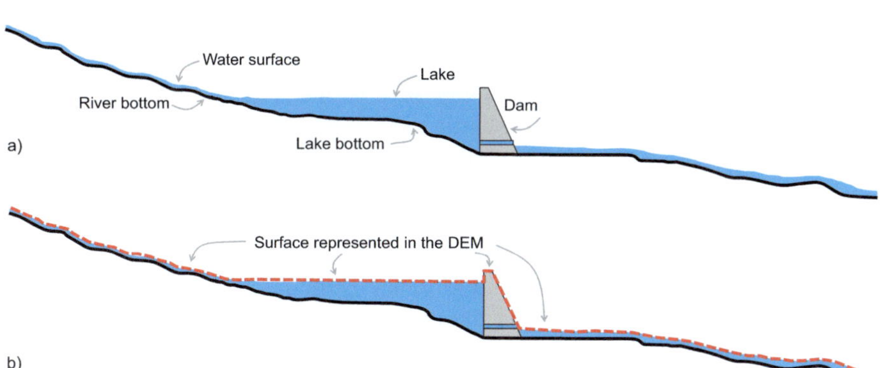

Fig. 7.11 **a** Lateral cross-section of a river reach containing a dam; **b** Topographic representation in a DEM whose data were acquired at the time of this water level in the dam

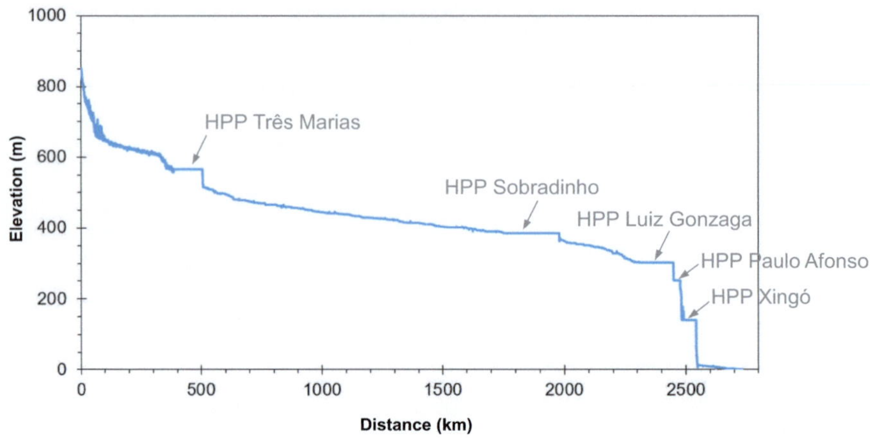

Fig. 7.12 Longitudinal topographic profile derived from the SRTM 30 DEM, highlighting the effect of five hydroelectric power plants (HPP)

of vegetation marginal to the watercourses can also be present in the topographic representation of the DEMs, as discussed in Chap. 5, which examines the cross-sections.

For example, five hydroelectric power plants' lakes are noticeable in the São Francisco River profile derived from SRTM 30 DEM (Fig. 7.12). As the acquisition of the SRTM data was carried out after the existence of these hydroelectric power plants, and as the type of survey does not capture information below the water level, the surface of these large lakes appears as flat areas in the profile. For instance, the lake of the Sobradinho hydroelectric power plant exceeds an extension of 200 km.

7.9.4 Dimensionless Longitudinal Topographic Profile

Another way to work a longitudinal topographic profile is to consider it dimensionless. On the *x*-axis, distances relative to the total length of the analyzed watercourse are adopted. On the *y*-axis, relative elevations are also adopted, obtained using the transformation indicated to elaborate dimensionless hypsometric curves (Chap. 5).

The primary purpose of presenting a profile in a dimensionless way is to be able to compare different rivers directly. For example, the longitudinal topographic profiles of the São Francisco (Fig. 7.13a) and Uruguay (Fig. 7.13b) rivers can be better compared when considering both in the dimensionless format (Fig. 7.13c). Each river has a specific extension, with the reach of the São Francisco River almost 1000 km longer than the reach of the Uruguay River. Dimensionless longitudinal profiles overlay the topographic variation from headwater to outlet of each reach relative to distance positioning, aligning the headwaters and outlets of both rivers at the exact location along the *x*-axis.

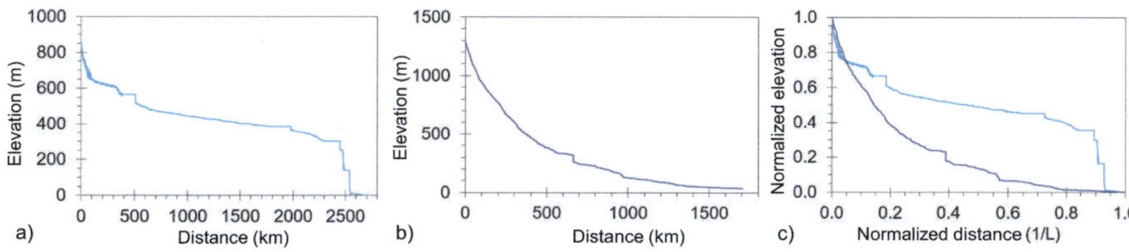

Fig. 7.13 **a**, **b** Longitudinal topographic profiles derived from the SRTM 90 DEM for the São Francisco and Uruguay rivers, respectively; **c** Dimensionless longitudinal topographic profiles of the São Francisco (light blue) and Uruguay (dark blue) rivers

7.10 Preprocessing a DEM to Incorporate an Existing River Network (Stream Burning)

7.10.1 Discrepancy Between an Existing Vector River Network and the DEM-Derived River Network

An assumed reliable vector river drainage network may be available for some regions or watersheds. However, when processing the DEM and obtaining a river drainage network according to the procedures described in the previous items, this network certainly does not fully coincide with the already available vector network (Fig. 7.14a). The uncertainties and errors involved in this processing are discussed in Chap. 14. Different data sources and ways of obtaining them lead to distinct networks, and the degree of agreement between the networks varies in each case.

Two questions usually emerge for the user. The first is conceptual and refers to the doubt of needing to process the DEM and obtain the river drainage network, given that a vector drainage network is already available. Suppose processing the DEM is solely to get a vector drainage network. This processing may be unnecessary because the user can work directly with the available vector. However, most situations need other data layers generated in this process, such as flow directions, accumulated drainage areas, and the drainage network itself in raster format. For example, this is a fundamental input to hydrological simulation models.

The second question refers to the justified perception of wasting valuable information. A previous effort was to generate and make a vector river drainage network available. In contrast, the user, who needs other data layers, must process the DEM, and the vector network that is already available does not contribute. If this occurs, there is a waste of not taking advantage of this valuable information that is the existing vector network. Another way is to take advantage of this vector network and increase the quality of DEM processing.

The available vector network can be incorporated to preprocess the DEM so that the river drainage network generated by processing this modified DEM is as close as possible to the available vector network.

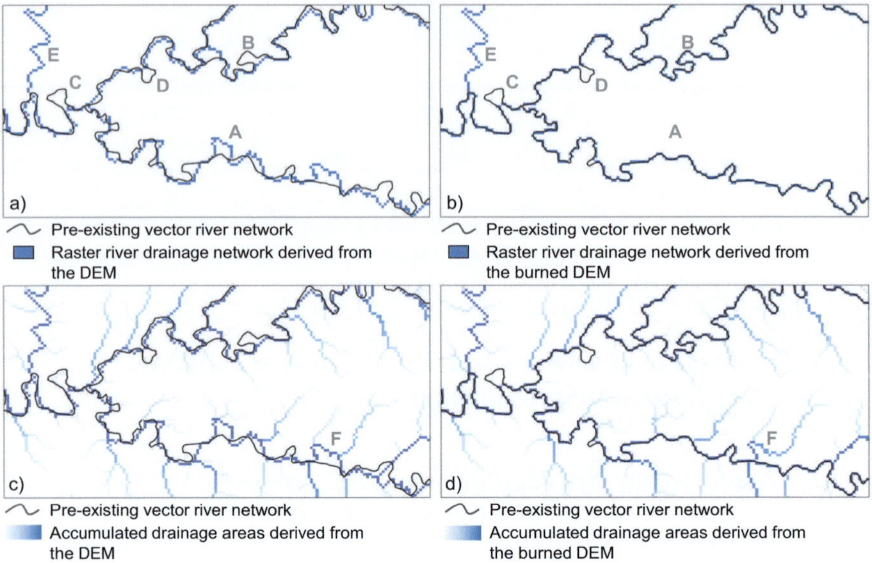

Fig. 7.14 **a** Example of a river drainage network derived from the DEM with divergences relative to a pre-existing vector network; **b** Higher convergence of the drainage network derived from the DEM preprocessed by stream burning (burned DEM) with the vector network; **c, d** Comparison of flow paths indicated by drainage accumulation areas derived from the original DEM and the burned DEM, respectively, with the vector stream network

7.10.2 The Concept of Stream Burning

The preprocessing of the DEM to incorporate an existing vector network is known as stream burning. This term reflects the central idea of the operation, which is to excavate the DEM in the pixels located along the available vector drainage network. In other words, these DEM pixels have their elevation lowered.

Stream burning alludes to a metal engraving technique based on chemical corrosion by an acid (Lindsay, 2016). In general terms, this engraving technique creates a protective film resistant to acid, which is pierced according to the design desired to be engraved on the metal. This protective film is applied to a metal plate for the acid treatment. The pierced areas of the film allow the acid to burn the metallic surface, excavating it in the shape of that film design.

According to Lindsay (2016), the exact origin of the stream burning technique is not definitively known, and evidence points to widespread dissemination from work developed at the Center for Research in Water Resources (CRWR) at the University of Texas in the second half of the 1990s. Some of the works produced by the group above and available for consultation are Graham et al. (1999), Saunders (1999), Maidment (1996), and Hellweger and Maidment (1997).

The modified DEM or burned DEM is used for the sequence of standard processing procedures described in this chapter and the previous chapter: depression removal, flow directions assignment, accumulated drainage areas determination, and river drainage network extraction. The difference is that the burned DEM induces the algorithms to identify the artificially lowered pixels from the burning process as a preferential flow path, since the reduction increases the slope from neighboring pixels toward the burned pixels. This also leads to surrounding flow paths converging toward this artificially lowered pathway.

As a result, the river drainage network derived from the modified DEM tends to agree more with the available vector network than the drainage network derived from the original DEM. For this reason, some authors refer to the burned DEM as a hydrologically consistent DEM. There is no consensus on the appropriateness of this term, which can be interpreted more broadly and generate some confusion. Other authors refer to this expression for the DEM processed for depression removal, as commented in Chap. 4.

For example, in Fig. 7.14b, the sections of the vector river network indicated by points A and B were well-represented in the network generated considering the burned DEM, but poorly represented in the network derived from the original DEM.

There may still be river segments where discrepancies remain between the vector stream network and the one derived from the burned DEM (segments C and D in the referenced figure). In contrast, segment E is nearly identical in both the network generated from the burned DEM and that derived from the original DEM, as there is no corresponding segment in the pre-existing vector network—in other words, the DEM was not carved along that reach. Additionally, when examining the flow paths beyond the river drainage network (Fig. 7.14c compared to Fig. 7.14d), further changes can be observed as a secondary effect of the DEM lowering along the primary drainage path, such as in segment F in the example.

Performing stream burning also contributes significantly to reducing the number of depressions in the DEM that result from artifacts of the data acquisition process and elaboration of the DEM (Burrough & McDonnell, 1998) since this type of artifact occurs frequently in valley bottoms and along the river drainage network.

7.10.3 Stream Burning Approaches

The most straightforward stream burning procedure constantly lowers the DEM by the same value for all pixels along an existing vector river drainage network's path. This procedure is technically equivalent to increasing the elevations of the entire land adjacent to the rivers (Lindsay, 2016). However, this approach requires more caution to avoid flow path discontinuities outside the area with raised elevations, for instance, ensuring that the increase is carried out for the entire basin. Lowering the elevations according to the vector river drainage network is more straightforward and involves modifying fewer pixels than increasing the elevations of the adjacent terrain (Lindsay, 2016).

It is recommended that the elevations be reduced by a value large enough, like 50 or 100 m, for example, so that the modified DEM results in the drainage network

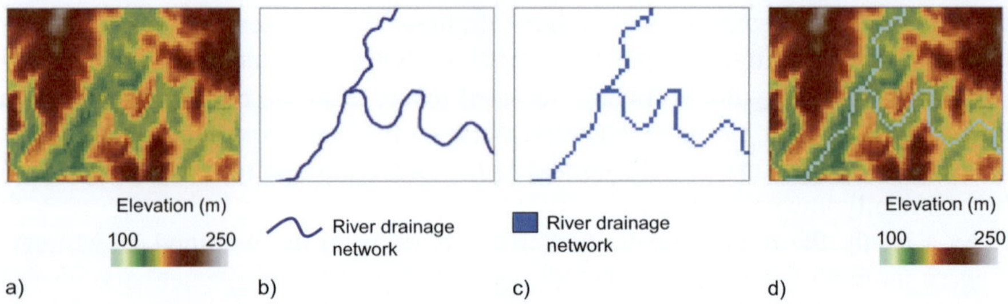

Fig. 7.15 **a** Digital elevation model; **b** Pre-existing vector river drainage network, **c** Pre-existing river drainage network converted to raster, and **d** DEM modified by lowering by 100 m the pixels along the pre-existing river drainage network (stream burning)

being obtained in the desired way. Small reductions like 1 m may not cause the desired effect. This is because there may already be neighboring pixels with this difference in elevations of around 1 m or more. DEMs with relatively coarse spatial resolution, like those from global surveys and freely available, may present elevations with this characteristic, for example.

It is essential to remember that the artificial lowering of the DEM is carried out to improve the subsequent result of the derivation of flow directions, accumulated drainage areas, and river drainage networks. The burned DEM should not be used as a topographic representation for other purposes since the elevation changes were made artificially and somewhat arbitrarily in terms of values.

For carrying out this type of stream burning procedure, it is enough to convert the vector of the pre-existing river drainage network to the raster format, where drainage pixels have attribute 1 and the remaining pixels have attribute zero, with the exact spatial resolution of the DEM (Fig. 7.15). Ideally, the raster representation of the river drainage network can be refined so that a one-pixel width represents each river reach flow path. If a river's reach in the raster is represented by a width of two or more pixels, there may be a tendency to generate parallel drainage flows when processing the burned DEM to obtain the flow paths.

Then, a mathematical operation is performed in which the modified DEM is generated by the difference between the original DEM and k times the drainage raster, where k is the value of the lowering (50 or 100 m, for example).

Figure 7.16 schematically exemplifies lowering a DEM according to the river drainage network. As the pixel highlighted by the blue rectangle is located on the drainage network, its elevation was reduced by 50 m compared to the original DEM since the operation was $114 - 50 \times 1 = 64$. For the pixel highlighted by the gray rectangle, which is not located on the drainage network, the operation was $111 - 50 \times 0 = 111$, maintaining the same elevation in the burned DEM compared to the original DEM.

Alternatively, a slope can be imposed during the lowering of DEM pixels to induce the direction of flow along the main pixel path during DEM processing. There are also

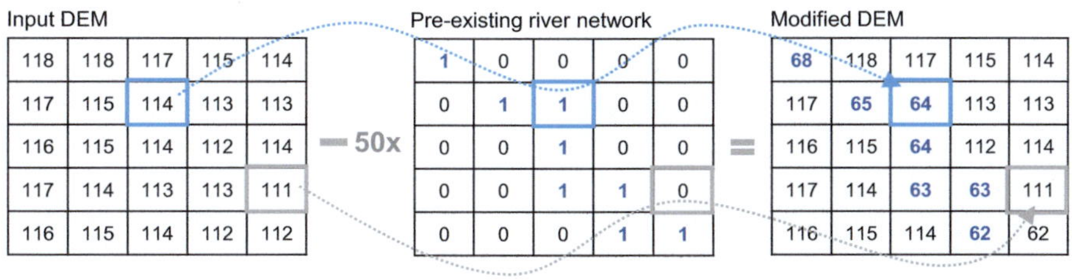

Fig. 7.16 Illustration of the operation of lowering the DEM by 50 m according to pixels of a raster river drainage network: pixels that correspond to the raster drainage network (attribute 1) have their elevation reduced by 50 m in the modified DEM compared to the original DEM, like the pixel indicated by the dotted blue line; pixels that do not correspond to the drainage raster network (attribute 0) remain at the same elevation as the original DEM (as example of the pixel indicated by the dotted gray line)

more sophisticated approaches to incorporate a pre-existing vector stream network into the DEM preprocessing stage.

Some stream burning methods proposed in the literature, for example, seek to account not only for river paths but also for lakes and wetlands, apply gradual lowering across extensive floodplains adjacent to rivers, consider sub-grid information (i.e., features smaller than the DEM pixel size), and incorporate topological attributes of the stream network into the procedure, among several other variations of the approach (Daniels et al., 2011; Getirana et al., 2009; Kenny & Matthews, 2005; Lindsay, 2016).

7.10.4 *Quality of the Vector River Network Used for Stream Burning*

Regardless of how stream burning is performed, it is essential to investigate the validity of the available vector drainage network for this procedure. This data layer can be provided by official institutions such as government agencies that manage and provide geographic databases. In this case, evaluating how the vector drainage network was developed is recommended, including the survey, whether editing and quality verification were performed, etc.

Another possibility to obtain a vector river drainage network for stream burning is manual digitization from satellite images, such as Landsat, or available from platforms like Google Earth. Care should be taken with the georeferencing of the images, attention should be paid to the line tracing during digitization, and the view's zoom level should be adjusted. Care should also be taken with the editing of the vectors to avoid discontinuity between sections and ensure the continuity of the vector segments.

In very wide rivers, like the reach of the Uruguay River indicated by A in Fig. 7.17, on the border between Brazil and Argentina, the river trace should follow along the

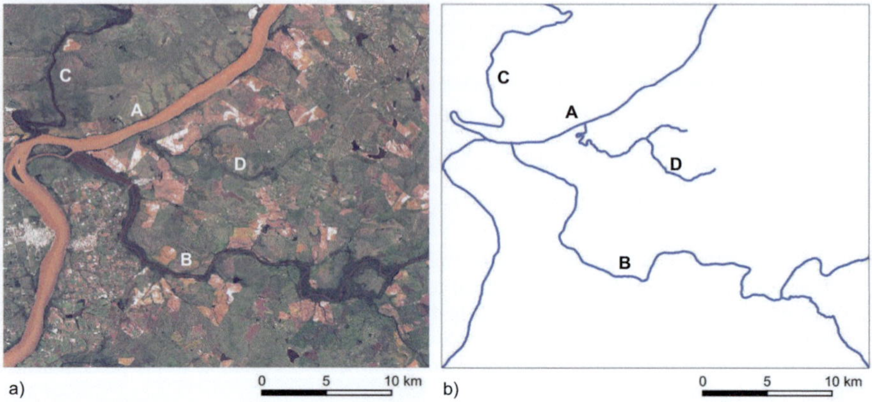

Fig. 7.17 **a** Color composition of Landsat image (path 225, row 81) of the region on the border between Brazil, Argentina, and Uruguay, highlighting a reach of the Uruguay River (A) and three tributaries (B, C and D); **b** Sections of rivers of the Uruguay river basin manually digitized from the Landsat image

center of the cross-section, and a zigzag effect within the area between the river banks should be avoided.

In relatively narrower rivers than the previous example, like the reach of the Quaraí River on the border between Brazil and Uruguay (section B in Fig. 7.17), the likelihood of the operator introducing a zigzag pattern during digitization decreases. In contrast, the effort required to accurately follow all the river's meanders increases.

For river segments with even narrower widths, such as the segment labeled C in the referenced figure, the limited visual ability to discern the river's course virtually eliminates the possibility of a zigzag pattern within the main channel. However, the risk of failing to trace all the river's meanders accurately increases significantly in such cases.

On the other hand, the river identified as D represents a situation in which the river's course is not visible due to its small width relative to the image's spatial resolution and the presence of riparian vegetation. In this case, one must weigh whether the greater uncertainty associated with digitizing the river is acceptable for the stream burning process. This decision should be guided by the intended purpose of the DEM processing and the end use of the derived products.

7.11 Scale Change in the River Drainage Network

In some situations, the user needs a river drainage network with a coarser spatial resolution than the spatial resolution of the available DEM. An example is making the river drainage network compatible with other data layers. Another example is the need for input data to hydrological models, for which there is no interest in working with a more refined spatial resolution. A third situation is circumventing the limitations of processing and storing large amounts of data.

There are two approaches to generating a river drainage network at a coarser spatial resolution than the original DEM. The first is the most intuitive, straightforward, and widely used. It consists of coarsening the DEM's resolution and processing this coarsened DEM to obtain the coarse river drainage network.

The other approach, flow direction upscaling, is more complex and requires specific algorithms. It involves obtaining the flow paths at the original DEM resolution and, from this information, deriving the flow directions and the river drainage network at coarser resolutions. Both approaches are detailed in the following items.

7.11.1 DEM Coarsening Approach

In this approach, the river drainage network at a spatial resolution coarser than the DEM is obtained mainly by coarsening the spatial resolution of the DEM itself (Fig. 7.18). Procedures such as averaging, interpolation methods, or nearest neighbor, for example, are used, and a DEM is obtained with elevations determined based on the elevations of the original DEM (as described in Chap. 4). From this coarsened DEM, the usual methods of DEM processing are applied to obtain the flow directions, the accumulated drainage areas, and the river drainage network, as described in this chapter and the previous one.

All the procedures involved are commonly available in geoprocessing and GIS software. Notably, the direct operation related to scale change is DEM coarsening, i.e., the elevations are transferred from one spatial resolution to another. This procedure is most recommended for relatively small spatial resolution intervals due to its convenience, ease of execution, and consistency of results.

When the level of spatial resolution coarsening is relatively large, with a considerable jump between the finest and coarsest scales, the loss of information that occurs with the DEM coarsening is very intense. Processing this coarsened DEM can result in a low-quality river drainage network that poorly represents the spatial patterns of flow paths observed in the fine-resolution river drainage network. The resulting drainage network tends to show apparent inconsistencies. It may require some time for manual editing and correction or lead to inconsistent analyses and errors in studies that use such information.

An outstanding example of inconsistency resulting from the aggregation of topographic information and subsequent derivation of the river drainage network is illustrated for the São Francisco river basin (drainage area of about 640,000 km^2; Fig. 7.19), based on results obtained by Saraiva and Paz (2014). The SRTM DEM with 90 m spatial resolution was aggregated to a spatial resolution of 50 km. The usual methods were applied to determine the river drainage network and the watershed delineation (a concept worked on in the next chapter) at this coarser resolution. The result is a significant error in the lower part of the basin, with the river draining to the north instead of curving to the east.

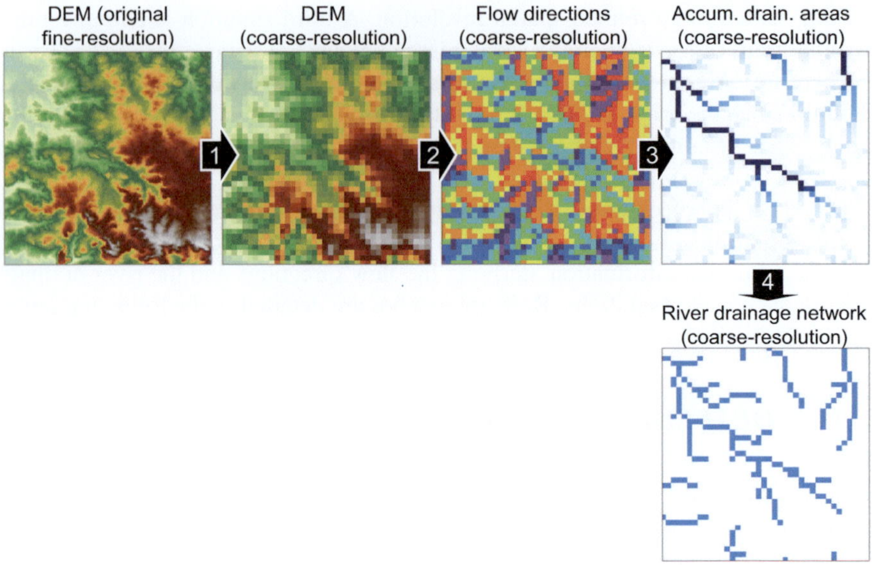

Fig. 7.18 Approach to obtaining a river drainage network at a spatial resolution coarser than the available DEM: coarsening of the DEM (step 1) and, sequentially, obtaining the flow directions, accumulated drainage areas, and river drainage network at the coarse resolution (steps 2, 3, and 4)

Throughout the basin in this example, the comparison between the layout of the drainage network at the 50 km resolution and the layout derived directly from the original SRTM resolution shows other inconsistent sections that should be corrected for better compatibility between these river drainage networks. The drainage network at the 50 km resolution (blue lines) should reproduce the patterns of the fine-resolution drainage network (white lines) as much as possible. Despite the inherent difficulty imposed by such a coarse grid, the network derived by the flow direction upscaling procedure (using the algorithm called COTAT+, described in the following items) achieves a much better result than the network obtained from the DEM aggregation. The section of the upper part of the basin highlighted in the enlargement of the maps demonstrates this difference in quality between the layouts of the 50 km spatial resolution networks.

7.11.2 Flow Direction Upscaling Approach

The approach of performing flow direction upscaling was initially conceived to reduce the number of pixels to be worked on in studies such as hydrological modeling. The available DEM presents a spatial resolution that configures a level of detail that is unwanted or even prohibitive for the development of subsequent studies, while maintaining this same grid.

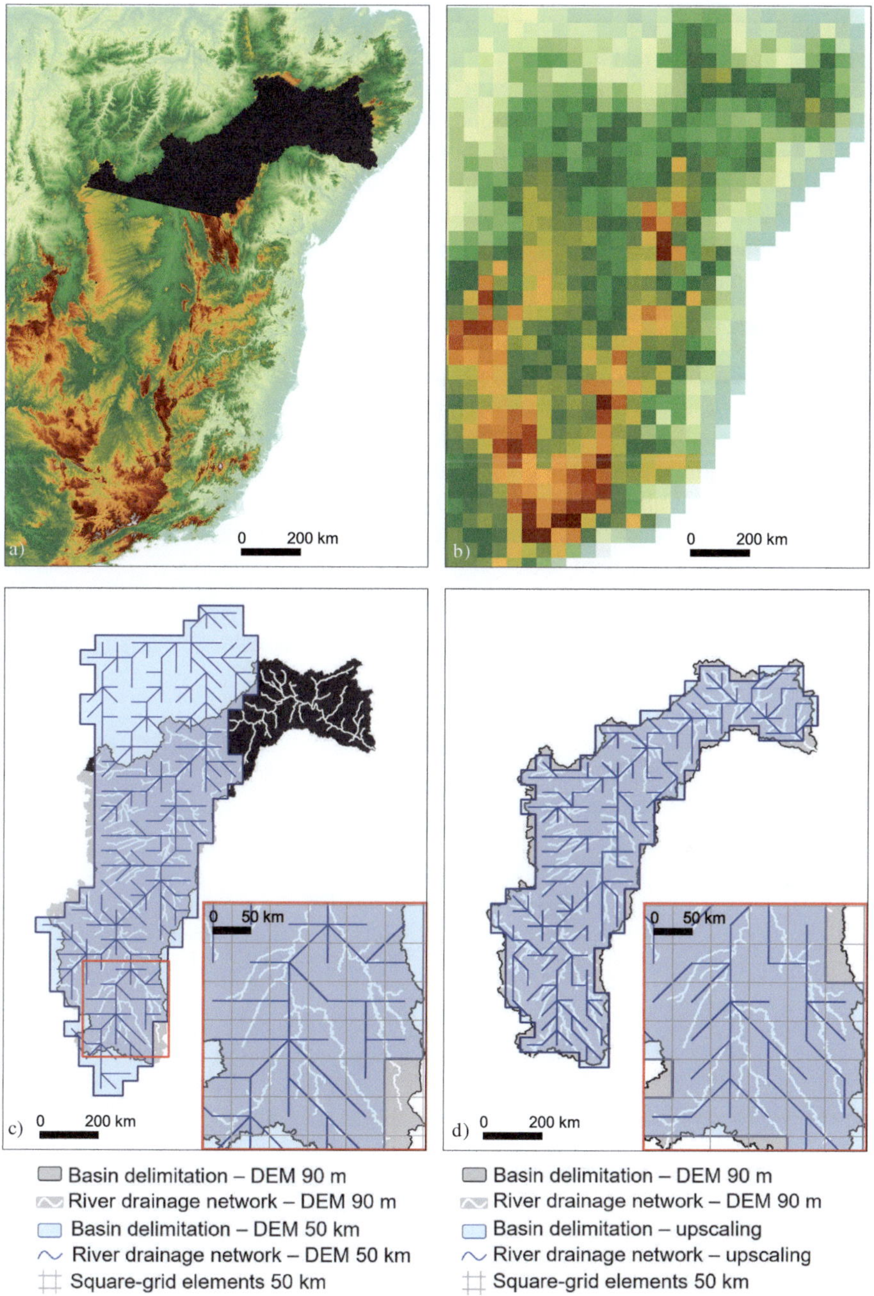

Fig. 7.19 **a** SRTM 90 DEM for the region that includes the São Francisco river basin; **b** 50 km spatial resolution DEM obtained by aggregating the SRTM 90 data; **c** Comparison between basin delineation and river drainage network obtained from the aggregated 50 km DEM relative to results obtained at the 90 m spatial resolution; **d** Comparison between basin delineation and river drainage network obtained by the flow direction upscaling algorithm COTAT+ and results obtained at the 90 m spatial resolution

In some cases, upscaling was applied to obtain flow directions and river drainage networks at a spatial resolution tens or hundreds of times coarser than the original resolution of the available DEM. This need occurs, for example, to obtain input information for large-scale hydrological models discretized in a regular grid of elements of 5, 10, or more kilometers in size. Global DEMs with an approximate spatial resolution of 30 or 90 m provide information assumed to be highly refined in this case. Coarsening this type of DEM to the resolution of the hydrological model represents a considerable scale jump. For example, when going from 90 m to 10 km, there is a jump between spatial resolutions of more than 100 times.

To perform the upscaling, the DEM in the original spatial resolution is first processed to obtain the flow directions and the accumulated drainage areas, adopting the usual methods to perform such operations, as seen in the previous chapter and the initial part of this chapter. These two products represent and characterize the flow paths at the spatial resolution of the DEM. Having this information as input data, an upscaling algorithm decides the flow directions of pixels at a coarsened spatial resolution (Fig. 7.20). Following the standard procedures, these coarse-resolution flow directions are input data for obtaining the coarse-resolution accumulated drainage areas and river drainage networks.

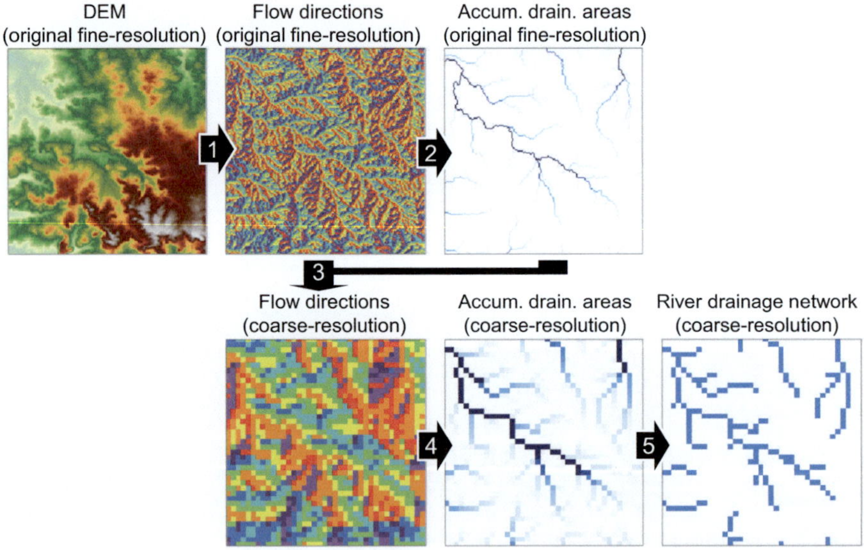

Fig. 7.20 Approach to obtaining a river drainage network at a spatial resolution more coarsened than the available DEM: obtaining flow directions (step 1) and accumulated drainage areas (step 2) at the exact DEM spatial resolution; upscaling of flow directions (step 3); obtaining the accumulated areas and the river drainage network at the coarsened resolution (steps 4 and 5)

7.11.3 Methods for Flow Direction Upscaling

Several upscaling methods have been proposed in the scientific literature, and works that describe some of these methods are listed in Table 7.3. The table also presents the original and coarse spatial resolutions and the corresponding scale jumps performed in each work.

Davies and Bell's (2009) work presents a comparative evaluation of results obtained with the methods of Fekete et al. (2001), Olivera and Raina (2003), and Paz et al. (2006). The comparison with results obtained by the most straightforward approach, aggregating the topographic information of the DEM and deriving the river drainage network, is also evaluated.

7.11.4 The COTAT+ Algorithm

One of the upscaling algorithms considered to be of excellent performance, according to Davies and Bell (2009), is the COTAT+ (Cell Outlet Tracing with an Area Threshold Plus) algorithm. This algorithm is described in Paz et al. (2006) and detailed in this item. It is an improvement of the COTAT algorithm (*Cell Outlet Tracing with an Area Threshold*) proposed by Reed (2003). The method proposed by Yamazaki et al. (2009) follows the COTAT+ very similarly.

The COTAT+ method has been successfully used to obtain river drainage networks at coarse spatial resolutions in watersheds with different characteristics in Brazil (Pereira et al., 2017; Saraiva & Paz, 2014) and in other parts of the world, such as in the United Kingdom (Davies & Bell, 2009; Davies et al., 2023), West Africa (Dadson et al., 2010), Italy (Rosim et al., 2015) and even for the whole globe (Riddick et al., 2018). The independent study of Davies and Bell (2009) attested to the quality of the COTAT+ algorithm compared to other procedures.

More recently, this algorithm has also proved helpful in scaling changes involving river drainage networks derived from LiDAR data. River drainage networks were obtained at resolutions of 30, 100, and 200 m from the flow paths information derived from the original resolution of 1 m of the LiDAR DTM (Almeida et al., 2024).

To better understand how the COTAT+ algorithm works, consider the coarser spatial resolution grid elements called cells. In contrast, the elements of the original (finer) spatial resolution grid are called pixels. Each cell has a set of pixels in its geographical coverage. For example, in Fig. 7.21, the upscaling is between spatial resolutions that represent a coarsening of 10 times. Each cell of the coarsened spatial resolution comprises 10×10 pixels of the original spatial resolution.

The COTAT+ algorithm consists of three main steps. The first step is the selection of the outlet pixel for each cell. The outlet pixel is selected as that pixel inserted in the area of the cell, which is located on the edge of the cell and represents the exit of the main flow path of pixels passing through the cell via the application of two

Table 7.3 List of scientific works that proposed upscaling flow direction methods, showing the spatial resolutions and scale changes considered in their study cases

References	Fine and coarse resolutions of the flow direction upscaling	Scale change
Kao (1992)	40–200 m	5×
Arora and Boer (1999)	100–420 km	4.2×
	100–310 km	3.1×
O'Donnell et al. (1999)	1–28 km	28×
Wang et al. (2000)	30–120 m	4×
	30–150 m	5×
Fekete et al. (2001)	1–10 km	10×
	1–50 km	50×
Döll and Lehner (2002)	1–20 km	20×
	10–60 km	6×
	20–60 km	3×
Olivera et al. (2002)	1–310 km	310×
Reed (2003)	30 m–1 km	33×
	30 m–4 km	133×
	400 m–4 km	10×
Olivera and Raina (2003)	1–100 km	100×
	1–310 km	310×
Shaw et al. (2005a)	30 m–3 km	100×
	200 m–90 km	450×
	1–230 km	230×
Shaw et al. (2005b)	1–230 km	230×
Paz et al. (2006)	200 m–10 km	50×
	250 m–10 km	40×
Arora and Harrison (2007)	50–156 km	3.1×
	50–210 km	4.2×
Yamazaki et al. (2009)	1–50 km	50×
	1–100 km	100×
Wu et al. (2011)	1 km to 7.5 up to 240 km	7.5× up to 240×
Saraiva and Paz (2014)	100 m–10 km	100×
	100 m–30 km	300×
	200 m–50 km	250×
Eilander et al. (2021)	90 m–1 km	11×
	90 m–10 km	111×
	90 m–25 km	278×
Almeida et al. (2024)	1–30 m	30×

(continued)

Table 7.3 (continued)

References	Fine and coarse resolutions of the flow direction upscaling	Scale change
	1–100 m	100×
	1–200 m	200×

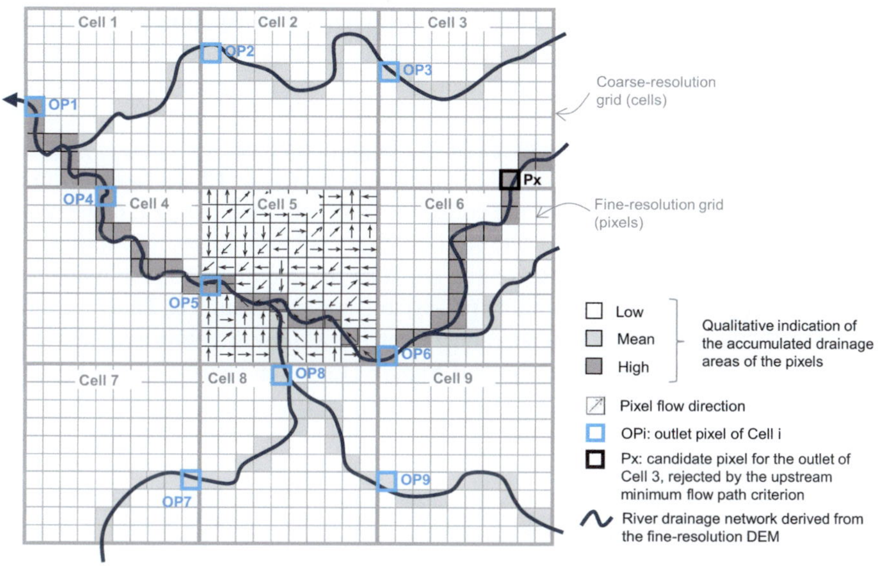

Fig. 7.21 Definition of the outlet pixel of each cell, as a step of the COTAT+ algorithm for upscaling flow directions, used to derive flow directions on a coarser grid than the spatial resolution of the available DEM

criteria: (i) highest accumulated drainage area; (ii) extension of the flow path within the cell higher than a minimum threshold.

The first criterion establishes that the first candidate for the outlet pixel is the one with the highest accumulated drainage area among those contained in the area of the cell. The second criterion establishes that this candidate from the first criterion must be evaluated for the extent of the upstream flow path from it within the cell. If such a path is higher than a pre-established value, a parameter of the algorithm called Minimum Upstream Flow Path (MUFP), this candidate is accepted as the outlet pixel.

If the candidate fails the MUFP criterion, the next candidate for the outlet pixel is the one with the second-largest accumulated drainage area within the cell. The second criterion must be checked, and if it is met, the definition of the outlet pixel is finalized. If it fails, proceed to the next, sequentially in descending order of larger accumulated drainage area, until you find one that meets the MUFP criterion. The

MUFP value recommended by Paz et al. (2006) is 1/5 of the dimension of the low spatial resolution cell.

The outlet pixels defined for the cells in Fig. 7.21 illustrate how the two criteria work. For all cells except for cell 3, the defined outlet pixel is precisely the largest accumulated drainage area. For cell 3, the pixel with the largest accumulated drainage area (pixel Px) was not accepted by the MUFP criterion. The primary flow path upstream from Px and inside the cell is too short, less than the MUFP. The idea of this criterion is to prevent main flow paths (with larger contribution area) that pass only marginally through the interior of the cell from being considered representative for defining the flow direction of the cell. It makes more sense that the flow path to which pixel PE3 belongs represents the primary path for defining cell 3 of the example (and not the flow path that passes through pixel Px).

After defining the outlet pixel of all cells, the next step of the COTAT+ algorithm is determining the flow direction of the cells. For each cell, the flow path downstream of the cell's outlet pixel is followed until another cell's outlet pixel is found. The difference in accumulated drainage area between these outlet pixels is quantified. Suppose this difference is higher than a pre-established value, an Area Threshold (AT) parameter. In that case, the trace ends, and the cell has its flow direction assigned in the direction of the cell where the trace ended. Suppose the calculated difference between accumulated areas is less than AT. In that case, the trace continues until another outlet pixel is found, again quantifying the incremental difference in accumulated area and checking if it is higher than AT.

Suppose the trace leaves the 3×3 window of the cell's immediate neighborhood without finding an outlet pixel that meets the AT criterion. In that case, the trace is interrupted, and the cell's direction is assigned to the direction of the last neighboring cell visited on the trace.

In the example of Fig. 7.21, to define the flow direction of cell 5, the traced flow path follows downstream from PE5 until it finds PE4. If the difference between the accumulated areas of PE4 and PE5 is higher than the pre-established AT value, cell 5 has an orthogonal direction to the left, draining to cell 4 (Fig. 7.22a).

If the difference is less than AT, tracing the path downstream from PE4 continues until PE1 is found. If the difference between the accumulated areas of PE1 and PE5 is higher than AT, the flow direction of cell 5 is upper right diagonal, draining to cell 1 (Fig. 7.22b). If this difference also does not meet the minimum established value, the flow path that would be continued would leave the immediate neighborhood of cell 5, and, as the last visited cell was cell 1, the flow direction of cell 5 would also be in the direction of this cell.

The AT parameter influences the occurrence of orthogonal and diagonal flow directions. The higher the AT value, the higher the requirement for incrementing the accumulated area between outlet pixels. This induces longer paths to travel until this criterion is met or the cell neighborhood is left, increasing the chance of diagonal flow directions.

The lower the AT value, the easier it is to meet this criterion, tending to have a higher quantity of orthogonal flow directions. The AT value Paz et al. (2006) recommended equals the low spatial resolution grid cell area.

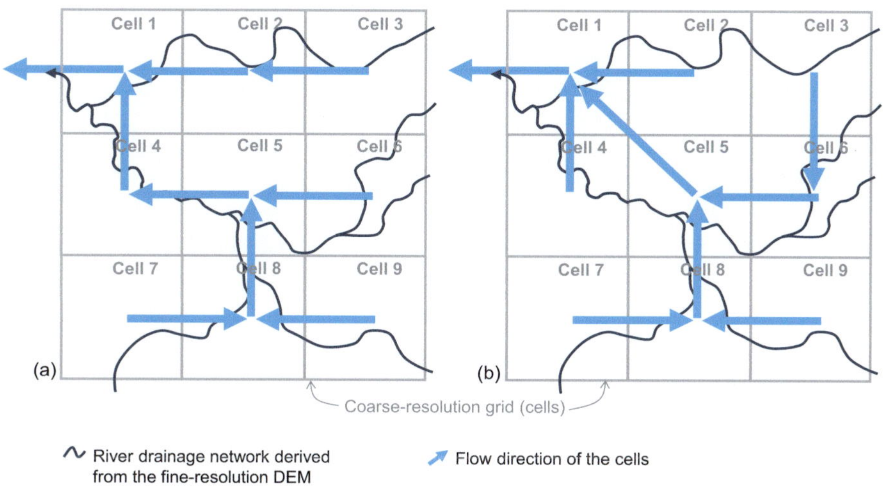

Fig. 7.22 Assignment of flow directions as part of the COTAT+ algorithm for upscaling flow directions

In the example of Fig. 7.21, if the outlet pixel of cell 3 were pixel Px, the direction of this cell would be in the direction of cell 6 or cell 5 (Fig. 7.22b), depending on the result of the verification of increments of accumulated drainage area. However, if PE3 is the outlet pixel of cell 3, the flow direction of this cell is in the direction of cell 2 (Fig. 7.22a), forming a flow path more coherent with the high-resolution drainage network. This illustrates the impact of the MUFP parameter on the definition of the choice of the outlet pixel on the assignment of flow directions.

7.11.5 Upscaling with Gradual Change of Spatial Resolution

When the scale jump is enormous, an alternative that tends to improve the results of upscaling flow directions is the application of gradual spatial resolution changes, as Saraiva and Paz (2014) proposed, referred to as a multi-step upscaling approach.

In the standard procedure, the transition is made directly from the initial refined resolution to the desired coarse resolution. The upscaling with gradual changes proposes to carry out sequential changes to intermediate spatial resolutions until reaching the coarsest one. This coarsest resolution is the final resolution of the process (Fig. 7.23).

For example, instead of applying the upscaling to jump from the resolution of 100 m to the resolution of 50 km (a jump of 500×), as is the standard procedure, the multi-step approach can make the jump from 100 m to 1 km, then jump from 1 to 10 km and then make the final jump from 10 to 50 km. The result of each change in spatial resolution serves as input for the subsequent change until the desired final resolution

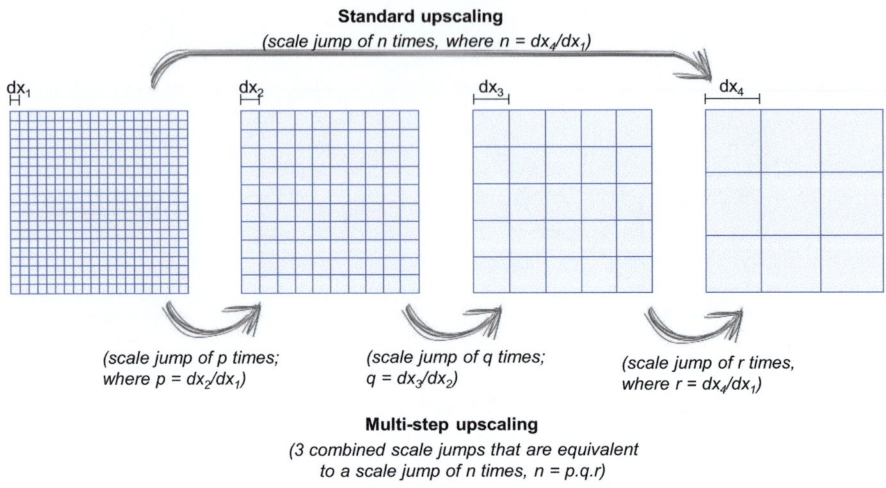

Fig. 7.23 Scheme of standard upscaling from an original spatial resolution dx_1 to obtain results at the coarse spatial resolution dx_4, and of gradual or multi-step upscaling, with intermediate scale jumps from dx_1 to dx_2, then from dx_2 to dx_3, and, finally, from dx_3 to dx_4

is reached. In this example, three shorter jumps (10×, 10×, and 5×, respectively) substitute for the single giant jump of 500×.

The main idea is that a sequence of smaller jumps favors preserving the spatial patterns of flow paths between spatial resolutions, unlike a single jump that represents a very intense change in spatial resolution. In this case, the refined resolution provides a pixel and flow path density level that increases the complexity of interpreting and determining the main flow paths to be represented at the coarse resolution. Results obtained by Davies and Bell (2009) in applying three distinct flow direction upscaling algorithms using the traditional single-step approach clearly illustrate how the upscaling performance worsens with the jump increase.

The user defines the number of partial jumps and the intermediate spatial resolutions in the multi-step upscaling approach. Two primary considerations are recommended for this definition. The first is to have a certain balance between the dimensions of the jumps. The second is that each jump represents a scale change of more than three times. Jumps less than this can lead some upscaling algorithms to not work properly, given that there would be a significantly reduced amount of pixels from the initial spatial resolution within the scope of the pixel of the final spatial resolution.

References

Almeida, A. A. R. D., Mendonça, R. L., Silva, N. M. M., & Paz, A. R. D. (2024). Evaluation of LiDAR-derived river networks coarsening with spatial patterns preservation. *Computers and Geosciences, 189*, Article 105639. https://doi.org/10.1016/j.cageo.2024.105639

Arge, L., Chase, J. S., Halpin, P., Toma, L., Vitter, J. S., Urban, D., & Wickremesinghe, R. (2003). Efficient flow computation on massive grid terrain datasets. *GeoInformatica, 7*, 283–313.

Arora, V. K., & Boer, G. J. (1999). A variable velocity flow routing algorithm for GCMs. *Journal of Geophysical Research, 104*, 30965–30979. https://doi.org/10.1029/1999JD900905

Arora, V. K., & Harrison, S. (2007). Upscaling river networks for use in climate models. *Geophysical Research Letters, 34*, L21407. https://doi.org/10.1029/2007GL031865

Barbedo, R., Siqueira, V., & Collischonn, W. (2022). Topographic Position-based Stream definition (TPS): A simple method to address spatial variability of drainage density in stream networks. *Hydrological Sciences Journal, 67*, 858–869. https://doi.org/10.1080/02626667.2022.2047190

Barnes, R. (2017). Parallel non-divergent flow accumulation for trillion cell digital elevation models on desktops or clusters. *Environmental Modelling and Software, 92*, 202–212. https://doi.org/10.1016/j.envsoft.2017.02.022

Burrough, P. A., & McDonnell, R. A. (1998). *Principles of geographical information systems.* Oxford University Press.

Clubb, F. J., Mudd, S. M., Milodowski, D. T., Hurst, M. D., & Slater, L. J. (2014). Objective extraction of channel heads from high-resolution topographic data. *Water Resources Research, 50*, 4283–4304. https://doi.org/10.1002/2013WR015167

Dadson, S. J., Ashpole, I., Harris, P., Davies, H. N., Clark, D. B., Blyth, E., & Taylor, C. M. (2010). Wetland inundation dynamics in a model of land surface climate: Evaluation in the Niger inland delta region. *Journal of Geophysical Research, 115*, 2010JD014474. https://doi.org/10.1029/2010JD014474

Daniels, M. H., Maxwell, R. M., & Chow, F. K. (2011). Algorithm for flow direction enforcement using subgrid-scale stream location data. *Journal of Hydrologic Engineering, 16*, 677–683. https://doi.org/10.1061/(ASCE)HE.1943-5584.0000340

Dantas, A. A. R., & Paz, A. R. (2021). Use of HAND terrain descriptor for estimating flood-prone areas in river basins. *Revista Brasileira De Ciências Ambientais, 56*, 501–516. https://doi.org/10.5327/Z21769478892

Davies, H. N., & Bell, V. A. (2009). Assessment of methods for extracting low-resolution river networks from high-resolution digital data. *Hydrological Sciences Journal, 54*, 17–28. https://doi.org/10.1623/hysj.54.1.17

Davies, H. N., Rameshwaran, P., Bell, V. A., & Dadson, S. (2023). Spatially consistent physical characteristics of UK rivers: 1-km data. *Geoscience Data Journal.* https://doi.org/10.1002/gdj3.209

De Jong, K., Panja, D., Karssenberg, D., & Van Kreveld, M. (2022). Scalability and composability of flow accumulation algorithms based on asynchronous many-tasks. *Computers and Geosciences, 162*, Article 105083. https://doi.org/10.1016/j.cageo.2022.105083

Do, H.-T., Limet, S., & Melin, E. (2011). Parallel computing flow accumulation in large digital elevation models. *Procedia Computer Science, 4*, 2277–2286. https://doi.org/10.1016/j.procs.2011.04.248

Döll, P., & Lehner, B. (2002). Validation of a new global 30-min drainage direction map. *Journal of Hydrology, 258*, 214–231. https://doi.org/10.1016/S0022-1694(01)00565-0

Eilander, D., Van Verseveld, W., Yamazaki, D., Weerts, A., Winsemius, H. C., & Ward, P. J. (2021). A hydrography upscaling method for scale-invariant parametrization of distributed hydrological models. *Hydrology and Earth Systems Science, 25*, 5287–5313. https://doi.org/10.5194/hess-25-5287-2021

Fan, F., Collischonn, W., Sorribas, M., & Pontes, P. (2013). Sobre o Início da Rede de Drenagem Definida a Partir dos Modelos Digitais de Elevação. *Revista Brasileira De Recursos Hídricos, 18*, 241–257. https://doi.org/10.21168/rbrh.v18n3.p241-257

Fekete, B. M., Vörösmarty, C. J., & Lammers, R. B. (2001). Scaling gridded river networks for macroscale hydrology: Development, analysis, and control of error. *Water Resources Research, 37*, 1955–1967. https://doi.org/10.1029/2001WR900024

Getirana, A. C. V., Bonnet, M.-P., Rotunno Filho, O. C., & Mansur, W. J. (2009). Improving hydrological information acquisition from DEM processing in floodplains. *Hydrological Processes, 23*, 502–514. https://doi.org/10.1002/hyp.7167

Graham, S. T., Famiglietti, J. S., & Maidment, D. R. (1999). Five-minute, 1/2°, and 1° data sets of continental watersheds and river networks for use in regional and global hydrologic and climate system modeling studies. *Water Resources Research, 35*, 583–587. https://doi.org/10.1029/1998WR900068

Hellweger, F., & Maidment, D. R. (1997). *AGREE—DEM surface reconditioning system.* University of Texas.

Kao, J.-J. (1992). Determining drainage pattern using DEM data for nonpoint-source water quality modeling. *Water Science and Technology, 26*, 1431–1438. https://doi.org/10.2166/wst.1992.0586

Kenny, F., & Matthews, B. (2005). A methodology for aligning raster flow direction data with photogrammetrically mapped hydrology. *Computers and Geosciences, 31*, 768–779. https://doi.org/10.1016/j.cageo.2005.01.019

Kotyra, B., Chabudziński, Ł, & Stpiczyński, P. (2021). High-performance parallel implementations of flow accumulation algorithms for multicore architectures. *Computers & Geosciences, 151*, Article 104741. https://doi.org/10.1016/j.cageo.2021.104741

Lindsay, J. B. (2016). The practice of DEM stream burning revisited: The practice of DEM stream burning revisited. *Earth Surface Processes and Landforms, 41*, 658–668. https://doi.org/10.1002/esp.3888

Maidment, D. R. (1996). GIS and hydrologic modeling—An assessment of progress. Presented at the the Third International Conference on GIS and Environmental Modeling, Santa Fe (EUA).

Montgomery, D. R., & Dietrich, W. E. (1988). Where do channels begin? *Nature, 336*, 232–234.

Montgomery, D. R., & Dietrich, W. E. (1992). Channel initiation and the problem of landscape scale. *Science, 255*, 826–830. https://doi.org/10.1126/science.255.5046.826

Montgomery, D. R., & Foufoula-Georgiou, E. (1993). Channel network source representation using digital elevation models. *Water Resources Research, 29*, 3925–3934. https://doi.org/10.1029/93WR02463

O'Donnell, G., Nijssen, B., & Lettenmaier, D. P. (1999). A simple algorithm for generating streamflow networks for grid-based, macroscale hydrological models. *Hydrological Processes, 13*, 1269–1275.

Olivera, F., Lear, M. S., Famiglietti, J. S., Asante, K. (2002). Extracting low-resolution river networks from high-resolution digital elevation models: extracting low-resolution river networks from DEMs. *Water Resources Research, 38*, 13-1–13-8. https://doi.org/10.1029/2001WR000726

Olivera, F., & Raina, R. (2003). Development of large scale gridded river networks from vector stream data. *Journal of the American Water Resources Association, 39*, 1235–1248. https://doi.org/10.1111/j.1752-1688.2003.tb03705.x

Passalacqua, P., Tarolli, P., & Foufoula-Georgiou, E. (2010). Testing space-scale methodologies for automatic geomorphic feature extraction from lidar in a complex mountainous landscape: testing feature extraction methodologies. *Water Resources Research.* https://doi.org/10.1029/2009WR008812

Paz, A. R., Collischonn, W., & Lopes Da Silveira, A. L. (2006). Improvements in large-scale drainage networks derived from digital elevation models: technical note. *Water Resources Research.* https://doi.org/10.1029/2005WR004544

Pelletier, J. D. (2013). A robust, two-parameter method for the extraction of drainage networks from high-resolution digital elevation models (DEMs): Evaluation using synthetic and real-world DEMs. *Water Resources Research, 49*, 75–89. https://doi.org/10.1029/2012WR012452

Pereira, F., Farinosi, F., Arias, M., Lee, E., Briscoe, J., & Moorcroft, P. (2017). Technical note: A hydrological routing scheme for the Ecosystem Demography model (ED2CR) tested in the Tapajós River basin in the Brazilian Amazon. *Hydrology and Earth System Sciences, 21*, 4628–4648. https://doi.org/10.5194/hess-21-4629-2017

Qin, C.-Z., & Zhan, L. (2012). Parallelizing flow-accumulation calculations on graphics processing units—From iterative DEM preprocessing algorithm to recursive multiple-flow-direction algorithm. *Computers & Geosciences, 43*, 7–16. https://doi.org/10.1016/j.cageo.2012.02.022

Reed, S. M. (2003). Deriving flow directions for coarse-resolution (1–4 km) gridded hydrologic modeling: Deriving flow directions for coarse resolution grids. Water Resources Research 39. https://doi.org/10.1029/2003WR001989

Riddick, T., Brovkin, V., Hagemann, S., & Mikolajewicz, U. (2018). Dynamic hydrological discharge modelling for coupled climate model simulations of the last glacial cycle: The MPI-DynamicHD model version 3.0. *Geoscientific Model Development, 11*, 4291–4316. https://doi.org/10.5194/gmd-11-4291-2018

Rosim, S., Ortiz, J., Oliveira, J., Jardim, A., & Abreu, E. (2015). Drainage network definition for low resolution DEM obtained from drainage network extracted from high resolution DEM using upscaling processes. In G. Lollino, M. Arattano, et al. (Eds.), *Engineering geology for society and territory—Volume 3: River basins, reservoir sedimentation and water resources*. Springer International Publishing. https://doi.org/10.1007/978-3-319-09054-2

Saraiva, A. G. S., & Paz, A. R. (2014). Multi-step change of scale approach for deriving coarse-resolution flow directions. *Computers Geosciences, 68*, 53–63. https://doi.org/10.1016/j.cageo.2014.04.002

Saunders, W. (1999). Preparation of DEMs for use in environmental modeling analysis. Presented at the Esri User Conference, San Diego (EUA).

Shaw, D., Martz, L. W., & Pietroniro, A. (2005a). Flow routing in large-scale models using vector addition. *Journal of Hydrology, 307*, 38–47. https://doi.org/10.1016/j.jhydrol.2004.09.019

Shaw, D. A., Martz, L. W., & Pietroniro, A. (2005b). A methodology for preserving channel flow networks and connectivity patterns in large-scale distributed hydrological models. *Hydrological Processes, 19*, 149–168. https://doi.org/10.1002/hyp.5765

Sten, J., Lilja, H., Hyväluoma, J., Westerholm, J., & Aspnäs, M. (2016). Parallel flow accumulation algorithms for graphical processing units with application to RUSLE model. *Computers & Geosciences, 89*, 88–95. https://doi.org/10.1016/j.cageo.2016.01.006

Su, C., Yu, W., Feng, C., Yu, C., Huang, Z., & Zhang, X. (2015). An efficient algorithm for calculating drainage accumulation in digital elevation models based on the basin tree index. *IEEE Geoscience and Remote Sensing Letters, 12*, 424–428. https://doi.org/10.1109/LGRS.2014.2345561

Tarboton, D. G., Bras, R. L., & Rodriguez-Iturbe, I. (1991). On the extraction of channel networks from digital elevation data. *Hydrological Processes., 5*, 81–100. https://doi.org/10.1002/hyp.3360050107

Tarboton, D. G., Bras, R. L., & Rodriguez-Iturbe, I. (1992). A physical basis for drainage density. *Geomorphology, 5*, 59–76. https://doi.org/10.1016/0169-555X(92)90058-V

Wang, M., Hjelmfelt, A. T., & Garbrecht, J. (2000). DEM aggregation for watershed modeling. *Journal of the American Water Resources Association, 36*, 579–584. https://doi.org/10.1111/j.1752-1688.2000.tb04288.x

Wu, H., Kimball, J. S., Mantua, N., & Stanford, J. (2011). Automated upscaling of river networks for macroscale hydrological modeling: upscaling of global river networks. *Water Resources Research*. https://doi.org/10.1029/2009WR008871

Yamazaki, D., Oki, T., & Kanae, S. (2009). Deriving a global river network map and its sub-grid topographic characteristics from a fine-resolution flow direction map. *Hydrology and Earth Systems Science, 13*, 2241–2251. https://doi.org/10.5194/hess-13-2241-2009

Yao, Y., & Shi, X. (2015). Alternating scanning orders and combining algorithms to improve the efficiency of flow accumulation calculation. *International Journal of Geographical Information Science, 29*, 1214–1239. https://doi.org/10.1080/13658816.2015.1027209

Zhou, G., Wei, H., & Fu, S. (2019). A fast and simple algorithm for calculating flow accumulation matrices from raster digital elevation. *Frontiers in Earth Sciences, 13*, 317–326. https://doi.org/10.1007/s11707-018-0725-9

Chapter 8
Watersheds

A watershed can be understood as the natural catchment area of precipitation whose resulting flows are directed to a certain point on the ground, the outlet of the basin (Fig. 8.1). All the runoff resulting from the transformation of precipitation into flow in the watershed area has the outlet as its only exit point from this system. Therefore, this area is often referred to as a drainage basin, drainage area, or contributing area, with the point taken as the outlet.

In other words, it can be established that, given a point chosen as the outlet, the delineation of the basin corresponding to this point is defined as identifying all portions of the land whose flows pass through it. But how do you know if the runoff from a specific region passes through the point chosen as the outlet?

The usual approach to get this answer is to analyze the region's topography. The former method for this analysis was by visual interpretation from topographic maps with contour lines. Today's most commonly used procedure is automated topographic analysis, with specific algorithms proposed for DEM processing and watershed delineation.

This chapter describes how the DEM is processed to obtain the watershed delineation and to estimate characteristics of the basin, such as area size, perimeter, drainage density, and geomorphological indices. It also presents the characterization of the stream order along a basin and the discretization of the basin into sub-basins according to three methods: confluence criterion, confluence criterion with size restriction, and Pfafstetter method (Ottobasins).

A. Rolim da Paz, *Digital Elevation Models for Environmental Studies*,
https://doi.org/10.1007/978-3-032-04523-2_8

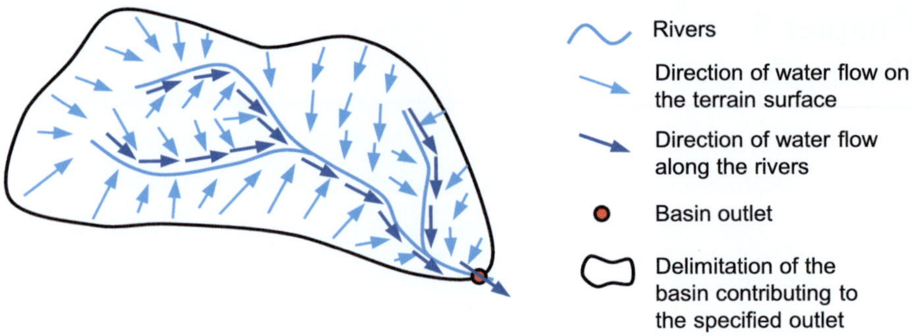

Fig. 8.1 Representation of the watershed concept, considering the contributing area to a certain point chosen as the outlet

8.1 DEM-Based Watershed Delineation

8.1.1 Conceptual Basis

As discussed in previous chapters, the interpretation of the topography represented by the DEM allows us to infer the general water flow paths on the surface. For example, one can visually and approximately infer where the flows originating from specific points on the surface, such as points A, B, C, and D in Fig. 8.2a, go. It is deduced that the flows originating from B and C converge into a single flow that continues until it reaches point Q (Fig. 8.2b). It is also deduced, for example, that the flow paths originating from A and D do not reach point Q.

Extending this analysis throughout the DEM, it is possible to perceive a set of areas whose flows follow paths that reach point Q (Fig. 8.2c). If a polygon is drawn bringing together all these areas with flow converging to Q, the watershed contributing to such a point is obtained (Fig. 8.2d).

The visual interpretation of flow patterns and the estimate of the contributing area to a certain point, as carried out in this example, are conceptual exercises. Once the topography is represented in the DEM, computational methods should be applied to delimit the watershed instead of a manual procedure, which presents inherent subjectivity and gross errors.

8.1.2 DEM Processing to Watershed Delineation

In processing the DEM to obtain the delineation of the basin, the key point is the determination of the flow directions. Obtaining this product is extensively described in Chap. 6 and is the only procedure involved in basin delineation that may be considered relatively complex. Among the procedures involved, the flow directions derivation is also the only step that can yield different results from distinct algorithms.

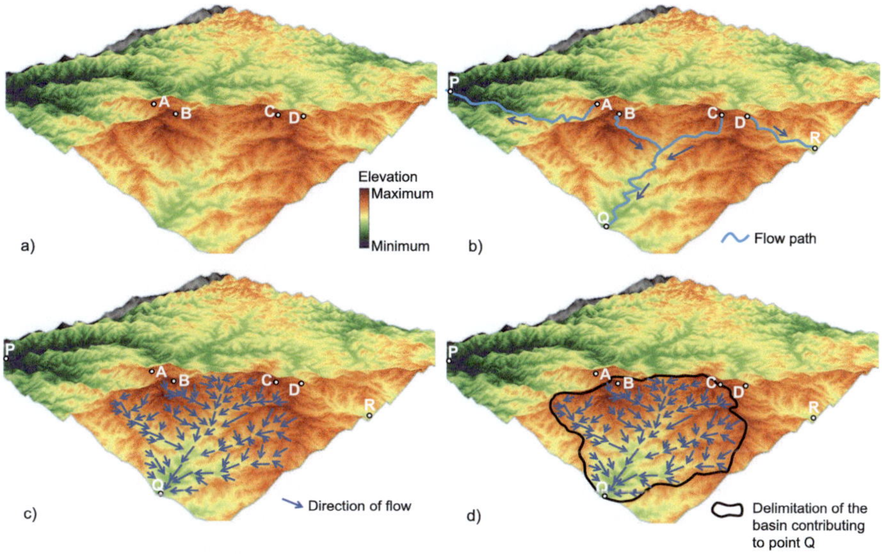

Fig. 8.2 **a** 3D representation of a DEM with indication of four points A, B, C, and D; **b** Estimate of the flow paths downstream of A, B, C, and D drawn manually, reaching points P, Q, and R; **c** Identification of all flows that drain to point Q; **d** Delineation of the contributing area (watershed) to point Q

This is because there are various rules and decisions among the algorithms proposed in the literature and used in software regarding the treatment of depressions, flat areas, and other definitions for the assignment of flow directions, as detailed in Chap. 6. In other words, when using different software and the same DEM as input, delineating a given watershed can have distinct results solely and exclusively due to differences in the flow directions.

The delineation of the watershed becomes trivial when the flow directions are defined (Fig. 8.3). Consider that the pixel chosen as the outlet of the basin (referred to here as the outlet pixel) has the light blue border highlighted in Fig. 8.3a. To know whether a pixel in the image belongs to the watershed whose outlet is this point, walk through the raster downstream of the pixel following the flow directions. Along this path, if the flow reaches the outlet pixel, then the pixel of origin of this path is part of the basin contributing to this outlet. In the example of this figure, the pixel highlighted with a dark blue border fits this case since the flow downstream from it reaches the outlet pixel of the basin (Fig. 8.3a).

Suppose the flow traced downstream from any pixel reaches the edge of the image without passing through the outlet pixel. In that case, the pixel of origin of the path does not belong to the area contributing to the outlet pixel. In the example, the pixel highlighted with a black border fits this situation.

By following the flow paths downstream from each pixel in the image and evaluating whether these paths reach the outlet pixel, all the pixels belonging to the basin

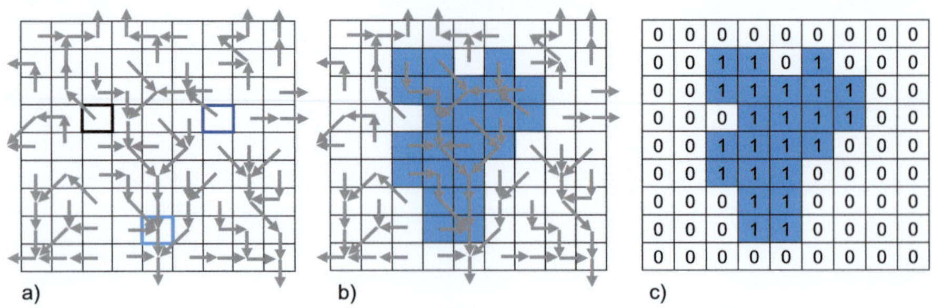

Fig. 8.3 **a** Flow directions derived from the DEM, represented by arrows; **b** Indication in blue of the pixels that are part of the basin contributing to the pixel highlighted in the same color in (**a**); **c** Raster of this contributing basin (pixels with attribute 1 are part of the basin; pixels with attribute zero are not part of it)

contributing to the selected outlet pixel are identified. For example, the pixels painted in light blue in Fig. 8.3b are part of the basin contributing to the outlet pixel indicated in Fig. 8.3a.

To delineate a basin as a data layer, the GIS software typically generates a raster in which each pixel is assigned a value of 1 or 0, indicating whether it belongs to the drainage area of the defined basin or not, respectively (Fig. 8.3c). This raster can subsequently be converted to a vector polygon format using the standard raster-to-vector conversion procedure (see Chap. 2).

Similarly, this same procedure can simultaneously delimit a set of basins, each with its outlet pixel. In this case, the result is a raster with attribute 0 when the pixel does not belong to any of these basins and an attribute from 1 to N when the pixel belongs to one of the N delimited basins.

In the example of Fig. 8.4, four pixels are indicated as outlets, and different colors and numbers indicate the basins contributing to each. It should be noted that when there is more than one outlet pixel, and one is downstream of another, there is a delineation of incremental basins. In the figure of the example, it can be seen that outlet pixel 4 is downstream of all the other three outlet pixels and, therefore, conceptually, the basin contributing to pixel 4 includes the basins contributing to the others. However, software usually identifies areas that do not overlap as contributors to the various outlet pixels. In the example, the basin labeled as 4 was left with only the area downstream of the other basins (i.e., the incremental contributing area to pixel 4).

The complete DEM processing for watershed delineation involves the following operations:

i. Obtaining and cropping the DEM for the area of interest
ii. Treatment of terrain depressions and determination of flow directions
iii. Determination of accumulated drainage areas
iv. Definition of the river drainage network
v. Selection of the pixel corresponding to the basin outlet

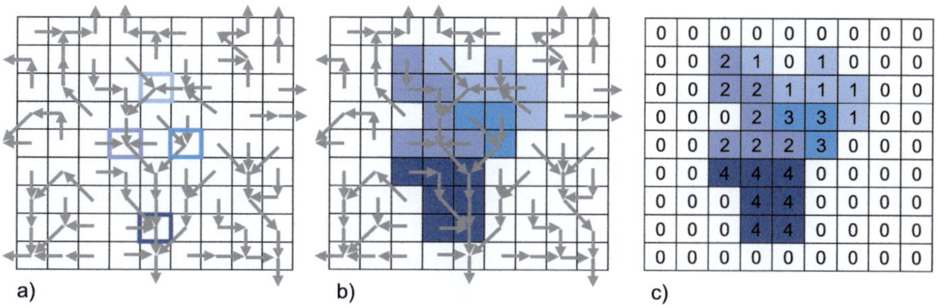

Fig. 8.4 **a** Flow directions derived from the DEM, represented by arrows; **b** Indication in different shades of blue of the pixels that are part of the incremental basins contributing to the pixels highlighted in the same color in (**a**); **c** Raster of the delineation of the contributing basins (pixels with attribute 1, 2, 3, and 4 are part of the basin of corresponding numbering; pixels with attribute zero do not belong to any of the basins)

vi. Identification of the areas belonging to the basin.

Step i is an essential stage of any DEM processing, obtaining the DEM with a geographical scope broader than the study area and cropping it to a window around the basin to be delimited. The caveat is to preserve a margin in the extension of this window relative to the basin so as not to have part of the basin outside the image. If this happens, the processing of the other stages can be carried out without any error being reported by the software, and the user may not notice the inconsistency. This type of error is detected more prominently when visualizing that part of the basin delineation is touching the edge of the image, as already illustrated in Chap. 4 (item 4.8).

Step ii involves treating the terrain depressions and determining the flow directions for this DEM. This is the most essential part of the procedure, for which different algorithms and distinct results exist.

Steps iii and iv are auxiliary and conceptually considered optional, but in practice, they are strongly recommended to be carried out. The primary advantage of obtaining the accumulated drainage areas and the river drainage network within the process of basin delineation is to ensure success in step v, which is the selection of the outlet pixel. The last step is identifying and associating the pixels of the watershed to generate its delineation.

8.1.3 Outlet Pixel Selection

In a real-world case, the DEM and its derived products comprise thousands of pixels. For this reason, the exact and correct choice of the outlet pixel to generate the watershed delineation by DEM processing can be made unfeasible if the accumulated drainage areas or the drainage network derived from this DEM are not used to guide this selection.

A widespread situation is for the user to have the geographical coordinates of the basin outlet they want to delimit or vector files of basin delimitation and river tracing from official databases, such as governmental environmental agencies. All this information is helpful in the choice of the outlet pixel to ensure that the delimitation of the basin from the DEM has as an outlet point equivalent to that available in official databases. However, this information should not be used in isolation, as it can be relatively coarse for this purpose and is insufficient to correctly select the exact outlet pixel. Visualizing the pixel-by-pixel flow path derived from the DEM is necessary to define the outlet pixel accurately.

Figure 8.5 exemplifies this situation, taking the Taperoá river basin in Paraíba as an example. We have the basin delineation and the tracing of the main river obtained from official sources (Fig. 8.5a). In Fig. 8.5b, there is a zoom of the outlet region of the basin and the indication of the pixel chosen for the outlet based solely on the available vector data (pixel P1, where the available river tracing leaves the available basin delineation).

When using pixel P1 as an outlet to delimit the referred basin from the processing of the SRTM 30 DEM, the result is a delineation that covers a minimal fraction of the basin intended to be identified (Fig. 8.5c). This happened because pixel P1, according to the flow paths derived from the SRTM 30 DEM using a particular algorithm, corresponds to a secondary flow, which flows into the path of the main river derived from the DEM.

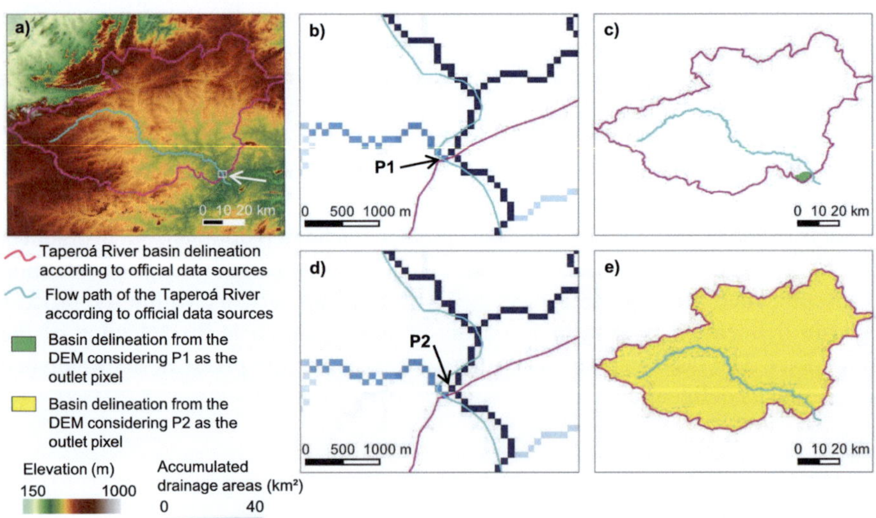

Fig. 8.5 **a** SRTM 30 DEM for the region of the Taperoá river watershed, in Paraíba, with basin delineation and tracing of the main river obtained from official geographic data sources; **b**, **c** Selection of pixel P1 as an outlet to delimit the basin from the DEM processing and resulting delineation, respectively; **d**, **e** Selection of pixel P2 as an outlet to delimit the basin from the DEM processing and resulting delineation, respectively

In this case, the correct pixel to be chosen as the outlet pixel is P2 (Fig. 8.5d), representing the point equivalent to the mouth of the main river tracing derived from the DEM. The result of the basin delineation, taking pixel P2 as an outlet, is coherent (Fig. 8.5e).

8.2 Drainage Area

The size of the watershed area is the quantification of its drainage area, whose value corresponds to the flat area between the topographic divides projected vertically. It is one of the most relevant quantities for hydrological studies since the primary water input for a basin is precipitation, represented as a water layer falling over the basin area. The multiplication of the basin area by precipitation represents the volume of water that enters the basin's water balance.

Given a watershed's delineation from the DEM, the corresponding drainage area can be quantified in two ways: one using this delineation in raster format and the other in vector format.

With the raster of the basin delineation, one can raise the number of pixels that integrate the basin and add up the surface areas of these pixels, following the general idea of calculating areas in a raster according to the different attributes (see Chap. 3). The sum of the areas of the pixels representing the basin is the size of the basin area. In the example of Fig. 8.6a, the area of the delimited basin corresponds to the sum of the surface areas of the 23 pixels that integrate it. Considering that each pixel has an area of 25 m^2, the area of the mentioned basin is 575 m^2.

When a projected coordinate and reference system is used, all pixels have the same surface area. In this situation, it is enough to quantify the number of pixels that are part of the basin and multiply this value by the surface area of one pixel, which is given by the square of the spatial resolution. When using a system of geographic coordinates, due to the variation of equivalence between degrees and planar units such as meters according to the latitude and longitude of the pixel, the surface area

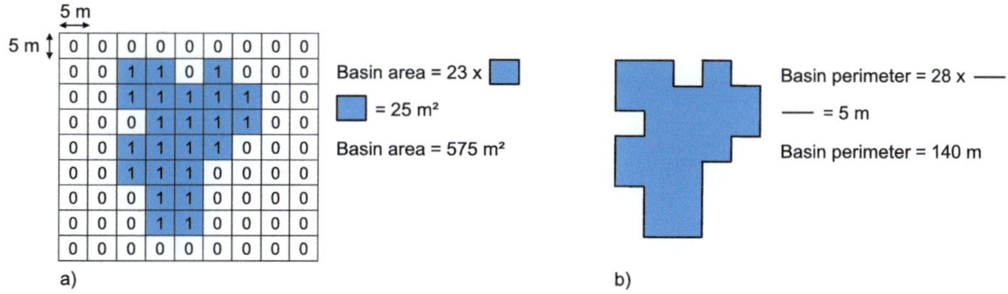

Fig. 8.6 **a** Identification of the pixels belonging to a watershed (attribute 1) and calculation of its area, formed by 23 pixels; **b** Identification and calculation of the basin perimeter, formed by 28 sections of size equal to the pixel dimension

among the pixels of the basin is variable, as detailed in Chap. 3. In this case, the basin area must be determined directly by quantifying the sum of the individual pixels' projected surface areas.

Alternatively, one can convert the raster basin delineation to a vector format of the polygon type and quantify the area of this polygon. Quantifying the area in raster and vector formats is a standard operation in any GIS software.

8.3 Perimeter

The perimeter of the watershed is defined as the extension of the contour that defines the basin's delineation. The basin's delineation in raster format obtained by processing the DEM can be converted to vector format (polygon). The basin's perimeter is obtained by quantifying the extension of a polygon's contour, a standard operation in GIS software.

For the basin of the example in Fig. 8.6, the perimeter is the extension of its contour and corresponds to 28 sections of dimension equal to the spatial resolution. As a result, the perimeter of this basin is 140 m.

8.4 Drainage Density

The drainage density of a watershed is conceptually defined as the relationship between the total length of all watercourses and the drainage area of a watershed (Eq. 8.1). It is a characteristic of the basin often used as a variable in discharge regionalization studies and is associated with understanding the dominant erosive processes in the region. Areas with high drainage density tend to have lower infiltration, higher surface runoff, moderately erodible surface materials, and a higher degree of channel abrasion for a given rainfall than areas with lower drainage density (Harsha et al., 2020).

$$D_{\mathrm{d}} = \frac{\sum_{i=1}^{n_{\mathrm{L}}} L_i}{A_{\mathrm{basin}}} \tag{8.1}$$

where L_i is the length of each river section in the watershed, n_{L} is the number of river sections in the basin, A_{basin} is the drainage area of the basin, and D_{d} is the drainage density of the basin.

The drainage density from a DEM can be obtained in two main steps. The first is the DEM processing for deriving the drainage network and subsequent quantification and sum of the lengths of all watercourses in the basin. The derivation of the drainage network was detailed in Chap. 7. River length quantification can be done by computing the distances pixel by pixel along the drainage network, as explained in the same chapter. Alternatively, this drainage network can be converted from raster

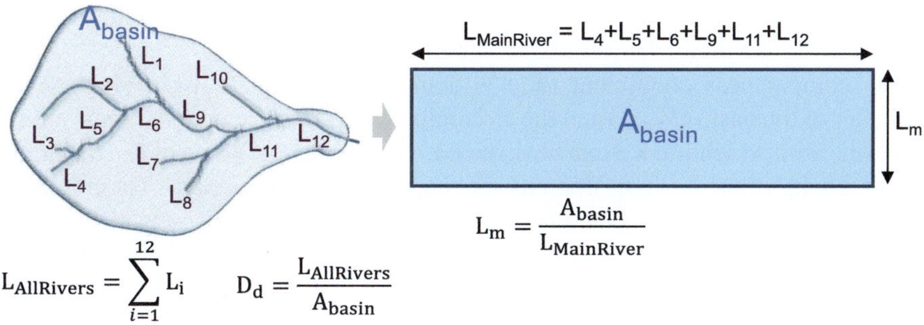

Fig. 8.7 Watershed of drainage area A_{basin} and numbering of sub-sections of watercourses, with equivalence of the basin in rectangular format of the same area and indication of the calculation of the sum of the lengths of all rivers ($L_{\text{AllRivers}}$) of the basin, of the drainage density (D_{d}), of the length of the main river represented by the sub-sections L4, L5, L6, L9, L11, and L12 and of the equivalent average width (L_{m})

format to vector-type lines (as described in Chap.2). Then, the lengths of each line and their sum (Fig. 8.7) are calculated.

As this chapter shows, the second primary step is the basin delineation and area quantification. Equation 8.1 is applied to obtain the drainage density by summing the lengths of the watercourses and quantifying the basin area.

As seen in the previous chapter, defining where rivers begin is key in determining the river drainage network from the DEM. This directly influences the length of the watercourses and, consequently, the drainage density. The earlier the start of the watercourses, the higher the length of these rivers, which leads to a higher drainage density.

8.5 Geomorphological Indices

Geomorphological indices can be used to characterize the watershed's physiography. Characterizing the basin's shape helps understand how a basin responds to rainfall events by transforming this water input into output flow. Geomorphological indices such as shape factor and compactness coefficient are used mostly to characterize the basin's shape and physiography.

The shape factor (K_{f}) is defined as the relationship between the average width of the basin (L_{m}) and the length of the main watercourse (L_{river}). The average width of the basin is established by idealizing the basin as a rectangle of the same area as the basin and with the longest side equal to the length of the main river (Fig. 8.7).

As L_{m} can be calculated by the relationship between the basin area and the length of the main river, the shape factor can be obtained by the quotient between the basin area and the square of the length of the main watercourse (Eq. 8.2). A higher value of K_f implies a basin with a higher tendency to flood than another with the same characteristics (climate, vegetation, soils, topography, etc.).

The compactness coefficient (K_c) is defined as the relationship between the perimeter of the basin (P_{basin}) and the circumference of a circle with the same area as the basin, from which Eq. 8.3 can be deduced. The smaller the value of K_c, the higher the propensity of the basin to flood as the shape approaches circular. The minimum conceptual value for this coefficient is $K_c = 1$ when the basin shape is a circle.

$$K_f = L_m/L_{river} \therefore K_f = A_{basin}/L_{river}^2 \quad (8.2)$$

$$K_c = 0.28 \cdot P_{basin}/\sqrt{A_{basin}} \quad (8.3)$$

8.6 Stream Order

The ordering of the streams is a characterization of the watershed related to the river drainage network. This is a qualitative way of denoting the degree of branching of the river drainage network, numerically classifying each river's reach between two confluences or between the headwaters and the downstream confluence.

The most widely adopted method for stream ordering is following Strahler's criteria as follows:

(a) All headwater river reaches are identified, i.e., those reaches that do not receive any contribution from another river upstream;
(b) These headwater river reaches are considered of order 1;
(c) The river drainage network is followed downstream from each order 1 reach;
(d) Reaches that are immediately downstream of the confluence of two order 1 streams receive order 2;
(e) Reaches that are formed downstream of the confluence between rivers of order 1 and order 2 receive order 2;
(f) Reaches that are downstream of the confluence between two rivers of order 2 receive order 3;
(g) The procedure continues until the basin's outlet is found, with these two primary rules: the junction of reaches of the same order results in a reach of the immediately higher order; the junction of reaches of different orders results in a reach of order equal to the highest order of the reaches upstream of the junction.

Figure 8.8 illustrates this Strahler-based ordering procedure. The more branched the river drainage network, the more likely the main river of the basin is of a higher order.

When the river drainage network results from DEM processing, the definition of the beginning of this network has a total influence on the ordering of the streams; the simplified criterion of adopting a minimum accumulated drainage area to denote

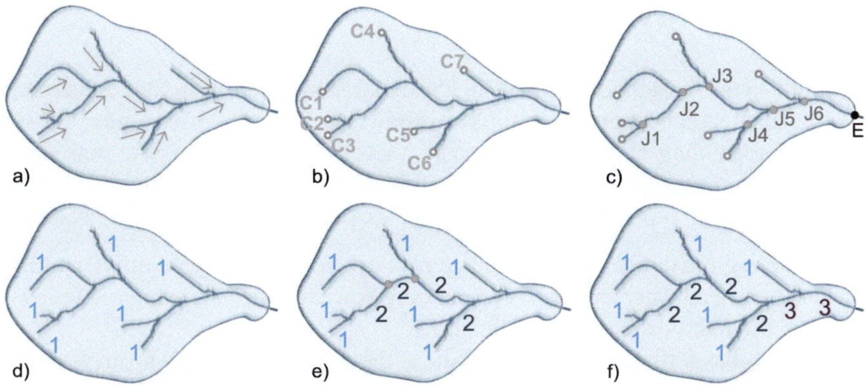

Fig. 8.8 **a** Watershed with an indication of the flow direction of the watercourses; **b** Identification of the points of the seven headwaters (points C1 to C7) of the drainage network; **c** Identification of the six junctions (points J1 to J6) and the outlet (point E); **d–f** Identification of the streams of order 1, 2, and 3, respectively, according to Strahler's rule

channel initiation is often used, as seen in Chap. 7. The larger this minimum area threshold, the less branched the network is; therefore, the lower the maximum ordering obtained in the basin. In contrast, the smaller the minimum area parameter for channel initiation, the more branched the network becomes, and the ordering reaches a higher value.

8.7 Automatic Division into Catchments

In a previous item of this chapter, we dealt with delineating one or more basins in the image area from the DEM processing, with the indication of the outlets corresponding to such basins. Another way of delimiting basins is by establishing criteria to automatically define the division of an area into a set of basins. In general, the user already has the delineation of a particular watershed and intends to subdivide it into sub-basins, without worrying about pointing out where each one's outlets would be precisely, but with some logical rule of subdivision criteria.

Three examples of this type of subdivision of a watershed are detailed below: according to the drainage network's confluences criterion, according to a modified version of this criterion by restricting sub-basin size, and according to the Pfastetter method (Ottobasins).

8.7.1 Confluence Criterion

The confluence criterion for establishing the division into sub-basins has as its central idea that each stream segment in the river drainage network corresponds to a sub-basin, understanding a stream segment as a river reach between two consecutive confluences or from the headwaters to the first downstream confluence.

In practice, this type of subdivision can be carried out in the following way:

i. Assuming the defined river drainage network, each starting point is identified, i.e., each headwater and the corresponding downstream watercourse is followed up to the first downstream confluence;
ii. The sub-basin that drains to this point of the river immediately upstream of the confluence is delimited;
iii. The river continues to be followed from this confluence to the next downstream confluence, and the incremental sub-basin draining to the river reach traced is delimited;
iv. The same is done for the other river segment that is part of the confluence;
v. This procedure is repeated throughout the entire river drainage network.

This type of automatic division into sub-basins is performed by the GRASS—QGIS algorithm r.watershed and by the Arc Hydro extension of the ArcGIS software, for example.

Within this subdivision method based on river confluences, the number of resulting sub-basins is entirely linked to the density of the drainage network. The definition of the beginning of the river reaches along the DEM processing has a total influence on the subdivision level. If the criterion of the minimum threshold of the accumulated area is used to establish the river drainage network (recall this method in Chap. 7), the smaller the value of this parameter, the more river segments exist in the network; consequently, the subdivision results in more sub-basins, generally smaller in size. In contrast, the larger the threshold value, the fewer segments and sub-basins the river drainage network has, most of which are larger.

Figure 8.9a shows a real-world example of this division into sub-basins for Pernambuco's upper part of the Sirinhaém river basin. A river drainage network was established from SRTM 30 DEM with a minimum drainage area threshold of 4.7 km^2. The result is the division into 65 sub-basins, with areas ranging from 0.28 to 20.89 km^2. For instance, the largest basin is about 75 times larger in area than the smallest basin.

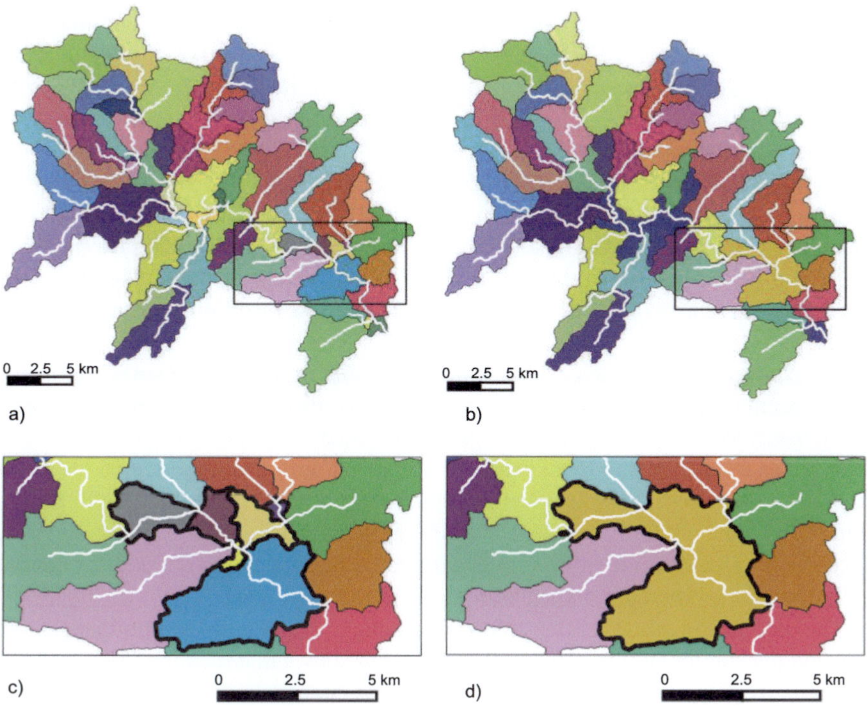

Fig. 8.9 **a** Subdivision of the upper part of the Sirinhaém river basin (Pernambuco) according to the confluence criterion, considering a minimum drainage area threshold of 4.7 km^2 as the definition of the beginning of the drainage network; **b** Subdivision according to this same criterion but additionally with a minimum basin size threshold of 4.7 km^2; **c**, **d** Zoom of the subdivisions shown in (**a**) and (**b**)

8.7.2 Confluence Criterion with Size Restriction

The confluence criterion for automatic division of a basin into sub-basins generally results in an extensive range of variation in sub-basin size. Along the DEM-processing steps, adjusting the minimum drainage area threshold to generate watercourses controls the number of sub-basin elements. It influences the headwater sub-basin size, which includes those contributing to the river's reach upstream of any confluence. However, it is only an influence and not a control of the size of the sub-basins since the position of the drainage network's confluences is not controllable by this minimum drainage area threshold parameter.

In some situations, the division into sub-basins with a considerable variation in size may be undesirable, whether from the point of view of water resources management at the sub-basin level or from a more technical point of view, such as in the application of distributed hydrological models. The tiny size of a sub-basin implies a small extension of the main river reach used for the flow routing downstream. This can cause problems with the spatial discretization of these river reaches and numerical instability when applying flow routing methods. Standard geographic data editing tools can agglomerate smaller sub-basins and overcome this issue manually or via some automatic procedure. This approach is more straightforward when considering sub-basins' delineation in vector format.

However, when the intention is to control the size of the sub-basins in automatic delineation, whether minimum and/or maximum values, and avoid the need for subsequent manual editing, an alternative is to impose additional criteria for the division into sub-basins. In this sense, different approaches can be taken to restrict both headwater basins and intermediate basins regarding the minimum and maximum sizes.

An example of this type of algorithm is available in the TerrSet software (Watershed command), which allows the user to specify the minimum sub-basin size of the division to be made. In the case of the Sirinhaém river basin worked on in the previous example, the subdivision by this algorithm with a minimum area restriction of 4.7 km^2 resulted in 48 sub-basins with areas ranging from 4.82 to 20.89 km^2 (Fig. 8.9b)—the choice of a 4.7 km^2 minimum drainage area was arbitrary and is not necessarily related to the minimum accumulated drainage area defined for channel initiation (also coincidentally set as 4.7 km^2 in this example).

The comparison between the subdivision obtained only with the confluence criterion and the subdivision by the TerrSet algorithm shows that the distinction between these subdivisions, in this case, is the grouping of intermediate sub-basins of the first subdivision approach to comply with the minimum size imposition. For example, in the highlight of Fig. 8.9c, there is a set of six sub-basins of the subdivision based on the confluence criterion that were agglomerated into a single sub-basin to meet the size restriction (Fig. 8.9d). Note that most downstream elements in this set of six sub-basins already met the minimum size, but the other five are so small that even grouping does not meet. It was necessary to agglomerate all six sub-basins to meet this requirement.

8.7.3 *Pfafstetter Coding System (Ottobasins)*

Another approach to dividing a basin into sub-basins is according to criteria that more emphatically aim at the final topology of this subdivision and the coding associated with the division's elements. The idea is that the subdivision and corresponding coding add more practical benefits from the point of view of the organization of hydrological data and water resources management.

A significant example of this idea is the division according to the criteria conceived by Otto Pfasftetter when acting as an engineer at the National Department of Sanitation Works (DNOS) in Brazil. This approach for basin dividing and coding (referred to as Ottobasins) is adopted by the Brazilian National Water and Sanitation Agency (ANA) for the management of water resources in Brazil (ANA, 2006) and is also internationally adopted (Stein, 2018; Verdin & Verdin, 1999).

In summary, the division into sub-basins according to the Pfafstetter method is structured by initially delimiting the basin into nine sub-basins, each of which is also divided into nine elements, and so on, recursively, for each sub-basin of the previous division until the desired level of subdivision is reached.

For the division of a basin element into nine sub-basins, the following rules are applied:

i. Identify the main river of the basin and its four main tributaries;
ii. Delimit the sub-basin contributing to each of these four main tributaries;
iii. These sub-basins receive numerical codes 2, 4, 6, and 8 in order from the downstream tributary to the upstream along the main river;
iv. Delimit the sub-basin contributing to the reach of the main river downstream of tributary 1 to the basin's outlet, and this sub-basin receives code 1;
v. Delimit the sub-basins of the reaches of the main river between the confluences of tributaries 2 and 4, 4 and 6, and 6 and 8, which receive numerical codes 3, 5, and 7, respectively;
vi. Delimit the sub-basin that contributes to the main river in the reach upstream of the confluence of sub-basin 8, receiving code 9.

When moving from one level of division to another, the digits 1 to 9 of the current division are added to the right of the numerical code already assigned to the element in the previous division. The first division, level 1, divides the basin into sub-basins 1 to 9. In the second division level, sub-basin 1 is subdivided into another nine elements, now coded as 11 to 19, and similarly for sub-basins 2 to 9. In the third level, for example, sub-basin 15 is divided into elements 151 to 159, similar to the division of the other sub-basins of the previous level. This procedure repeats until the desired level of subdivision is reached.

A significant benefit of this division and coding is the logic of the resulting numerical code. This coding allows for automatic recognition of addressing and spatial positioning of each division element. For example, sub-basin 149 is a headwater basin at the third level of division, as indicated by the final digit 9, and it belongs to the sub-basin of the river's second-largest tributary (indicated by the digit 4 at level 2). Additionally, it is known that this level-2 sub-basin, to which sub-basin 149 belongs, flows directly into the main river of the entire basin near its outlet, as suggested by the digit 1 at the first level of the 149 code.

The previous examples illustrate the division into Ottobasins of the upper part of the Sirinhaém river basin in Fig. 8.10. It shows the first and second levels of complete division, and at the third level only for sub-basin 14.

Limitations to the Pfafstetter division and coding system have been pointed out in the literature, with propositions for improvements, as in the works of Teixeira et al. (2022), Stein (2018), and Verdin and Verdin (1999).

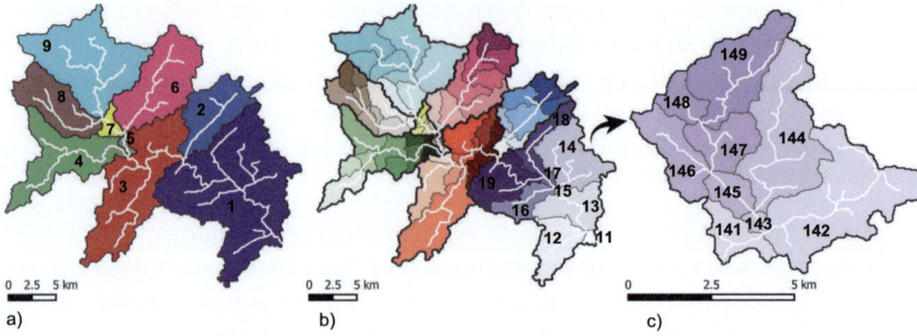

Fig. 8.10 **a** Subdivision of the upper part of the Sirinhaém river watershed (Pernambuco) according to Pfafstetter's criterion (Ottobasins) at the first level; **b** Subdivision of basin 1 at the second level; **c** Subdivision of basin 14 at the third level

References

ANA. (2006). Topologia Hídrica: método de construção e modelagem da base hidrográfica para suporte à gestão de recursos hídricos (No. versão 1.11). Agência Nacional de Águas.

Harsha, J., Ravikumar, A. S., & Shivakumar, B. L. (2020). Evaluation of morphometric parameters and hypsometric curve of Arkavathy river basin using RS and GIS techniques. *Applied Water Science, 10*, 86. https://doi.org/10.1007/s13201-020-1164-9

Stein, J. L. (2018). An enhanced Pfafstetter catchment reference system. *Water Resources Research, 54*, 9951–9963. https://doi.org/10.1029/2018WR023218

Teixeira, A. D. A., Silva, A. M., Pruski, F. F., Davis, C., Jr., Roig, H. L., & Diniz, N. C. (2022). Improvements on the Pfafstetter basin coding system proposal. *Revista Brasileira De Recursos Hídricos, 27*, Article e6. https://doi.org/10.1590/2318-0331.272220210120

Verdin, K. L., & Verdin, J. P. (1999). A topological system for delineation and codification of the Earth's river basins. *Journal of Hydrology, 218*, 1–12. https://doi.org/10.1016/S0022-1694(99)00011-6

Chapter 9
Lake Formed by Constructing a Dam

Water storage reservoirs along the river drainage network are built and used for various purposes, such as human supply, hydroelectricity, irrigation, flood control, and downstream flow regulation. A particular river point is chosen, and a dam is built at this cross-section, allowing for water accumulation as a lake (Fig. 9.1).

Reservoirs play a key role in strategically planning and managing water resources. Therefore, they are a driving factor for social and economic development, mainly related to energy, water, and food security (Avisse et al., 2017; Nguy-Robertson et al., 2018). Reservoirs can also contribute to increasing resilience to the impacts of climate change.

Before constructing a reservoir, the appropriate selection of the dam site is necessary. Several criteria and factors are involved in this selection process and are mainly related to the purpose of the reservoir construction and environmental, economic, and legal issues. Two main technical factors commonly considered in the choice of location are: (i) the estimate of the maximum flooded area with the dam's construction, and (ii) the area-volume-elevation curve of the lake.

The information on the maximum flooded area helps analyze potential environmental impacts, such as those associated with the relocation of people, biodiversity of the flooded area, removal of vegetation, and flooding of legally protected areas, such as archeological sites and indigenous lands, to evaluate costs related to land expropriation and to estimate losses by evaporation, among other purposes.

The area-volume-elevation curve establishes the relationship between water level, lake surface area, and storage volume for different points from the minimum to the maximum water accumulation. This information is functional for sizing and operating the reservoir and assessing the suitability of the chosen location for reservoir construction.

A. Rolim da Paz, *Digital Elevation Models for Environmental Studies*,
https://doi.org/10.1007/978-3-032-04523-2_9

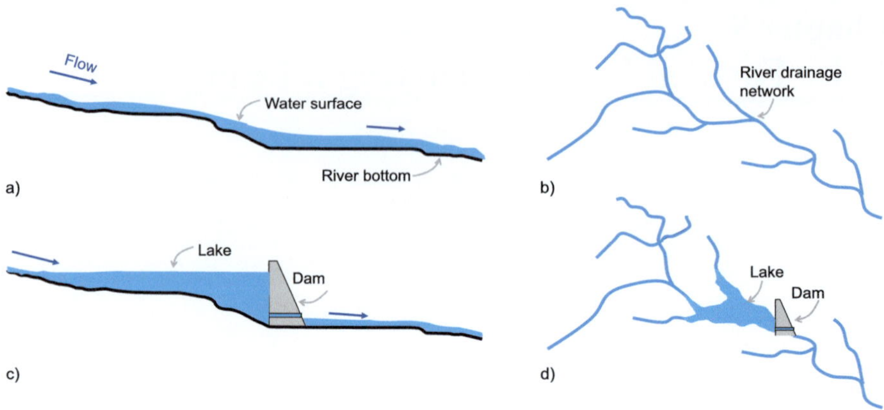

Fig. 9.1 **a** Longitudinal profile of a river reach; **b** River drainage network; **c**, **d** Formation of a lake due to the construction of a dam, with a view in the longitudinal profile of the river reach and in the mapping of the river drainage network, respectively

9.1 Area of the Reservoir Lake

A relatively quick and low-cost way to estimate the flooded area with the possible construction of a reservoir at a certain point in the river drainage network is from the DEM processing. The estimate of the lake formed by damming the river drainage network is based on identifying all the points that will be part of this lake. For any point on the ground to be part of this lake, two criteria must be met: (a) it belongs to the basin contributing to the dam location point, and (b) it has an elevation lower than the maximum water accumulation level at the dam.

Applying the first criterion requires DEM-derived products, which were described in previous chapters. The second criterion involves only the analysis of the DEM itself. Operationally, the estimate of the area to be flooded with the construction of a dam based on these two criteria can be detailed as follows (Fig. 9.2):

i. Obtaining and windowing the DEM for the area of interest (with enough margin so that the lake to be delimited is wholly contained within the DEM coverage area);
ii. Removal of depressions from the DEM;
iii. Determination of flow directions;
iv. Determination of accumulated drainage areas;
v. Extraction of the river drainage network;
vi. Definition of the exact cross-section of the dam location, according to user criteria;
vii. Delineation of the watershed draining to this section of the river (pixels that are part of the basin receive attribute 1, while the remaining pixels have attribute 0);

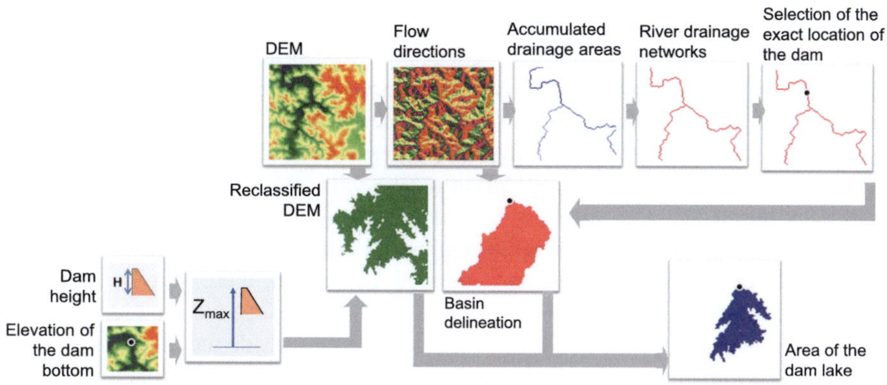

Fig. 9.2 Flowchart of main steps for estimating the area of the lake to be formed with the construction of a dam from the DEM and derived products

viii. Definition of the water level corresponding to the maximum water accumulation in the dam, which we simplistically call here the height of the dam (H; the user chooses this value depending on the purpose of the reservoir);
ix. Determination of the elevation corresponding to the height of the dam ($Z_{\max}$) by the sum between the elevation of the dam location point in the DEM (Z_{base}; base elevation of the dam) and the height of the dam: $Z_{\max} = H + Z_{\text{base}}$;
x. Reclassification of the DEM: Pixels lower than the elevation of the maximum water level in the dam ($Z_{\max}$) receive attribute 1 (they can be part of the lake). In contrast, pixels with elevation equal to or higher than such a value receive attribute 0 (they cannot be part of the lake).
xi. Multiplication between the rasters resulting from steps (vii) and (x) and obtaining, in the resulting raster, the delimitation of the lake area as the grouping of all pixels with attribute 1.

The purpose of steps ii, iii, iv, and v is to obtain the river drainage network derived from the DEM. In some software, the drainage network extraction is performed directly by inputting the DEM. Removing depressions, determining flow directions, and calculating accumulated drainage areas are executed internally without being explicit to the user. Similarly, the delineation of the contributing basin is conceptually done from the flow directions. However, some software may require only the DEM as input and internally determine the flow directions.

Determining the river drainage network from the DEM and choosing the exact location of the dam in this drainage network are fundamental to avoid inconsistencies that can occur, for example, when wanting to define such a location based on a vector drainage network from another data source. There may not be a coincidence of the network flow paths derived from the DEM with the layout of this vector network precisely at the point chosen for the dam. A similar problem was described in Chap. 8 regarding the choice of the representative point of the outlet for a basin delineation.

The definition of the beginning of the river drainage network from the raster of accumulated drainage areas does not affect the result of the lake estimate. On the other hand, in reclassifying the DEM to determine the areas that meet the second criterion, the user can choose between using the original DEM without having the terrain depressions removed or using the depressionless DEM. Using the original DEM has a great chance of resulting in a lake with small portions disconnected from the larger water mass at the end of the procedure. This happens because slight elevation differences between neighboring pixels at the lake boundaries distinguish whether or not the second criterion is met. In this region, there may be situations where upstream pixels have a lower elevation (and therefore were included in the lake) than downstream neighboring pixels (considered outside the lake).

With the removal of terrain depressions, this lake discontinuity tends to be eliminated, as this procedure seeks precisely to modify the DEM to establish a topographic gradient always descending from upstream to downstream. As seen in Chap. 6, several methods for treating and removing DEM depressions modify the DEM differently. Using the DEM modified by different procedures can result in different lake delimitations estimated for a hypothetical reservoir construction at a given point along the river drainage network.

A real-world example of estimating the lake to be formed with the hypothetical construction of a dam is presented in Fig. 9.3 for a section on the Ijuí River, a tributary of the Uruguay River, in southern Brazil. Several products were extracted from the DEM processing (Fig. 9.3a), and the dam construction's hypothetical location (Fig. 9.3b) was defined on the river drainage network. For this dam location, the corresponding contributing basin was determined (Fig. 9.3c), which is not fully shown in the figure. This dam location has an elevation in the DEM equal to 166 m, and a dam height of 30 m was arbitrated, resulting in a maximum water accumulation corresponding to an elevation of 196 m.

The reclassification of the original DEM indicated all the image pixels with an elevation of up to 196 m (Fig. 9.3d), and overlaying this raster with the basin delimitation resulted in the delimitation of the lake that would be formed with the construction of the 30 m dam at the indicated point (Fig. 9.3e). This lake has an area of approximately 90.54 km^2. It extends longitudinally for almost 76 km, from point A at the dam site to point B at the most downstream point of the lake delimitation. However, this definition of the lake extension's end is confused with the watercourse itself. The zoom of the lake delimitation illustrates the discontinuity at some points due to the terrain depressions (Fig. 9.3f).

When using a depressionless DEM generated by a depression filling approach to identify the pixels with an elevation lower than 196 m (Fig. 9.3g), the resulting lake has continuity. However, it has reduced the extension of its ends (Fig. 9.3h and i). Filling depressions causes the elevation of the pixels in these areas to rise due to neighboring pixels, which tends to reduce the upstream extremes of the lake area. In this case, the resulting lake has an area of 85.95 km^2, about 5% less than the area obtained with the original DEM.

When a DEM modified by a procedure that treats depressions primarily by lowering or carving the pixels neighboring the depressions (Fig. 9.3j) was used,

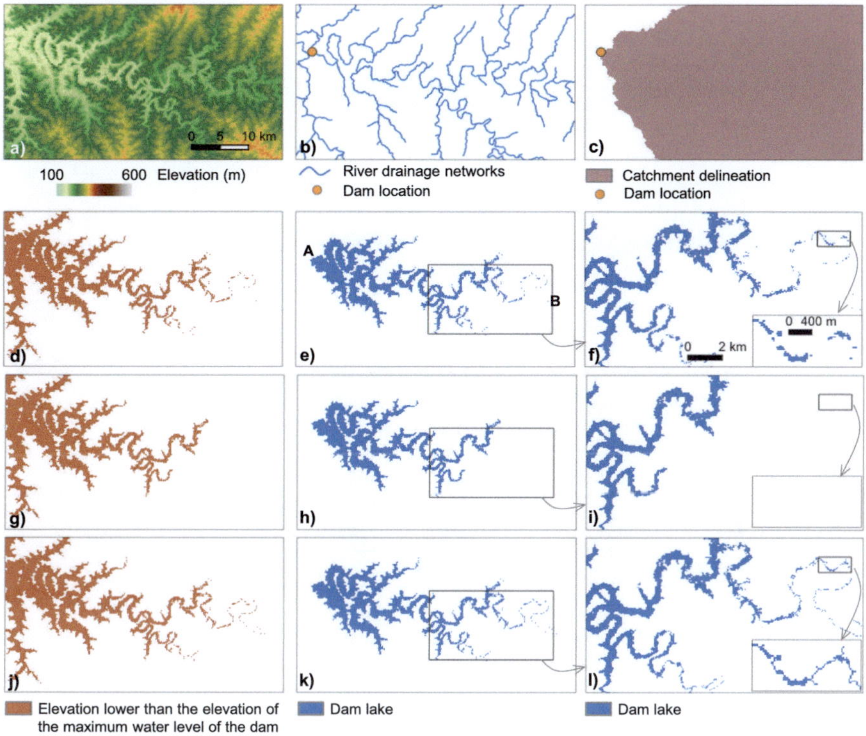

Fig. 9.3 **a** SRTM 30 DEM for an area of the Ijuí river basin (RS); **b** River drainage network derived from the DEM with a minimum area threshold of 4.5 km^2 and indication of the location of the hypothetical dam (adopted with a height of 30 m), in the municipality of Vitória das Missões; **c** Contributing basin to the hypothetical dam (partial view); **d** Identification of areas with elevation lower than the elevation of the maximum water accumulation level at the dam (196 m); **e, f** Estimate of the lake formed by the dam; **g, h, i** Analogous to **d**, **e**, **f**, but considering the DEM modified by a depression filling algorithm; **j, k, l** Analogous to **d**, **e**, and **f**, but considering the DEM modified by an algorithm that prioritizes the lowering of elevations in the treatment of terrain depressions

the effect was the opposite. There is an extension of the upstream ends of the lake relative to the use of the original DEM, as well as continuity of the flooded area (Fig. 9.3k and l). The area of the resulting lake in this example is 91.21 km^2, or 0.7% higher than the area obtained when using the original DEM.

9.2 Volume of the Reservoir Lake

Considering the maximum level of water accumulation in a dam, there is a spatial variation of depths along the lake according to the topography of the flooded terrain. The maximum depth tends to occur near the dam and decreases as it moves upstream

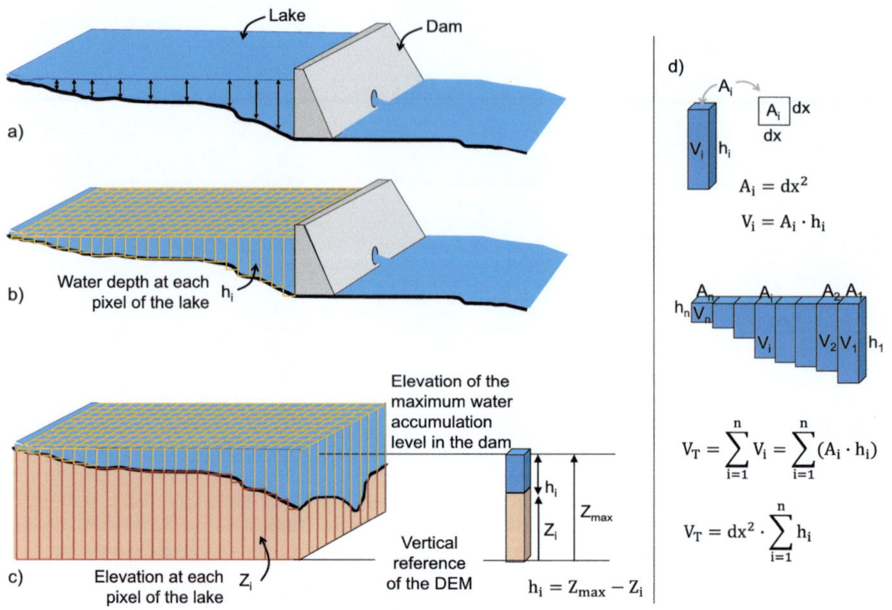

Fig. 9.4 **a** Spatial variability of depths along a lake formed by the construction of a dam; **b** Representation of the lake depths in each pixel; **c** Indication of the elevations and depths in each pixel, whose sum is constant and equal to the elevation of the maximum water accumulation level in the dam (Z_{max}); **d** Deduction of the calculation of the lake volume, from the sum of the volume in each pixel

(Fig. 9.4a and b). It is assumed that the water level at the dam is horizontal; therefore, the sum of the depth and the elevation of each pixel is constant and equals the elevation of this water level (Fig. 9.4c). Thus, the spatial variability of depths in the lake reflects the spatial variability of the bottom topography.

The estimated volume of the lake formed by a dam to be constructed is derived in Fig. 9.4d based on the DEM processing. For any given pixel *i* within the lake, the stored water volume *Vi* corresponds to the volume of a quadrangular prism with base area *Ai* and height *hi* ($Vi = Ai \times hi$). The variable *hi* represents the water depth at each pixel. The quantity *Ai* is the pixel's surface area, which is considered constant and equal to dx^2, where *dx* is the pixel dimension.

The lake's total volume is the sum of the volumes stored in each pixel. All pixels have the same surface area, so this area can be highlighted outside the sum. Therefore, calculating the lake volume boils down to adding the depths of all pixels and multiplying the result by the pixel's surface area (in this case, this area is dx^2).

Operationally, based on the DEM and using GIS software, the volume of a lake to be formed by constructing a dam can be determined by combining standard operations available in this type of software. There are distinct approaches for combining available operations and achieving the same result. However, a relatively quick and easy-to-execute alternative is illustrated in Fig. 9.5 and depicted as follows:

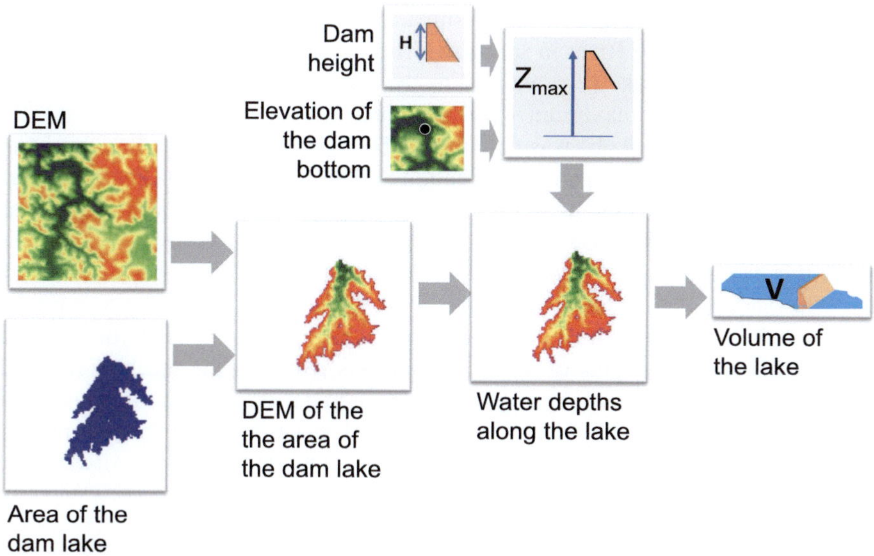

Fig. 9.5 Workflow of the main steps for estimating the water volume accumulated in the lake to be formed by constructing a dam, based on the DEM and its derived products

i. The DEM and the lake boundary (determined as described in the previous item in this chapter) generate a DEM with valid values only within the lake area. This can be done using tools that clip a raster based on the lake boundary vector. Alternatively, this step can be performed by multiplying the DEM by the lake boundary raster, which is assigned a value of 1 for pixels inside the lake and 0 for those outside.
ii. The user should define the dam's height, and depending on the exact location of the dam, the elevation of this pixel in the DEM should be identified. The sum of the height of the dam (H) and the elevation of the dam location results in the elevation of the maximum water accumulation level (Z_{max});
iii. With an operation like subtraction between a scalar and a raster, the difference between Z_{max} and the DEM generated in step i is made. The result is a raster of lake depths, whose attribute in each pixel is the lake depth considering the water level Z_{max};
iv. The last step consists of first summing up the attributes of the lake depth raster. This is commonly done with operations that quantify statistics of a raster, with the choice of sum as the statistic to calculate. Next, this resulting sum must be multiplied by the surface area of a pixel (the square of the pixel dimension, dx^2), obtaining the lake volume in m^3 (assuming DEM with elevations in meters and H and dx taken in meters). This number can be huge, being more practical to work in units like the cubic hectometer (hm^3), which is obtained by dividing the calculated value in m^3 by 1,000,000.

This approach was applied for the hypothetical reservoir worked on in the previous item, located on the Ijuí River in southern Brazil, and the DEM was cropped to the area surrounding the lake (Fig. 9.6a). For the dam height considered to be 30 m and the corresponding maximum water accumulation level (elevation 196 m), the lake depth raster (Fig. 9.6b) was calculated. This hypothetical lake has depths of up to 34 m, following the spatial distribution presented in the map. The maximum depth is higher than the dam height because the original DEM was used with terrain depressions present.

For this example, three cross-sectional depth profiles were developed to characterize the variation of depths along the reservoir section (Fig. 9.7). The A-B profile is located closer to the dam (less than 500 m upstream), and the reservoir is about 1.8 km wide and has depths of up to 30 m in this reach. In the reach of the C-D profile, located almost 17 km upstream in longitudinal distance from the axis of the dam, the width is about 800 m, and the depths still reach 28 m. In the E–F profile, located almost 46 km upstream from the dam, the width is less than 400 m and depths

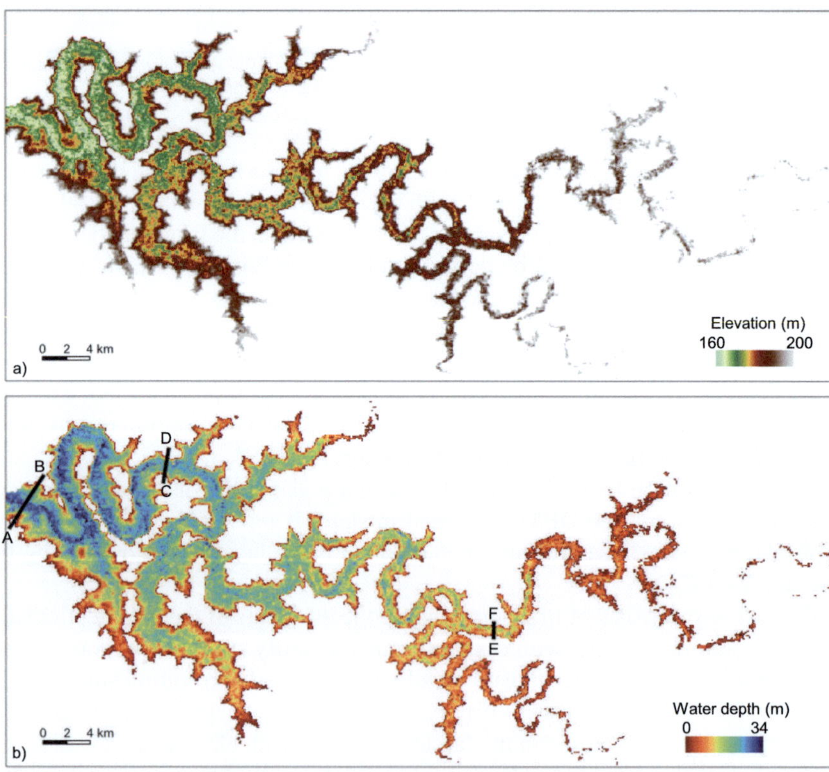

Fig. 9.6 **a** SRTM 30 DEM for the area of the lake formed by a hypothetical dam on the Ijuí River (RS); **b** Raster of lake depths of the dam (considering dam height of 30 m)

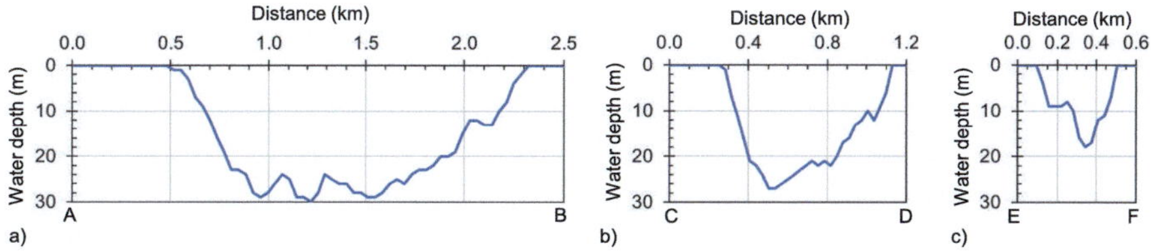

Fig. 9.7 **a**, **b**, **c** Cross-sectional depth profiles drawn along the dam lake, following the transects indicated in the previous figure

up to 18 m. In the example, the horizontal and vertical scales are equal among the three profiles, allowing for better relative comparison.

9.3 Area-Volume-Elevation Curve of a Flooded Area

Each water level in a reservoir corresponds to a specific surface area and accumulated water volume of the lake (Fig. 9.8). The so-called area-volume-elevation (AVE) curve of a reservoir establishes the relationship between the lake's area and volume and the elevation of the water level.

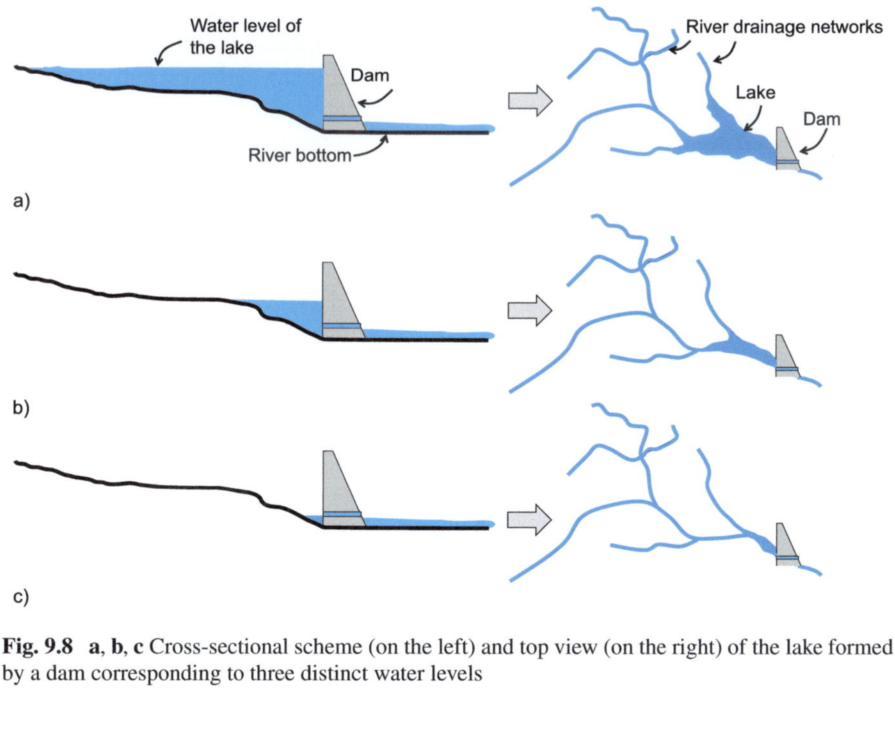

Fig. 9.8 **a**, **b**, **c** Cross-sectional scheme (on the left) and top view (on the right) of the lake formed by a dam corresponding to three distinct water levels

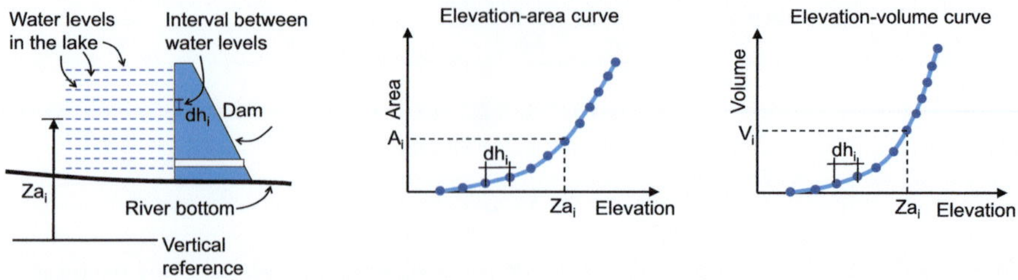

Fig. 9.9 Indication of discretization of the water levels of a dam with interval dh_i for the elaboration of the area-volume-elevation curve, where each water level has elevation Za_i that corresponds to an area A_i and volume V_i

The AVE can be expressed in three ways: (i) as a set of points presented in tabular format; (ii) in graphical format, with curves that relate elevation versus area and elevation versus volume (Fig. 9.9); (iii) as mathematical equations, also relating elevation versus area and elevation versus volume.

The AVE curve is essential for the sizing and operation of the reservoir. It is relatively cheap and straightforward to monitor the reservoir's water level and, from the AEV, estimate the corresponding area and volume. The field survey of the lake's area and volume would be high cost and unfeasible to be redone with the necessary frequency.

With the reservoir built and in operation, information on water level, surface area, and volume at each moment is key for operational procedures and decision-making. This data supports improved reservoir management following design criteria and objectives, helps minimize conflicts arising from multiple water uses, and optimizes operational rules. The relationship between water level, surface area, and volume is fundamental to various reservoir operation simulation studies, including those developed for economic forecasts, energy demand projections, and climate change scenarios.

Topographic surveys and photogrammetry usually provide data to prepare the AVE curve before dam construction. After the lake exists, this curve should be systematically reviewed and updated over time due to the deposition and accumulation of sediments at the bottom. Sediments tend to reduce the reservoir's water accumulation capacity. Bathymetric surveys are preferably used for this purpose, with recent advances in methods based on remote sensing.

Before the dam's construction and in a preliminary way, including as a criterion for defining the exact location of the dam, the AVE curve can be prepared from the DEM processing. The procedures for estimating the maximum area and volume of the lake that would result from the hypothetical construction of a dam, described in the previous sub-items, can be repeated analogously to estimate the area and volume of this lake for any water level in the dam. In this case, such procedures should be repeated for a set of water levels that range from the minimum to the maximum level of the dam, with a specific regular spacing between them (for example, 50 cm or

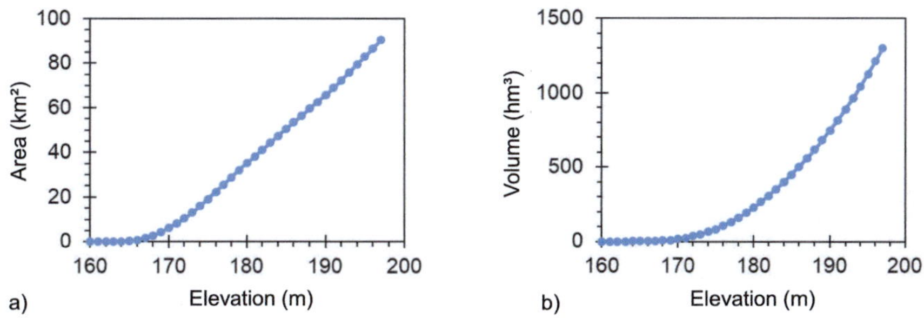

Fig. 9.10 **a**, **b** Area-elevation and volume-elevation curves in graphic format for a hypothetical reservoir on the Ijuí River (southern Brazil), at the point indicated in Fig. 9.6, obtained by DEM processing

1 m). The result is a set of points of the corresponding lake's water level, area, and volume, that is, the preliminary AVE curve of the planned reservoir.

The procedure for estimating the AVE curve from the DEM is straightforward. However, the repetition process to obtain the set of points can be laborious, as you must calculate each water level's area and volume individually.

An alternative is to automate the procedure with modeling tools available in some software so that the sequence of operations is automatically performed from the input data. This is particularly recommended if the generation of AVE curves will be carried out for several reservoirs. Another alternative is to take advantage of specific commands, plug-ins, or scripts already available in some software that prepare the AVE curve automatically.

An intermediate alternative between automation and manual repetition of operations uses a command that generates the hypsometry of a watershed, which is available in some software. In this case, the extent of analysis should be restricted to the maximum area of the lake, and the resulting hypsometry indicates the set of points of the flooded area and the corresponding water level. The numerical integration of the elevation-area curve can deduce the elevation-volume curve.

The area-volume-elevation curve generated for the lake of the hypothetical dam of the example in Fig. 9.6, on the Ijuí River in Rio Grande do Sul, obtained from the SRTM 30 DEM, is presented in Fig. 9.10, in graphical format. Meter by meter, there is a relationship between the elevation of the water level of the lake and the corresponding area and volume. According to each new water level measurement, this information could be used operationally, in real time, to determine the area and volume of water stored at that moment.

References

Avisse, N., Tilmant, A., Müller, M. F., & Zhang, H. (2017). Monitoring small reservoirs storage from satellite remote sensing in inaccessible areas (preprint). *Water Resources Management/ Remote Sensing and GIS*. https://doi.org/10.5194/hess-2017-373

Nguy-Robertson, A., May, J., Dartevelle, S., Birkett, C., Lucero, E., Russo, T., Griffin, S., Miller, J., Tetrault, R., & Zentner, M. (2018). Inferring elevation variation of lakes and reservoirs from areal extents: Calibrating with altimeter and in situ data. *Remote Sensing Applications: Society and Environment, 9*, 116–125. https://doi.org/10.1016/j.rsase.2018.01.001

Chapter 10
Distances Related to Water Flow

In studies of various areas of knowledge, in addition to the topography represented in the DEM and the layout of the drainage network, there is interest in computing distances related to the flow of water in a watershed. This chapter presents ways to compute these distances, with real-world examples and didactic explanations about interpreting the results.

Given any point on the terrain, one can calculate the shortest straight-line distance to the nearest watercourse, which is the Euclidean distance, or determine the distance from this point to reach the watercourse according to the layout dictated by the topography, that is, according to the flow paths.

Another distance that can be calculated is from each point of the watershed to reach the outlet, with reaches accounted for over the land and along watercourses. If the reference is fixed on each head pixel, that is, each pixel that does not receive runoff from any other, one can account for the distance traveled along the route to the outlet.

Analogously, it is possible to compute distances exclusively for the pixels part of the drainage network: the distances traveled from each headwater pixel in the direction of water flow to the outlet and the distances measured from the outlet along these paths (i.e., following the river upstream).

10.1 Distance to the Drainage Network in a Straight Line (Euclidean)

In several situations, the distances from any point on the surface to the nearest watercourse need to be accounted for. For example, to select suitable areas for installing a landfill or to delimit permanent preservation areas. These distances are determined in a straight line to the closest point of the drainage network. This approach of distance accounting corresponds to the so-called Euclidean distance.

A. Rolim da Paz, *Digital Elevation Models for Environmental Studies*,
https://doi.org/10.1007/978-3-032-04523-2_10

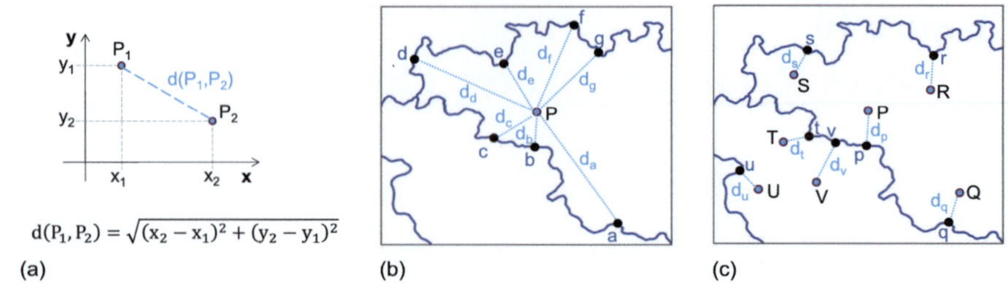

Fig. 10.1 **a** Calculation of Euclidean distance in two-dimensional (*xy*) space between two points (P_1 and P_2); **b** Indication of Euclidean distances (d_a to d_g) between any point on the surface (P) and eight points along the drainage network (*a* to *g*); **c** Indication of Euclidean distances (d_p to d_v) between seven surface points (P to V) and the corresponding closest points on the drainage network (*p* to *v*)

Figure 10.1a shows the mathematical definition of the Euclidean distance in two-dimensional space, proven by the Pythagorean Theorem. Theoretically, infinite distances can be calculated for the infinite points chosen on the network for distances between the surface and the drainage network. However, the interest is usually in identifying the minimum distance from this point to the drainage network, i.e., the distance from this point to the network's closest point along a straight line.

In the example of Fig. 10.1b, the distances established in a straight line between point P on the surface and eight points of the drainage network are indicated. If this were replicated for all points of the drainage network, the smallest of the distances would be the Euclidean distance from point P to the network. Such distance is d_b since point *b* is the point of the drainage network located closest to P in a straight line. Similarly, in Fig. 10.1c, there is an indication of the shortest distance from each of the seven surface points relative to the drainage network.

By calculating the Euclidean distance from each point on the surface to the nearest drainage network and storing this information as an attribute of the pixel corresponding to this point, you get a raster that represents the surface of Euclidean distances. In this case, the distance is computed from the centroid of each pixel to the point of the nearest drainage network. The reference drainage network for the computation of distances can be a line vector or a raster, depending on the software and command used.

Determining a raster of Euclidean distances to a drainage network, also represented as a raster, is illustrated didactically in Fig. 10.2. It is noted that pixel A is located two pixels away from the nearest drainage network pixel. Assuming a pixel size of 1 m, this distance (d_A) corresponds to 2 m. For pixels B and C, the straight-line distances between their centroid and the centroid of the nearest drainage network pixel are determined by the Pythagorean Theorem, as indicated in the figure.

By repeating this calculation procedure for all image pixels, you get the raster of Euclidean distances. The pixels that represent the drainage network have zero distance.

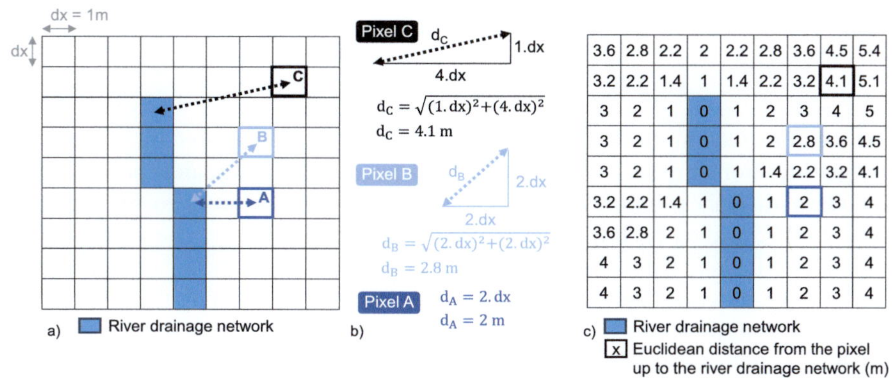

Fig. 10.2 **a** Indication of Euclidean distance between three pixels (A, B, C) and the corresponding nearest drainage network pixels to each one; **b** Calculation of the Euclidean distances of pixels A, B, and C to the drainage network; **c** Surface of Euclidean distances relative to the drainage network

A real-world example is shown to the Icamaquã river basin in Rio Grande do Sul (Fig. 10.3), with distances obtained as a raster with a spatial resolution of 90 m relative to a vector drainage network determined by processing the SRTM 90 DEM and considering a minimum drainage area threshold of 10 km^2. The spatial patterns of the distance surface are noticeable in buffers around the watercourses. Also clearly identified are the areas farthest from any watercourse, reaching around 6 km, such as the region where point A is located.

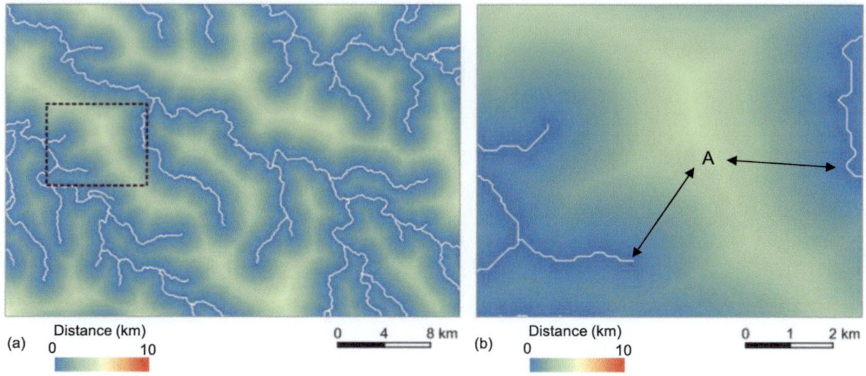

Fig. 10.3 **a** Surface of Euclidean distances relative to the drainage network (white line) of part of the Icamaquã river basin (RS), obtained from the processing of the SRTM 90 DEM; **b** Enlargement of the area indicated by the rectangle indicated in **a**, with A being a point equidistant in a straight line from the drainage networks at the points indicated by the arrows

10.2 Distance to the Drainage Network Along the Flow Paths

The previous section shows that the Euclidean distance is measured by supposing a straight line from any surface point to the drainage network's nearest point. Unlike Euclidean distances, another approach to account for distances to the drainage network is according to flow paths. Given a point on the surface, one moves downstream from it along the flow path established by the DEM processing until it reaches the drainage network. The length of this route is the distance along the flow path to the drainage network.

The conceptual difference in accounting for these two types of distances to the drainage network for two pixels (P_1 and P_2) is exemplified in Fig. 10.4. In the Euclidean distance, the reference point of the drainage network for calculating the distance is the one that provides the shortest straight-line distance for each case (points R_1 and R_2 in Fig. 10.4a). In the distance accounted for along the flow paths, the reference point of the drainage network is the one to which the downstream flow of the pixel converges—points R_1 and R_2 in Fig. 10.4b.

By definition, for a given point, the distance along the flow paths is always equal to or higher than the Euclidean distance since the straight-line distance is the shortest distance between two points. The equality between these distance measures only occurs when two conditions co-occur: (i) the flow path follows a straight path to the drainage network, and (ii) this path coincidentally reaches the nearest network point.

Analogous to Euclidean distances, one can account for the distances along the flow to the drainage network for all points and obtain a distance surface represented

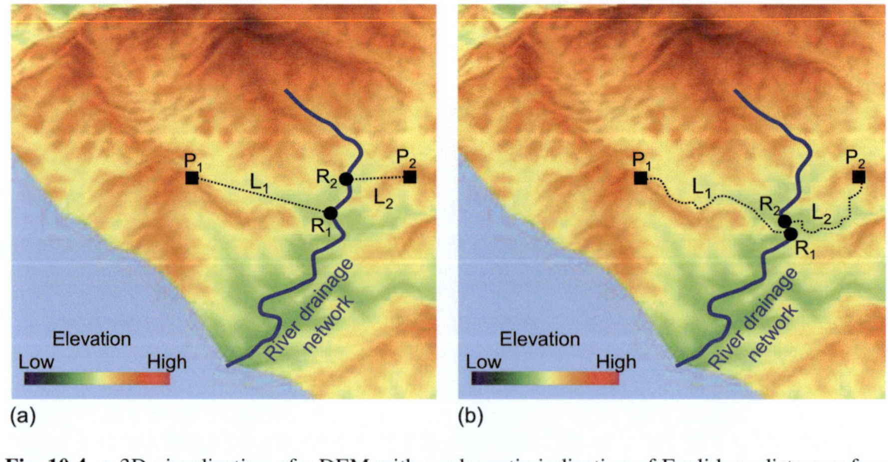

Fig. 10.4 **a** 3D visualization of a DEM with a schematic indication of Euclidean distances from pixels P_1 and P_2 to the drainage network, corresponding to measures L_1 and L_2 and to points R_1 and R_2 of the network, respectively; **b** Analogous to the previous one, but with distances accounted for along the flow paths from pixels P_1 and P_2 to the drainage network

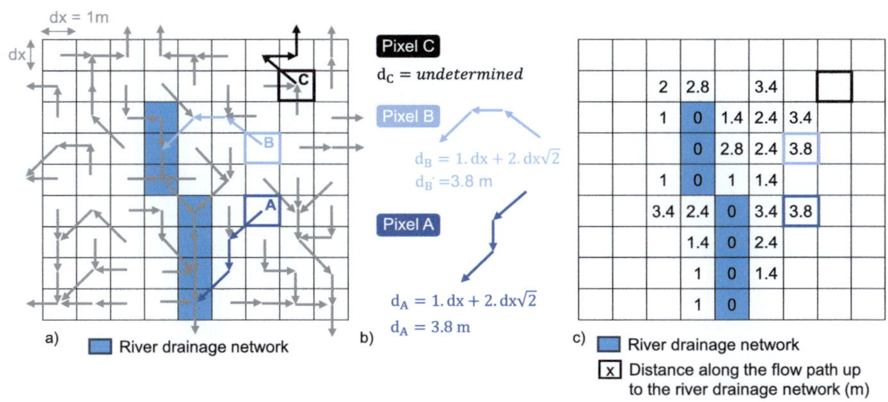

Fig. 10.5 **a** Flow directions, drainage network and indication of distance along the flow paths from three pixels (A, B, C) to the corresponding pixels of the drainage network to which they connect; **b** Calculation of the distances of pixels A, B, and C to the drainage network along the flow paths; **c** Surface of distances along the flow paths to the drainage network (pixels without value do not have determined distance as they do not connect to some drainage reach in the image)

as a raster. The drainage network, taken as a reference for computing distances, can be in vector or raster formats, the latter being more common.

In the example of Fig. 10.5, the same structure and drainage network of Fig. 10.2 illustrate Euclidean distances. For pixels A and B, there is an indication of the flow path traveled to reach the drainage network and the quantification of the extension of the corresponding route. The standard rule for measuring distance is to compute an extension equal to the pixel size (dx) for each orthogonal step and an extension equal to $\sqrt{2} \cdot dx$ for each diagonal step. For pixel C, the downstream flow from it leaves the image without reaching any part of the drainage network, resulting in an indeterminacy of the distance along the flow path to the drainage network for this pixel. Repeating the procedure for all pixels, the raster of the distance surface is obtained, as illustrated in the figure.

The application of the calculation of distances along the flow paths to the drainage network for a real-world case is exemplified in Fig. 10.6 for a reach of the Icamaquã river basin in Rio Grande do Sul; the same example worked for Euclidean distances (Fig. 10.3).

This example illustrates how the spatial patterns of the surface of distances along the flow paths are very distinct from those obtained on the surface of Euclidean distances. On the surface of Euclidean distances, geometric patterns such as buffers occur around watercourses. On the surface of distances along the flow paths, the boundaries of the contributing areas to each reach of drainage are visibly observed, that is, the topographic divides between sub-basins contributing to river reaches.

In the area highlighted in Fig. 10.6, the region between Points A and B presents distance attributes approximately 10 km apart. This difference occurs because this region has the flow paths draining to the watercourse on the left in the image, although

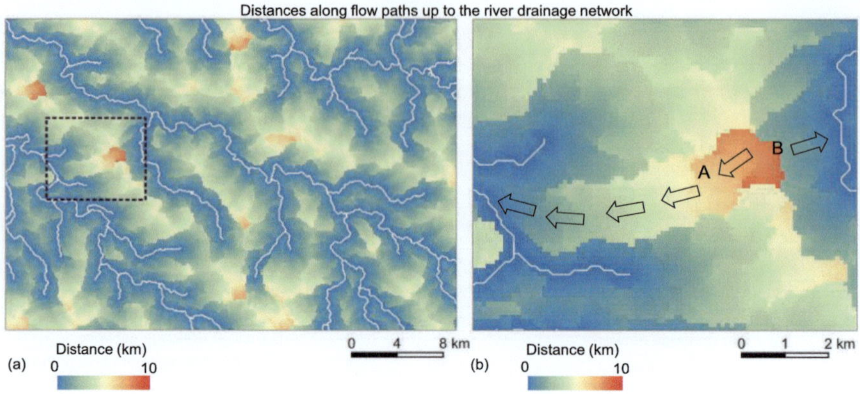

Fig. 10.6 **a** Surface of distances along the flow paths to the drainage network for a section of the Icamaquã river basin (RS), obtained from the processing of the SRTM 90 DEM; **b** Enlargement of the area indicated by the rectangle in **a**, with A being the same point indicated in Fig. 10.3a and b point in the region that separates the area on the left, whose distances refer to the section of the network on the left, and the area on the right, whose distances refer to the section of the network on the right, as indicated by the arrows; in **a** and **b**, the white line represents the reference drainage network for distance computation

there is a watercourse on the right closer to this region in a straight line (at a distance of less than 2 km).

10.3 Distances Along the Flow Paths to the Outlet

Another way to account for distances along the flow paths is relative to the basin outlet. For each pixel in the basin, the distance of the flow path downstream is quantified until it reaches the pixel selected as the basin outlet. The procedure is analogous to the one described in the previous item for distances to the drainage network, but the final reference is changed to the basin outlet.

For each pixel, one moves downstream according to the flow directions, accounting by the standard rule of distance measurement an extension equal to the pixel size (dx) for each orthogonal step and an extension equal to $\sqrt{2} \cdot dx$ for each diagonal step along this traced path. For pixels not part of the drainage network (for example, pixels A and B in Fig. 10.7), the route covered includes the section to reach the drainage network and the section along the drainage network to the outlet. For pixels that are part of the drainage network (for example, pixel D in Fig. 10.7), the traced route is formed only by the section along the network itself.

By storing the measured distance for each pixel as an attribute in a raster, the surface of distances along the flow paths to the outlet is formed. This raster has no distinction between pixels that are part of the drainage network and those that are not. For all pixels, there is distance information to the outlet according to the flow

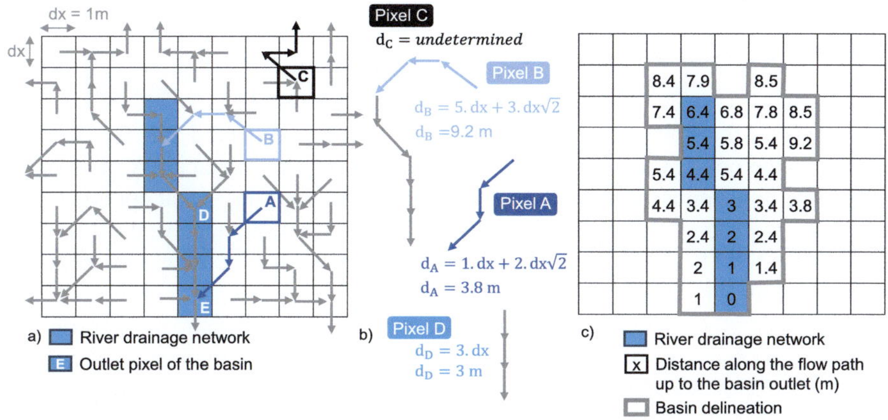

Fig. 10.7 **a** Flow directions, drainage network and an indication of distance along the flow paths from four pixels (A, B, C, D) to the basin outlet pixel; **b** Calculation of the distances from pixels A, B, C and D to the basin outlet; **c** Surface of distances along the flow paths to the outlet (pixels without value do not have a determined distance because they do not have a flow that drains to the basin outlet, that is, they do not belong to the basin)

paths. Pixels not belonging to the basin contributing to the chosen outlet do not have a determined distance, like pixel C in Fig. 10.7.

A real-world example of a surface of distances along the flow paths relative to the outlet is indicated for the Icamaquã river basin in Rio Grande do Sul, obtained from the processing of the SRTM 90 DEM (Fig. 10.8). The spatial pattern of the obtained surface illustrates how the distances to the outlet vary as one moves from downstream to upstream, toward the basin's headwaters. The spatial variation of the distances follows the pattern of the spatial configuration of the drainage network and reaches distances of up to 270 km. The areas equidistant to the outlet are noticeable, such as the regions around points A, B, C, and D, located approximately 210 km from the basin outlet. On the other hand, the area of point E is geographically adjacent to the region of point D, but is located almost 60 km further from the outlet than point D due to the different flow paths that follow from each of these points.

Fig. 10.8 Surface of distances along the flow paths to the outlet of the Icamaquã river basin (RS), with superposition of the drainage network (white line), both derived from the SRTM 90 DEM

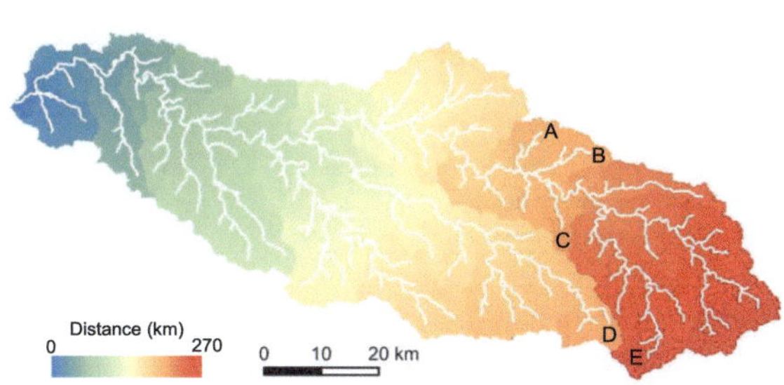

10.4 Distances Along the Flow Paths from the Flow Initiation Point

According to the flow paths, another distance can be accounted for: the distance relative to the point of flow initiation. DEM processing shows that every pixel in the image has flow connectivity with the others, forming flow paths that continue until they leave the image. However, some pixels in the image do not receive the flow from any other pixel. These pixels are characterized as flow initiation points (for example, A, B, and F in Fig. 10.9).

From each of these flow initiation pixels, one can move downstream according to the flow directions, accounting for the distance covered since the start of the route, with the same standard rule of accounting for an extension equal to the pixel size (dx) for each orthogonal step and an extension equal to $\sqrt{2} \cdot dx$ for each diagonal step.

From pixel F to pixel G in Fig. 10.9, for example, a distance of 1 dx (1 m, in this case) is covered, and at pixel H, the distance covered from F is 2 dx (2 m). The pixel downstream of H is pixel I. Unlike the previous ones, pixel I receives not only the flow coming from F but also from four other points of origin of the flow. The distance traveled from each of these flow initiation points to reach the point I have specific values, which may or may not coincide with each other. In this case, it is usual to consider the highest of these distances as the value to be assigned to pixel I, corresponding to the flow path originating at pixel B (6.2 m, having one orthogonal section and two diagonal sections).

Therefore, it is understood that the distance stored in a pixel corresponds to the distance measured relative to the farthest upstream start of the flow. Repeating this procedure for all the pixels in the image, a raster is obtained that represents the

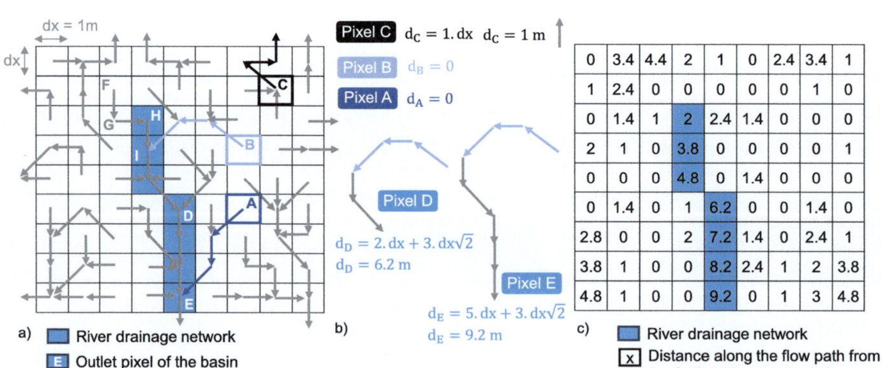

Fig. 10.9 **a** Flow directions represented by arrows and indication of the drainage network and some pixels taken as an example; **b** Calculation of distances along the flow paths from the farthest upstream start of flow to reach pixels A, B, C, D, and E; **c** Surface of distances along the flow paths from the farthest upstream start of flow

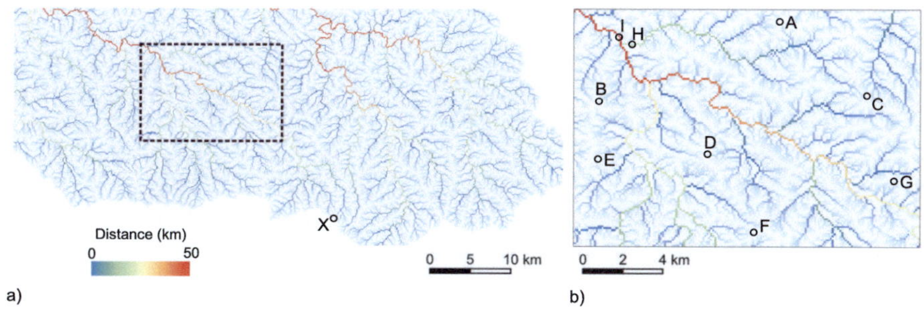

Fig. 10.10 **a** Surface of distances along the flow paths from the farthest start of the flow, for the upper part of the Icamaquã river basin (RS), obtained from the processing of the SRTM 90 DEM; **b** Enlargement of the area indicated by the rectangle in **a**

surface of distances along the flow paths accounted for from the farthest start of the flow.

Figure 10.10 illustrates a real-world example of a distance surface of this type for a section of the Icamaquã river basin in Rio Grande do Sul. Very short-distance regions are distributed throughout the area, as there are starting points of flow throughout the area (like points A to G, for example), which should not be confused with the starting points of the drainage network.

As the flows are traversed, the distances from the start of the path increase, highlighting the main flow paths that coincide with sections of the drainage network, like points H and I, for example. However, the reader should notice that, although H and I are spatially very close to each other and I is downstream of H, they present distances from distinct farthest upstream points. This is because the distance recorded at H refers to the flow traced from the point around C, which is the farthest start of flow that passes through H. In contrast, the distance measured for point I refers to a flow path from a point located upstream of the area enlarged in Fig. 10.10b (point X indicated in Fig. 10.10a).

10.5 Distances Along the Drainage Network from the Headwaters

After processing the DEM, the drainage network can be derived in raster format, and specific distances can be accounted for exclusively for the pixels that belong to this network. The two primary ways are the distances calculated along the drainage network from the headwaters and the distances calculated relative to the basin's outlet. This sub-item deals with the first approach, while the second is described in the following sub-item.

For each start of the drainage network, the pixel that is part of the network and is located farthest upstream can be identified and considered as the headwater of that

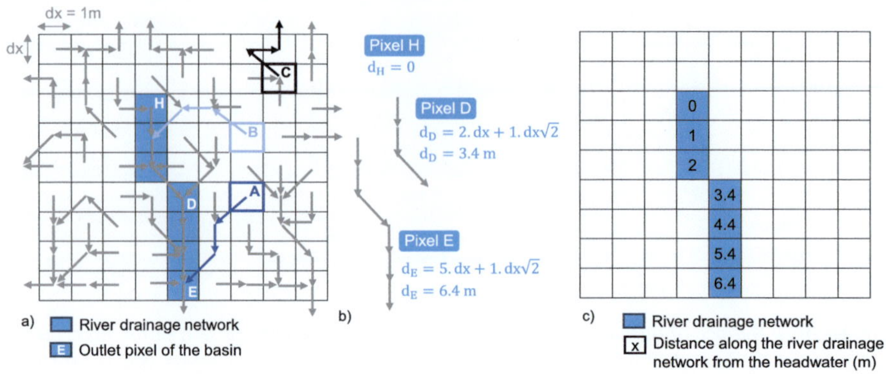

Fig. 10.11 **a** Flow directions represented by arrows and an indication of the drainage network and some pixels taken as an example; **b** Calculation of distances along the drainage network from the headwater H to reach pixels D and E; **c** Raster of distances along the drainage network from the headwater

section (pixel H in the example of Fig. 10.11). Starting from this pixel, the network is traversed downstream according to the flow directions, and the accumulated distance covered from the start point (headwater) is quantified. At each orthogonal step, the pixel size (dx) is added to the distance covered, and at each diagonal step, $\sqrt{2} \cdot dx$ is added. In each pixel of the drainage network, the information regarding the distance covered from the headwater (Fig. 10.11) is stored.

As the network is followed downstream, there are confluences of river reaches. For a given network pixel, there may be several headwaters upstream. In this case, the information to be stored is the one with the longest accumulated distance, i.e., the distance relative to the headwater located farthest from the pixel.

By performing this procedure for all the pixels in the network, a raster is obtained whose attribute is the distance along the drainage network from the farthest headwater. Pixels not belonging to the network do not have a determined value (for example, pixels A, B, and C and many others in Fig. 10.11).

A real-world case of a raster of distances along the drainage network from the headwaters is illustrated in Fig. 10.12 for the upper part of the Icamaquã River basin in southern Brazil. In the displayed region, the calculated distances reach about 30 km. In the section indicated by the letter G, the distances are around 24 km. Upstream of G is the confluence of two river arms, whose distances from the headwaters are distinct. The river arm that passes through E is about 13 km from the headwater, which is point A. The river arm that passes through F has three headwaters upstream, indicated by points B, C, and D, and is 22 km from the farthest headwater (point B).

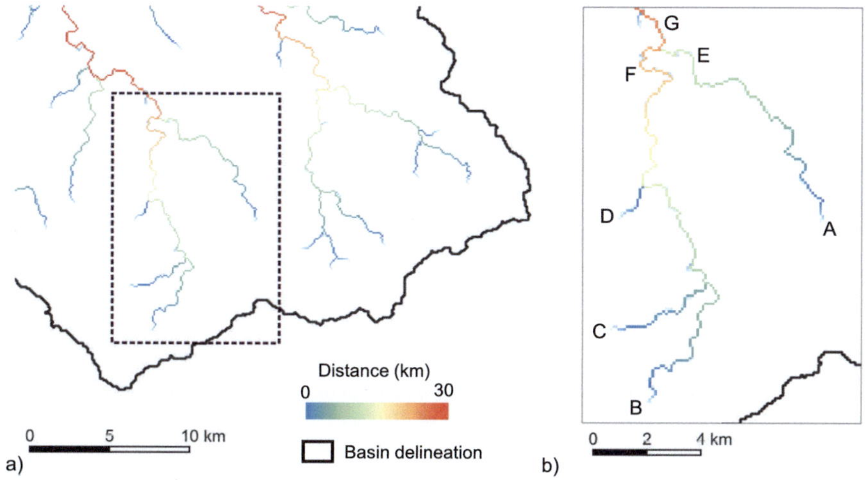

Fig. 10.12 **a** Distances along the drainage network from the headwaters for the upper part of the Icamaquã river basin (RS), obtained from the processing of the SRTM 90 DEM; **b** Enlargement of the area indicated by the rectangle in **a**

10.6 Distances Along the Drainage Network from the Outlet

Instead of measuring distances relative to the headwaters, as in the previous sub-item, distances along the drainage network can be accounted for relative to the outlet. In this case, one should start from the outlet and walk upstream along the drainage network. The standard rule for computing distances is employed, adding to the distance traveled the pixel (dx) size at each orthogonal step and adding $\sqrt{2} \cdot dx$ at each diagonal step.

For example, pixel D in Fig. 10.13 is 3 m away from the outlet defined as pixel E, while pixel H is at a distance of 6.4 m. Pixels that do not integrate the drainage network do not have a distance to be calculated, as in the case of pixels A, B, and C in the example.

Upon reaching a confluence or junction in the drainage network, one arm of the river is traversed upstream to the headwater. Then, it returns to the junction and continues upstream along the other river arm. By repeating this procedure for all the pixels in the drainage network, a raster is obtained whose attribute indicates the distance along the drainage network relative to the outlet.

A real-world example of this type of raster is shown in Fig. 10.14 for the Icamaquã river basin (RS). A spatial pattern of increasing distances is perceived as one walks from the mouth along the different watercourses toward the headwaters to the southeast. Points located on different equidistant watercourses relative to the mouth can also be identified. For example, points A, B, and C are about 60 km away from the mouth, while points D, E, F, and G are around 130 km from the mouth, points H,

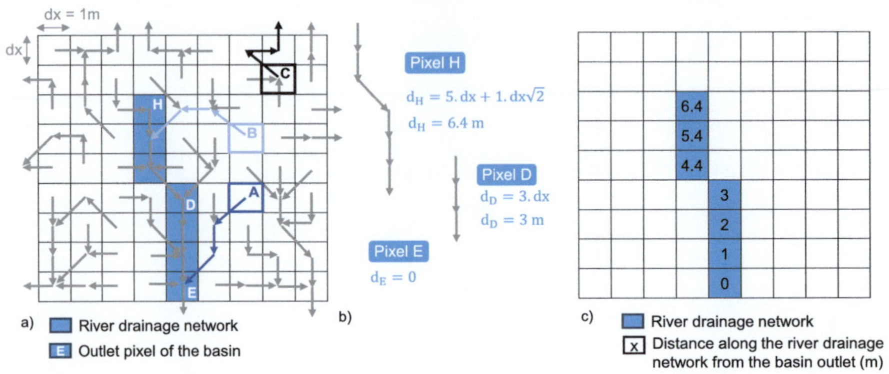

Fig. 10.13 **a** Flow directions represented by arrows and an indication of the drainage network and some pixels taken as an example; **b** Calculation of distances along the drainage network from the outlet to pixels H, D, and E; **c** Raster of distances along the drainage network from the outlet

I, and J are around 190 km, and points K, L, and M are very close to the farthest headwaters from the mouth (more than 250 km away from the basin outlet, which is point X).

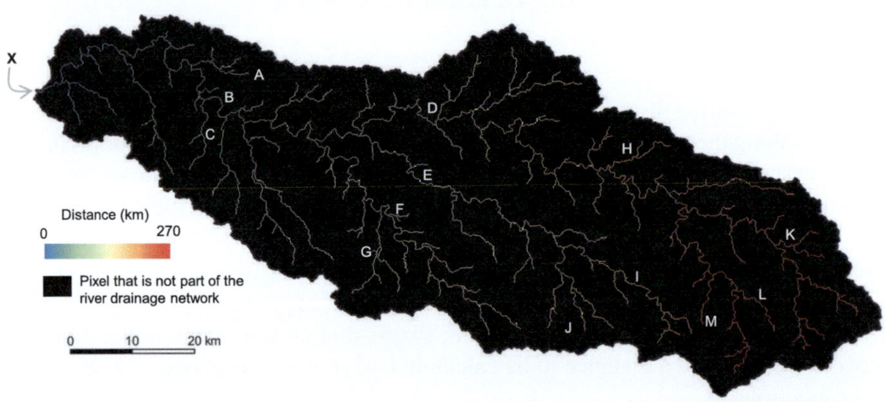

Fig. 10.14 Raster of distances along the drainage network relative to the basin outlet (point X) for the Icamaquã river basin (RS), obtained from the processing of the SRTM 90 DEM

Chapter 11
Topographic Indices—Part 1: Neighborhood Relative Elevations and Terrain Roughness

A versatile DEM-derived type of product is the so-called topographic index. It is a way to explore and relate the information on spatial variation of topography to help explain different patterns and processes in knowledge areas such as hydrology, ecology, geography, renewable energies, agronomy, and climatology (Amatulli et al., 2018; Florinsky, 2017). Some of these indices are obtained directly and solely from the DEM. Other indices may require further DEM-derived products or even DEM-independent auxiliary information.

The general idea of topographic indices is to take advantage of the key role of topography in the environment and the enormous influence that topography has as a primary or secondary explanatory variable on numerous processes in various themes. Another favorable aspect is the relative ease of obtaining and processing topographic data, including automating the processing and generation of analyses with the development of computational algorithms. Topographic data can also be applied to various spatial scales. In contrast, specific variables related to these environmental processes have limited or non-existent availability of observations.

Most topographic indices are considered a composite type of attribute derived from the DEM, as they involve the combination of primary attributes directly derived from the DEM, such as elevation and slope (Moore et al., 1991). Therefore, some of these indices describe or characterize the spatial variability of specific processes that occur in the landscape, such as soil moisture distribution and sheet erosion potential. Other indices can be auxiliary to investigate the occurrence and spatial distribution of very complex geomorphological, hydrological, and ecological processes (Moore et al., 1991).

In ecology, for example, topographic features portrayed by indices can be used to study the difficulty of species movement in the environment, the ability of the habitat to provide shelter from extreme weather conditions, the ability to help escape from predators or prey, and the spatial variability of plant and animal species diversity, among other purposes. For example, topographic diversity has been pointed out as a

A. Rolim da Paz, *Digital Elevation Models for Environmental Studies*,
https://doi.org/10.1007/978-3-032-04523-2_11

key piece for the resilience of protected areas in the face of climate change (Lawrence et al., 2021).

Another field of application for topographic indices is analyzing the environment's exposure to wind. Topographic wind exposure refers to the degree of protection or shelter the topography offers to a particular terrain location. Understanding and spatial mapping of this issue contribute to the study of the spatial distribution of snow, optimal location of wind farms, ecological processes such as spatial distribution of plant species and the transition from seedlings to trees, spatial patterns of air pollution, and potential wind impact on plantations and trees, for example.

An extensive set of topographic indices is presented in the literature, some of which are improvements proposed from previously established indices. This chapter and the following two chapters present topographic indices grouped into three major classes. This chapter presents 10 topographic indices related to neighborhood relative elevations or the terrain roughness. Each index's conception, formulation, and purpose of application are described, followed by a real-world example.

11.1 Topographic Position Index (TPI)

The Topographic Position Index (TPI) is defined as the difference between the terrain elevation at a point and the average elevations of a pre-established neighborhood around that point (Eq. 11.1 and Eq. 11.2) as a function of a specific search radius (Weiss, 2001; Fig. 11.1). This index is equivalent to the difference from mean elevation (or simply DIFF), a measure of relative topographic position presented by Wilson and Gallant (2000). The unit of the TPI is the same as the elevations.

$$\mathrm{TPI}_{\mathrm{P}} = Z_{\mathrm{P}} - \overline{Z_n} \tag{11.1}$$

$$\overline{Z_n} = \frac{1}{n}.\sum_{i=1}^{n} Z_i \tag{11.2}$$

where $\mathrm{TPI}_{\mathrm{P}}$ is the value of the topographic position index at point P, Z_{P} is the elevation at P, $\overline{Z_n}$ is the average of the elevations of the pixels in the neighborhood of P, n is the number of pixels in this neighborhood, Z_i is the elevation of pixel i in this neighborhood.

The TPI is mainly useful in distinguishing topographic features such as hilltop areas, valley bottoms, lowlands, ridges, etc. For example, this topographic characterization was used as an auxiliary variable for automatically detecting the channel initiation points when deriving a river drainage network from a DEM (Barbedo et al., 2022).

By definition, the determination of the TPI only depends on the DEM itself. Once the search radius is chosen, all pixels located in this neighborhood are identified and

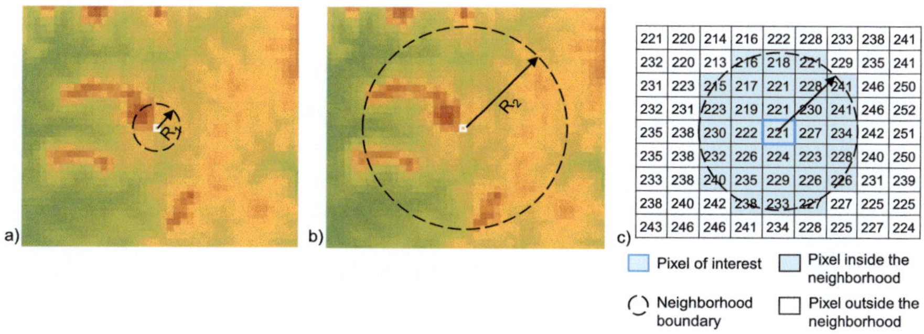

Fig. 11.1 **a**, **b** Example of neighborhood definition based on radii R1 and R2, respectively, for determining the TPI of the DEM pixel highlighted in white; **c** DEM with elevations in meters and identification of pixels located inside and outside the neighborhood for determining the TPI of a target pixel

used to calculate the average of the elevations, which is then used to determine the difference relative to the central pixel under analysis.

The larger the search radius, the higher the number of pixels that make up the neighborhood for calculating the TPI (Fig. 11.1a and b). Whether or not a pixel belongs to the neighborhood defined by the search radius can be decided, for example, by checking whether the centroid of the pixel is located or not, respectively, within the circle established by the radius (Fig. 11.1c).

The standard procedure uses the arithmetic mean to obtain the neighborhood's average elevation. Alternatively, some software allows the adaptation of the TPI calculation by a weighted average of the neighborhood elevations. In this case, the idea is to use weights to distinguish between points closer and farther from the central point, with higher weights for the former. Weighting methods include the inverse of the distance raised to a power, an exponential function, or a Gaussian approximation.

When working with a raster DEM, the neighborhood can be defined as a square matrix window instead of specifying a radius, with the same number of rows and columns around the central pixel under analysis. For example, one can consider neighborhood moving windows of 3 rows by 3 columns (3×3), 5×5, 7×7, or even larger dimensions, always maintaining an odd number of rows and columns. This window traverses the entire image, centered on each pixel analyzed each time, configuring an operation similar to spatial filtering.

Figure 11.2 illustrates calculating the TPI for a hypothetical area given the DEM and considering a 3×3 window as a neighborhood. For example, for pixel P (whose elevation is 215 m), the average elevation of the 9 pixels that make up the neighborhood of P is 218.6 m, considering the standard arithmetic mean approach. As this value is higher than the elevation of P, this pixel gets a negative TPI of − 3.6 m.

A negative TPI does not imply that all pixels in the neighborhood have a higher value than the pixel under analysis. Instead, this pixel has an elevation lower than the neighborhood average. In the example, there is a pixel in the neighborhood of pixel P with an elevation lower than that of pixel P. Pixel Q has a positive TPI of 2.3 m.

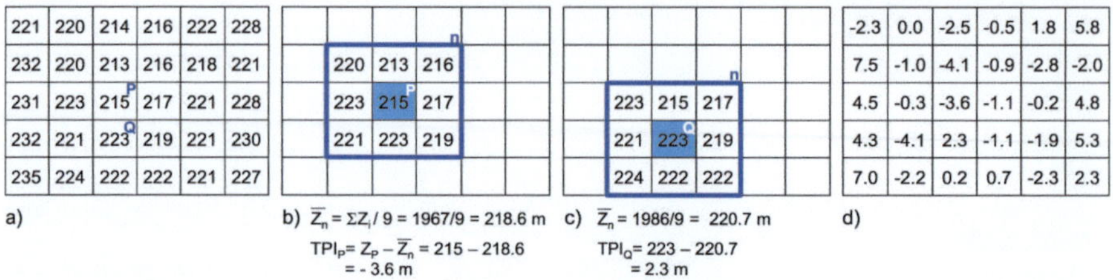

Fig. 11.2 **a** DEM with elevation in meters; **b** Calculation of topographic position index (TPI) for pixel P considering neighborhood (n) of size 3×3; **c** Same for pixel Q; **d** TPI raster (meters)

This indicates that it is an area of higher elevation relative to the neighborhood, but not necessarily all neighbors are lower than it. In calculating the example, a simplification was adopted that pixels located at the edge of the image had the TPI determined considering only the existing neighboring pixels within the 3×3 window.

In some software, there is a specific command to determine the TPI, having the DEM as input data. However, it is possible to determine this index by combining standard operations found in most GIS software. One approach is by combining the following two steps:

i. Apply a spatial average filter to the DEM with the desired neighborhood window size (see Chap. 3). This operation results in a raster of elevations, where each pixel has the average of the elevations of the defined neighborhood as an attribute.
ii. Subtract the DEM raster from the raster of the average elevations obtained in the previous step. The result of this subtraction is a raster whose attribute in each pixel is the TPI index.

In general, negative TPI values can be associated with valleys, while positive TPI values can indicate tops or ridges, as in the case of points P_5 and P_8, respectively, in Fig. 11.3. The intensity of the positive value can be used to distinguish more prominent ridges in the landscape (point P_8) relative to softer hilltops (point P_5).

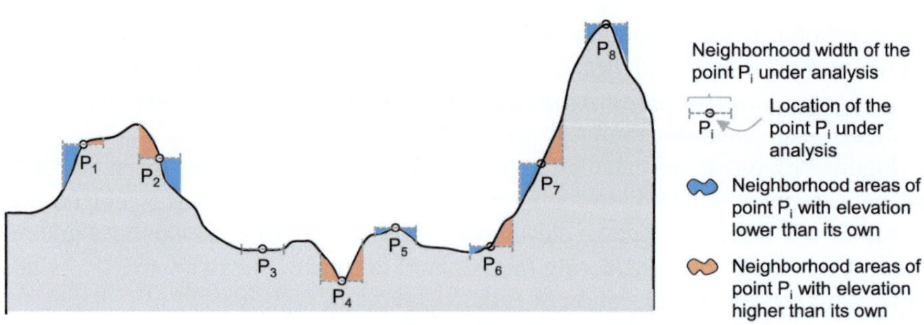

Fig. 11.3 Topographic profile indicating 8 points (P_1 to P_8) with assignment of equal width of the neighborhood window and distinction of neighboring areas of higher elevation and lower than the central point, in different situations: hilltop or peak (P_8 and P_5); flat areas (P_3); areas of constant slope (P_2 and P_7); base of hill (P_6); bottom of valley (P_4); hill edge (P_1)

A positive TPI can also be obtained for hill edge areas (point P_1, for example), indicating a sharp topographic transition with a predominance of lower neighboring areas. However, higher neighboring areas are also present. The opposite situation, where neighboring areas are predominant with higher elevation than the central point (still presenting neighboring areas of lower elevation), is typical of hill base regions (point P_6).

Null or minimal TPI values may indicate flat areas with slight topographic variation relative to the neighborhood (point P_3). These values can also correspond to regions of approximately constant slope, whether steep or not (points P_7 and P_2).

Due to the influence of slope in interpreting TPI results, it is recommended that slope raster analysis be combined with TPI analysis to make this type of distinction of geomorphological features (Weiss, 2001).

The relief from the TPI can also be characterized by considering thresholds of standard deviation quantity relative to the elevation of the point under analysis, instead of considering thresholds of the absolute values of the index. In this way, the analysis incorporates the variability of the elevations of the neighborhood (Weiss, 2001), making it so that points with the same TPI value can be classified differently in different areas, depending on the variability of the elevations of the neighborhood of each point. For a more in-depth discussion on this aspect, the reader is invited to read the work of Weiss (2001).

The neighborhood size considered for the TPI calculation significantly influences the index's resulting value and, consequently, the topographic characterization estimated from it (Salinas-Melgoza et al., 2018). The increase in the search radius causes more pixels to be part of the neighborhood, as in the case of R_2 relative to R_1 in Fig. 11.1, and covers higher distances. Increasing this distance increases the tendency for more significant topographic variation within the neighborhood relative to the central point. Tiny neighborhoods tend to lower TPI values due to a higher chance of proximity between the elevations of neighboring pixels and the central pixel.

For example, for the hypothetical relief schematized in Fig. 11.4a, the neighborhood defined by n_1 around point P results in an approximately zero TPI for this point. With the increase in the size of the neighborhood, with values such as n_2 (Fig. 11.4b) and n_3 (Fig. 11.4c), the TPI value at P becomes positive, and the value increases since, for such neighborhoods, the average of the elevations is lower than the elevation at P (Z_P). However, due to the region's topographic variation pattern, when considering an even broader neighborhood around P, such as size n_4 (Fig. 11.4d), there is a predominance of points of elevation higher than Z_P to the detriment of points of lower elevation. In this case, the average elevation of the neighborhood is higher than the elevation at P; therefore, the TPI becomes negative at P.

In an analogous and reverse sense, the TPI sign for point Q goes from almost null to negative and then to positive with the increase of the neighborhood (Fig. 11.4e to h). There may be a specific window width between n_3 and n_4 that results in an approximately zero TPI for Q, as a balanced compensatory effect of points with higher and lower elevation than Q. If this occurs, from the point of view of TPI

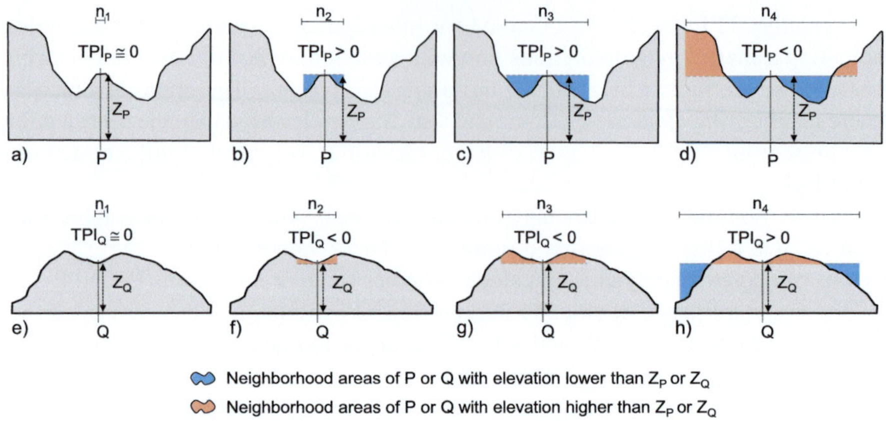

Fig. 11.4 **a**, **b**, **c**, **d** Hypothetical area with an indication of the determination of the TPI value for point P, whose elevation is Z_P, as a function of different neighborhood sizes: n_1, n_2, n_3, and n_4, respectively; **e**, **f**, **g**, **h** Another hypothetical area with indication of the determination of the TPI for point Q (elevation Z_Q) for neighborhoods of size n_1, n_2, n_3, and n_4, respectively

analysis with this window size, the region around Q would be flat. The same can happen for an intermediate window between n_3 and n_4 for pixel P (Fig. 11.4c and d).

More generally, the neighborhood size adopted in the TPI calculation determines the scale at which the landscape is characterized regarding relief. A larger neighborhood indicates that the TPI analysis aims to represent the main features of the relief, considering a general view of the landscape. A smaller neighborhood suggests that the study focuses on the more localized variations of the relief.

The definition and influence of neighborhood size vary according to the characteristics of the study area and the purpose of the index's application, with no indication of absolute values. Zwolinski and Stefanska (2015) indicate that the more significant the heterogeneity of a region's relief, the smaller the neighborhood should be for calculating the TPI.

An interesting recommendation is to combine TPI results applied at two scales (one larger and one smaller), which facilitates the identification of relief classes (Weiss, 2001). The relationship between these scales also influences the results.

In addition, the literature also uses the approach of standardizing TPI values, that is, transforming the obtained values into a new raster with a zero mean and variance equal to one. This is equivalent to taking the TPI raster, subtracting the mean value of the raster from each pixel, and then dividing by the raster's standard deviation.

Ideally, field visits and knowledge of the area help to find a better configuration for calculating the TPI through trial and error and enhance the understanding of the values obtained from this index and the consequent characterization of the relief.

An example of determining the TPI index in a real-world case is illustrated in Fig. 11.5 for an area in the municipality of Franca, located in the northeast of the state of São Paulo, from the SRTM 30 DEM. It illustrates how the TPI determined on a smaller scale identifies the localized topographic variations and has more significant

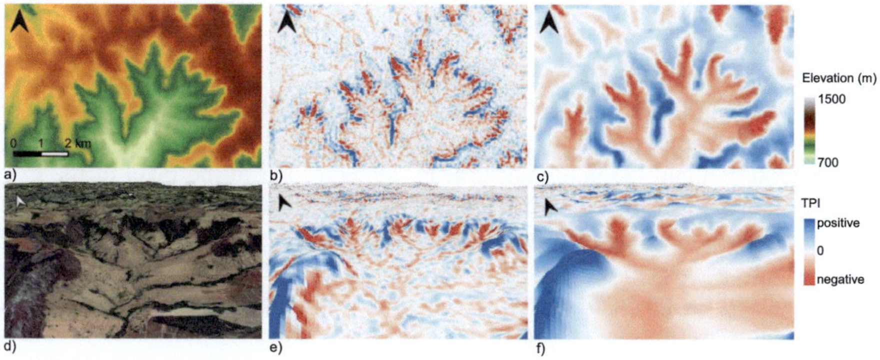

Fig. 11.5 **a** SRTM 30 DEM for an area in the municipality of Franca (SP); **b** TPI calculated with local neighborhood analysis (0 to 100 m); **c** TPI calculated with extended neighborhood analysis (250 to 1000 m); **d**, **e**, **f** 3D view generated from the DEM with an overlay of Google Earth satellite image of the TPI in the local analysis and the TPI in the extended analysis, respectively

spatial heterogeneity, with more remarkable alternation between positive and negative TPI values. On the other hand, the TPI calculated for the larger scale presents a higher spatial continuity of positive or negative TPI and characterizes the main features of the relief.

One way to apply the TPI index to watersheds is to extract and analyze the values of this index for a marginal strip to the watercourses (a buffer), as Weiss (2001) suggested. Considering the division of the basin into sub-basins, the average TPI value in the marginal buffer to the rivers in each sub-basin is then obtained. The interpretation of the results is as follows: sub-basins with higher negative average values have a higher proportion of watercourses in relatively deeper and narrower valleys. Again, the neighborhood size considered in obtaining the TPI defines the scale of analysis. For a smaller neighborhood, the analysis reflects how embedded the watercourse is and how narrow the valley is. For a larger neighborhood, the analysis demonstrates the broader morphology of the valley. For example, Weiss (2001) used 300 m and 2000 m neighborhoods for this analysis.

Other topographic indices have been proposed in the literature similar to the TPI index, in the sense of evaluating the elevation of a point regarding the comparison with the elevation of the neighborhood or elevations of another reference in the image. These indices can be called local topographic position metrics (Newman et al., 2018). These indices can generally be subdivided into two groups, representing absolute or relative values.

The TPI is classified within a class of indices working with absolute values. It is the most used of all metrics for evaluating the local topographic position. As already seen, the TPI quantifies the elevation difference of a point relative to the average of the neighborhood elevations. A variant of the TPI in this same category is the index of topographic position referred to as the median. In this case, the average of the neighborhood elevations is replaced by the median, as described in the following item in this chapter.

The terrain descriptor HAND (see item 12.7 of Chap. 12) can also be classified as a variant of an index that evaluates the local topographic position. However, in this case, the reference is not some attribute of elevation of the geometrically defined neighborhood around the pixel but, instead, the elevation of the pixel of the river drainage network to which the pixel under analysis connects.

The category of relative metrics of local topographic evaluation gathers indices that evaluate the elevation of each pixel of the DEM relative to the distribution of elevations of the pixels of a geometrically defined neighborhood or according to another reference criterion. This set of indices includes elevation percentile, deviation from mean elevation, percentage of elevation range, and relative topographic position, which are described in the following items.

11.2 Topographic Position Index Referred to the Median Elevation (TPI^{med})

A TPI index variant quantifies the elevation difference of a point relative to the median of the neighborhood elevations (TPI^{med}; Eq. 11.3), configuring a TPI that takes as reference the median and not the average of the elevations. The purpose of applying the TPI^{med} index is the same as that of the TPI.

An advantage of this approach relative to the original TPI is that the median is less sensitive than the average to extreme values, which can be irregularities or anomalies in the DEM due to noise. The occurrence of unreal extreme elevation values tends to disturb the average much more than the median, impacting the value of the TPI and leading to mistaken interpretations. Like the TPI, the TPI^{med} unit is the same as the elevations.

$$TPI_{P}^{med} = Z_{P} - Z_{n}^{med} \tag{11.3}$$

where TPI_{P}^{med} is the value of the topographic position index referred to the median at point P, Z_P is the elevation at P, and Z_n^{med} is the median of the elevations in the neighborhood of P.

Figure 11.6 illustrates the calculation of TPI^{med} for a DEM. For point P, the median elevation of the 3×3 neighborhood is 219 m, making TPI^{med} negative with a value of – 4 m. Point Q has a neighborhood with a median elevation of 222 m, resulting in a TPI^{med} of 1 m.

Operationally, the same two steps indicated for determining the TPI index may be used to obtain the TPI^{med} index in geoprocessing or GIS software:

i. Application of spatial filter to the DEM
ii. Subtraction operation between the DEM and the result of the filtering.

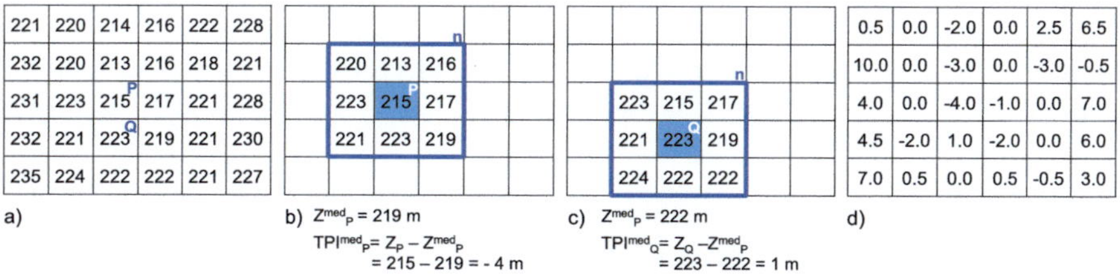

Fig. 11.6 **a** DEM with elevation in meters; **b** Calculation of the topographic position index referred to the median (TPI^{med}) for pixel P, considering a neighborhood (*n*) of size 3 × 3; **c** Same for pixel Q; **d** Raster of TPI^{med} (meters)

The difference in determining the TPI^{med} compared to the TPI is the type of spatial filtering, which becomes the median in this case of the TPI^{med}.

As an example of results from a real-world area, Fig. 11.7 shows the TPI^{med} obtained from the SRTM 30 DEM with a 7 × 7 neighborhood window for a region between the municipalities of Franca and Patrocínio Paulista in the northeast of São Paulo. The standard TPI, obtained with the arithmetic mean and the same window size, is also compared. The spatial pattern of the TPI^{med} is very similar to the TPI. A higher concentration of more intense negative values is most notably observed in the valley bottom areas, where watercourses flow.

Among the results of the TPI and TPI^{med} of this example, a systematic difference is the higher agglomeration of pixels of the same sign and close values, whether negative or positive, in the TPI. As a result, there are relatively uniform patches

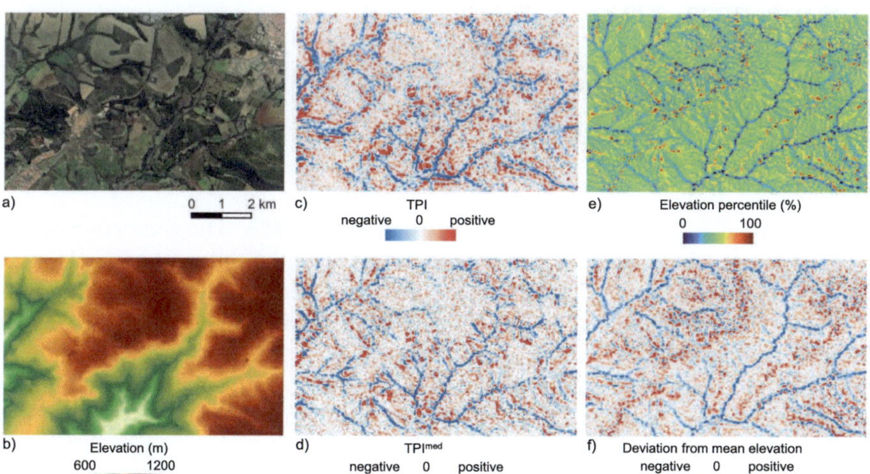

Fig. 11.7 **a** Google Earth satellite image of an area of the municipalities of Patrocínio Paulista and Franca (SP); **b** SRTM 30 DEM; **c, d, e, f** Topographic indices TPI, TPI^{med}, elevation percentage and relative deviation to the average elevation, respectively, calculated for 7 × 7 window type neighborhood

with larger dimensions in the TPI raster than in the TPI^{med}. This effect results from the arithmetic mean operator, which is more influenced by extreme values than the median.

11.3 Elevation Percentile (EP)

One of the relative local topographic position metrics is the elevation percentile (EP). The literature provides two definitions for this same topographic index. It can be defined as the percentage of points in the neighborhood with a lower or higher elevation than the point under analysis.

Wilson and Gallant (2000) and other authors, such as Walker et al. (2020) and Gallant and Dowling (2003), adopt the consideration of points with lower elevation. Other authors use the EP index to represent the percentage of neighboring points whose elevation is higher than that of the point under analysis, such as Newman et al. (2018) and as adopted in the WhiteBox GAT software (Lindsay, 2016).

In this book, we adopt the version of the EP index defined in terms of the percentage of points in the neighborhood with elevation lower than that of the point under analysis. The definition of the neighborhood of a pixel P can be based on a search radius or a matrix window (a given number of rows and columns) around P. The counting of pixels in the neighborhood that have an elevation lower than the elevation of P is made, and the value of EP is obtained by dividing the result of this count by the total number of pixels that make up the neighborhood, then multiplying by 100 to get percentage values (Eq. 11.4).

$$EP_P = \frac{\text{Counting}_{i \in n}(Z_i < Z_P)}{n} \cdot 100 \tag{11.4}$$

where EP_P is the elevation percentile index at point P, Z_p is the elevation at P, Z_i is the elevation at pixel i of the neighborhood of P, and n is the number of pixels in the neighborhood of P.

The value of EP can assume values between 0 and 100%. The zero value occurs when the pixel under analysis has the lowest elevation of the entire set of pixels in the neighborhood. The maximum of 100% is when this pixel has the highest elevation. A value of EP equal to 30%, for example, indicates that the analyzed pixel has an elevation higher than 30% of the pixels in the neighborhood.

The EP index has applications, for example, in evaluating the spatial distribution of plant species and assisting in archeological investigations (Llobera, 2001; Wilson & Gallant, 2000).

Figure 11.8 shows an example of EP calculation from a DEM, considering a neighborhood defined by a 3 × 3 window. Pixel P has an elevation of 215 m and only one pixel in the neighborhood (8 pixels) with an elevation lower than it. The EP of pixel P is 12.5%. For pixel Q, there are 7 of the 8 neighbors with lower elevation, which results in an EP value of 87.5% in this case. In the example of the figure, the

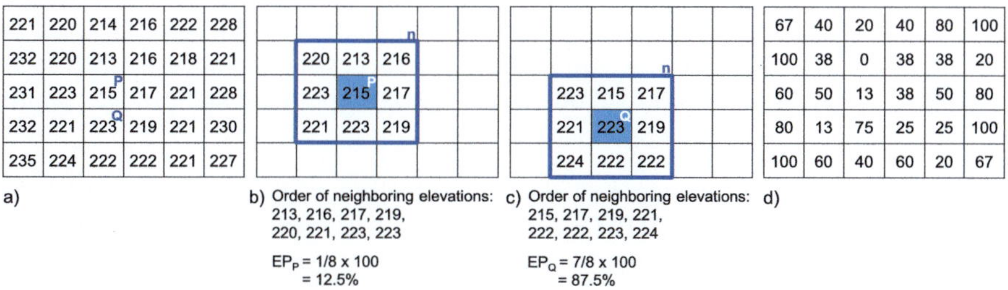

Fig. 11.8 **a** DEM with elevation in meters; **b** Calculation of the elevation percentile metric (EP) for pixel P, considering a neighborhood (n) of size 3 × 3; **c** Same for pixel Q; **d** Raster of EP (%)

value of the EP for the edge pixels was calculated considering the number of valid neighbors in each case.

Due to its formulation, the EP index is considered a robust metric for assessing local topographic position. However, it has the disadvantage of requiring more computational processing time than other indices due to the counting process (Newman et al., 2018).

The determination of the EP requires only the DEM as input. Some software, such as WhiteBox GAT, provides a specific tool. Alternatively, the EP index can be calculated involving the following operations:

i. Counting pixels in the neighborhood with an elevation lower than the central pixel under analysis (n_{viz}). Some software may not provide a direct operation to perform this count, which can be circumvented by developing a specific algorithm. At the end of this step, there is a raster whose attribute is the number of pixels in the neighborhood with an elevation lower than the central pixel.
ii. The second operation is the division of the raster obtained in the previous step by n_{viz} and multiplication by 100. This operation would, in principle, be a mathematical operation between a raster and two scalars (n_{viz} and 100), which is available in almost any GIS software. The additional issue is related to the treatment of pixels on the edge of the DEM or close to it, a situation in which there is not the standard number of valid neighbors—that is, the value of n_{viz}, in this case, differs from the other pixels not influenced by proximity to the edge. Three approaches can overcome this situation. The first would be to restrict calculating the EP index only for the DEM pixels with precisely the number of valid n_{viz} neighbors, making the calculation undefined for the other pixels. Another approach would be to adjust the value of n_{viz} to the exact number of neighbors found in each case. In these two alternatives, conditional operators must operate with the raster using tools available in the software or by developing a specific computational routine. On the other hand, a third approach would be for the user to maintain the standard operation for the entire image, using the constant value of n_{viz} for all DEM pixels, including the edge ones. But, recognizing that the pixels near the edge have an inconsistent index value (since there are no n_{viz} neighbors to them), only the EP results for the pixels not influenced by the edge would be

considered for further analyses. This can be implemented by selecting a DEM subset for the EP index calculation that extends beyond the area of interest. In this way, the regions discarded due to their proximity to the image edges would lie outside the study area boundaries.

Figure 11.7e shows an example of the determination of the elevation percentile topographic index from the SRTM 30 DEM with a 7 × 7 neighborhood window for a region northeast of the São Paulo state. The index's spatial pattern shows lower values, generally below 30% and even reaching null values, in the areas at the bottom of the valley, where the watercourses flow. Near the peaks and watershed divides, values are predominantly above 60% and approach 100%.

11.4 Deviation from Mean Elevation (DEV)

The deviation from mean elevation topographic index (DEV) is similar to TPI, as it quantifies the deviation of the pixel's elevation relative to the average elevation of the neighborhood. The distinction is that the DEV normalizes this difference by the standard deviation of the distribution of elevations in the neighborhood (Eq. 11.5). The DEV is, therefore, a dimensionless topographic index.

$$\mathrm{DEV_P} = \frac{Z_\mathrm{P} - \overline{Z_n}}{\sigma_n} \tag{11.5}$$

where $\mathrm{DEV_P}$ is the relative deviation to the average elevation index at point P, Z_p is the elevation at P, $\overline{Z_n}$ is the average elevation of the neighborhood of P, and σ_n is the standard deviation of the elevations in this neighborhood.

Analogous to TPI, the DEV index assumes negative values when the pixel under analysis has an elevation lower than the average elevation of the neighborhood and presents positive values when the elevation of this pixel is higher than the average elevation of the neighborhood. In other words, it is understood that negative and positive values of DEV correspond to negative and positive values of TPI, respectively, since it is the difference in elevation of the pixel relative to the average of the neighborhood that determines the sign. However, as the standard deviation in the denominator of the DEV formulation is a specific value for each neighborhood, this index is not merely a scale change of the TPI. The relative differences between the DEV values along the image do not proportionally follow the differences found for the TPI, given an equal window definition in both indices.

The magnitude of the DEV value indicates the relative distancing to the distribution of elevations in the neighborhood (Newman et al., 2018). This characteristic implies that topographic variability is better incorporated in DEV than in TPI, making DEV a more suitable metric to characterize smooth variations in topography (De Reu et al., 2013). A disadvantage of DEV is implicitly assuming that the elevation data of

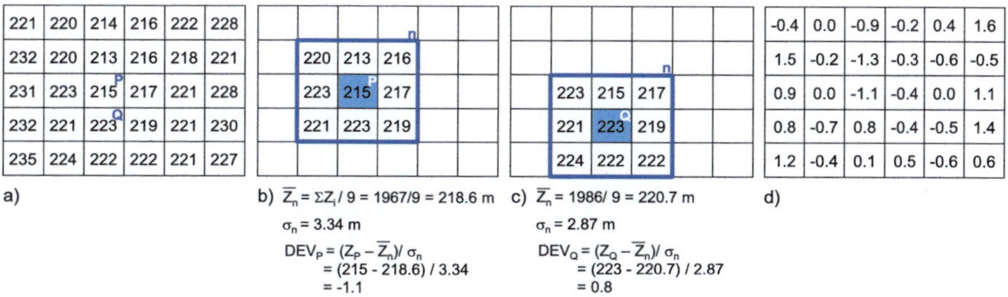

Fig. 11.9 **a** DEM with elevation in meters; **b** DEV calculation for pixel P considering a neighborhood (n) of size 3 × 3; **c** Same for pixel Q; **d** Raster of the DEV index (dimensionless values)

the neighborhood set of points follows a normal distribution (Newman et al., 2018), which does not necessarily occur.

The value of DEV is dimensionless and assumes values mainly in the interval from − 1 to 1 (Wilson & Gallant, 2000), tending to limit itself to values between − 3 and 3 (Lindsay et al., 2015). Some of these values that are further from the − 1 to 1 range may indicate anomalies in the DEM, that is, point elevations outside the expected range for the region (Wilson & Gallant, 2000).

As an example of DEV calculation, the pixel P in Fig. 11.9 has a neighborhood with an average elevation of 218.6 m and a standard deviation of 3.34 m, considering the neighborhood defined by a 3 × 3 window. As a result, the DEV value at P is − 1.1. Similarly, the DEV value for pixel Q is 0.8. The example in this figure is the same one depicted as an example for the calculation of the TPI in Fig. 11.2. As expected, the sign of the DEV and TPI raster values between these examples remains the same. However, the proportionality and relativity between the internal values in each raster are different.

Like other topographic indices, the determination of DEV requires only the DEM as input data and involves the following main steps:

i. A spatial average filter is applied to the DEM, resulting in a raster whose attribute is the average elevation of the neighborhood defined around each pixel.
ii. A spatial filter determined by the DEM's standard deviation is applied, resulting in a raster whose attribute is the standard deviation of the elevations of the neighborhood centered on the pixel.
iii. Mathematical operation between rasters to obtain the DEV index at each pixel according to Eq. (11.5): subtraction of the DEM by the average filter result and then division by the standard deviation filter result.

A real-world example of DEV calculation is shown in Fig. 11.7f, along with other topographic indices. The spatial pattern obtained for the DEV follows the one identified for the TPI, TPI^{med}, and elevation percentile indices: more intense negative values along the watercourses and valley bottoms, and more intense positive values on hilltops. Although there is a similar spatial pattern of positive and negative values between TPI and DEV, the effect of the scaling of DEV values due to the variable standard deviation for each neighborhood is evident. In addition, there is a bit more continuity of patches of high-intensity negative values in the DEV raster than in the TPI and TPI^{med}.

The window size adopted for the DEV calculation affects the type of result obtained: smaller windows result in more significant variability of results and detection of more local features. In comparison, larger windows result in higher spatial uniformity of results and focus on larger-scale features. This analysis can be illustrated with the example worked on in the following section of this chapter.

11.5 Maximum Elevation Deviation (DEVmax)

The concept of the DEV index, deviation from mean elevation, was expanded by Lindsay et al. (2015) in the conception of another topographic index called maximum elevation deviation (DEVmax). The DEVmax index is the maximum DEV value at each pixel calculated for different spatial scales.

The corresponding DEV is obtained for each neighborhood size, considering a specific range of spatial scales (Fig. 11.10). All DEV rasters are analyzed together to define, pixel by pixel, the maximum value among the various DEVs and result in the DEVmax index.

In the example in Fig. 11.10, there are n DEV values for pixel P, one for each window size that defines the neighborhood at multiple scales. In absolute terms,

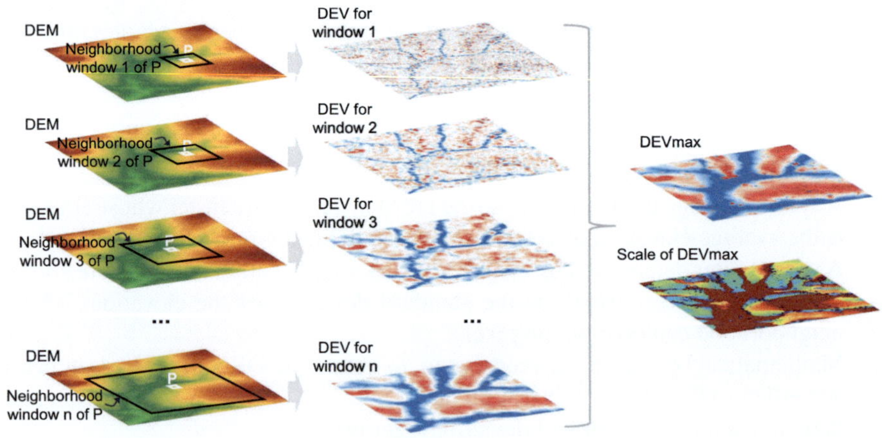

Fig. 11.10 Scheme for determining the topographic index DEVmax: from the DEM, the determination of the DEV index is repeated for n neighborhood window sizes; for each pixel, the highest absolute value obtained in the various DEVs is identified to obtain the DEVmax raster, with the corresponding positive or negative sign; an auxiliary raster is generated with the information of the neighborhood size (scale) that provided the DEVmax for each pixel

the largest of these values is identified and represents the DEVmax of this pixel, maintaining the positive or negative sign corresponding to this absolute maximum.

In the DEVmax raster, each pixel has as an attribute the value of the DEV at the specific scale that provided the highest relative deviation to the neighborhood. The maximum value or maximum relative deviation is computed in absolute terms. Then, a negative or positive sign is adopted for each raster attribute, indicating whether the deviation represents an area lower or higher than the neighborhood.

Lindsay et al. (2015) introduced the DEVmax proposal to represent terrain features referring to multiple scales in a single topographic index. In other words, the DEVmax index represents patterns of local topographic position considering the heterogeneity of scales between the pixels since, for each pixel, the optimal scale was obtained in terms of higher prominence relative to the landscape.

When calculating the DEVmax from the various DEVs obtained at multiple scales, it is also possible to determine a raster that indicates the scale, or neighborhood size, that provided the maximum DEV for each pixel. The scale information of the DEVmax assists in interpreting and analyzing the topographic characterization provided by this index.

Operationally, to obtain the DEVmax, one can use specific commands available in software like Whitebox GAT. An alternative is to use a set of standard operations present in most GIS software, such as the following procedures:

i. The DEV index is determined for the smallest neighborhood size of interest in the analysis. If there is no specific DEV command, it can be obtained through spatial filtering operations and raster arithmetic operations, as described in the previous section of this chapter.
ii. Repeat the previous step *n*-1 times, once for each incremental neighborhood size, up to the highest desired size.
iii. Take the n DEV rasters calculated in the previous steps, from the smallest to the highest neighborhood size, with absolute values, and apply a maximum operation to the series of images. For each maximum absolute value found, it is necessary to identify the corresponding sign in the original value. The maximum absolute value and the sign make up the attribute stored for each pixel, forming the DEVmax raster.

To illustrate the concept of DEVmax with a real-world case, Fig. 11.11 shows this topographic index obtained from the SRTM 30 DEM for an area of the state of São Paulo between Franca and Restinga municipalities. The DEV indices obtained for five different window sizes show that the larger the neighborhood window, the higher the deviation relative to the average, forming spatial patterns within relatively larger homogeneous areas.

The DEVmax index aggregates the information obtained from multiple windows. It is noticeable in the DEVmax raster of the example that the DEV pattern obtained in the larger windows presents a higher extension of the relatively homogeneous areas. But it is also noticeable in the DEVmax the occurrence of areas of smaller dimensions that portray local variability of the information, which was detected in the DEV obtained for the smaller windows. The raster that indicates the window size

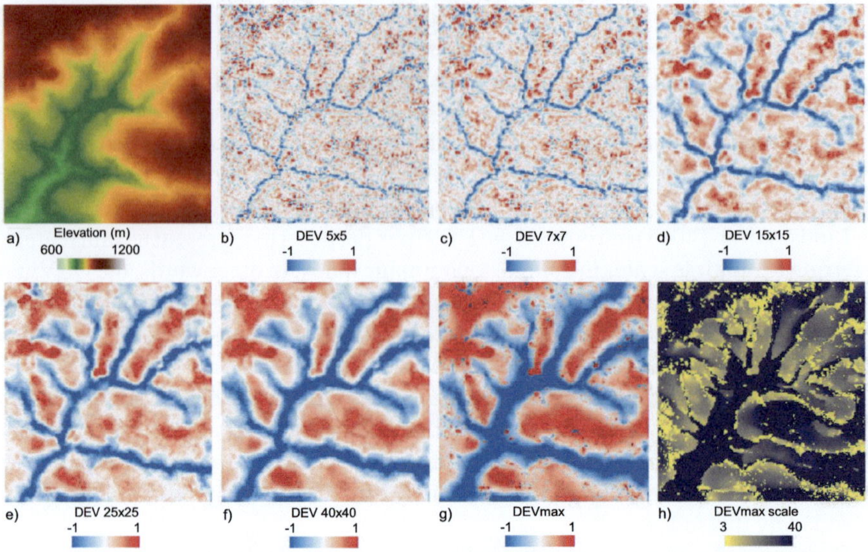

Fig. 11.11 **a** SRTM 30 DEM for an area between the municipalities of Franca and Restinga (São Paulo state); **b**, **c**, **d**, **e**, **f** Topographic index DEV for different neighborhood window sizes (5 × 5, 7 × 7, 15 × 15, 25 × 25, and 40 × 40 pixels, respectively; **g** Topographic index DEVmax obtained from the analysis of DEV in the scale range 3 × 3 to 40 × 40); **h** DEVmax scale (the attribute indicates the window size in pixels of the corresponding DEV used to compute the DEVmax for each pixel)

of the DEV of each pixel considered to integrate the DEVmax raster shows how the DEVmax results from DEV values in different windows.

11.6 Percent Elevation Range (PER)

The percent elevation range (PER) quantifies the local topographic position of a point's elevation in terms of how much higher it is than the minimum elevation within the neighborhood, relative to the elevation range in that neighborhood—that is, relative to the difference between the maximum and minimum elevation values (Eq. 11.6). This ratio is then multiplied by 100 to express it as a percentage.

$$\mathrm{PER_P} = \frac{(Z_\mathrm{P} - Z_{\min})}{(Z_{\max} - Z_{\min})} \cdot 100 \tag{11.6}$$

where $\mathrm{PER_P}$ is the percentage of elevation range index at point P, Z_P is the elevation at P, and $Z_{\min}$ and $Z_{\max}$ are the minimum and maximum elevations in the neighborhood of P.

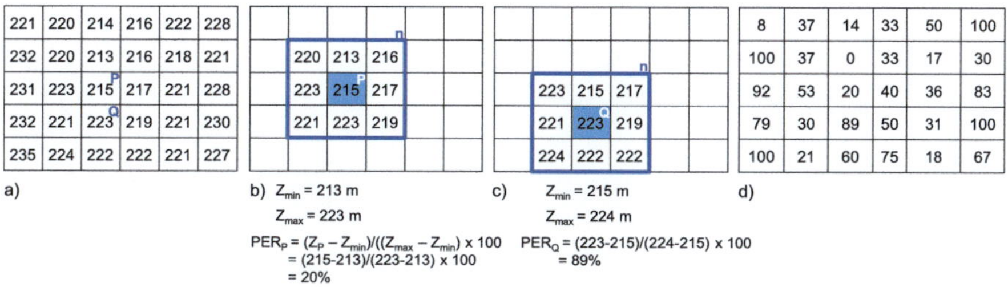

Fig. 11.12 **a** DEM with elevation in meters; **b** Calculation of the percent elevation range (PER) for pixel P considering neighborhood (*n*) of size 3 × 3; **c** Same for pixel Q; **d** Raster of the PER index (%)

The interpretation of PER is as follows: a PER value of 20%, for example, indicates that this point has an elevation situated 20% between the distance from the minimum elevation to the maximum elevation of the neighborhood. This is the value, for example, of the PER at point P in Fig. 11.12. The neighborhood of P has a minimum elevation of 213 m and a maximum of 223 m, representing a range of 10 m. Twenty percent of this range equals 2 m; therefore, an elevation of 215 m has a PER index of 20%, as with P, since the sum of the minimum value and the increment of 2 m results in 215 m. In this figure, point Q has an elevation very close to the maximum elevation of the neighborhood, implying that the PER index has a higher value in Q than in P. In fact, by the calculation performed, PER in Q equals 89%.

The values of the PER index can vary from 0, when the elevation of the pixel under analysis is the minimum of the set of points in the neighborhood, up to 100%, when the elevation of the analyzed pixel is the maximum of the neighborhood.

The determination of PER is quickly performed via standard operations commonly present in GIS software, as follows:

i. Applying the minimum filter to the DEM results in a raster whose attribute at each pixel is the minimum elevation value of the neighborhood defined around the pixel.
ii. Apply the maximum filter to the DEM to obtain a raster with the maximum elevation of the neighborhood centered on the pixel as an attribute.
iii. Algebraic operation between rasters (the DEM, the result of the minimum filter, and the result of the maximum filter) and multiplication by a scalar (100) according to the mathematical expression of Eq. (11.6).

This metric offers advantages regarding the convenience of determination and ease of interpretation of the results. On the other hand, considering the maximum and minimum filters following the established formulation means that evaluating the local topographic position is based on a linear interpolation between these extremes of the neighborhood without an explicit reference to the distribution of elevations (Newman et al., 2018).

Figure 11.13 shows a real-world example of determining the PER topographic index for an area between the Restinga and Franca municipalities in São Paulo. The

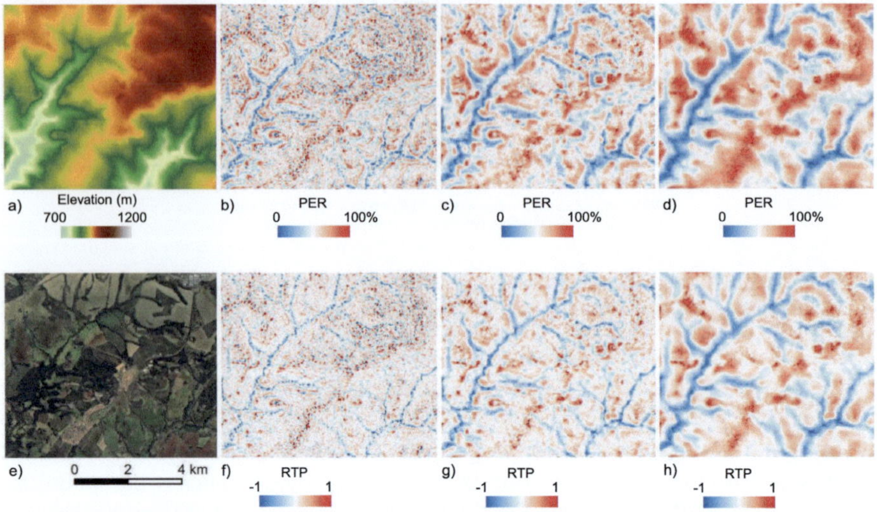

Fig. 11.13 **a** SRTM 30 DEM for an area between the municipalities of Restinga and Franca (SP); **b**, **c**, **d** Percent elevation range (PER) for neighborhood windows of 5 × 5, 11 × 11 and 25 × 25, respectively; **e** Google Earth satellite image; **f**; **g**, **h** Relative Topographic Position (RTP) topographic index for windows of 5 × 5, 11 × 11 and 25 × 25, respectively

SRTM 30 DEM is the topographic data source, and three distinct sizes of neighborhood windows are considered. The lowest PER values are concentrated in the areas where the watercourses flow, ranging from less than 30% to null values. With the increase of the analysis neighborhood, these areas become broader, and there is a growing gradient of PER in the direction of where the channels would be traced toward the marginal strips, still with values below 50% (shades of blue in the figure). This gradient continues to increase toward the watershed divides but is now above 50% (shades of red).

11.7 Relative Topographic Position (RTP)

The topographic index called Relative Topographic Position (RTP) was proposed to modify the PER index to circumvent the deficiency that the PER establishes a linear interpolation and does not incorporate the distribution of elevations (Newman et al., 2018). The formulation of the RTP is based on a two-part linear function, depending on the position of the point's elevation relative to the average elevation of the neighborhood (Eq. 11.7). In addition to considering the maximum and minimum values of the neighborhood in the calculation, as adopted for the PER index, the RTP index also incorporates the average elevation of the neighborhood.

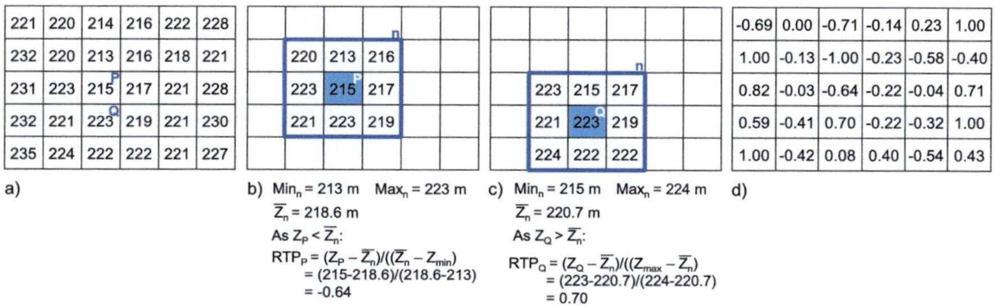

Fig. 11.14 **a** DEM with elevation in meters; **b** Calculation of the relative topographic position (RTP) for pixel P considering neighborhood (n) size 3 × 3; **c** Same for pixel Q; **d** RTP index raster (dimensionless)

$$\mathrm{RTP_P} = \left\{ \begin{array}{l} \frac{(Z_P - \overline{Z_n})}{(\overline{Z_n} - Z_{\min})}, \text{ if } \; Z_P < \overline{Z_n} \\ \frac{(Z_P - \overline{Z_n})}{(Z_{\max} - \overline{Z_n})}, \text{ if } \; Z_P > \overline{Z_n} \end{array} \right\} \quad (11.7)$$

where $\mathrm{RTP_P}$ is the relative topographic position index at point P, Z_P is the elevation at P, $Z_{\min}$, $Z_{\max,}$ and $\overline{Z_n}$ are the minimum, maximum, and average elevations, respectively, in the neighborhood of P, n is the number of pixels in this neighborhood.

The values of RTP vary in the interval from − 1 to 1. Negative values indicate that the pixel's elevation is below the average elevation of the neighborhood, and positive values indicate that the pixel's elevation is above this average, as is the case for the TPI and DEV indices. However, the relative proportionality of values across the RTP raster differs from that observed for the TPI and DEV indices. A didactic example of calculating the RTP topographic index is presented in Fig. 11.14.

Some software, such as Whitebox GAT, allows you to determine the RTP index via a specific command. Alternatively, you can determine RTP by operating spatial filters and algebraic operations between rasters, but with one more procedure. This additional procedure verifies the relative condition of the pixel's elevation in the window center regarding the neighborhood average, since the RTP formulation involves a function defined by parts.

The sequence of operations to obtain the RTP index can be summarized as follows:

(i) Application of the average filter to the DEM to obtain a raster whose attribute is the average of the neighborhood elevations;
(ii) Application of the maximum filter to the DEM, resulting in a raster whose attribute is given by the maximum value of the neighborhood at each pixel;
(iii) Application of the minimum filter to the DEM, obtaining the raster of minimum values of the neighborhood at each pixel;
(iv) Application of a conditional operation that relates the pixel's elevation with the average elevation of the neighborhood and, depending on the result, uses one of the two parts of the function that defines the RTP via algebraic operation of subtraction and division between the involved rasters (already determined

in the previous steps). The result of this sequence of steps is a raster whose attribute is the value of the RTP index at each pixel.

A real-world example of RTP index determination is illustrated in Fig. 11.13 for an area in the northeastern region of São Paulo State, based on the SRTM 30 DEM. The results obtained using three different neighborhood sizes reveal a consistent spatial pattern, similar to that observed for the PER index. In valley bottoms and along watercourses, RTP values are negative and approach -1 as proximity to the river course increases. It is important to note, however, that the main river channels have widths smaller than 30 m, which corresponds to the spatial resolution of the DEM.

In this example, the increase in the analyzed neighborhood makes RTP patterns relatively more homogeneous regarding the concentration of negative or positive values. Positive RTP values predominate in areas corresponding to watershed divides. However, the smaller the analysis window, the higher the occurrence of positive values representing local conditions in a more spatially distributed manner.

11.8 Terrain Ruggedness Index (TRI)

Another characteristic associated with the topography of a region is the concept of topographic ruggedness, widely used to study geomorphological processes (Lindsay et al., 2019), geological characteristics (Grohmann et al., 2011), and landslides (McKean & Roering, 2004). This issue is also helpful to characterize the environment from an ecological point of view regarding habitat diversity, difficulty of movement of animal species, habitat suitability to provide protection against extreme weather conditions, and even the ease of escaping from predators (Dilts et al., 2023).

More interesting than knowing whether a particular region is more rugged or smoother topographically than another region is understanding how rugged it is and where higher or lower ruggedness occurs, that is, quantifying the ruggedness and its spatial variability (Grohmann et al., 2011). This highlights the value of analyzing topographic ruggedness from the DEM.

Topographic ruggedness is associated with two topographic properties: local variability of elevations and surface complexity (Lindsay et al., 2019). The local variability of elevations can be represented by metrics such as the range of neighborhood elevations, while surface complexity is related to topographic texture. A metric associated with this latter aspect is the terrain ruggedness index (TRI). The TRI was conceived by Riley et al. (1999) as a measure of the terrain surface's heterogeneity, which can be quantified quickly and objectively.

The TRI is determined solely from the DEM as a function of the differences in elevation between each pixel and its eight neighbors in a 3×3 neighborhood window (Eq. 11.8 and Fig. 11.15). These differences are squared and added, with the square root applied to the sum result. This formulation considers that the TRI quantifies the total elevation difference of any terrain point relative to the immediate neighborhood (Sappington et al., 2007). Some authors expanded the original idea to calculate the

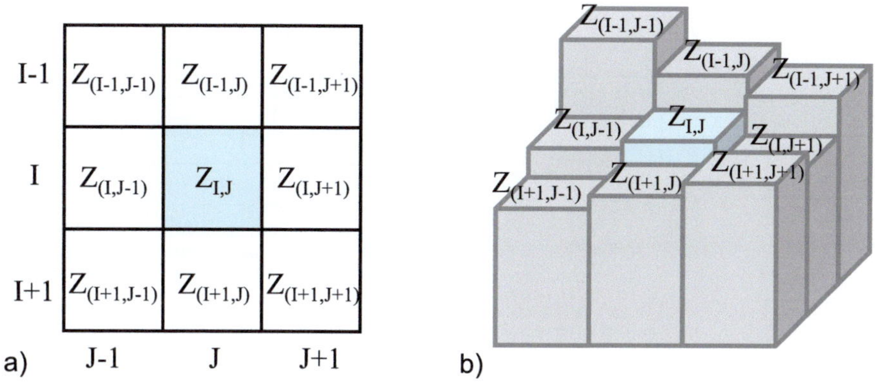

Fig. 11.15 3 × 3 neighborhood window around a pixel with the identification of the elevations used to calculate the topographic index TRI: **a** matrix definition; **b** 3D view

TRI by considering larger neighborhood windows defined by a search radius.

$$\mathrm{TRI}_{I,J} = \sqrt{\sum_{i=I-1}^{I+1} \sum_{j=J-1}^{J+1} \left(Z_{i,j} - Z_{I,J}\right)^2} \tag{11.8}$$

where $\mathrm{TRI}_{I,J}$ is the terrain ruggedness index at the pixel of row I and column J, $Z_{I,J}$ is the elevation at this pixel of coordinates (I, J), $Z_{i,j}$ is the elevation at a pixel of the neighborhood at coordinates i, j.

The calculation of the TRI is detailed in Fig. 11.16's example. For point P, the TRI value is 21.4 m, while for point Q, it is equal to 3.9 m. For the DEM of the example, the TRI ranges from 6.0 to 21.4 m. In this example, the TRI was not calculated for the edge pixels, as not all eight valid neighbors are available.

Some software, such as QGIS and SAGA GIS, have a specific function for determining the TRI, with the DEM as the input information. These software allow variations of the original formulation of TRI in the sense of flexibility in the choice of the size of the neighborhood window and the weighting of the calculated differences relative to each neighbor according to different criteria.

If a specific software function is not used, developing a new algorithm is an alternative to determining the TRI. This algorithm is relatively simple: scanning the DEM matrix and identifying the eight neighbors for each pixel, computing the square of the differences, making the sum of these, and then applying the square root. A specific treatment must be used to avoid calculating the TRI for edge pixels.

A real-world case of TRI index determined from the SRTM 30 DEM is shown in Fig. 11.17 for an area in the municipality of Cláudio, Minas Gerais. The spatial pattern of the TRI is similar to the spatial distribution of local slope values, as shown in the figure. In general, it is observed that an imaginary line joining points D and E separates two very distinct regions: above this line, an area of relief with lower

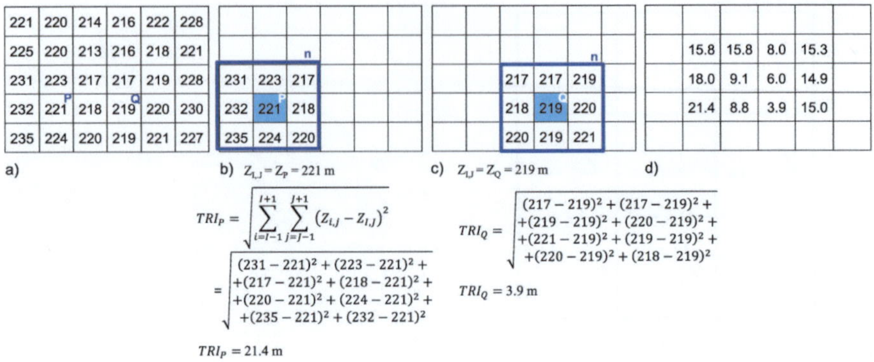

Fig. 11.16 **a** DEM (elevation in meters); **b, c** Highlight for the neighborhood n of size 3 × 3 around pixel P and pixel Q, respectively, for determining the TRI index (Terrain Ruggedness Index); **d** Raster of the topographic index TRI (in meters)

slopes and more reduced TRI values; below this line, the topography presents higher gradients, with higher slopes and higher TRI values.

The strong correlation between TRI and slope is expected due to the mathematical formulation of these variables. Such correlation is seen by some authors as a disadvantage of the TRI as an auxiliary variable in the study of ecological aspects related to topography, as discussed by Sappington et al. (2007). According to these authors, the TRI does not distinguish steep and uniform terrains from steep and irregular topography, for example. Both terrains have steep slopes. However, the second type presents higher topographic ruggedness than the first; therefore, distinct ecological relationships occur in these two environments.

In the result of the example in Fig. 11.17, areas such as below point A are seen, marked by a high slope and presenting a high TRI—in this case, above 120 m. The area to the left of B is another interesting example: there is a kind of local hill, with slopes reaching up to 30° and TRI up to 115 m. On the other hand, valley bottom areas, such as where point C is located, are areas of more uniform and flat topography, with very low slopes (generally less than 3°) and significantly reduced TRI, less than 5 m.

11.9 Surface-Area to Planar-Area Ratio (SAPR)

The topographic index known as the Surface-Area to Planar-Area Ratio (SAPR) represents how much larger the surface area, accounting for the three-dimensionality of the terrain, is compared to the area measured as a two-dimensional planar projection (Jenness, 2004). It is a ruggedness terrain metric (Dilts et al., 2023).

Usually, the planar area is considered when quantifying a region's area. However, when considering resource availability to support animal species in the environment,

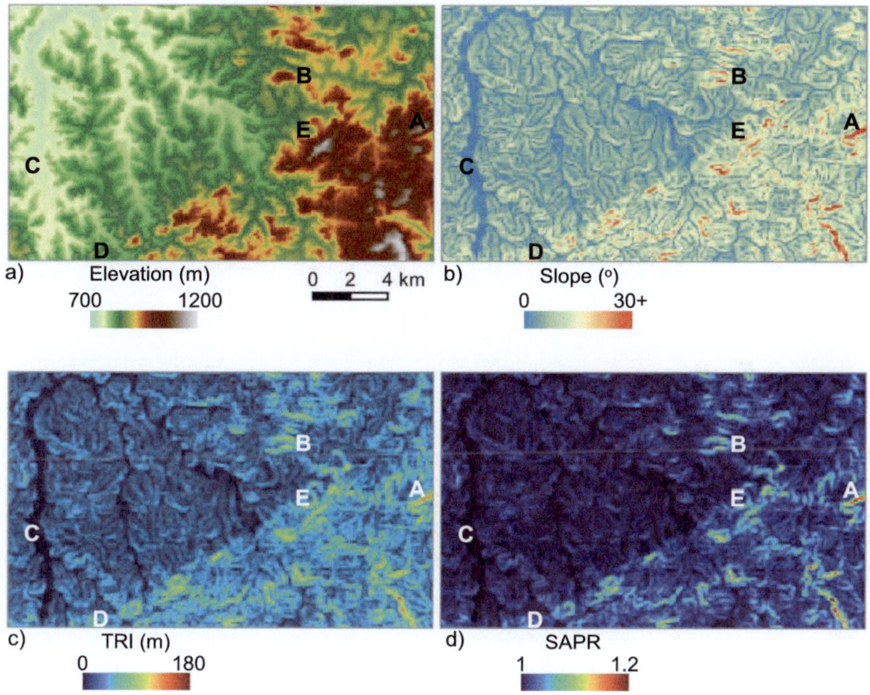

Fig. 11.17 **a** SRTM 30 DEM of an area in the municipality of Cláudio (Minas Gerais state); **b** Local slope; **c** Terrain roughness index (TRI); **d** Surface-Area to Planar-area Ratio (SAPR) topographic index

the surface area quantification more realistically portrays this availability than the planar area (Jenness, 2004).

A practical and relatively easy way to calculate the surface area from a DEM was developed by Jenness (2004), according to which the surface area of a pixel is determined based on the elevations of its eight neighbors in a 3 × 3 window, in addition to its elevation. This method can be summarized with the following step-by-step:

i. For a given pixel P, the eight neighbors of the 3 × 3 window and the corresponding elevations are identified (Fig. 11.18a and f);
ii. Considering the centroids of each pixel in this window positioned at the corresponding pixel elevation, lines are drawn connecting the centroid of each pixel to the centroid of the central pixel and the centroid of the lateral neighbors (Fig. 11.18b and g);
iii. Eight three-dimensional triangles are identified, whose sides correspond to the lines drawn in the previous step (Fig. 11.18c and h, triangles identified by A to H);

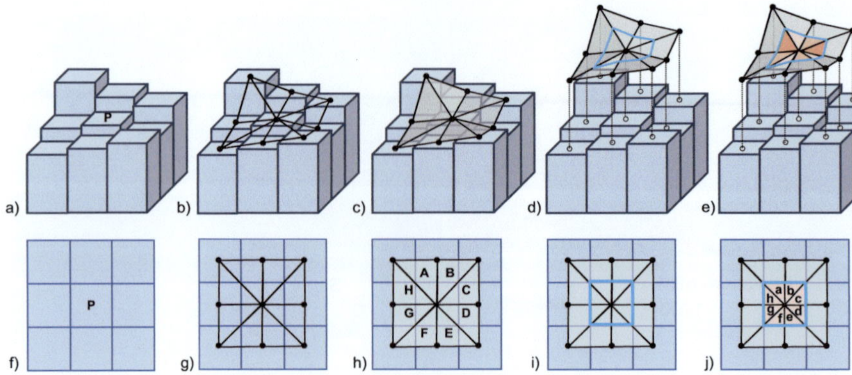

Fig. 11.18 Synthesis of the approach proposed by Jenness (2004) for determining the surface area of a pixel P: **a, f** Identification of pixel P and its eight neighbors in the 3 × 3 window, in 3D visualization and the plane, respectively; **b, g** Drawing of lines connecting centroids of neighboring pixels, with the centroids located in 3D at the pixel elevation; **c, h** Identification of eight 3D triangles whose sides correspond to the lines that join the centroids of the pixels; **d, i** Drawing of lines that join the midpoints of the sides of the triangles identified in the previous step; **e, j** Identification of eight smaller triangles from the lines drawn joining the midpoints of the sides of the larger triangles (*Source* based on Jenness, 2004)

iv. The length of the three sides of each of these eight triangles is quantified based on the difference in elevations and the planar distance between neighboring pixels (orthogonal neighbors have a distance equal to the pixel size; diagonal neighbors have a distance equal to this size multiplied by $\sqrt{2}$). To illustrate this step, consider the triangle formed by the lines that join the centroids of pixels P, Q, and R in Fig. 11.19. Each line has a planar distance $L_{2D,}$ as indicated, and there is a difference in elevation ΔZ between each pair of pixels. The size of each side of the triangle computed in a three-dimensional way (L_{tri}) is calculated according to the Pythagorean theorem as the hypotenuse of the triangle with corresponding sides L_{2D} and ΔZ.
v. The midpoints of each side of these triangles are identified, which constitute the limits of these triangles corresponding to the planar area of the central pixel; the neighboring midpoints are joined by lines laterally (Fig. 11.18d and i);
vi. Eight smaller three-dimensional triangles are identified, formed by the lines that joined the midpoints of the sides of the larger triangles (Fig. 11.18e and j, which are the triangles identified by *a* to *h*);
vii. The extension of the three sides of each of the triangles is smaller (up to *h*), as being half the length of the corresponding larger side;
viii. The area of each triangle from *a* to *h* is determined, based on the three corresponding sides, by Eq. (11.9) and Eq. (11.10), which constitute Heron's Theorem in mathematics;
ix. The surface area of the central pixel P is quantified by the sum of the areas of the eight smaller triangles (from *a* to *h*).

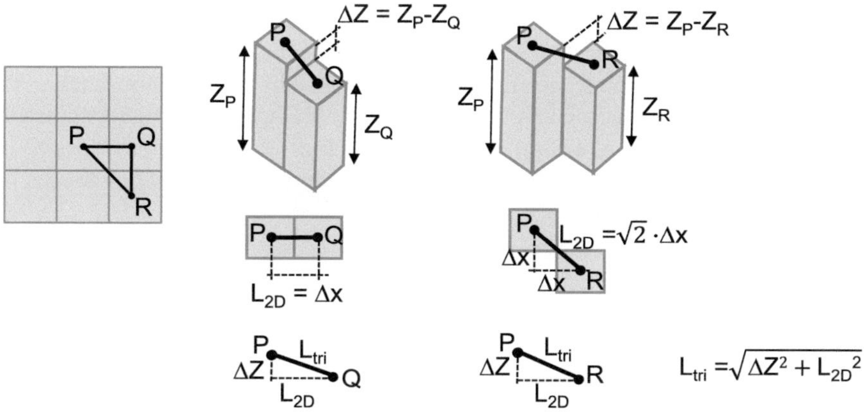

Fig. 11.19 Scheme of the determination of the elevation difference (ΔZ) and planimetric distance (L_{2D}) for the calculation of the length of the sides of the triangles formed by the connection of the centroids of the neighboring pixels, as part of the method of obtaining the surface area of a pixel from the DEM: the distinction between the situation of orthogonal neighboring pixels (P and Q) and the situation between diagonal neighbors (P and R) is highlighted

x. Move the 3 × 3 window and repeat this procedure for each new central pixel.

$$A_{tri} = \sqrt{p \cdot (p-a) \cdot (p-b) \cdot (p-c)} \tag{11.9}$$

$$p = \frac{a+b+c}{2} \tag{11.10}$$

where A_{tri} is the area of a triangle with sides of size *a*, *b*, and *c*; *p* is the semi-perimeter of this triangle.

For each pixel, the topographic index SAPR is given by the quotient between the pixel's surface area obtained according to the previous script and the planimetric area, which corresponds to Δx^2. This index has a minimum value of 1 when the surface corresponding to the pixel is strictly flat, and the surface area equals the planimetric area. The higher the SAPR value, the larger the surface area is proportional to the planimetric area, indicating a higher terrain roughness.

Determining the SAPR index from the DEM requires a specific algorithm, as it is infeasible to obtain it just by combining standard operations available in GIS software. Some software, like Whitebox GAT, provides a particular function for obtaining the SAPR.

Figure 11.17d shows a real-world example of a raster of the SAPR topographic index for an area in Cláudio municipality, in Minas Gerais, determined from the SRTM 30 DEM. This example shows how the spatial variability pattern of SAPR follows the spatial distribution of slopes, as pointed out in the literature, and for the TRI index.

In the example, in areas of higher topographic gradient, such as below A and to the left of B, there are higher slopes and the largest surface areas, which deviate more from the value of the planimetric area. The SAPR value in this region reaches up to 1.2, indicating that the surface area, in this case, is 20% larger than the planimetric area. On the other hand, along valley bottoms, such as along the area where C is located, there are lower slopes, and flatter regions are configured, which results in SAPR values close to 1.

11.10 Vector Ruggedness Measure (VRM)

Sappington et al. (2007) presented the Vector Ruggedness Measure (VRM). It corresponds to the roughness factor described in Etzelmüller (2000) and is based on Mark's work (1975).

The VRM is a metric that quantifies the dispersion of vectors orthogonal to the terrain surface (Sappington et al., 2007). Conceptually, for a given pixel P, the VRM is obtained by analyzing the normal vector to the surface of each pixel in a neighborhood window of the DEM around P (Fig. 11.20). This analysis consists of decomposing each vector in the axes of the three-dimensional Cartesian system (x, y, z) according to trigonometric operators and considering each pixel's local slope and aspect.

Therefore, the VRM at a point is a roughness metric measured based on all the pixels within the considered neighborhood. For example, nine pixels are analyzed if the adopted neighborhood window is of the 3 × 3 type.

In areas of more continuous topography, with less roughness, the orthogonal vectors of the neighborhood window have directions closer to each other (McKean & Roering, 2004), as in point P of Fig. 11.20. In areas of higher roughness, like point Q of the figure, the orthogonal vectors have a higher variation in orientation between them. Areas of lower topographic roughness have less dispersed vectors, while regions of higher roughness have more dispersed vectors (Grohmann et al., 2011).

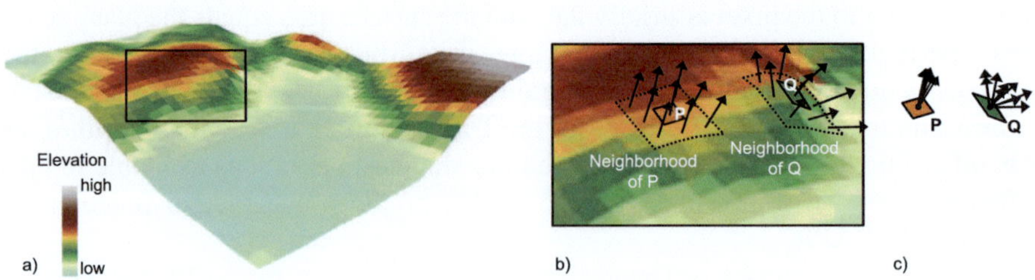

Fig. 11.20 **a** 3D visualization of a DEM; **b** Detail of part of the 3D DEM showing the orthogonal vectors to the terrain surface in the pixels of the 3 × 3 neighborhood window around pixels P and Q; **c** Grouping of all orthogonal vectors at the centroid of the central pixel of the neighborhood window, for better visualization of the dispersion between them

The calculation of the VRM is according to Eq. (11.11), which includes the division by the number of pixels belonging to the neighborhood and the subtraction of 1. As a result, the VRM index is dimensionless and assumes values ranging in the interval from 0 to 1. The value zero corresponds to the situation of a flat area, and the closer to 1, the higher the indication of the roughness of the region. The detail of the orthogonal vector decomposition resulting in Eq. (11.11) is presented in Etzelmüller (2000) and Sappington et al. (2007).

$$\mathrm{VRM_P} = 1 - \frac{\sqrt{\left[\sum(\sin\beta\cdot\sin\gamma)\right]^2 + \left[\sum(\sin\beta\cdot\cos\gamma)\right]^2 + \left[\sum(\cos\beta)\right]^2}}{n} \tag{11.11}$$

where $\mathrm{VRM_P}$ is the vector roughness index at point P, β is the local slope at P, γ is the terrain aspect at P, and n is the number of pixels in the neighborhood of P.

The proposal of the VRM, according to Sappington et al. (2007), was to quantify the topographic roughness in a way not correlated to the slope. These authors point out that other indices describing topographic roughness, such as TRI and SAPR, have the disadvantage of being strongly correlated with slope, as discussed in previous sections of this chapter. When this occurs, it becomes difficult to distinguish the separate effects of slope and roughness (Dilts et al., 2023). For example, applying a regression model for habitat suitability analysis using TRI and slope as explanatory variables becomes problematic due to their collinearity (Sappington et al., 2007).

The VRM index incorporates both the slope and the aspect in its mathematical formulation. Although it has a relationship with the slope, the formulation conceived leads to results with less correlation with the slope than other similar indices (Sappington et al., 2007).

Three approaches can be used to obtain the VRM from the DEM. The first is to take advantage of a specific operation in some GIS software, such as SAGA GIS. The second approach is to develop your computational routine involving the aspect and local slope rasters according to the mathematical formulation of the VRM.

The third approach is to combine a series of standard operations available in most GIS software in the following way, for example:

i. Obtain the local slope raster β (in radians) from the DEM using a specific software command, as described in Chap. 5;
ii. Obtain the terrain aspect raster γ (in radians) from the DEM using a specific software command (also covered in Chap. 5);
iii. Determine three auxiliary rasters (R1, R2, and R3) via a raster calculator command according to the mathematical operations of Eqs. (11.12), (11.13), and (11.14)

$$\mathrm{R1} = \sin\beta\cdot\sin\gamma \tag{11.12}$$

$$\mathrm{R2} = \sin\beta\cdot\cos\gamma \tag{11.13}$$

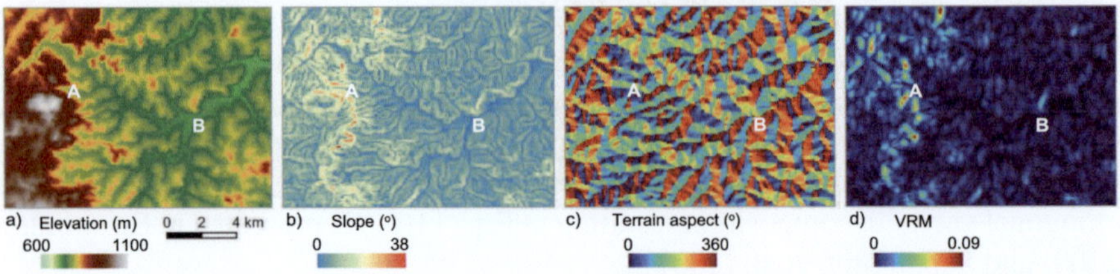

Fig. 11.21 **a** SRTM 30 DEM for an area in the municipality of Divinópolis (Minas Gerais state); **b** Local slope; **c** Terrain aspect; **d** Vector roughness index (VRM)

$$\mathrm{R3} = \cos\beta \tag{11.14}$$

iv. Apply a spatial sum filter to each of the auxiliary rasters generated in the previous step, obtaining the rasters $\mathrm{R1_f}$, $\mathrm{R2_f}$ and $\mathrm{R3_f}$; in this filtering, each pixel is replaced by the value corresponding to the sum of the attributes of all the neighboring pixels, including itself, within a neighborhood window (window of the desired size for the VRM);

v. Via a raster calculator command, the final step is to obtain the VRM raster via mathematical formulation given by Eq. (11.15)

$$\mathrm{VRM} = 1 - \frac{\sqrt{[\mathrm{R1_f}]^2 + [\mathrm{R2_f}]^2 + [[\mathrm{R3_f}]]^2}}{n} \tag{11.15}$$

An example of determining the VRM index for a real-world case is illustrated in Fig. 11.21 for an area in the municipality of Divinópolis in Minas Gerais. A VRM was obtained from the SRTM 30 DEM, reaching a maximum of 0.09. The areas of highest topographic roughness correspond to the areas with the steepest gradient and topographic variability on the left side of the region shown in the figure, as to the left of A, for example. Like regions to the left of B, the valley bottom areas have VRM close to zero. There are similarities in the general spatial pattern between the VRM and the local slope, but they are less expressive than observed for the TRI and SAPR indices in the example of Fig. 11.17.

References

Amatulli, G., Domisch, S., Tuanmu, M.-N., Parmentier, B., Ranipeta, A., Malczyk, J., & Jetz, W. (2018). A suite of global, cross-scale topographic variables for environmental and biodiversity modeling. *Scientific Data, 5*, Article 180040. https://doi.org/10.1038/sdata.2018.40

Barbedo, R., Siqueira, V., & Collischonn, W. (2022). Topographic position-based Stream definition (TPS): A simple method to address spatial variability of drainage density in stream networks. *Hydrological Sciences Journal, 67*, 858–869. https://doi.org/10.1080/02626667.2022.2047190

De Reu, J., Bourgeois, J., Bats, M., Zwertvaegher, A., Gelorini, V., De Smedt, P., Chu, W., Antrop, M., De Maeyer, P., Finke, P., Van Meirvenne, M., Verniers, J., & Crombé, P. (2013). Application of the topographic position index to heterogeneous landscapes. *Geomorphology, 186*, 39–49. https://doi.org/10.1016/j.geomorph.2012.12.015

Dilts, T. E., Blum, M. E., & Shoemaker, K. T. (2023) Improved topographic ruggedness indices more accurately model fine-scale ecological patterns. *Landscape Ecology 38*, 1395–1410. https://doi.org/10.1007/s10980-023-01646-6

Etzelmüller, B. (2000). On the quantification of surface changes using grid-based digital elevation models (DEMs). *Transactions in GIS, 4*, 129–143. https://doi.org/10.1111/1467-9671.00043

Florinsky, I. V. (2017). An illustrated introduction to general geomorphometry. *Progress in Physical Geography: Earth and Environment, 41*, 723–752. https://doi.org/10.1177/0309133317733667

Gallant, J. C., & Dowling, T. I. (2003). A multiresolution index of valley bottom flatness for mapping depositional areas: multiresolution valley bottom flatness. *Water Resources Research, 39*. https://doi.org/10.1029/2002WR001426

Grohmann, C. H., Smith, M. J., & Riccomini, C. (2011). Multiscale analysis of topographic surface roughness in the midland valley, Scotland. *IEEE Transactions on Geoscience and Remote Sensing, 49*, 1200–1213. https://doi.org/10.1109/TGRS.2010.2053546

Jenness, J. S. (2004). Calculating landscape surface area from digital elevation models. *Wildlife Society Bulletin, 32*, 829–839. https://doi.org/10.2193/0091-7648(2004)032[0829:CLSAFD]2.0.CO;2

Lawrence, A., Hoffmann, S., & Beierkuhnlein, C. (2021). Topographic diversity as an indicator for resilience of terrestrial protected areas against climate change. *Global Ecology and Conservation, 25*, Article e01445. https://doi.org/10.1016/j.gecco.2020.e01445

Lindsay, J. B. (2016). Whitebox GAT: A case study in geomorphometric analysis. *Computers and Geosciences, 95*, 75–84. https://doi.org/10.1016/j.cageo.2016.07.003

Lindsay, J. B., Cockburn, J. M. H., & Russell, H. A. J. (2015). An integral image approach to performing multi-scale topographic position analysis. *Geomorphology, 245*, 51–61. https://doi.org/10.1016/j.geomorph.2015.05.025

Lindsay, J. B., Newman, D. R., & Francioni, A. (2019). Scale-optimized surface roughness for topographic analysis. *Geosciences, 9*, 322. https://doi.org/10.3390/geosciences9070322

Llobera, M. (2001). Building past landscape perception with GIS: Understanding topographic prominence. *Journal of Archaeological Science, 28*, 1005–1014. https://doi.org/10.1006/jasc.2001.0720

Mark, D. M. (1975). Geomorphometric parameters: A review and evaluation. *Geografiska Annaler. Series A, Physical Geography, 57*, 165. https://doi.org/10.2307/520612

McKean, J., & Roering, J. (2004). Objective landslide detection and surface morphology mapping using high-resolution airborne laser altimetry. *Geomorphology, 57*, 331–351. https://doi.org/10.1016/S0169-555X(03)00164-8

Moore, I. D., Grayson, R. B., & Ladson, A. R. (1991). Digital terrain modelling: A review of hydrological, geomorphological, and biological applications. *Hydrological Processes, 5*, 3–30. https://doi.org/10.1002/hyp.3360050103

Newman, D. R., Lindsay, J. B., & Cockburn, J. M. H. (2018). Evaluating metrics of local topographic position for multiscale geomorphometric analysis. *Geomorphology, 312*, 40–50. https://doi.org/10.1016/j.geomorph.2018.04.003

Riley, S. J., DeGloria, S. D., & Elliot, R. (1999). A terrain ruggedness index that quantifies topographic heterogeneity. *Intermountain Journal of Sciences, 5*, 23–27.

Salinas-Melgoza, M. A., Skutsch, M., & Lovett, J. C. (2018). Predicting aboveground forest biomass with topographic variables in human-impacted tropical dry forest landscapes. *Ecosphere, 9*. https://doi.org/10.1002/ecs2.2063

Sappington, J. M., Longshore, K. M., & Thompson, D. B. (2007). Quantifying landscape ruggedness for animal habitat analysis: A case study using bighorn sheep in the Mojave desert. *Journal of Wildlife Management, 71*, 1419–1426. https://doi.org/10.2193/2005-723

Walker, S. J., Wilkinson, S. N., Van Dijk, A. I. J. M., & Hairsine, P. B. (2020). A multi-resolution method to map and identify locations of future gully and channel incision. *Geomorphology, 358*, Article 107115. https://doi.org/10.1016/j.geomorph.2020.107115

Weiss, A. (2001). *Topographic position and landforms analysis*. Presented at the ESRI Users Conference, San Diego (EUA).

Wilson, J. P., & Gallant, J. C. (2000). *Terrain analysis: Principles and applications*. Wiley.

Zwolinski, Z., & Stefanska, E. (2015). Relevance of moving window size in landform classification by TPI. *Geomorphometry for Geosciences*, 273–277.

Chapter 12
Topographic Indices—Part 2: Flow Convergence, Soil Moisture, and Flooding

This chapter continues the previous chapter by covering DEM-derived topographic indices. Eight topographic indices are presented, with scope related to flow convergence and the influence of topography on soil moisture or flooding.

This type of topographic characterization has been used, for example, to help identify and explain patterns of soil moisture and biological and ecological characteristics, such as vegetation patterns, species richness and composition, net primary productivity, and forest biomass, in addition to applications related to agricultural management and silviculture operations.

Some indices presented in this chapter specifically refer to erosive processes, which help characterize spatial vulnerability to erosion and distinguish areas with predominant erosion or sediment deposition behaviors.

This chapter also highlights the topographic index HAND, which is widely used to estimate floodable areas and various other applications.

12.1 Convergence Index (CI)

The convergence index (CI) was proposed to characterize the terrain by distinguishing local flow convergence areas from local flow divergence areas. The calculation of CI from the DEM generally considers a moving window 3×3 around each pixel under analysis. In this case, the CI of the central pixel is determined based on the analysis of the eight neighboring pixels. A version of CI calculation considers only the orthogonal neighboring pixels within the 3×3 window, effectively reducing the analysis to a 2×2 configuration.

Mathematically, the CI determines the degree of agreement between the direction of the aspect of the neighboring pixels and the direction of each pixel toward the central pixel. As seen in Chap. 5, the aspect is measured in degrees, varying clockwise,

A. Rolim da Paz, *Digital Elevation Models for Environmental Studies*,
https://doi.org/10.1007/978-3-032-04523-2_12

from zero when pointing north and increasing clockwise, reaching 180º in the south direction and up to the value of 360º again when it returns to the north.

In the situation of maximum agreement, all neighboring pixels have an aspect with a direction toward the central pixel. This central pixel represents a point of maximum convergence—all neighboring pixels have direction converging to this central pixel (Fig. 12.1a and d). The opposite situation is when there is maximum disagreement, i.e., the direction of the aspect of each pixel is precisely in the opposite direction to the direction pointed to the central pixel. This situation configures the case of maximum divergence (Fig. 12.1b and e).

The CI value for a given pixel P is calculated as the mean of the angles θi, representing the deviation between each neighboring pixel's aspect direction and the direction toward the central pixel P. It can be computed using Eq. (12.1) (Kiss, 2004; Thommeret et al., 2010).

$$\mathrm{CI_P} = \left(\frac{1}{8}\sum_{i=1}^{8}\theta_i\right) - 90 \tag{12.1}$$

where $\mathrm{CI_P}$ is the convergence index at P.

In more detail, the calculation of the CI can be understood with the following steps (Fig. 12.2):

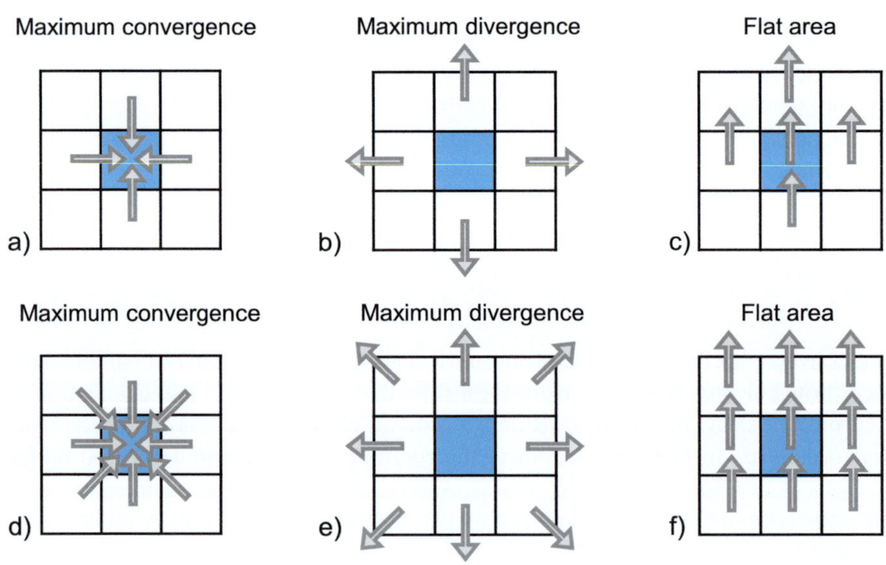

Fig. 12.1 Illustrative schemes of situations of maximum convergence **a**, maximum divergence **b** and flat area **c**, considering neighborhood 2 × 2 around the pixel (only orthogonal neighbors); **d**, **e**, **f** situations analogous to the previous ones, but for a 3 × 3 neighborhood around the pixel (orthogonal and diagonal neighbors)

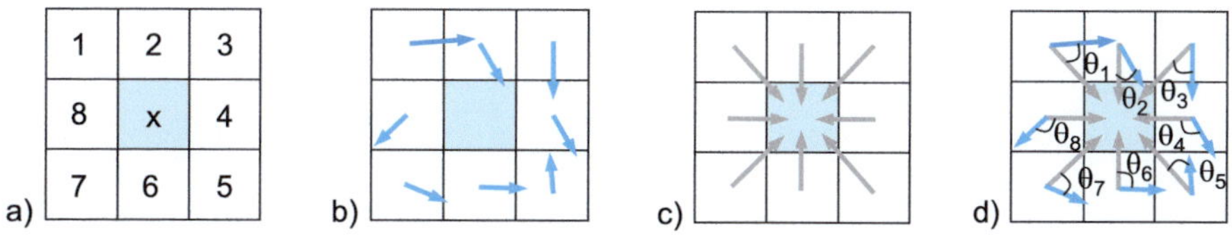

Fig. 12.2 Determination of the convergence index CI for pixel x: **a** Identification of the eight neighbors to pixel x in the 3 × 3 window around it; **b** Direction of the aspect of each neighboring pixel to x; **c** Direction of maximum convergence for each pixel in the neighborhood; **d** Identification of the angle between the direction of the aspect and the direction of maximum convergence for each pixel in the neighborhood (θ_i)

i. For each pixel, identify the pixels in the neighborhood in the 3 × 3 window around it (Fig. 12.2a);
ii. Determine the direction of the aspect for each pixel of this neighborhood, that is, the direction toward the steepest slope (Fig. 12.2b);
iii. Determine for each neighboring pixel the direction that provides the maximum convergence, which is toward the central pixel (Fig. 12.2c);
iv. Identify the angle of deviation between the directions of the aspect and maximum convergence for each pixel (Fig. 12.2d);
v. Calculate the arithmetic mean of these deviation angles of the eight neighbors;
vi. Calculate the CI index by the difference between the mean and 90º angles.

The CI value calculated by Eq. (12.1) varies from − 90º to 90º. Negative values indicate a convergence area. The lower limit value of − 90º corresponds to the situation of maximum convergence when the direction of the aspect of the neighborhood pixels points toward the central pixel (Fig. 12.1d). The average of the angles θ_i, in this case of maximum convergence, is zero.

Positive CI values refer to divergent areas. The upper limit case of the value 90º corresponds to the situation in which the direction of the aspect of all neighboring pixels is opposite to the central pixel. In this case, the average of the angles θ_i is 180º, and the divergence is maximum (Fig. 12.1e).

When it comes to flat areas or flat slopes, that is, areas without curvature, the CI value is zero since the average of the angles θ_i is 90º (Fig. 12.1f). In some software, the value obtained by a specific function for the CI by Eq. (12.1) is multiplied by the factor 10/9 to convert the variation scale of this index to the interval − 100 to 100 for the convenience of interpretation.

Figure 12.3 shows an example of calculating the CI from the raster of aspect direction. For pixel P, taking a 3 × 3 neighborhood window, the calculation of the deviation angles θ_i is detailed based on the direction of the aspect and the direction toward maximum convergence. The average of these angles is 84.85°, which results in a CI value of − 5.15°. This value indicates that pixel P is a point of convergence, but relatively small. Other pixels in the example also have negative values, and some have positive values, indicating divergence areas. Edge pixels do not have all valid neighbors; in this case, the CI value is indeterminate.

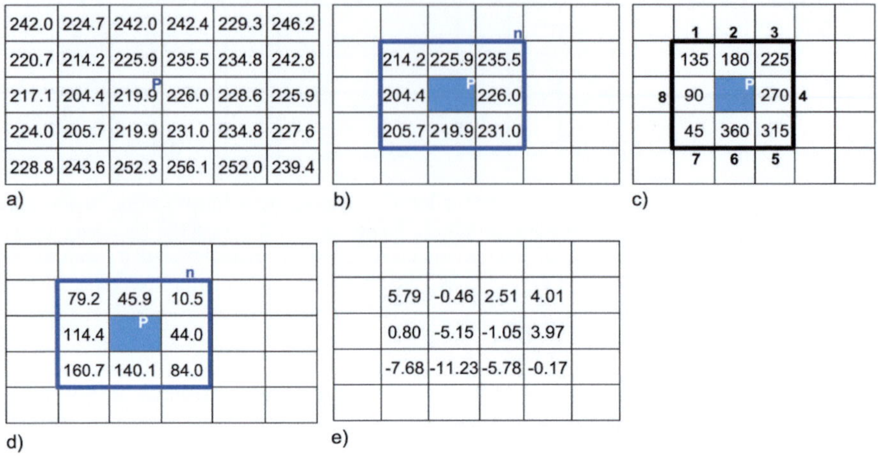

$\theta_1 = |214.2 - 135| = 79.2^{\circ}$ $\quad \theta_5 = |231.0 - 315| = 84.0^{\circ}$

$\theta_2 = |225.9 - 180| = 45.9^{\circ}$ $\quad \theta_6 = |219.9 - 360| = 140.1^{\circ}$

$\theta_3 = |235.5 - 225| = 10.5^{\circ}$ $\quad \theta_7 = |205.7 - 45| = 160.7^{\circ}$

$\theta_4 = |226.0 - 270| = 44.0^{\circ}$ $\quad \theta_8 = |204.4 - 90| = 114.4^{\circ}$

$\frac{1}{8}\sum_{i=1}^{8} \theta_i = (79.2 + 45.9 + 10.5 + 44 + 84 + 140.1 + 160.7 + 114.4)/8 = 678.8/8 = 84.85^{\circ}$

$$CI_P = \left(\frac{1}{8}\sum_{i=1}^{8} \theta_i\right) - 90 = 84.85 - 90$$

$$CI_P = -5.15^{\circ}$$

Fig. 12.3 **a** Aspect direction (degrees) obtained from the DEM; **b** Highlight for the 3 × 3 size neighborhood n around pixel P for the determination of the CI index; **c** Aspect angles of maximum convergence of each neighboring pixel 1 to 8 in the direction of pixel P (degrees); **d** Deviation angle (θ_i) between aspect direction and maximum convergence direction, for pixel P (degrees); **e** CI raster (edge pixels have undefined value)

The input data for determining the CI is the raster with aspect direction derived from the DEM. Some software has an explicit function for calculating the CI, generally taking the DEM as input data and internally determining the aspect direction raster.

Without using a specific command designed for obtaining the CI, a set of standard operations in GIS software can be performed to calculate this index, as follows:

i. Obtaining the raster with aspect direction for each pixel from the DEM, as described in Chap. 5;
ii. Apply a mathematical operation involving the aspect direction raster and the maximum convergence directions to compute the sum and average of the deviation angles. This operation requires formulating an algorithm using programming modules available in the software or via a computational routine elaborated externally to some GIS software. At this step, the raster indicating the average deviation angle for each pixel is obtained as a product.
iii. Finally, the raster of the average deviation angle is subtracted by the scalar 90, resulting in the raster of the convergence index for each pixel.

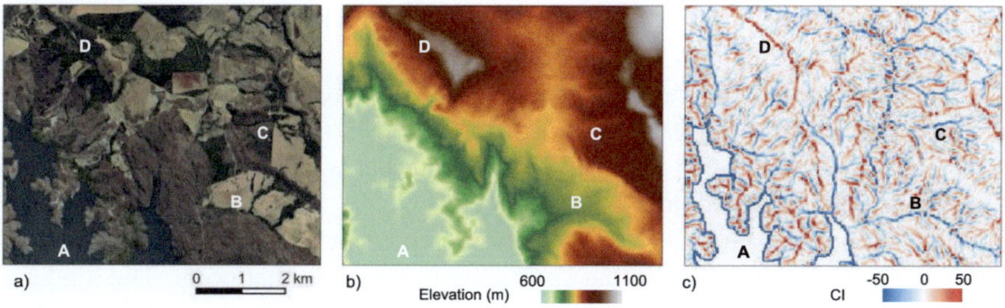

Fig. 12.4 **a** Google Earth satellite image for an area of the municipality of Sacramento (Minas Gerais state); **b** SRTM 30 DEM; **c** CI calculated from this DEM considering 3 × 3 window and scaled values for intervals from − 100 to 100, but color scale saturated for limits − 50 and 50 for better contrast and visualization of spatial patterns

A real-world example is shown in Fig. 12.4, regarding the CI obtained for a region south of the municipality of Sacramento, in the southwest of Minas Gerais. The spatial pattern of the CI shows similarities with recognizable features from the satellite image or the DEM. Some points of higher convergence (areas shaded in blue) coincide with the courses of rivers visible in the satellite image or with flow paths identifiable through the topographic pattern represented in the DEM.

For example, below the point indicated by letter B and above point C, rivers pass through, and their paths stand out as areas of strong convergence. On the other hand, hilltops show greater divergence (areas shaded in red), such as to the right and above point D. The area around point A is homogeneous in terms of CI, with a value near zero, as it corresponds to a large flat area—specifically, part of the lake formed by the Luiz Carlos Barreto de Carvalho Hydroelectric Power Plant, also known as Estreito.

12.2 Topographic Wetness Index (TWI)

The Topographic Wetness Index (TWI) describes the tendency or potential of a point on the terrain to accumulate water. It is the topographic index most used in hydrological studies (Mattivi et al., 2019). The initial idea present in the TWI index was conceived by Beven and Kirkby (1979), associated with the proposal of the TOPMODEL hydrological model and assuming the condition of permanent state (not variable over time) of the precipitation rate and the spatially uniform condition for precipitation and soil infiltration.

Less commonly, this index is also referred to by other terms, such as Compound Topographic Index (CTI; for example, in Dindaroglu et al., 2023, and Thompson et al., 2001), Topographic Convergence Index (TCI; for example, in Das et al., 2015), Topographic Index (TI; Wilson & Gallant, 2000), and Soil Wetness Index (SWI; for example, in Murphy et al., 2009). The TWI nomenclature is most used and recommended as it is more readily associated with the index and avoids confusion with other indices.

The central concept of the TWI is a mass balance, relating for a given point on the terrain two contrasting tendencies: (i) the tendency of this point to receive water and (ii) the tendency of this point to drain water, that is, for the gravitational force to move the water downstream.

The tendency of a point to receive water is represented by its drainage area (A), that is, the total upstream areas that contribute to runoff to that point. The tendency of a point to drain water is represented by two factors: the flow width (L) and the local slope (β). Considering the division of A by L as the specific drainage area (α), also referred to as the contribution area per unit of flow width, the expression for the calculation of TWI is as per Eq. (12.2), with the TWI being a dimensionless value.

$$\mathrm{TWI_P} = \ln\left(\frac{\alpha}{\tan\beta}\right) \tag{12.2}$$

where $\mathrm{TWI_P}$ is the topographic wetness index at point P.

High TWI values mean areas that are more prone to soil saturation. According to the TWI formulation, high values can result from large drainage areas or significantly reduced slopes. Similarly, reduced TWI values correspond to regions of lower propensity for soil saturation according to this index, which can originate from having small contribution areas or presenting steep slopes.

The TWI can also be seen as a topographic index representing the predisposition for subsurface runoff accumulating at a point in the terrain, increasing soil moisture and the frequency of saturated conditions (Meles et al., 2020). By applying the TWI to all terrain points, we have spatial information about water distribution in the soil.

Studies have found a correlation between the actual distribution of soil moisture and the TWI, mainly in small basins. Given this relationship and the concept that prolonged exposure to water influences soil properties, the TWI has also been related to soil characteristics such as organic matter content, sand and silt, depth, texture, and nutrient availability.

The association between soil moisture and TWI is limited because the index formulation does not capture the effect of seasonal variations and other variables not directly related to topography, but that influence soil moisture, such as soil characteristics, land cover, land use, solar radiation, vegetation characteristics, and evapotranspiration (Mattivi et al., 2019; Raduła et al., 2018).

Due to the relationship between TWI and the spatial distribution of soil water, this index has also been applied to spatially investigate hydrological processes within a watershed and in studies of a biological or ecological nature. In this second case, the index has been related to vegetation patterns, species richness and composition, net primary productivity, and forest biomass, among other variables.

The raw information needed to determine the TWI is the topography, commonly represented as the DEM. However, the calculation of the TWI involves variables derived from the DEM, not the elevation values themselves, as seen in Eq. (12.2): drainage area, flow width, and local slope.

Some software has a specific command to calculate the TWI. However, the following procedures and considerations can be adopted for the determination of the TWI index from the DEM, involving standard operations that are usually available in GIS software:

i. Removal of depressions from the DEM, as seen in Chap. 6;
ii. Determination of flow directions from the depressionless DEM, which is also detailed in Chap. 6;
iii. Calculation of accumulated drainage areas. For each pixel of the area of interest, the drainage area is equivalent to the size of the watershed that contributes to this pixel (Fig. 12.5a), as detailed in Chap. 7. Determining the drainage area values for all pixels is an operation resulting from the DEM processing, which follows from the flow directions. However, depending on the software, the input data may not be the raster of flow directions but the depressionless DEM or even the original DEM before removing depressions. Internally, the software performs the expected processing sequence, removing the terrain depressions, determining the flow directions, and only then calculating the accumulated drainage areas. Suppose the software calculates the accumulated drainage areas as the number of draining pixels. In that case, this raster should be multiplied by the surface area of a pixel in m^2 to generate a new raster of accumulated drainage areas in m^2.
iv. One approach to the flow width L is to assume two possible values depending on the flow direction of each pixel under analysis. In this approach, L has a value equal to the pixel size when the flow direction of this pixel is orthogonal (the pixel drains to one of its upper, lower, right, or left neighbors). When the pixel has a diagonal flow direction, the value of L equals $\sqrt{2}$ times the pixel size (as seen in Chap. 6, Fig. 6.6). Another approach to the definition of L adopted by some authors is to assume a constant value equal to the pixel size, regardless of whether the downstream flow is orthogonal or diagonal. Whether by one approach or another, dividing the raster of accumulated drainage areas (in m^2)

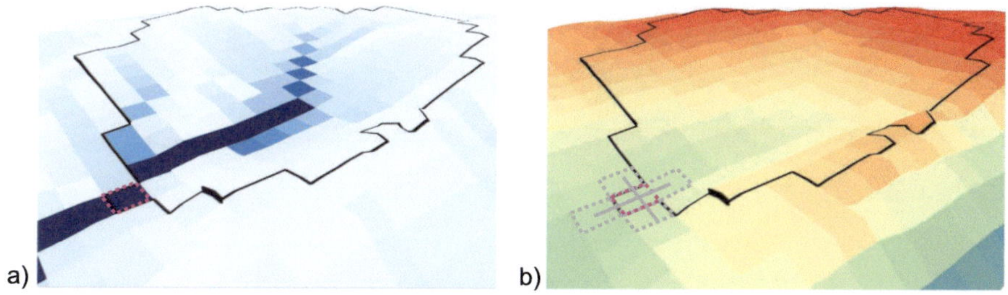

Fig. 12.5 **a** 3D visualization of the DEM with an overlay of accumulated drainage areas and delimitation of the area contributing to the pixel highlighted in pink; **b** 3D visualization of the DEM emphasizing the four orthogonal neighboring pixels (in gray border) to the pixel highlighted in pink, whose elevations are used for the determination of local slope in this pink pixel

by the corresponding value of L in each pixel (taken in meters) results in the raster of specific drainage areas.

v. The local slope variable for a given pixel is most commonly obtained as the average slope in two directions (horizontal and vertical), calculated based on the elevations of the four orthogonal neighbors (upper, lower, right, and left neighbor; Fig. 12.5b), as detailed in Chap. 5. This slope is usually obtained in percentage or degrees, with the second option necessary for calculating the TWI. However, the user should check the software used if the input argument of the tangent function is the angle in degrees or radians. Most software requires angles in radians to calculate trigonometric functions like the tangent. In this case, the slope obtained in degrees should be converted to radians by multiplying by π (3.1415…) and dividing by 180 for input into the TWI calculation (Eq. 12.2).

vi. As a final step, the mathematical operation involving the logarithmic function applied to the result of the quotient between the raster of specific drainage areas and the raster of the tangent of the local slope is performed. With this processing, the raster of the TWI index is obtained, where each pixel has the value of this index as an attribute.

Figure 12.6 illustrates the TWI index obtained for part of the municipality of Delfinópolis, in the southwest of the Minas Gerais state, as an example of a real-world case study. It also shows the input topographic information (SRTM 30 DEM) and the intermediate products necessary for the TWI: the local slope and the accumulated drainage areas.

In the example, the slopes are steeper on the hillsides (points A and B, for example) and flatter in the areas where the flows converge (points C and D, for example), with larger accumulated drainage areas. The hillsides are the regions with the least tendency to accumulate water and present lower TWI values.

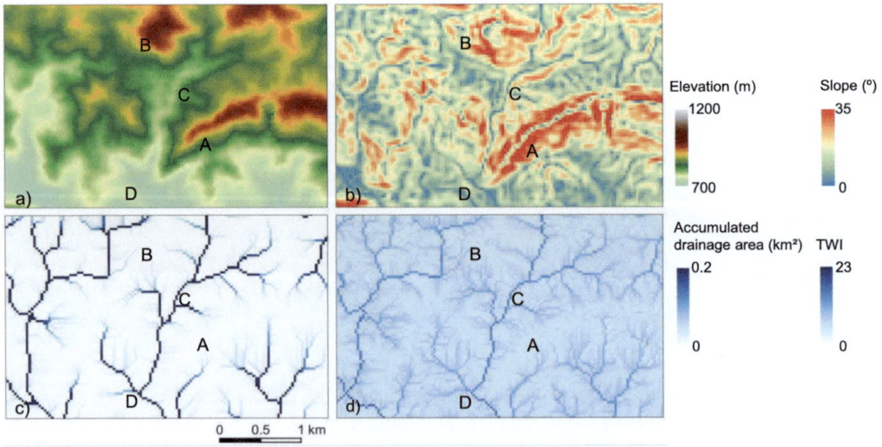

Fig. 12.6 **a** SRTM 30 DEM for part of the municipality of Delfinópolis (Minas Gerais); **b** Local slope; **c** Accumulated drainage areas; **d** TWI index

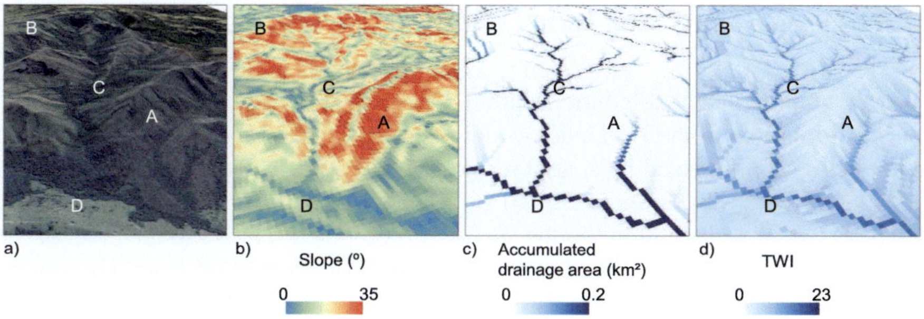

Fig. 12.7 3D visualization of part of the area shown in Fig. 12.6, with an overlay of Google Earth satellite image **a**, local slope raster **b**, accumulated drainage areas raster **c**, and TWI index **d**; the points A, B, C and D indicated in the figure are the same as those shown in the previous figure

The highest values of this index are strongly related to the spatial pattern of the preferred flow paths represented in the raster of accumulated drainage areas. The understanding of the spatial patterns and the effect of topographic variation on the intermediate variables (slope and drainage areas) and on the final TWI product is highlighted in the 3D visualization shown in Fig. 12.7, in which the same points A, B, C, and D from the previous figure are identified.

A disadvantage of the TWI is its dependence on the spatial resolution of the DEM. Several studies have reported that the variation of the resolution of the DEM used to derive the TWI influences the results and the spatial patterns found. An interesting research investigating this influence is the work of Riihimäki et al. (2021).

To understand in more depth the different methodological alternatives for determining the variables necessary for the calculation of the TWI and the impact of this choice on the values of this index, the reader is invited to visit the works of Kopecký et al. (2021), Kopecký and Čížková (2010), Mattivi et al. (2019), and Sørensen et al. (2006).

For example, Kopecký and Čížková (2010) point out that the determination of the TWI with the approach of multiple flow directions is recommended for ecological applications. The study by Lanni et al. (2011) evaluates variants of the TWI considering variable accumulated drainage area over time and smoothed by the average of the 3 × 3 neighborhood around the pixel.

Finally, for a discussion and deepening of the limitations of the scope of the application of the TWI index for relative evaluation of soil saturation areas, the recommended readings are the works of Meles et al. (2020) and Murphy et al. (2009).

12.3 Depth-To-Water Index (DTW)

The topographic index called the Depth-To-Water (DTW) was proposed to circumvent the limitations of the TWI index. Studies indicate that the DTW has the advantage of not being as influenced by spatial resolution in estimating spatial patterns of soil moisture as the TWI (Ågren et al., 2014).

The DTW of a given point on the surface is computed relative to the pixel of the river drainage network identified as most possibly connected to it from a hydrological point of view, considering for this the search in the direction of the minimum accumulated slope from the pixel to the drainage network (Murphy et al., 2009). This index considers the effect of each pixel's horizontal and vertical distance on the surface of the river drainage network (Oltean et al., 2016).

For each pixel on the surface, the path from this pixel to some point on the drainage network that provides the least accumulation of slope weighted by distance is identified. For a given pixel on the surface, there are numerous possible paths to the river drainage network, according to different pixel-to-pixel paths that consider as possible both orthogonal steps (from one pixel to one of the four upper, lower, left, or right neighbors) and diagonal steps (from one pixel to one of the four diagonal neighbors of the 3×3 window).

For each of these alternative paths from the pixel under analysis to some point on the drainage network, one can calculate and associate the value referring to the accumulation of slopes weighted by the incremental distance traveled. Along the route, the slope value between them is accumulated at each incremental step from one pixel to another but weighted by a factor (called 'a'). Each orthogonal step has the slope multiplied by the factor 'a' with value 1, and each diagonal step has the slope multiplied by 'a' with value $\sqrt{2}$. This distinguishes diagonal neighbors from orthogonal neighbors due to the higher relative distance between the former.

The path that provides the minimum accumulation can be identified among all possible paths from the pixel to some point in the river drainage network. The DTW index for this pixel is calculated as the value of this accumulation multiplied by the pixel size in meters (Eq. 12.3).

$$\mathrm{DTW_P} = \left[\sum \frac{dz_i}{dx_i} a\right] \cdot \Delta x \tag{12.3}$$

where $\mathrm{DTW_P}$ is the depth-to-water index at point P, $\frac{dz_i}{dx_i}$ is the slope at an incremental step of the flow path from P to the watercourse, a is a multiplier factor, and Δx is the spatial resolution of the DEM.

The DTW value is assumed to be zero for pixels belonging to the river drainage network and tends to increase as they move away from the network due to the increase in both distance and accumulated slope. The DTW value is inversely proportional to the tendency of soil moisture: lower DTW values are related to higher soil moisture, and the higher the DTW, the lower the predisposition for soil moisture.

According to the conceived formulation, the DTW increases more rapidly in steeper terrains and more slowly in flatter terrains (Murphy et al., 2009), indicating that soil moisture tends to decrease more quickly with the increase of more abrupt elevation as it moves away from water bodies.

Studies have adopted a threshold value for the DTW to distinguish the result into two classes: areas of wet soil (DTW below the threshold) and areas of dry soil (DTW above the threshold). However, this threshold is variable for each situation, mainly due to the local soil characteristics, topography, and climatic conditions (Mohtashami et al., 2022).

As an indicator of areas with higher soil moisture potential, DTW has been used in a variety of applications, as listed by Mohtashami et al. (2022), including: planning of harvesting operations in forestry areas; development of soil trafficability models; identification of suitable locations for stream crossings; and planning of agricultural management practices, such as fertilizer application.

The DTW index proposes to relate each pixel to the pixel of the drainage network to which the former would have a significant chance of being hydrologically connected (Murphy et al., 2009). However, in a strict sense, the path of minimum slope accumulation weighted by the incremental distance does not necessarily lead to the pixel of the drainage network to which it is hydrologically connected. This hydrological connection could be identified differently if the downstream flow directions from the pixel were followed pixel by pixel until reaching the drainage network. Determining flow directions has, as one of the main rules, the higher slope of a pixel relative to its neighbors, but other conditions exist mainly in the case of neighbors that provide the same slope or in the case of flat areas, as seen in Chap. 6. The occurrence of these situations and how they are treated can make the path given by the minimum accumulation of slopes differ from the path established by the flow directions.

One way to obtain the DTW index is to use minimum cost path tools. You need the local slope raster and the river drainage network as input data. The slope raster is obtained from the DEM processing, as described in Chap. 5. Attention should be paid to the unit required by the minimum cost path procedure—whether it requires slopes in degrees or radians, for example. The river drainage network is also derived from the DEM, as described in Chap. 7. It is essential to define the beginning of the river drainage network coherently since there is total dependence on the result of the DTW based on the channel initiation definition and the density of rivers in the study area, as illustrated in the results of Mohtashami et al. (2022).

In the minimum cost analysis, it is considered that the river drainage network is the goal or target to be reached. Starting from each point of the terrain, the surface of the terrain slopes represents the difficulty to be overcome, i.e., the cost. There are infinite paths to be followed from a point on the terrain to some point in the river drainage network. Each path, configured by the path taken and the destination point reached, presents as accumulated cost the accumulation of slopes, considering the distance traveled over the slope surface. Minimum cost path tools can identify which point in the river drainage network and which path represents the lowest accumulated cost for each point on the terrain. In this case, the value of this accumulated cost corresponds to the DTW index.

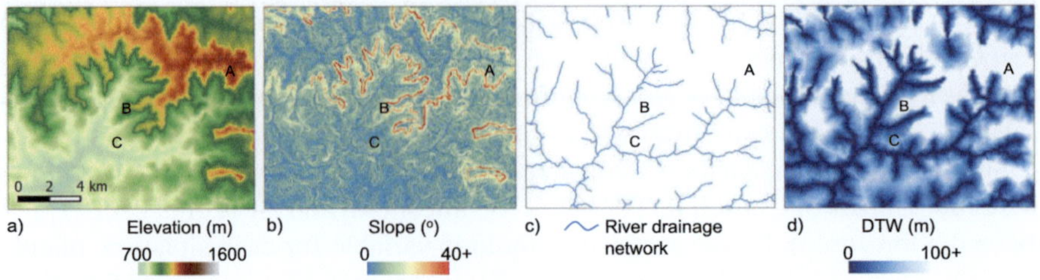

Fig. 12.8 **a** SRTM 30 DEM of the area between the municipalities of Franca and Patrocínio Paulista (São Paulo state); **b** Local slope; **c** River drainage network; **d** DTW topographic index

Figure 12.8 shows the raster of the topographic DTW index obtained for an area between the municipalities of Franca and Patrocínio Paulista in São Paulo from the SRTM 30 DEM. As some studies point out, using DEMs with finer spatial resolution, such as those from LiDAR surveys with 1 or 2 m resolution, is recommended to better identify the spatial patterns of DTW estimation. Nonetheless, the example presented in Fig. 12.8 clarifies the concept of DTW and illustrates the general nature of the results obtained for a real-world study case.

In this example, the spatial pattern of DTW is linked to the layout of the river drainage network, starting from zero exactly where the rivers flow, and increasing the value of this index as it moves away from them. In areas with a steeper topographic gradient, there is a higher accumulation of slopes, resulting in the highest values of DTW, as at point A. Points B and C illustrate regions of almost the same distance and configuration relative to the river drainage network, as they are between river traces in an approximately equidistant manner. However, there are very distinct DTW patterns between points B and C due to the spatial pattern of slopes: at point C, gentler slopes predominate, resulting in lower DTW values than those obtained around B, which has steeper slopes.

12.4 Wetness Index Based on Landscape Position and Topography (WILT)

The Wetness Index based on Landscape position and Topography (WILT) was motivated by the limitations of the TWI index, like the DTW index seen in the previous item. Meles et al. (2020) proposed modifying the TWI to improve the prediction of relative soil moisture in watersheds with a higher contribution from groundwater flow, moderate slopes, and high subsurface hydraulic conductivity. In these areas, surface water bodies such as rivers, lakes, and even wetlands mark relatively continuous positions of the water table. The soil is generally wetter near these water bodies and drier in other areas (Meles et al., 2020).

The topographic convergence represented in the TWI by the accumulated drainage areas and by the local slope also acts to favor soil saturation. However, the idea of Meles et al. (2020) was to combine this factor with the distance to the water table. These authors proposed to modify the original formulation of the TWI index to incorporate two variables that relate to the position of the point under analysis relative to the water bodies and the water table (Eq. 12.4).

One of the variables is ΔC, which represents the horizontal distance (Euclidean distance, in a straight line) of the point under analysis relative to the nearest water body. The other variable is ΔL, which represents the vertical distance of the point relative to the water table. In the calculation of WILT, the contribution area is considered per se and not the value of this magnitude divided by the flow width, as occurs for the TWI index.

$$\mathrm{WILT_P} = \ln\left(\frac{A}{\Delta C \cdot \Delta L \cdot \tan\beta}\right) \tag{12.4}$$

where $\mathrm{WILT_P}$ is the WILT index at point P, A is the contribution area at P, and β is the local slope at P.

According to the WILT index proposal, soil moisture is inversely proportional to ΔC and ΔL. The higher the horizontal distance from a point to the nearest watercourse/water body or the vertical distance to the water table, the lower the predisposition for soil moisture. Similarly, the closer to a water body or the water table, the higher the tendency for soil moisture. The scheme indicated in Fig. 12.9a, based on Meles et al. (2020), illustrates the computation of distances ΔC and ΔL for two points (P_1 and P_2) along a topographic cross-section that includes a river and the explicit position of the water table. According to the WILT index, point P_1 tends to have higher soil moisture due to its proximity to the watercourse and higher proximity to the water table than point P_2. According to WILT, if point P_1 is located along a primary flow path, it also has a large accumulated drainage area, contributing to higher soil moisture.

However, points in the region of P_1 with a small contribution area still have high soil moisture according to the estimate of the WILT index due to their proximity to the watercourse and the water table. Conversely, points around P_2 with a large accumulated drainage area tend to have relatively low soil moisture, according to WILT, due to their higher distance to the watercourse and the water table.

In the absence of data on the height or depth of the water table, the vertical distance ΔL in the formulation of the WILT index can be adapted to refer to the elevation difference of each point under analysis relative to the nearest watercourse (Fig. 12.9b; Meles et al., 2020).

The determination of the WILT index for each pixel of a DEM can be carried out via a set of standard geoprocessing operations that are usually available in GIS software, as follows:

i. From the DEM, the operation of depression removal is performed (as per Chap. 6);
ii. The depressionless DEM is processed to determine the flow directions, as described in Chap. 6 and following specific commands in the software;

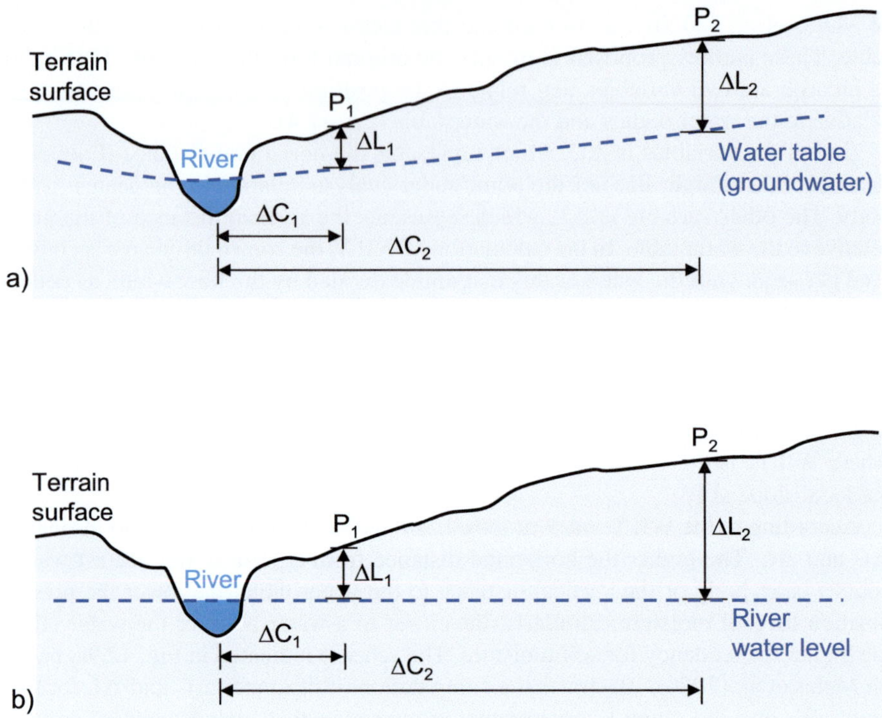

Fig. 12.9 **a** Illustrative scheme of the distances of two points (P_1 and P_2) to the nearest watercourse (ΔC_1 and ΔC_2) and the water table (ΔL_1 and ΔL_2) for the calculation of the WILT topographic index; **b** Adaptation of the calculation of ΔL_1 and ΔL_2 using the nearest watercourse as a reference for the water level, when data on the water table depth is unavailable (*source* Based on Meles et al., 2020)

iii. The raster of accumulated drainage areas is determined, in units of m^2, based on the flow directions (see Chap. 7);

iv. From the DEM, the local slope for each pixel is calculated based on the elevations of its four orthogonal neighbors (as per Chap. 5);

v. To determine ΔC distances, a layer of information regarding the location and layout of the watercourses is needed. Data from reliable sources, such as governmental portals of geographic databases, can be used, or the river drainage network can be extracted from the DEM processing, as seen in Chap. 7. In the first case, the primary concern is that this database represents the region's watercourses and is consistent with the topographic information present in the DEM. In the second case, there are two main concerns: that the definition of the beginning of the rivers is realistically carried out, compatible, and consistent with field information gathered from satellite images; that the river drainage network incorporates the location and extent of existing lakes in the region.

vi. From the definition of the river drainage network, a distance operator is used to compute the Euclidean distance (in a straight line) from each pixel to the nearest point of this network (as seen in Chap. 10);

vii. There are two main ways of obtaining the vertical distances ΔL, depending on the available information:

   a. If there is data on the depth of the water table at irregular points along the area of interest, these data can be interpolated to generate a surface as a raster with the exact spatial resolution and geographic coverage as the DEM. In this case, the calculation of ΔL for each pixel occurs by directly subtracting the DEM from this raster of water table depths.
   b. The second case is when there is no information on the water table, and it is decided to simplify by computing vertical distances relative to the elevation of the nearest watercourse (situation illustrated in Fig. 12.9b). The operation involves identifying which point of the river drainage network is closest to each pixel of the surface in a straight line and then calculating the difference in elevation between the surface pixel and this nearest pixel of the network. Some software has specific functions for these operations. Alternatively, a computational algorithm could be developed in a programming language with the rasters of the river drainage network and the DEM as input.

viii. The final step involves performing the mathematical operation involving the four rasters obtained in the previous steps (accumulated drainage areas, horizontal distances to watercourses, vertical distances to the water table, and local slope) according to the mathematical expression of Eq. (12.4). The result is a raster whose attribute is the value of the WILT index in each pixel.

Figure 12.10 shows a real-world example of the WILT index calculated for an area in southeastern Brazil. The SRTM 30 DEM used in determining the index and its derived products necessary for WILT are also shown to understand the spatial pattern of the results better. The value of WILT ranged from − 8 to 15.

Negative values correspond to the situation where the argument of the logarithm is less than 1. This means the accumulated drainage area A is less than the product of ΔL, ΔC, and $\tan \beta$. This occurs, for example, in the area below region A indicated in Fig. 12.10, which corresponds to a headwater region with a reduced drainage area and large slope, distance, and elevation relative to the river drainage network. The southeast area of region B is located near the main water flows but has a reduced drainage area and large slope and elevation relative to the drainage network, resulting in negative WILT. The value of C is zero in the drainage network pixels, like those to the left of B. In this case, there is a mathematical indeterminacy in the calculation of WILT.

The spatial pattern of WILT is strongly influenced by the spatial pattern of the accumulated drainage areas for two reasons. The first is obvious, as it directly includes this information in the calculation. The second reason is that the distances and the elevation relative to the drainage network have, by definition, a spatial pattern conditioned by the positioning and layout of the river drainage network. However, there are

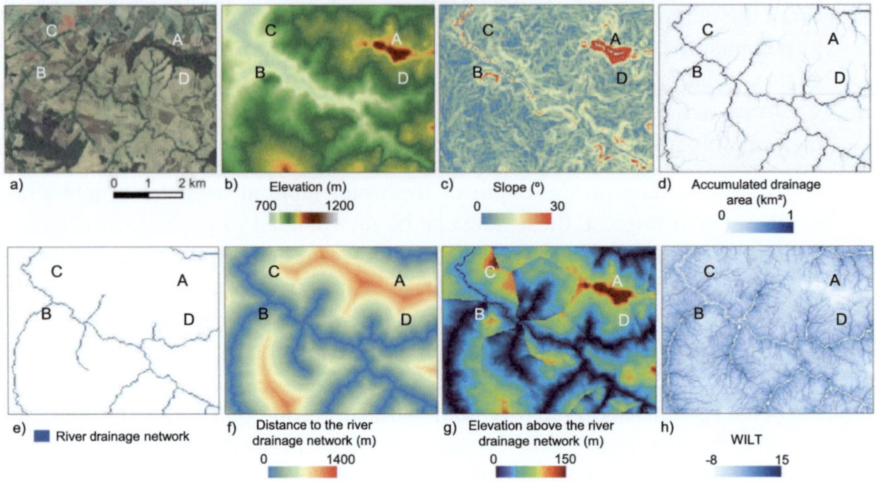

Fig. 12.10 **a** Google Earth satellite image of an area between the municipalities of Cristal Paulista (São Paulo state) and Claraval (Minas Gerais state); **b** SRTM 30 DEM; **c** Local slope; **d** Accumulated drainage areas; **e** River drainage network; **f** Euclidean distance to the drainage network; **g** Elevation above the drainage network; **h** WILT topographic index

areas of flow accumulation that do not correspond to the formation of the drainage network itself and appear in the WILT results. An example of this case is the areas near C and D, where there are flow accumulations without corresponding sections of the river drainage network.

12.5 Stream Power Index (SPI)

The Stream Power Index (SPI) is a topographic index that measures the erosive power of water flow (Moore et al., 1988) or, in other words, the energy available for surface runoff to cause soil erosion and particle transport (Moore & Nieber, 1989). Generally, as water flows downstream, potential energy is gradually converted into kinetic energy, part of which is responsible for causing erosion and sediment transport.

The SPI index comes from the concept of stream power or fluvial power (ω; Eq. 12.5) of a watercourse, understood as the ability of the watercourse to modify the geomorphology from the joint action of the hydraulic gradient and flow (Bizzi & Lerner, 2015). The SPI is directly proportional to ω (Burrough & McDonnell, 1998). For a deeper understanding of this concept of fluvial power, the reader is referred to the works of Howard (1994) and Lague (2014).

$$\omega = \rho \cdot g \cdot q \cdot \tan \beta \tag{12.5}$$

where ρ is the water density, g is the acceleration of gravity, q is the surface runoff flow per unit of width, and β is the local slope.

The original formulation of the SPI index is given by the multiplication of the specific drainage area by the tangent of the local slope (Eq. 12.6). It is assumed that the flow is proportional to the specific drainage area. Some works consider the application of the natural logarithm to the product indicated in Eq. (12.6), resulting in Eq. (12.7) for the calculation of the SPI (for example, Galzki et al., 2011). This is a transformation in the scale of index values. The original version without the logarithmic function is the most commonly adopted in literature.

$$\mathrm{SPI_P} = \alpha \cdot \tan\ \beta \tag{12.6}$$

$$\mathrm{SPI_P} = \ln(\alpha \cdot \tan\ \beta) \tag{12.7}$$

where $\mathrm{SPI_P}$ is the stream power index at point P, α is the specific drainage area at P, and β is the local slope at P.

The variables involved in the calculation of the SPI are the same variables used for calculating the TWI index, as discussed in item 12.2, but organized under a different mathematical formulation. The specific drainage area refers to the accumulated drainage area or the contributing area upstream of the point under analysis, divided by the flow width. The local slope, in general, is taken as the average slope in both directions (horizontal and vertical), calculated based on the elevations of the four orthogonal neighbors (upper, lower, right, and left), as presented in Chap. 5.

The SPI index distinguishes areas with different behaviors in terms of erosion (Wilson & Gallant, 2000): (i) it estimates that net erosion occurs in areas that combine convexity in terms of profile curvature and concavity in terms of plan curvature, which configure acceleration of flow and convergence zones; (ii) it estimates that there is a net deposition of sediments in areas of concavity in terms of profile curvature, as these are zones of decreasing flow speed.

The higher the SPI value, the higher the vulnerability to erosion. With the increase in the accumulated drainage area and slope, there is an increase in the volume of water that comes as a contribution from upstream and an increase in the flow speed. Both factors increase the SPI value and, therefore, the erosion risk. The highest SPI values are concentrated along the watercourses and tend to decrease rapidly, moving away from them (Moore & Nieber, 1989).

The SPI has been identified as a valuable indicator for targeting areas where intensified efforts in soil conservation and surface runoff erosion control practices are needed (Moore & Nieber, 1989). For instance, Galzki et al. (2011) demonstrated the effectiveness of SPI analysis in identifying gullies and areas experiencing severe erosion, with results validated through field inspections.

The determination of the SPI from a DEM can be performed directly by a specific command available in some software, such as Whiteboxtools GAT and ArcGIS, or from a set of operations available in most software. In this second case, the same steps established for the calculation of the TWI index are necessary:

i. Removal of depressions from the DEM;
ii. Determination of flow directions from the depressionless DEM;
iii. Calculation of accumulated drainage areas based on the flow directions;
iv. Definition of flow width L and calculation of specific accumulated drainage areas by dividing the raster of accumulated drainage areas by the width L;
v. Determination of the local slope raster;
vi. Performing a mathematical operation involving the multiplication of the raster of specific accumulated drainage areas by the raster of the tangent of the local slope.

As an example of a real-world case study, Fig. 12.11 shows the SPI index determined from the SRTM 30 DEM for an area of the municipality of Pedregulho in São Paulo state. As expected, the spatial pattern of the SPI is strongly influenced by the pattern of accumulated drainage areas, with a higher index value along the accumulated flow paths, denoting higher vulnerability to erosion. Two examples of these areas are to the left of G and above H.

The slope's influence on the example's SPI results is also noticeable. Along the area between points A, B, and C, for example, there are the highest slopes in the region but reduced accumulated flows, as these are areas of initial flow formation. Even so, there are high SPI values, i.e., areas more prone to erosion. On the other hand, the areas above and below D and between E and F do not configure flow accumulation zones and have a slight slope. In these areas, the SPI has reduced values, reaching very close to the minimum value of zero.

In Fig. 12.11, the SPI index calculated considering the application of the logarithm is also shown, according to the formulation of Eq. (12.7). The spatial pattern is maintained relative to the original SPI. However, the logarithmic function helps narrow

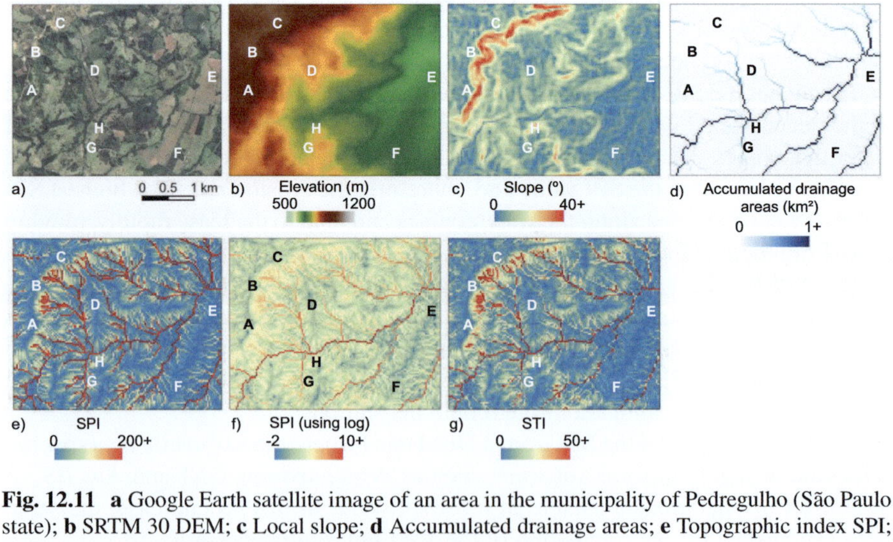

Fig. 12.11 **a** Google Earth satellite image of an area in the municipality of Pedregulho (São Paulo state); **b** SRTM 30 DEM; **c** Local slope; **d** Accumulated drainage areas; **e** Topographic index SPI; **f** SPI index calculated with logarithm; **g** Topographic index STI

the range of values. Consequently, the color association of the raster symbolism sweeps more entirely from the minimum to the maximum value. In the original SPI, the values range from close to zero to almost 580,000. In Fig. 12.11e, the maximum value of the color scale bar was saturated at 200 to identify spatial patterns better. The negative consequence of this adjustment in the color scale is that it does not distinguish among the values above this threshold.

When applying the logarithmic function (Fig. 12.11f), the SPI varies in the region of the example from close to − 3 to almost 14. This small range of values associated with the same color symbolism allows the visualization of more nuances of the spatial pattern, including the differentiation of SPI values related to higher erosive power. For example, the accumulated flow that passes to the left of G appears indistinguishable from the main flow above H in the original SPI. In the SPI with logarithm, the main flow above H is perceived as having higher erosive power than that to the left of G.

With the application of the logarithm in this example, the area between E and F is also distinguished as less vulnerable to erosion than in the region around D. This distinction was not perceived in the original version of the SPI. The change in the calculation of the SPI also makes negative values appear, which are due to the product $\propto \cdot \tan \beta$ with a value less than 1. This occurs when the slope is tiny, less than 1.

12.6 Sediment Transport Index (STI)

Moore and Burch (1986) conceived the Sediment Transport Index (STI), which some authors also call the Sediment Transporting Capacity Index (De Roo, 1998) or the Sediment Transport Capacity Index (Wilson & Gallant, 2000). The STI is defined by Eq. (12.8) (Burrough & McDonnell, 1998), which involves the variables specific accumulated drainage area and local slope. These are the same variables needed to calculate the TWI and SPI indices.

$$\mathrm{STI_P} = \left(\frac{\alpha}{22.13}\right)^{0.6} \cdot \left(\frac{\sin \beta}{0.0896}\right)^{1.3} \tag{12.8}$$

where $\mathrm{STI_P}$ is the sediment transport index at point P, α is the specific drainage area at P, and β is the local slope at P.

The exponent of the term that contains the specific accumulated drainage area appears in the literature predominantly with the value 0.6, as presented in Eq. (12.8) and the works of Burrough and McDonnell (1998), De Roo (1998), Wilson and Gallant (2000) and Jaafari et al. (2014), for example. Other works present 0.4 for such an exponent (examples: Ahmad et al., 2019; Mojaddadi et al., 2017; and the Whiteboxtools GAT software).

In some circumstances, the STI index is equivalent to the LS (Length-Slope) topographic factor of the Universal Soil Loss Equation when applied to three-dimensional surfaces (Wilson & Gallant, 2000). The STI is used to characterize erosion and sediment deposition processes, highlighting the effects of topography on soil loss (Burrough & McDonnell, 1998).

Software like Whiteboxtools GAT provides a specific tool for determining the STI, taking the raster of specific accumulated drainage areas and the raster of the local slope as input. However, in most GIS software, standard tools that are usually available can be combined to determine the STI from the DEM processing. For this, a set of steps is carried out similar to those necessary to obtain the TWI and SPI indices, as follows:

i. Removal of depressions from the DEM;
ii. Determination of flow directions from the depressionless DEM;
iii. Calculation of accumulated drainage areas based on flow directions;
iv. Calculation of specific accumulated drainage areas by dividing the raster of accumulated drainage areas by the flow width;
v. Determination of the local slope raster;
vi. Mathematical operation between rasters and scalars according to Eq. (12.8).

Figure 12.11g presents an example of the determination of the STI index for an area in the municipality of Pedregulho, in São Paulo, along with SPI index results. The spatial pattern obtained for the STI is similar to that described for the SPI index in the previous item. This is expected, as it involves the same variables of the specific accumulated drainage area and local slope in the numerator of the mathematical formulation. Higher STI index values indicate a greater erosion occurrence along the larger accumulated flow paths. In contrast, the index suggests a predominance of depositional processes in the areas between these flows and with lower slopes.

12.7 Height Above the Nearest Drainage (HAND)

12.7.1 Fundamentals

As already seen in Chap. 10, one can identify the flow path that passes through each point of the terrain and follow it until it reaches the river drainage network. Additionally, with this path traced, one can quantify the difference in elevation or altitude between the point of the terrain where the path originates and the point of the river drainage network reached.

Repeating this procedure for all points in a region, one obtains a raster of elevations relative to the river drainage network, a concept worked on by Crave and Gascuel-Odoux (1997) and Rodda (2005), for which Rennó et al. (2008) designated the name HAND (Height Above the Nearest Drainage). In summary, HAND is a topographic index whose attribute in each pixel represents the difference in elevation of this point relative to the point of the drainage network to which it connects according to the flow paths (Fig. 12.12). In other words, one can think of HAND as a DEM whose vertical reference varies for each pixel (Nobre et al., 2016) since, at each pixel, the elevation

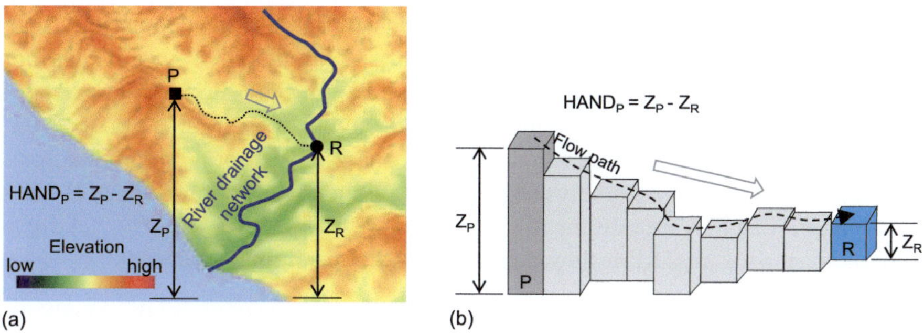

Fig. 12.12 **a** Illustrative scheme of the determination of the topographic index HAND for a point P, located at elevation Z_P, and whose flow path connects to point R of the river drainage network, located at elevation Z_R; **b** Detail of the flow path traveled from P to R and indication of the elevations used for the calculation of HAND

is relative to the point of the river to which it connects. The distance traveled along the flow path from the pixel under analysis to the point of the river drainage network is not accounted for in the HAND calculation; only the difference in elevation between them is considered.

Higher HAND values indicate a higher potential for drainage of the area, while lower HAND values indicate areas more prone to water accumulation and soil saturation (Nobre et al., 2011). HAND has been used for multiple purposes, notably for estimating flood-prone areas (Garousi-Nejad et al., 2019; Nobre et al., 2016), as exemplified later. Other applications of the HAND index include the establishment of terrain classes with distinct characteristics of soil moisture and vegetation cover (Rennó et al., 2008), the identification of areas with different hydrological functions (Gharari et al., 2011), understanding of spatial patterns of plant species diversity (Marca-Zevallos et al., 2022; Schietti et al., 2014), development of new approaches to hydrological modeling of the rainfallrunoff transformation (Gao et al., 2019), and identification of landslide occurrence patterns (Hong et al., 2018).

12.7.2 HAND Determination

Determining HAND for an area of interest requires three data layers: DEM, flow directions, and the river drainage network. The last two products are derived from the first (as seen in Chaps. 6 and 7). The depressionless DEM and the raster of accumulated drainage areas must also be generated as intermediate products to obtain them.

Some software may provide a direct and specific tool to obtain HAND. Other software presents operations that, when combined, allow the elaboration of HAND. The third way to obtain HAND is the development of a specific algorithm by the user in a programming language of preference.

Specific commands available in most GIS software remove depressions from the DEM, determine the flow directions, calculate accumulated drainage areas, and extract the river drainage network. However, there are usually no ready-made tools for identifying the pixel of the drainage network to which each pixel connects and computing the difference in elevation between them. Developing a computational routine for these tasks is relatively easy, given the simple concept involved in HAND.

The main idea of a routine for the calculation of HAND is as follows, taking as input the rasters of the DEM, flow directions, and river drainage network:

i. For each pixel in the study area that does not belong to the river drainage network, the elevation of this pixel is recorded. For example, for pixel A in Fig. 12.13, the elevation $Z_A = 254$ m;
ii. The flow path from each pixel to the river drainage network is traced according to the DEM-derived flow directions. In the case of pixel A, the flow continues until it reaches pixel P of the river drainage network;
iii. The elevation of the pixel of the river drainage network reached by tracing this flow path is recorded. In the example, it is found that $Z_P = 213$ m;
iv. The HAND value for the pixel under analysis is calculated as the difference between the elevation of this pixel and the elevation of the pixel of the river drainage network reached by the traced path. For pixel A, it is found that $HAND_A = Z_A - Z_P = 254 - 213 = 41$ m;

DEM (m)

a)

221	220	214	216	222	228	233	238	241	242
232	220	213	216	218	221	229	235	241	250
231	223	215	217	221	228	241	246	250	254
232	231	223	219	221	230	241	246	252	254
235	238	230	222	221	227	234	242	251	255
235	238	232	226	224	223	228	240	250	253
233	238	240	235	229	226	226	231	239	244
238	240	242	238	233	227	227	225	225	228
243	246	246	241	234	228	225	227	224	223

River drainage network

b)

0	0	1	0	0	0	0	0	0	0
0	0	1	0	0	0	0	0	0	0
0	0	1	0	0	0	0	0	0	0
0	0	0	1	0	0	0	0	0	0
0	0	0	0	1	0	0	0	0	0
0	0	0	0	0	1	0	0	0	0
0	0	0	0	0	0	1	0	0	0
0	0	0	0	0	0	0	1	0	0
0	0	0	0	0	0	0	0	0	0

c)

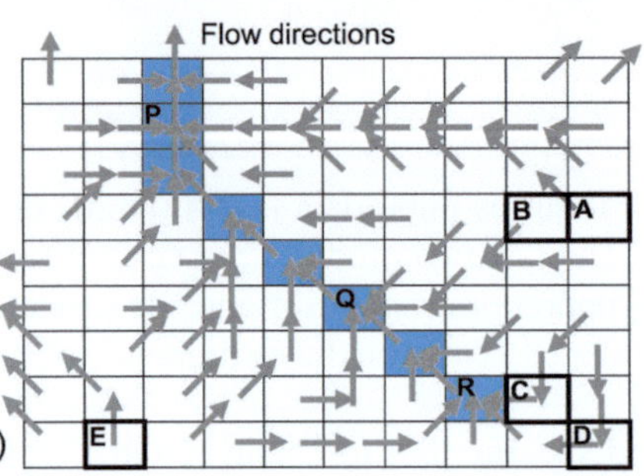

HAND (m)

d)

	6	0	2	8	15	20	25		
19	7	0	3	5	8	16	22	28	37
16	8	0	4	8	15	28	33	37	41
17	16	8	0	8	17	28	23	29	41
	23	11	3	0	6	11	19	28	32
	19	13	7	3	0	5	17	24	28
	19	21	16	8	3	0	5	14	19
		23	17	10	4	1	0	0	3
		25	16	9	3	0	2	-1	-2

Fig. 12.13 **a** DEM with elevation in meters; **b** Raster river drainage network (pixels that are part of the network or not have attribute 1 or 0, respectively); **c** Representation of flow directions; **d** HAND topographic index (in meters)

v. Pixels belonging to the river drainage network are assigned a zero value for the HAND.

In HAND, each pixel's elevation is measured relative to a specific point of the river drainage network. Situations can occur as illustrated by pixels A and B in Fig. 12.13. These pixels are neighbors and have elevations that differ by only 2 meters. However, pixel A has a flow that drains to the river drainage network at point P, while pixel B has a flow that reaches the river drainage network at point Q. As Q is 10 m above P, the HAND of pixel A is 12 m above the HAND of pixel B.

Other situations that can occur when calculating the HAND are:

a. Pixels with elevation equal to that of the pixel of the river drainage network to which they connect. This is the case of pixel C in the example of Fig. 12.13, which results in a HAND with a value of zero.
b. Pixels with elevation lower than that of the pixel of the river drainage network to which they connect. This is the situation of pixel D in Fig. 12.13, resulting in a HAND with a negative value.
c. Pixels whose flow path does not reach the river drainage network within the windowed image. This is the case of pixel E in Fig. 12.13 and others in its vicinity. In this case, the HAND is indeterminate as there is no reference to the elevation value of the pixel of the river drainage network.

Situation (b) occurs when the DEM is used as input information for calculating HAND without removing the depressions. Terrain depressions in a DEM are common and can consist of isolated pixels or a set of neighboring pixels. These depressions can be natural areas of the terrain with elevations lower than the surrounding neighborhood, or they can be the result of artifacts of the data acquisition and preparation process to elaborate the DEM, as already discussed in Chap. 6.

If the depressions are at an elevation below the river point to which the flow converges, the result is a negative HAND for the pixels that make up the depression. If the HAND is calculated with a depressionless DEM, all pixels have HAND with positive values or at least zero values (Dantas & Paz, 2021).

There is no consensus in the literature on whether to calculate the HAND using the DEM with or without depression removal treatment. For example, Garousi-Nejad et al. (2019) and Dantas and Paz (2021) recommend using the depressionless DEM to estimate flood areas. The second-mentioned work points out the advantage of removing depressions from the DEM, which would reduce the noise effect caused by marginal vegetation to rivers when working with global DEMs like the SRTM. On the other hand, Zheng et al. (2018a) recommend using the DEM with depressions, as this would allow identifying local flood areas, such as lakes not connected to the river drainage network. Removing the depressions would no longer represent these natural lakes in the DEM.

12.7.3 Real-World Example of HAND Determination

The HAND obtained for the Icamaquã river basin (located in Rio Grande do Sul state and tributary of the Uruguay River) from the SRTM 90 DEM is presented in Fig. 12.14. It illustrates a real-world example of this topographic index. In this case, the HAND ranges from − 3 to 165 m, calculated with the DEM without removing the terrain depressions. The spatial distribution shows a predominance of areas with higher HAND values in the upper parts of the basin, close to the headwaters, and lower values in the lower region of the basin, near the outlet. This is the general expected pattern of spatial variation of HAND in a watershed.

In the upper part of the basin, there is higher topographic variation, and the relief has higher slopes, so there is a higher difference in elevation between the points of the terrain and the points of the river drainage network to which they connect. The flatter areas are located in the lower part of the basin, with a lower topographic amplitude of the terrain relative to the river drainage network, which results in lower HAND values.

The HAND variation along line A-B, also shown in Fig. 12.14, highlights how the river drainage network is the reference for the elevations computed in this terrain descriptor. The cross-sectional profile drawn crosses two watercourses, River 1 and River 2, having a null value for the HAND at these crossing points of the watercourses. As indicated by C, the region between these points is more than 100 m in altitude

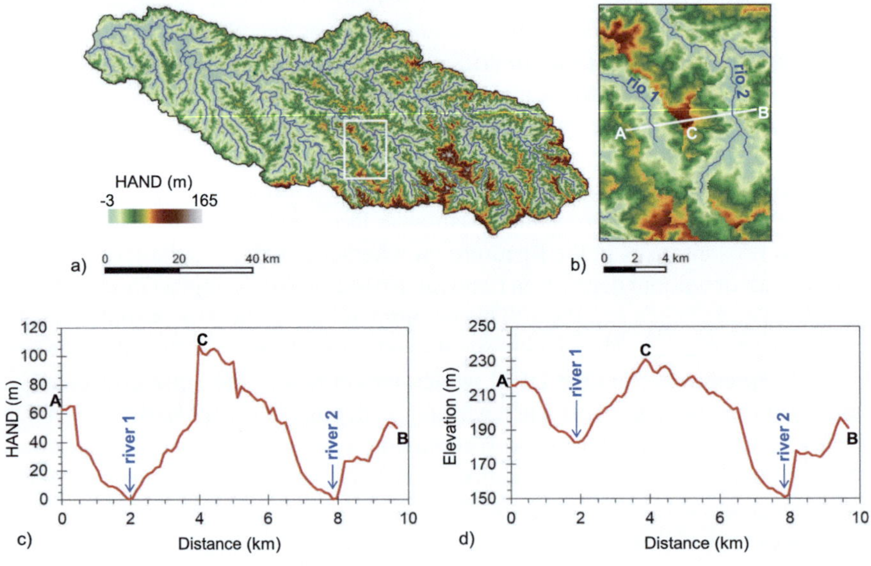

Fig. 12.14 **a** HAND generated for the Icamaquã river basin (RS) from the SRTM 90 DEM without removing the terrain depressions; **b** Enlargement of the area indicated by the rectangle in a); **c** Variation of the HAND along the line A-B indicated in b); **d** Cross-sectional topographic profile along A-B obtained from the SRTM 90 DEM

relative to the river drainage network to which it connects. Even without indicating the flow paths in this figure, it can be deduced that the areas to the left of C have flow directed to River 1, as HAND values gradually increase since the pixels of this river up to this region.

Similarly, the areas to the right of C have a flow that follows River 2. Thus, the areas to the left of C have River 1 as a reference for the HAND. In contrast, River 2 is a reference for the regions to the right of C. However, the crossing points of line A-B with Rivers 1 and 2 are not exactly where the flows from the areas around C converge. This convergence occurs for points of these rivers downstream of these crossings, located at a lower elevation.

The topographic profile along the A-B line obtained from the DEM shows that the elevation difference between that point on River 1 and region C in a straight line is less than 50 m. In comparison, the difference between the point on River 2 under the A-B line and region C is around 80 m. This highlights how HAND differs from the elevation relative to the drainage network based on Euclidean, straight-line distance.

12.7.4 Estimating Floodable Areas Based on HAND

One of the main applications of HAND is the estimation of floodable areas, using various approaches with different degrees of complexity. The most simplified procedure consists of obtaining the HAND and reclassifying this raster to distinguish floodable and non-floodable areas (Afshari et al., 2018; Nobre et al., 2016).

According to this approach, all pixels with HAND below a given threshold are floodable areas. It is assumed that the set of watercourses of the river drainage network has its water level increased equally and parallel to the level represented in the DEM (Dantas & Paz, 2021). The chosen uniform increase in water level (ΔH in Fig. 12.15) is the value of the HAND threshold considered to distinguish flooded and non-flooded areas.

In any GIS software, determining the floodable areas from the HAND according to this simplified approach is summarized in a reclassification operation. The schematic example of Fig. 12.16 illustrates this procedure in practice. Given the HAND and assuming a flood threshold of 15 m, all pixels whose HAND is below this value are identified. These pixels constitute the floodable area by this method and for this flood threshold value, receiving attribute 1. In contrast, the other pixels (those with HAND above the threshold) have a zero value (or undetermined, signaled by some typical value of *no data*).

This approach of identifying the floodable areas by a maximum HAND threshold assumed uniformly for all the rivers of the study area ignores the differentiated hydraulic and hydrodynamic behavior among the basin's rivers. Due to differences in the drainage areas' dimensions and characteristics and the river reaches' characteristics (slope, roughness, cross-section area), floods occur with different intensities and frequencies along these sections in a basin. This is disregarded when

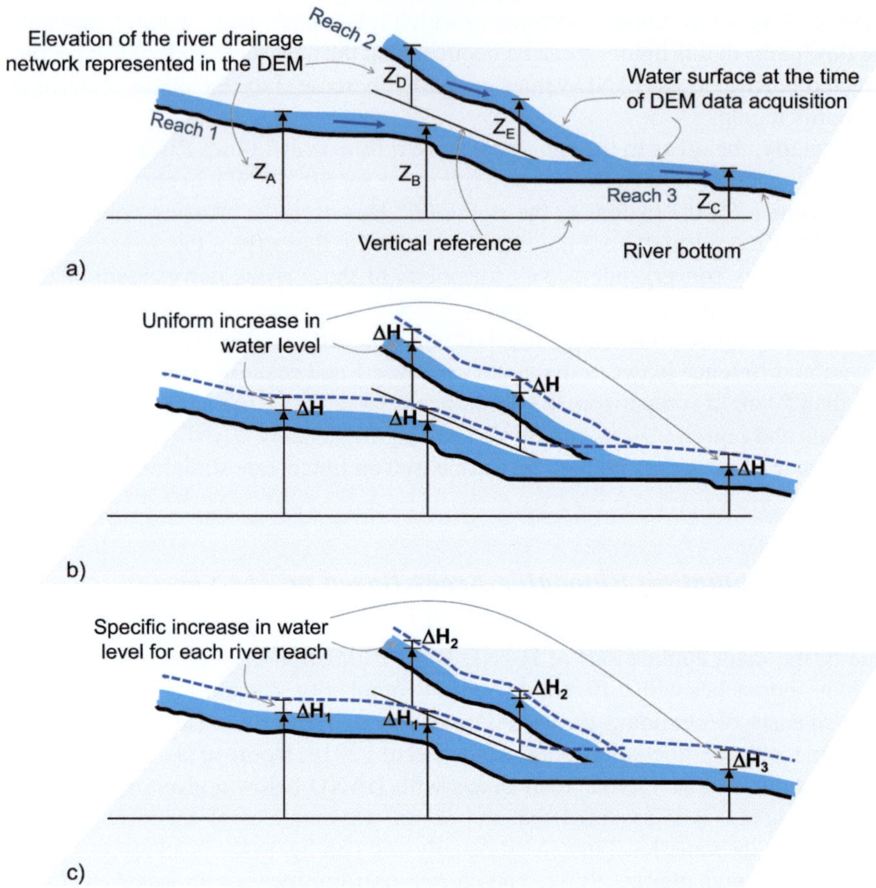

Fig. 12.15 **a** Perspective view of a river drainage network formed by three river reaches, highlighting the elevations (Z) represented in the DEM for five points (A, B, C, D, E); **b** Water surface considering a uniform increase equal to ΔH for all the sections; **c** Water surface considering a specific increase ΔH_1, ΔH_2 and ΔH_3 for the reaches 1, 2 and 3, respectively

adopting a uniform HAND threshold to distinguish floodable and non-floodable areas throughout the basin.

A more advanced approach than the previous one considers variable water level increases per reach of the river drainage network. One way to define how much the water level increases for each river reach is to estimate the maximum flow per reach using hydraulic methods (Garousi-Nejad et al., 2019; Zheng et al., 2018b). Synthetic key curves calculated from the DEM for each reach can be used to estimate the elevation of the water level corresponding to that maximum simulated flow.

The estimated water level increase for each reach is adopted as the HAND threshold to define the floodable areas connected to each reach. Each river reach i has a corresponding ΔH_i value. In practice, for each sub-basin contributing to a

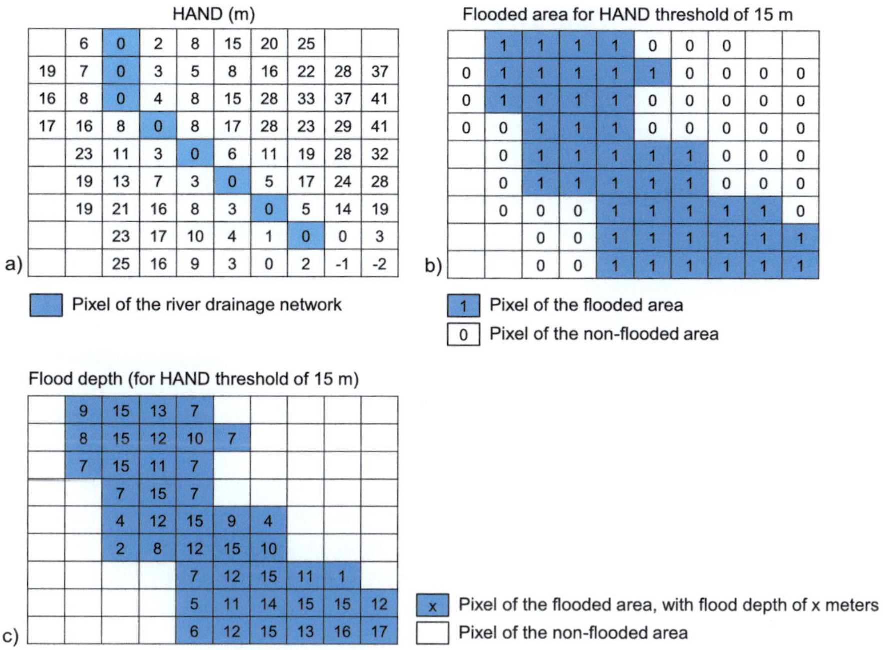

Fig. 12.16 **a** Raster of the HAND topographic index; **b** Identification of the pixels that make up the floodable area assuming a HAND threshold of 15 m; **c** Raster of flood depths for the floodable area assuming HAND threshold of 15 m; In the three rasters, pixels with undetermined HAND (because they do not drain to the indicated drainage network) do not present values

specific river section, the HAND is reclassified considering the sub-basin threshold (ΔH_i), and the mapping of floodable and non-floodable areas is obtained. In the example of Fig. 12.15c, the HAND thresholds were defined with ΔH_1, ΔH_2, and ΔH_3 values for river Sects. 12.1, 12.2, and 12.3, respectively.

In this second approach, we advance in considering the heterogeneity of flooding conditions along the basin's rivers, compared to the initial, more simplified approach that standardizes all reaches. However, this second approach has as its main disadvantage the occurrence of discontinuities in the floodable areas near the river junctions. In the example of Fig. 12.15c, it can be seen that at the confluence of the rivers, the uneven increase in water level causes discontinuity in the water surface. The corresponding mapping of floodable areas per reach amplifies this continuity and generates inconsistencies in the results. This problem is reported by Aristizabal et al. (2023), who propose adapting the method to reduce this occurrence.

The scientific literature proposes other variations of approaches to estimating floodable areas from HAND and evaluates these estimates against field data or hydrodynamic modeling results. The reader is recommended to read the works of Speckhann et al. (2018); Afshari et al. (2018); McGrath et al. (2018); and Aristizabal et al. (2023).

12.7.5 Estimating Flood Depths Based on HAND

It is possible to estimate flood depths when mapping flooded areas from HAND. This estimate is direct and operationally trivial when using the approach to determine floodable areas more simply, in which all pixels whose HAND is below a maximum HAND threshold (ΔH) are considered flooded.

To explain this concept, consider the schematic example of Fig. 12.17. All indicated pixels have flow directed to the river pixel identified by R. Therefore, the HAND of each pixel is the difference between its elevation and the elevation of R, as exemplified for pixels P and Q (Fig. 12.17a). Adopting ΔH as the HAND threshold to distinguish floodable and non-floodable areas, it is found that all pixels with HAND above ΔH do not flood, as is the case with P (Fig. 12.17b). Pixels with HAND below ΔH are flooded, like pixel Q. Pixel R, by definition, always floods since it belongs to the river itself.

In this example, the flood depths in the floodable areas estimated from HAND correspond to the difference between the adopted flood threshold (ΔH) and the HAND value in each pixel. For example, for pixel Q, the flood depth is given by $\text{depth}_Q = \Delta H - \text{HAND}_Q$ (Fig. 12.17c). At point R, belonging to the river and taken as the reference in the HAND calculation, there is a zero value for this topographic index. Therefore, by definition, the depth in the river is equal to the adopted flood threshold $\left(\text{depth}_R = \Delta H\right)$.

Operationally, in any GIS software, the determination of the raster of estimated flood depths from HAND can be obtained as follows:

i. Mathematical operation of subtraction between a scalar (adopted HAND threshold, ΔH) and the HAND raster (obtained as described in the previous sub-item);
ii. Reclassification of the result of the previous step so that negative values are converted to zero (if value < 0, change to value = 0). Negative values correspond

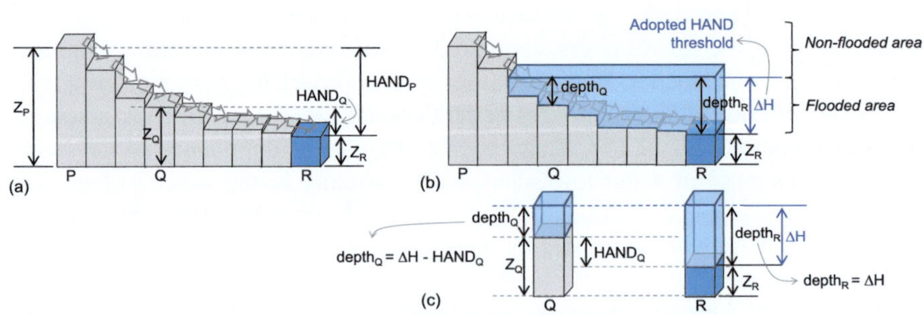

Fig. 12.17 **a** Elevation scheme in a set of pixels whose flow is directed to a particular pixel R of the river (of elevation Z_R), indicating the HAND for points P and Q, whose elevations are Z_P and Z_Q, respectively; **b** Indication of the floodable area referring to a HAND threshold adopted as ΔH, highlighting the depths of the flood in R (depth_R) and in Q (depth_Q); **c** Calculation of flood depths in R and Q

to pixels that are not flooded since they have a HAND value higher than the flood threshold. The remaining values are zero or positive. These positive values indicate the depths at each flooded pixel.

iii. An alternative to the previous step: Multiplication of the result from step (i) by the floodable areas raster obtained from HAND (described in the last subitem). This raster has 0 (non-flooded pixels) and 1 (flooded pixels). The multiplication ensures that the difference raster calculated in step (i) remains with nonzero values only for the pixels that are part of the flood. These values will always be positive and indicate the flood depth at each pixel.

When the HAND is determined with the depressionless DEM, all pixels have a positive or null value for this attribute. Consequently, the flood depths raster presents values limited to the maximum, defined as the HAND threshold used to identify the floodable area.

If the HAND was determined with the DEM without removing the depressions, some pixels may have negative HAND values. If these pixels are part of the floodable area, they have a depth higher than the HAND threshold chosen to distinguish flooded and non-flooded regions.

A didactic example of determining the flood depth raster is shown in Fig. 12.16, presented in the previous sub-item. A HAND threshold of 15 m was considered to denote the floodable areas, resulting in depths of up to 17 m. The pixels that are part of the river and others with a HAND value of zero have a depth equal to the threshold, 15 m. Two pixels have a HAND with a negative value because they represent terrain depressions, resulting in depths higher than such a threshold. The other pixels in the floodable area have depths less than 15 m.

12.7.6 Estimating Flood Volume Based on HAND

Given the determination of the flood depths raster, calculating the corresponding flood volume is conceptually simple. This operation is analogous to calculating the flood volume of the lake formed by constructing a reservoir along the river, as seen in Chap. 9. To illustrate, let's consider the schematic example of Fig. 12.18, which continues the example treated in the previous figure.

For any pixel-like pixel Q in Fig. 12.18a, the flood volume corresponds to the quadrangular prism whose base is the pixel's surface area and whose height is the flood depth in that pixel (Fig. 12.18b). The flooded volume in a pixel is, therefore, the result of the multiplication between the surface area of the pixel and the flood depth.

The flood volume corresponds to the sum of the flooded volumes in all pixels in the floodable area. Considering that all the pixels have the same surface area, this sum of volumes is simplified to the sum of the depths, the result of which is multiplied by the surface area of a pixel (Fig. 12.18c). This method is valid regardless of whether the HAND was determined from the DEM with or without depressions.

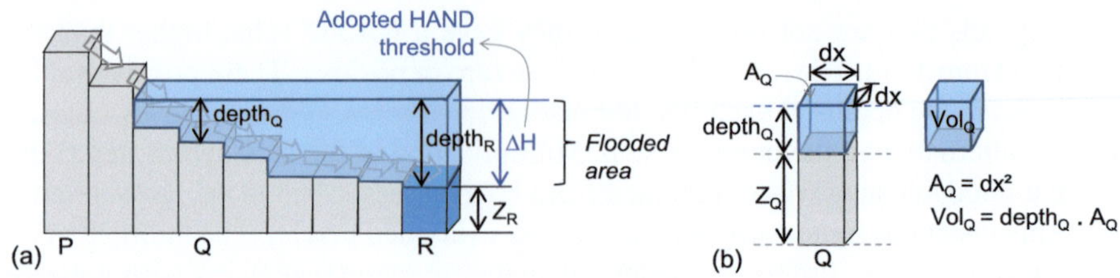

(c)
$$Vol_{total} = \sum_{i=1}^{n} Vol_i = \sum_{i=1}^{n} (depth_i.A_i) \Rightarrow Vol_{total} = \sum_{i=1}^{n} (depth_i.dx^2) \Rightarrow Vol_{total} = dx^2.\sum_{i=1}^{n} depth_i$$

Fig. 12.18 **a** Floodable area referring to a given HAND threshold (ΔH), highlighting the flood depths at the river reference point R (depth$_R$) and in Q (depth$_Q$); **b** Deduction of the flood volume in pixel Q (Vol$_Q$), based on its surface area (A_Q) and depth, with dx being the spatial resolution; **c** Deduction of the calculation of the total flooded volume (V_{total}) along the n pixels that make up the floodable area

In operational terms, in any GIS software, the determination of the flood volume can be performed in the following way:

i. Apply a command that generates raster statistics to perform the sum of the depth raster attributes. This input raster should have as an attribute the flood depth for each pixel belonging to the floodable area, zero for the non-flooded pixels, or of indeterminate value (*no data*), just like the raster indicated in Fig. 12.16c.
ii. Perform the mathematical calculation, on a calculator, of the product between the sum found in the previous step and the surface area of a pixel, given by the square of the spatial resolution in meters.

Returning to the didactic example of Fig. 12.16, the sum of the depths of the indicated floodable area results in a value of 444 m. If the pixel size is 30 × 30 m, the surface area of each raster element is 900 m^2. As a result, the flood volume of this example is obtained by the calculation 444 × 900, that is, 399,600 m^3.

12.7.7 *Real-World Example of HAND Application for Studying Floodable Areas*

A real-world example of determining floodable areas from HAND is presented in Fig. 12.19 for the Icamaquã river basin (drainage area of approximately 4629 km^2), a tributary of the Uruguay River, in southern Brazil. The HAND was determined from the SRTM 90 DEM processing, considering two distinct river drainage networks. Both networks were derived from the processing of this same DEM, with the same procedures and algorithms, except for the minimum drainage area value adopted to configure the beginning of the watercourses. When the value of 1 km^2 was adopted as the minimum drainage area, a much denser river drainage network was obtained

(0.28 km/km^2) than the network obtained with the minimum threshold of 5 km^2 (density of 0.13 km/km^2), as already discussed in Chap. 7.

The differences between the HANDs obtained from two river drainage networks of different densities occur for the areas that connect to the river reaches that exist in one network and do not exist in the other (Fig. 12.19c and d). The higher the drainage density, the shorter the downstream flow path distance from a pixel of these areas to reach the river drainage network. As rivers are the reference for calculating HAND elevations, the elevation difference between a pixel and the river it connects to decreases. Thus, higher drainage density implies lower HAND values.

In this example, floodable areas were identified with the more simplified approach of only considering a HAND threshold as a criterion, adopted with a uniform value

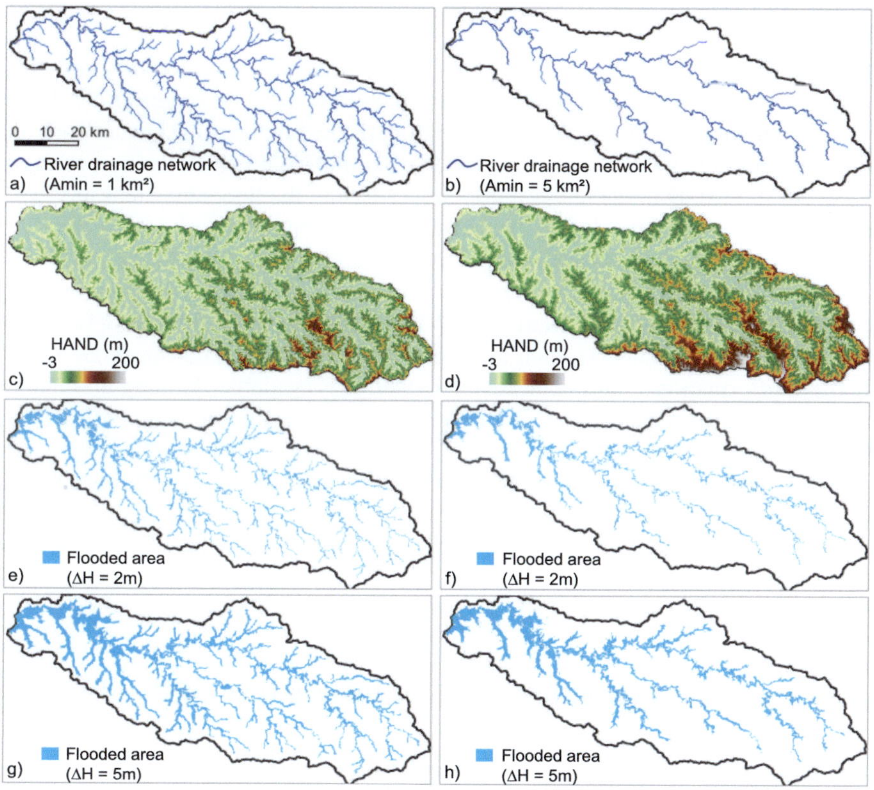

Fig. 12.19 **a**, **b** River drainage network derived from the SRTM 90 DEM for the Icamaquã river basin, a tributary of the Uruguay river (southern Brazil), considering minimum area thresholds of 1 and 5 km^2, respectively, for the formation of watercourses; **c**, **d** HAND derived from the SRTM 90 DEM considering the river drainage networks **a** and **b**, respectively; **e**, **f** Floodable area derived from the HAND presented in **c** and **d**, respectively, considering a HAND threshold of 2 m (ΔH); **g**, **h** Floodable area derived from the HAND presented in **c** and **d**, respectively, considering a HAND threshold of 5 m (ΔH)

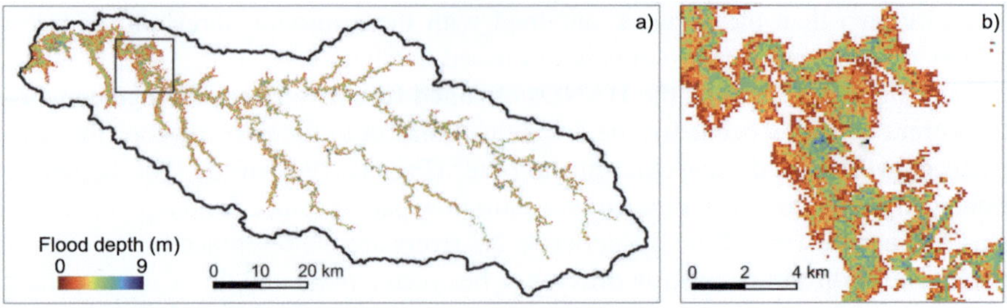

Fig. 12.20 **a** Raster of flood depths estimated from the HAND for the Icamaquã river watershed (southern Brazil), considering the flooded area referring to a drainage network with a minimum area of 5 km^2 and a HAND threshold of 5 m to denote flooding; **b** Enlargement of the area indicated by the rectangle in **a**

for the entire basin.When a HAND threshold of 2 m was considered to represent flooding, it resulted in about 388.3 km^2 of floodable areas, considering the denser drainage network (Fig. 12.19e). When adopting the less dense drainage network, the floodable areas total 259.3 km^2 (Fig. 12.19f), about 33% less than the first case.

For a 5 m HAND threshold as a flooding criterion, the floodable areas are 713.1 and 493.7 km^2 when considering the denser and less dense networks, respectively (a reduction of 31% in the second case compared to the first).

These results illustrate the relevance of adequately defining the beginning of the river drainage network for calculating HAND and subsequent estimation of floodable areas, as illustrated in works such as Dantas and Paz (2021) and Speckhann et al. (2018). If information regarding headwater location is available, defining the channel initiation based on this data is an alternative to eliminating this uncertainty (Liu et al., 2018).

The flood depths were determined for the floodable areas considering the less dense drainage network and a HAND threshold of 5 m (Fig. 12.20). Depths of up to 9 m were obtained, i.e., higher than the 5 m threshold, due to the DEM containing terrain depressions.

12.8 Reduced Dissipation Per Unit Length (rDUNE)

The topographic index called reduced Dissipation per UNit lEngth (rDUNE) was proposed by Loritz et al. (2019) as a reinterpretation and improvement of the HAND concepts. These authors used thermodynamic principles to conceive an index considering the potential energy difference and the accumulated dissipative energy loss along a flow path. The proposal was to relate geomorphic characteristics to a region's processes and hydrological structure.

When the rDUNE index was conceptualized, it was considered that most of the precipitation's potential energy is dissipated and not transformed into runoff's kinetic energy, and this dissipation occurs both in hillslope runoffs and along the river drainage network itself (Phillips, 2022).

The potential energy difference is represented by the topographic difference (h) between each surface point and the river drainage network point to which it connects, according to the DEM-derived flow paths. This topographic difference corresponds to the HAND. The extension of the flow path (l) that connects a given point on the surface to the point it connects in the river drainage network represents the accumulated dissipative energy loss.

The rDUNE index comprises the ratio between h and l, followed by two mathematical operations (Eq. 12.9; Loritz et al., 2019). First, the logarithmic function is applied to this result for convenience in the scale of values and to approximate them to a normal distribution. Then, it is multiplied by − 1 for ease of interpretation of results. The h/l ratio can be understood as the slope of the downstream flow path from a given point on the terrain until it meets the river drainage network. Schematically, the variables involved in determining the rDUNE are illustrated in Fig. 12.21.

$$\mathrm{rDUNE_P} = -\ln\left(\frac{\mathrm{h}}{\mathrm{l}}\right) \tag{12.9}$$

where $\mathrm{rDUNE_P}$ is the rDUNE index at point P.

According to the proposed formulation, the values of rDUNE are dimensionless and have no upper and lower limits, being able to vary from $-\infty$ to $+\infty$, with the following general interpretation (Loritz et al., 2019):

a. Negative values of rDUNE indicate that the h/l ratio is higher than 1. In this case, the vertical drop to the drainage network is higher than the flow path length.
b. Positive values of rDUNE signal that h/l is less than 1, that is, the flow path length is higher than the vertical drop to the river drainage network;
c. When rDUNE equals zero, the vertical drop and the flow path length are equal (h/l equals 1).

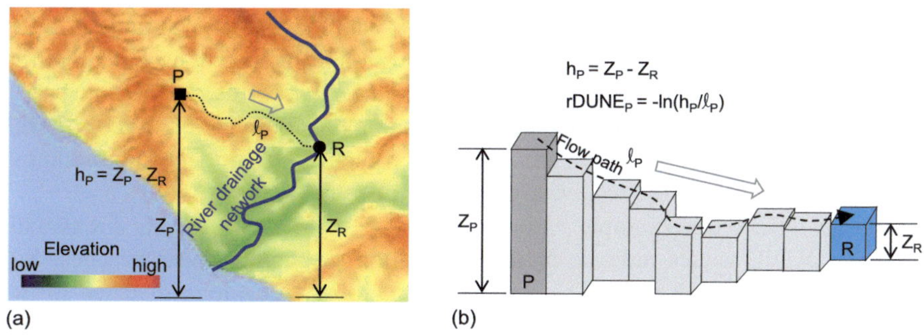

Fig. 12.21 **a** Illustrative scheme of the determination of the topographic index rDUNE for a point P, located at elevation Z_P, and whose flow path connects to the point R of the river drainage network, located at elevation Z_R; **b** Detail of the flow path traveled from P to R and an indication of the variables h_P and l_P for the calculation of rDUNE

Looking at it another way: for the same distance to the network, the higher the topographic drop, the higher the h/l ratio and the smaller the rDUNE, representing a higher dissipation of potential energy; for the same topographic drop, the higher the distance to the network, the smaller the h/l ratio and the higher the value of rDUNE, that is, less dissipation of potential energy.

Determining the rDUNE index is similar to HAND and requires the same input data, including all DEM-derived products: DEM, flow directions, accumulated drainage areas, and river drainage network. The rDUNE incorporates the HAND in its formulation since the HAND attribute equals the value of the topographic drop h inserted in the rDUNE. Therefore, obtaining the rDUNE involves obtaining the HAND and some additional operations, as follows:

i. Basic DEM processing operations for the removal of depressions, determination of flow directions, accumulated drainage areas, and extraction of the river drainage network;
ii. Calculation of the raster of the h attribute, a topographic drop of each pixel to the point of the river drainage network to which it connects—that is, calculation of the HAND index (described in the previous item of this chapter);
iii. Calculation of the distance along the flow path from each pixel to the point of the river drainage network to which it connects, as described in Chap. 10, resulting in the raster whose attribute in each pixel is the value of l required for the rDUNE;
iv. Mathematical operation involving the raster of the h attribute and the raster of the l attribute, according to the mathematical formulation of Eq. (12.9), resulting in the raster whose value in each pixel is the rDUNE index.

Figure 12.22 shows the rDUNE index obtained for an area in the municipality of Divinópolis in Minas Gerais. The river drainage network was derived from the SRTM 30 DEM, and the distances along the flow paths to reach the network were computed, representing the variable l in the rDUNE equation. The raster of these distances highlights the areas with the highest distance to the network, in some points with values up to more than 4 km.

The HAND was also derived from the DEM, which equals the variable h in the formulation of the rDUNE. It represents the elevation difference of each pixel to the point of the river drainage network to which it connects, according to the flow paths. The raster of the h/l ratio shows a spatial pattern that is a combination of the patterns of the variables h and l. However, this pattern becomes more evident by applying the logarithmic function to the values of the rDUNE itself.

The rDUNE in this example varied from just under 1 to almost 11. Areas with reduced rDUNE values indicate higher dissipation of potential energy. These are areas with a higher h/l ratio. This occurs for regions with a long distance to the network, such as around point A, whose distance is higher than 3 km, due to the drop of around 300 m relative to the drainage network ($h/l = 0.1$). As a result, rDUNE is approximately 2.3. However, it also occurs in regions to the left of B, with a shorter distance to the network (~660 m) but with an elevation drop of almost 130 m ($h/l = 0.2$), resulting in an rDUNE of around 1.6. This becomes even more intense

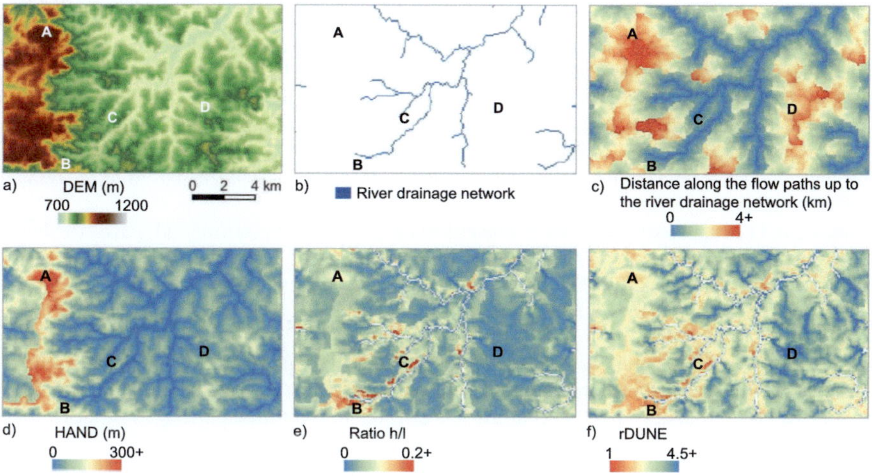

Fig. 12.22 **a** SRTM 30 DEM for an area of the municipality of Divinópolis (Minas Gerais state) and derived products: **b** River drainage network; **c** Distances along the flow paths to the drainage network; **d** HAND topographic index; **e** h/l ratio, given by the division between distances calculated in **c** and HAND elevations; **f** rDUNE topographic index

in other areas, like the point to the right of C, whose distance to the network is 210 m and elevation drop of only 60 m (h/l = 0.29), corresponding to a rDUNE of 1.25.

Areas around E present high rDUNE values, approximately equal to 5 in this case, configuring low potential energy dissipation. This is mainly due to the very close elevation to the river drainage network (less than 10 m) combined with a high distance (1650 m), which corresponds to h/l = 0.006.

References

Afshari, S., Tavakoly, A. A., Rajib, M. A., Zheng, X., Follum, M. L., Omranian, E., & Fekete, B. M. (2018). Comparison of new generation low-complexity flood inundation mapping tools with a hydrodynamic model. *Journal of Hydrology, 556*, 539–556. https://doi.org/10.1016/j.jhydrol.2017.11.036

Ågren, A. M., Lidberg, W., Strömgren, M., Ogilvie, J., & Arp, P. A. (2014). Evaluating digital terrain indices for soil wetness mapping—A Swedish case study. *Hydrology and Earth System Sciences, 18*, 3623–3634. https://doi.org/10.5194/hess-18-3623-2014

Ahmad, I., Dar, M. A., Teka, A. H., Gebre, T., Gadissa, E., & Tolosa, A. T. (2019). Application of hydrological indices for erosion hazard mapping using Spatial analyst tool. *Environmental Monitoring and Assessment, 191*, 482. https://doi.org/10.1007/s10661-019-7614-x

Aristizabal, F., Salas, F., Petrochenkov, G., Grout, T., Avant, B., Bates, B., Spies, R., Chadwick, N., Wills, Z., & Judge, J. (2023). Extending height above nearest drainage to model multiple fluvial sources in flood inundation mapping applications for the U.S. national water model. *Water Resources Research, 59*, e2022WR032039. https://doi.org/10.1029/2022WR032039

Beven, K. J., & Kirkby, M. J. (1979). A physically based, variable contributing area model of basin hydrology/Un modèle à base physique de zone d'appel variable de l'hydrologie du bassin versant. *Hydrological Sciences Bulletin, 24*, 43–69. https://doi.org/10.1080/02626667909491834

Bizzi, S., & Lerner, D. N. (2015). The use of stream power as an indicator of channel sensitivity to erosion and deposition processes: SP as an indicator of erosion and deposition. *River Research and Applications, 31*, 16–27. https://doi.org/10.1002/rra.2717

Burrough, P. A., & McDonnell, R. A. (1998). *Principles of geographical information systems.* Oxford University Press.

Crave, A., & Gascuel-Odoux, C. (1997). The influence of topography on time and space distribution of soil surface water content. *Hydrological Processes, 11*, 203–210.

Dantas, A. A. R., & Paz, A. R. (2021). Use of HAND terrain descriptor for estimating flood-prone areas in river basins. *Revista Brasileira De Ciências Ambientais, 56*, 501–516. https://doi.org/10.5327/Z21769478892

Das, A., Nagendra, H., Anand, M., & Bunyan, M. (2015). Topographic and bioclimatic determinants of the occurrence of forest and grassland in tropical montane forest-grassland mosaics of the western Ghats, India. *PLoS ONE, 10*, Article e0130566. https://doi.org/10.1371/journal.pone.0130566

De Roo, A. P. J. (1998). Modelling runoff and sediment transport in catchments using GIS. *Hydrological Processes, 12*, 905–922. https://doi.org/10.1002/(SICI)1099-1085(199805)12:6%3c905::AID-HYP662%3e3.0.CO;2-2

Dindaroglu, T., Tunguz, V., Babur, E., Menshov, O., & Battaglia, M. L. (2023). Determination of the relationship among compound Topographic Index (CTI), soil properties and land-use in karst ecosystems. *Physical Geography, 44*, 307–329. https://doi.org/10.1080/02723646.2021.1961361

Galzki, J. C., Birr, A. S., & Mulla, D. J. (2011). Identifying critical agricultural areas with three-meter LiDAR elevation data for precision conservation. *Journal of Soil and Water Conservation, 66*, 423–430. https://doi.org/10.2489/jswc.66.6.423

Gao, H., Birkel, C., Hrachowitz, M., Tetzlaff, D., Soulsby, C., & Savenije, H. H. G. (2019). A simple topography-driven and calibration-free runoff generation module. *Hydrology and Earth Systems Science, 23*, 787–809. https://doi.org/10.5194/hess-23-787-2019

Garousi-Nejad, I., Tarboton, D. G., Aboutalebi, M., & Torres-Rua, A. F. (2019). Terrain analysis enhancements to the height above nearest drainage flood inundation mapping method. *Water Resources Research, 55*, 7983–8009. https://doi.org/10.1029/2019WR024837

Gharari, S., Hrachowitz, M., Fenicia, F., & Savenije, H. H. G. (2011). Hydrological landscape classification: Investigating the performance of HAND based landscape classifications in a central European meso-scale catchment. *Hydrology and Earth Systems Science, 15*, 3275–3291. https://doi.org/10.5194/hess-15-3275-2011

Hong, H., Tsangaratos, P., Ilia, I., Liu, J., Zhu, A.-X., & Xu, C. (2018). Applying genetic algorithms to set the optimal combination of forest fire related variables and model forest fire susceptibility based on data mining models. The case of Dayu County, China. *Science of the Total Environment, 630*, 1044–1056. https://doi.org/10.1016/j.scitotenv.2018.02.278

Howard, A. D. (1994). A detachment-limited model of drainage basin evolution. *Water Resources Research, 30*, 2261–2285. https://doi.org/10.1029/94WR00757

Jaafari, A., Najafi, A., Pourghasemi, H. R., Rezaeian, J., & Sattarian, A. (2014). GIS-based frequency ratio and index of entropy models for landslide susceptibility assessment in the Caspian forest, northern Iran. *International Journal of Environmental Science and Technology, 11*, 909–926. https://doi.org/10.1007/s13762-013-0464-0

Kiss, R. (2004). Determination of drainage network in digital elevation models, utilities and limitations. *Journal of Hungarian Geomathematics, 2*, 16–29.

Kopecký, M., & Čížková, Š. (2010). Using topographic wetness index in vegetation ecology: Does the algorithm matter?: Topographic wetness index in vegetation ecology. *Applied Vegetation Science, 13*, 450–459. https://doi.org/10.1111/j.1654-109X.2010.01083.x

Kopecký, M., Macek, M., & Wild, J. (2021). Topographic wetness index calculation guidelines based on measured soil moisture and plant species composition. *Science of the Total Environment, 757*, Article 143785. https://doi.org/10.1016/j.scitotenv.2020.143785

Lague, D. (2014). The stream power river incision model: evidence, theory and beyond: Stream power incision model. *Earth Surface Processes and Landforms, 39*, 38–61. https://doi.org/10.1002/esp.3462

Lanni, C., McDonnell, J. J., & Rigon, R. (2011). On the relative role of upslope and downslope topography for describing water flow path and storage dynamics: A theoretical analysis: Upslope and downslope topography in water flow path. *Hydrological Processes, 25*, 3909–3923. https://doi.org/10.1002/hyp.8263

Liu, Y. Y., Maidment, D. R., Tarboton, D. G., Zheng, X., & Wang, S. (2018). A CyberGIS integration and computation framework for high-resolution continental-scale flood inundation mapping. *Journal of the American Water Resources Association, 54*, 770–784. https://doi.org/10.1111/1752-1688.12660

Loritz, R., Kleidon, A., Jackisch, C., Westhoff, M., Ehret, U., Gupta, H., & Zehe, E. (2019). A topographic index explaining hydrological similarity by accounting for the joint controls of runoff formation. *Hydrolology and Earth Systems Science, 23*, 3807–3821. https://doi.org/10.5194/hess-23-3807-2019

Marca-Zevallos, M. J., Moulatlet, G. M., Sousa, T. R., Schietti, J., Coelho, L. D. S., Ramos, J. F., Lima Filho, D. D. A., Amaral, I. L., De Almeida Matos, F. D., Rincón, L. M., Cardenas Revilla, J. D., Pansonato, M. P., Gribel, R. B., et al. (2022). Local hydrological conditions influence tree diversity and composition across the Amazon basin. *Ecography, 2022*. https://doi.org/10.1111/ecog.06125

Mattivi, P., Franci, F., Lambertini, A., & Bitelli, G. (2019). TWI computation: A comparison of different open source GISs. *Open Geospatial Data, Software and Standards, 4*, 6. https://doi.org/10.1186/s40965-019-0066-y

McGrath, H., Bourgon, J.-F., Proulx-Bourque, J.-S., Nastev, M., & Abo El Ezz, A. (2018). A comparison of simplified conceptual models for rapid web-based flood inundation mapping. *Natural Hazards, 93*, 905–920. https://doi.org/10.1007/s11069-018-3331-y

Meles, M. B., Younger, S. E., Jackson, C. R., Du, E., & Drover, D. (2020). Wetness index based on landscape position and topography (WILT): Modifying TWI to reflect landscape position. *Journal of Environmental Management, 255*, Article 109863. https://doi.org/10.1016/j.jenvman.2019.109863

Mohtashami, S., Eliasson, L., Hansson, L., Willén, E., Thierfelder, T., & Nordfjell, T. (2022). Evaluating the effect of DEM resolution on performance of cartographic depth-to-water maps, for planning logging operations. *International Journal of Applied Earth Observation and Geoinformation, 108*, Article 102728. https://doi.org/10.1016/j.jag.2022.102728

Mojaddadi, H., Pradhan, B., Nampak, H., Ahmad, N., & Ghazali, A. H. B. (2017). Ensemble machine-learning-based geospatial approach for flood risk assessment using multi-sensor remote-sensing data and GIS. *Geomatics, Natural Hazards and Risk, 8*, 1080–1102. https://doi.org/10.1080/19475705.2017.1294113

Moore, I. D., & Burch, G. J. (1986). Modelling erosion and deposition: Topographic effects. *Transactions of the ASAE, 29*, 1624–1630. https://doi.org/10.13031/2013.30363

Moore, I. D., Burch, G. J., & Mackenzie, D. H. (1988). Topographic effects on the distribution of surface soil water and the location of ephemeral gullies. *Transactions of the ASAE, 31*, 1098–1107. https://doi.org/10.13031/2013.30829

Moore, I. D., & Nieber, J. L. (1989). Landscape assessment of soil erosion and nonpoint source pollution. *Journal of the Minnesota Academy of Science, 55*.

Murphy, P. N. C., Ogilvie, J., & Arp, P. (2009). Topographic modelling of soil moisture conditions: A comparison and verification of two models. *European Journal of Soil Science, 60*, 94–109. https://doi.org/10.1111/j.1365-2389.2008.01094.x

Nobre, A. D., Cuartas, L. A., Hodnett, M., Rennó, C. D., Rodrigues, G., Silveira, A., Waterloo, M., & Saleska, S. (2011). Height above the nearest drainage—A hydrologically relevant new terrain model. *Journal of Hydrology, 404*, 13–29. https://doi.org/10.1016/j.jhydrol.2011.03.051

Nobre, A. D., Cuartas, L. A., Momo, M. R., Severo, D. L., Pinheiro, A., & Nobre, C. A. (2016). HAND contour: A new proxy predictor of inundation extent: Mapping flood hazard potential using topography. *Hydrological Processes, 30*, 320–333. https://doi.org/10.1002/hyp.10581

Oltean, G. S., Comeau, P. G., & White, B. (2016). Linking the depth-to-water topographic index to soil moisture on boreal forest sites in Alberta. *Forest Science, 62*, 154–165. https://doi.org/10.5849/forsci.15-054

Phillips, J. D. (2022). Store and pour: Evolution of flow systems in landscapes. *CATENA, 216*, Article 106357. https://doi.org/10.1016/j.catena.2022.106357

Raduła, M. W., Szymura, T. H., & Szymura, M. (2018). Topographic wetness index explains soil moisture better than bioindication with Ellenberg's indicator values. *Ecological Indicators, 85*, 172–179. https://doi.org/10.1016/j.ecolind.2017.10.011

Rennó, C. D., Nobre, A. D., Cuartas, L. A., Soares, J. V., Hodnett, M. G., Tomasella, J., & Waterloo, M. J. (2008). HAND, a new terrain descriptor using SRTM-DEM: Mapping terra-firme rainforest environments in Amazonia. *Remote Sensing of Environment, 112*, 3469–3481. https://doi.org/10.1016/j.rse.2008.03.018

Riihimäki, H., Kemppinen, J., Kopecký, M., & Luoto, M. (2021). Topographic wetness index as a proxy for soil moisture: The importance of flow-routing algorithm and grid resolution. *Water Resources Research, 57*. https://doi.org/10.1029/2021WR029871

Rodda, H. J. E. (2005). The development and application of a flood risk model for the Czech Republic. *Natural Hazards, 36*, 207–220. https://doi.org/10.1007/s11069-004-4549-4

Schietti, J., Emilio, T., Rennó, C. D., Drucker, D. P., Costa, F. R. C., Nogueira, A., Baccaro, F. B., Figueiredo, F., Castilho, C. V., Kinupp, V., Guillaumet, J.-L., Garcia, A. R. M., Lima, A. P., & Magnusson, W. E. (2014). Vertical distance from drainage drives floristic composition changes in an Amazonian rainforest. *Plant Ecology and Diversity, 7*, 241–253. https://doi.org/10.1080/17550874.2013.783642

Sørensen, R., Zinko, U., & Seibert, J. (2006). On the calculation of the topographic wetness index: Evaluation of different methods based on field observations. *Hydrology and Earth System Sciences.*

Speckhann, G. A., Borges Chaffe, P. L., Fabris Goerl, R., Abreu, J. J. D., & Altamirano Flores, J. A. (2018). Flood hazard mapping in Southern Brazil: A combination of flow frequency analysis and the HAND model. *Hydrological Sciences Journal, 63*, 87–100. https://doi.org/10.1080/02626667.2017.1409896

Thommeret, N., Bailly, J. S., & Puech, C. (2010). Extraction of Thalweg networks from DTMs: application to badlands. *Hydrology and Earth Systems Science, 14*, 1527–1536. https://doi.org/10.5194/hess-14-1527-2010

Thompson, J. A., Bell, J. C., & Butler, C. A. (2001). Digital elevation model resolution: Effects on terrain attribute calculation and quantitative soil-landscape modeling. *Geoderma, 100*, 67–89. https://doi.org/10.1016/S0016-7061(00)00081-1

Wilson, J. P., & Gallant, J. C. (2000). *Terrain analysis: Principles and applications.* Wiley.

Zheng, X., Maidment, D. R., Tarboton, D. G., Liu, Y. Y., & Passalacqua, P. (2018a). GeoFlood: Large-scale flood inundation mapping based on high-resolution terrain analysis. *Water Resources Research, 54*. https://doi.org/10.1029/2018WR023457

Zheng, X., Tarboton, D. G., Maidment, D. R., Liu, Y. Y., & Passalacqua, P. (2018b). River channel geometry and rating curve estimation using height above the nearest drainage. *Journal of the American Water Resources Association, 54*, 785–806. https://doi.org/10.1111/1752-1688.12661

Chapter 13
Topographic Indices—Part 3: Protection and Exposure to Wind

Continuing from the previous two chapters, this chapter deals with a third group of DEM-derived topographic indices. Seven indices related to topographic protection and exposure to wind are presented.

From topography, one can infer an area's exposure to or protection from wind action. This analysis is relevant for different environmental studies, such as analyses of the spatial distribution of atmospheric pollutants, the occurrence and distribution of plant species, the potential for wind energy utilization, spatial patterns of snow accumulation, and climate forecasts.

Topographic indices derived from a DEM that help analyze this degree of exposure or topographic protection to the wind action include the aspect of the terrain relative to the wind, horizon angle, wind fetch, directional relief, and terrain exposure to the sloped wind flux, among others. These topographic variables are determined in a wind-dependent manner and considered for analysis, that is, as a function of a wind flow acting in a specific direction.

This chapter discusses seven topographic indices related to wind protection and exposure, presenting concepts and formulations and providing a real-world example of determining and interpreting the results.

13.1 Aspect of the Terrain Relative to the Wind

The aspect of the terrain relative to the wind is a metric derived from the aspect, a fundamental concept that refers to the direction of maximum slope at a certain point of the terrain, as seen in Chap. 5. The aspect of the terrain relative to the wind is defined as the angle between the aspect of the terrain and the wind direction, both measured relative to the north (Fig. 13.1; Antonić & Legović, 1999). The terrain aspect is derived from the DEM, while the user chooses the wind direction for the analysis.

A. Rolim da Paz, *Digital Elevation Models for Environmental Studies*,
https://doi.org/10.1007/978-3-032-04523-2_13

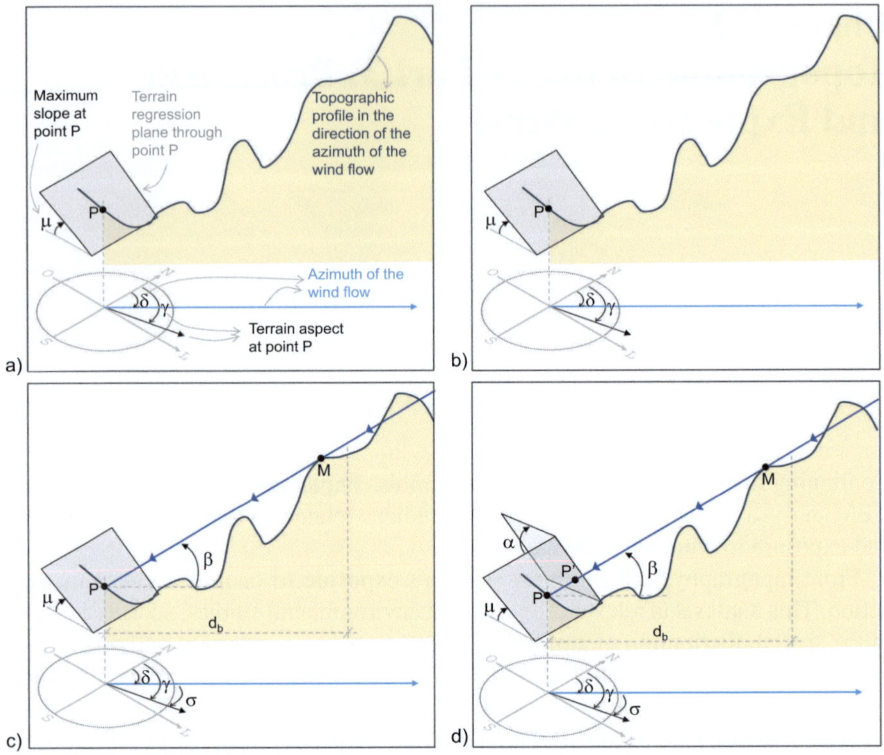

Fig. 13.1 **a** Cross-sectional topographic profile in the direction of the wind flow azimuth (δ), with indication of a plane adjusted to the local topography at point P, whose local slope is μ and terrain aspect is γ; **b** Indication of the terrain aspect relative to the wind (σ), given by the difference between terrain aspect and wind azimuth; **c** Determination of the horizon angle β, from the identification of point M in the topographic profile, defined within a search distance d_b in the wind direction (*Source* Based on Antonic and Legovic, 1999)

If the difference between the terrain aspect and the wind direction is less than or equal to 180°, the aspect relative to the wind is directly such difference (Fig. 13.2a). If this difference is higher than 180°, the aspect relative to the wind is adjusted to be equal to 360 minus such difference (Fig. 13.2b). Thus, the variable aspect relative to the wind is between 0° and 180°.

The value zero corresponds to the situation in which the point of the terrain is oriented in the windward position (Lindsay & Rothwell, 2008), that is, aligned in the direction to fully receive the wind, representing the maximum exposure to the considered wind. The value of 180° corresponds to the opposite situation, where the point is in the leeward position, opposite to the wind reception, configuring the maximum topographic protection for that adopted wind. Alternatively, one can work with the cosine of the terrain aspect relative to the wind (cos σ), which varies from 1 (windward, maximum wind exposure) to − 1 (leeward, maximum topographic wind protection) (Böhner & Antonić, 2009).

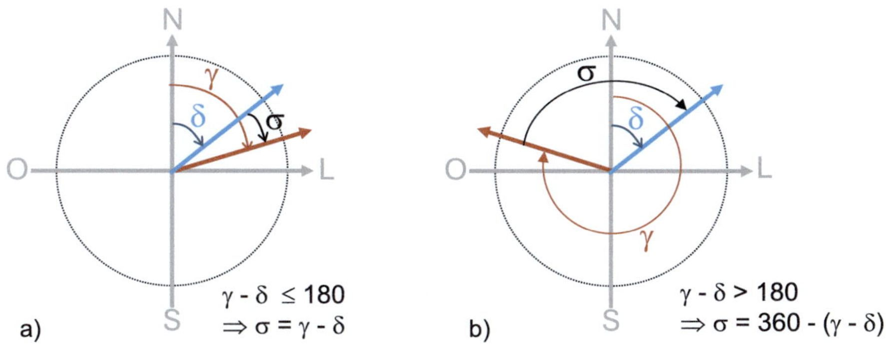

Fig. 13.2 Determination of the terrain aspect relative to the wind (σ), as a function of the difference between the terrain aspect (γ) and the azimuth of the wind flow in a specific direction (δ): **a** Situation in which $\gamma - \delta \leq 180°$; **b** Situation in which $\gamma - \delta > 180°$

A limitation of analyzing the degree of protection or topographic exposure to the wind of an area of interest based on the terrain aspect relative to the wind is that this topographic attribute contemplates only local information in the immediate 3×3 neighborhood around each pixel (Lindsay & Rothwell, 2008). The elevations of the eight neighboring pixels are analyzed to determine the terrain aspect and, by quantifying the difference with the angle of wind incidence, the estimate of the terrain aspect relative to the wind is obtained. The existence of possible topographic obstacles in the surroundings, outside the 3×3 neighborhood, is ignored in this estimate but can represent actual effects of protection against the action of the wind (Lindsay & Rothwell, 2008).

The determination of the raster of the terrain aspect relative to the wind can be performed by a specific function in some software that has this availability, such as Whitebox GAT. The input information is the DEM or the terrain aspect raster and the azimuth information of the wind direction.

It is also possible to determine the topographic attribute aspect of the terrain relative to the wind from a set of standard operations available in most GIS software, as follows:

i. Determination of the terrain aspect raster from the DEM, with angle measured relative to the north (azimuth), via a specific function;
ii. Arithmetic operation with conditional logic to calculate the difference between the terrain aspect at each pixel and the azimuth angle of the selected wind direction: this is a subtraction operation between a raster and a scalar, with a condition check as specified in Fig. 13.2. The operation decides whether to retain the result of the difference at the pixel (when $\gamma - \delta \leq 180$) or to replace it with a value obtained through another arithmetic operation between raster and scalars [$360 - (\gamma - \delta)$], if the condition $\gamma - \delta > 180$ occurs at that pixel.

Figure 13.3 illustrates a real-world case of determining the terrain aspect relative to the wind for an area of the municipality of Cristais Paulista, in the state of São Paulo,

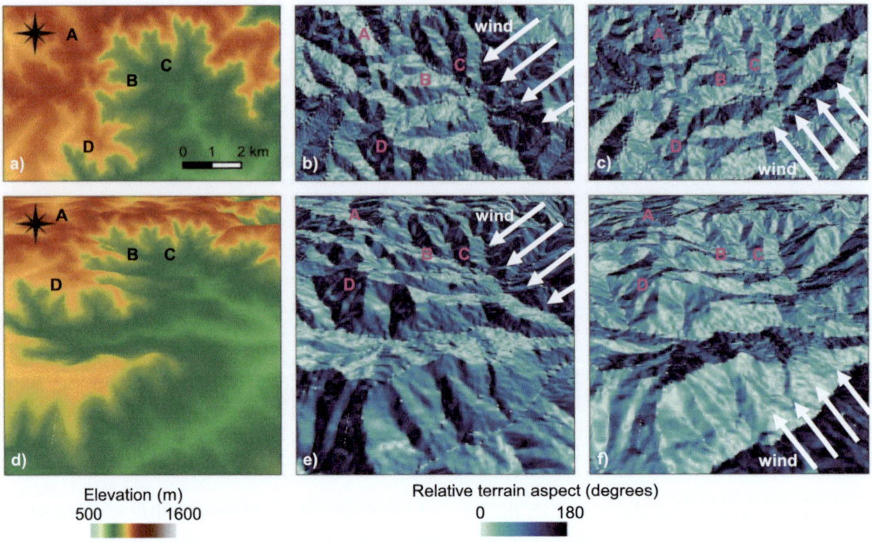

Fig. 13.3 **a** DEM derived from SRTM 30 data for an area of the municipality of Cristais Paulista (São Paulo state); **b**, **c** Terrain aspect relative to the wind, determined from the DEM for a wind with direction at 45° and 135°, respectively; **d**, **e**, **f** 3D view of part of the DEM and the raster of terrain aspect relative to the wind shown in **a**, **b** and **c**, respectively

from the SRTM 30 DEM. Two winds in directions distinct from each other by 90° were considered to compare results.

As expected from the definition of this topographic attribute, slope areas facing directly into the wind correspond to the lowest values—lighter areas in the images, such as points A (9°) and B (3°) for a wind direction of 45°. The areas less exposed to the wind appear in darker tones, such as points C (148°) and D (176°) for the same 45° wind. When considering a wind from a different direction, at 135°, the intensity of wind exposure—estimated indirectly by the terrain aspect relative to the wind—is altered, as can be seen by comparing points A, B, C, and D, which change from values of 9°, 3°, 148°, and 176° in the 45° wind analysis to 81°, 87°, 58°, and 86°, respectively, in the 135° wind analysis.

13.2 Horizon Angle

The horizon angle is another estimator of the degree of topographic exposure of an area to the wind. For a given point on the terrain, this topographic attribute is defined as the maximum vertical angle in the direction of the wind flow provided by the topography within a certain pre-established distance (Antonić & Legović, 1999).

Figure 13.1c illustrates the concept of the horizon angle. It shows the cross-sectional topographic profile along the direction of the wind. For the indicated point

P and assuming a search distance of d_b, the maximum possible slope is according to the line connecting point P to point M. This maximum slope configures the horizon angle.

This topographic attribute is also known as maximum elevation angle (Yokoyama et al., 2002) or maximum upwind slope, with examples of applications being the modeling of the effect of solar radiation and the study of spatial patterns of snow accumulation (Antonić & Legović, 1999; Winstral & Marks, 2002; Winstral et al., 2002).

The horizon angle at a given point on the terrain depends on the wind direction chosen for analysis and the search distance defined. For each wind direction, there is a specific topographic conformation of the terrain in that direction, and one cannot make inferences about the horizon angle without determining it. The search distance has a conceptually expected effect on the horizon angle. By increasing this distance, keeping all other conditions unchanged, the horizon angle can remain the same or increase, but not decrease.

For example, in Fig. 13.4, for the search distances d_{b1}, d_{b2}, and d_{b3}, sequentially increasing, the resulting horizon angle grows from situation 1 to situation 2 ($\beta_2 > \beta_1$) but remains the same from situation 2 to situation 3 ($\beta_3 = \beta_2$).

By increasing the search distance, points of higher elevation that provide higher slopes, considering a line drawn from point P, may be included. This occurs when passing from distance d_{b1} to d_{b2}, with point M_2 causing the angle β_2 to be higher than the angle that point M_1 caused for the distance $\mathrm{d}_{\mathrm{b}1}$.

However, when moving from distance d_{b2} to d_{b3}, the selected point M_3 for determining the horizon angle is the exact point M_2, identified at the distance d_{b2}, even though there are points like $M_3{}'$ of higher elevation than M_3. This illustrates how the increase in the analysis distance can increase or not the horizon angle, but cannot result in the reduction of the horizon angle, regardless of the topographic variability included in this extension of the search distance.

The mathematical formulation of the horizon angle for a given point is presented in Eq. (13.1) (Böhner & Antonić, 2009) as a function of the distance to the point

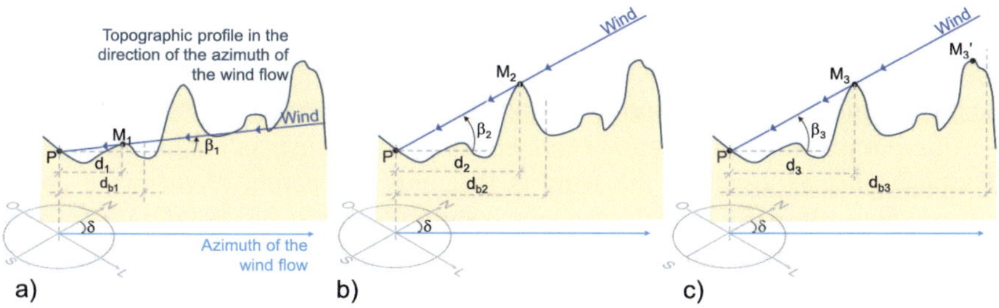

Fig. 13.4 **a, b, c** Determination of the horizon angle (β_1, β_2, β_3) for a point on the terrain P relative to a wind flow with azimuth δ and considering a certain search distance (d_{b1}, $d_{b2}$and d_{b3}): from point P and within the maximum distance along direction δ, the point (M_1, M_2 and M_3) that provides the maximum inclination, corresponding to β, is identified

identified as the one that provides the highest angle and the difference in elevation between the point of analysis and this identified point. In the case of Fig. 13.4a, for example, the calculation of the horizon angle is a function of the distance d_1 and the difference in elevation between M_1 and P. For the situations in Fig. 13.4b and c, the horizon angles are equal to each other, even with $d_{b3} > d_{b2}$, because the variables involved have the same value: $d_2 = d_3$, and the difference in elevation between M_2 and P is the exact difference between M_3 and P.

$$\beta_P = \arctan\left(\frac{\Delta z}{d}\right)_{\max} \tag{13.1}$$

where β_P is the topographic index horizon angle at point P, d is the horizontal distance from P to a point M, Δz is the difference in elevation between P and M, and M is the point that provides the maximum vertical angle in the direction of the wind flow provided by the topography within a search distance d_b.

The larger the horizon angle, the less topographic exposure or higher protection from wind action in the direction under analysis within the adopted search distance. For example, Fig. 13.4a shows a reduced horizon angle within a search distance of d_{b1}, indicating relatively large topographic exposure to the wind at this distance. For higher analysis distances, such as d_{b2} (Fig. 13.4c), the horizon angle is high, indicating reduced topographic exposure to the wind.

There is no consensus on the recommended value of distance to be adopted; it is usually the practice to test different values, although there is also in the literature the approach of not imposing a distance restriction in the analysis (Lindsay & Rothwell, 2008). In this last situation, the analysis goes along the direction of the wind, considered up to the edge of the scene. However, very high distances may invalidate the simplification of disregarding the effect of the Earth's curvature.

If the analysis indicates that all points have an elevation lower than P within the established search distance, this results in a horizon angle below the horizontal line, which corresponds to a negative value of this angle. However, the differentiation among the negative values has no meaning for this magnitude, and therefore, it is recommended to adopt a zero value in this situation (Antonić & Legović, 1999). Whether negative or zero values, this indicates areas exposed to the wind (within the analysis distance), while positive values represent areas protected against the wind; the higher this value of the horizon angle, the more protected they are.

A limitation of the association between the variable horizon angle and the degree of topographic exposure to the wind is that only the action of the wind and the topographic protection directly and precisely aligned with a specific direction are considered for analysis. The effects of deviation or change in wind direction are disregarded. One way to circumvent this is to calculate the horizon angle for a range of wind directions, for example, every 2° for directions between 30° and 50°, and average these values. This result would be more representative than just calculating for a particular direction (Winstral & Marks, 2002; Winstral et al., 2002).

Determining the horizon angle requires only input information from the DEM, the definition of the wind incidence angle measured relative to the north, and the

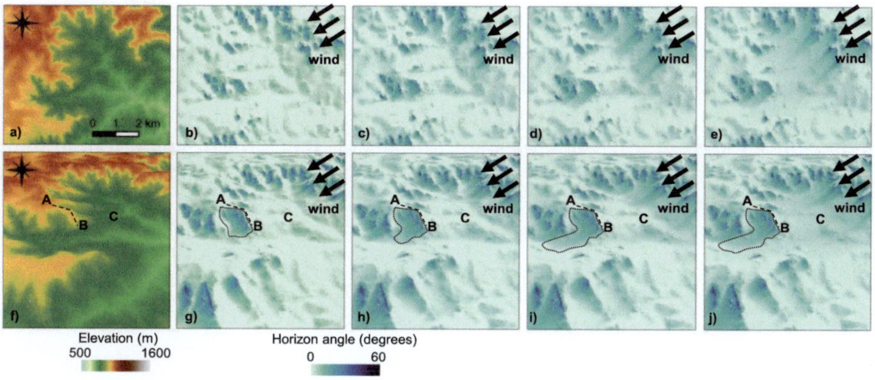

Fig. 13.5 **a** SRTM 30 DEM for an area of the municipality of Cristais Paulista (São Paulo state); **b**, **c**, **d**, **e** Horizon angle rasters determined from the DEM, considering a wind at 45° and search distances of 300 m, 800 m, 2000 m and limited only by the edge of the image (edge outside the figure), respectively; **f**, **g**, **h**, **i**, **j** 3D view of part of the DEM and the horizon angle rasters shown in **a**, **b**, **c**, **d**, and **e**, respectively

value defined for the search distance. Although the concept is simple, calculating the horizon angle requires a specific algorithm. Getting such a determination via standard operations available in GIS software is infeasible. Some software, such as Whitebox GAT, provides a particular function for this aim.

An example of a horizon angle raster determined from the SRTM 30 DEM is shown in Fig. 13.5 for an area of the municipality of Cristais Paulista in the state of São Paulo. Wind at 45° and four distinct distances were considered: 300 m, 800 m, 2000 m, and distance limited only by the edge of the image. The last condition represents the situation where the search occurs throughout the extension of the scene, according to the orientation of the wind angle. In this example, the entire processed scene (not entirely shown in the figure) has an edge more than 50 km away from the displayed region, according to a calculation at a 45° angle.

In the horizon angle rasters of the example, the light-toned areas are the most topographically exposed to the wind. In contrast, the dark areas are more protected, corresponding to the larger horizon angles. The 3D view of the DEM and the horizon angle rasters help to understand the results of the spatial pattern of areas more or less exposed to the wind.

The general pattern is that increasing the analyzed distance decreases the number of areas with higher exposure to the wind. As conceptually known, increasing the distance does not reduce the horizon angles; they may increase or remain the same.

To illustrate, consider the highest elevation region along the dashed line from A to B. For the wind selected for analysis, this region serves as topographic protection at a search distance of 300 m for the lower elevation area indicated by the dotted polygon. With the increase in the search distance in the calculation of the horizon angle, more areas southwest of A-B become protected by the topography of A-B, as evidenced by the polygons drawn for distances of 800 m and 2000 m. When moving from 2000 m to a situation of distance to the edge of the image, more than 50 km away, the dotted region protected by A-B almost does not change. This is because, from the limit of this dotted area in the southwest direction, the elevations begin to increase and even surpass the elevations of A-B at some point. In other words, either significantly reduced horizon angles are formed, an indicator of small topographic protection provided by A-B, or areas of high exposure are configured, with null values for the horizon angle.

13.3 Surface Openness

Yokoyama et al. (2002) conceived the surface openness index to characterize and represent the topographic surface, emphasizing the predominant concavity and convexity. The original motivation of these authors was to enhance the visualization of topography and facilitate the interpretation of the landforms of the Earth's surface, but recognizing that it also applies to analyses of the relief of the seabed, the surface of other planets, and even any irregular surface such as those produced by industrial procedures. The surface openness index represents the degree of dominance or enclosure of a point on the irregular surface where it is located (Prima et al., 2006).

Various applications of the surface openness index include, for example, the analysis of landslide areas (Görüm, 2019), support for semi-automated geomorphological mapping techniques (Anders et al., 2011), and the spatial distribution of soil characteristics such as organic carbon stocks (Ließ et al., 2016), among others.

The surface openness index is an angular measure composed of two distinct and non-complementary parts: positive and negative openness. Understanding the concept and representation of positive and negative openness becomes more manageable with the scheme shown in Fig. 13.6. It shows the cross-sectional topographic profile drawn in a specific direction with azimuth δ and passing through a point of interest P on the terrain. Considering a particular distance of search d_b, from P, one can identify which point provides the highest positive inclination (in the "above the surface" sense) in the topographic profile. This point is indicated by Ma and is located at a distance d_{Ma} from P in the horizontal direction. A line drawn from P to Ma creates an angle β with the horizontal line that is the maximum elevation angle, also known as the horizon angle, as described in a previous item of this chapter.

In contrast, one can identify the point that provides the highest negative inclination (in the "below the surface" sense) along the same topographic profile of Fig. 13.6 and within the search distance restriction. Mi indicates this point in the figure, located at a distance d_{Mi} from P. The inclination of the line P-Mi with the horizontal corresponds to the angle ϕ, which is called the minimum elevation angle.

Equations (13.2 and 13.3), respectively, can be used to mathematically obtain the maximum and minimum elevation angles. Equation (13.2) is the same as Eq. (13.1), which corresponds to the horizon angle.

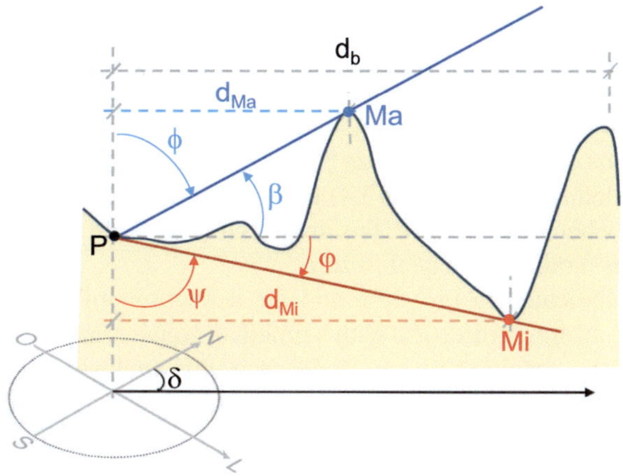

Fig. 13.6 Cross-sectional topographic profile and scheme for determining positive (φ) and negative (ψ) surface openness for a point P on the terrain considering a specific analysis direction with azimuth δ and search distance d_b: the point Ma, located at a horizontal distance d_{Ma} from P, is the one that provides maximum elevation angle β; and the point Mi, located at a horizontal distance d_{Mi} from P, is the one that provides minimum elevation angle ϕ (based on Yokoyama et al., 2002)

$$\beta_{\mathrm{P}} = \arctan\left(\frac{Z_{\mathrm{Ma}} - Z_{\mathrm{P}}}{d_{\mathrm{Ma}}}\right) \tag{13.2}$$

$$\varphi_{\mathrm{P}} = \arctan\left(\frac{Z_{\mathrm{Mi}} - Z_{\mathrm{P}}}{d_{\mathrm{Mi}}}\right) \tag{13.3}$$

where β_{P} and ϕ_{P} are the maximum and minimum elevation angles, respectively, at point P; Z_{P} is the elevation at P; Z_{Ma} and Z_{Mi} are the elevations at points Ma and Mi that provide the angles β_{P} and ϕ_{P}, respectively; d_{Ma} and d_{Mi} are the horizontal distances from points Ma and Mi to point P, respectively.

Equation (13.2) results in a positive angle β if point Ma has a higher elevation than P $\left(Z_{\mathrm{Ma}} > Z_{\mathrm{p}}\right)$ and a negative angle otherwise $\left(Z_{\mathrm{Ma}} < Z_{\mathrm{p}}\right)$. Similarly, Eq. (13.3) results in a negative angle ϕ if $Z_{\mathrm{Mi}} < Z_{\mathrm{p}}$ and in a positive angle ϕ for $Z_{\mathrm{Mi}} > Z_{\mathrm{p}}$.

From β and ϕ, one can define the angles φ and ψ, also shown in Fig. 13.6. The angle φ is the zenith angle formed between the vertical (zenith) of point P and the line of maximum positive inclination, which joins P and Ma. The angle ψ is the nadir angle, formed between the nadir at point P (the line diametrically opposite to the zenith) and the line of maximum negative inclination, which joins P and Mi. Mathematically, the angles φ and ψ are determined according to Eqs. (13.4) and (13.5), respectively.

$$\phi_{\mathrm{P}} = 90 - \beta_{\mathrm{P}} \tag{13.4}$$

$$\psi_{\mathrm{P}} = 90 + \varphi_{\mathrm{P}} \tag{13.5}$$

where φ_P and ψ_P are the zenith and the nadir angles at point P, respectively.

While the angles β and ϕ can take positive or negative values, depending on the difference in elevations between the chosen point (Ma for β and Mi for ϕ) and point P, the angles φ and ψ always result mathematically in positive values. When β is positive, it is limited to a maximum of 90°, as can be understood from the scheme of Fig. 13.6. Therefore, it is deduced from Eq. (13.4) that φ is limited to the minimum value of zero. Similarly, when ϕ is negative, this angle can reach − 90°. From Eq. (13.5), it is seen that ψ has a minimum equal to zero.

The scheme indicated in Fig. 13.6 and the deductions of the indicated angles correspond to an arbitrary direction with azimuth δ, considering a search distance of d_b. For point P and keeping this search distance, this analysis can be repeated for directions corresponding to the eight angles according to the cardinal and intercardinal directions: 0°, 45°, 90°, 135°, 180°, 225°, 270°, and 315°. The maximum and minimum elevation angles are identified, and the zenith angles φ and nadir ψ for each of those eight directions under analysis are determined. From these values, the positive openness index φ_P and the negative openness index ψ_P at point P are defined as equal to the arithmetic mean of the zenith and nadir angles in the eight directions, respectively (Eq. 13.6 and Eq. 13.7; Yokoyama et al., 2002).

$$\phi_P = \left(\frac{\phi_0 + \phi_{45} + \phi_{90} + \phi_{135} + \phi_{180} + \phi_{225} + \phi_{270} + \phi_{315}}{8} \right) \tag{13.6}$$

$$\psi_P = \left(\frac{\psi_0 + \psi_{45} + \psi_{90} + \psi_{135} + \psi_{180} + \psi_{225} + \psi_{270} + \psi_{315}}{8} \right) \tag{13.7}$$

As the angles φ and ψ are always positive in any direction, the metrics φ_P and ψ_P also assume only positive values. However, Yokoyama et al. (2002) adopted the convention of naming φ_P and ψ_P as positive openness index and negative openness index, respectively, by associating the nomenclature with the physical representation of the indices. The positive openness of the surface is more easily associated with the opening 'above the surface', while the negative openness refers to the opening 'below the surface' (Yokoyama et al., 2002). Therefore, surface openness is assumed to incorporate the line-of-sight principle, considering two viewer perspectives (Prima et al., 2006).

Figure 13.7 illustrates three distinct relief configurations and the corresponding determinations of positive and negative openness in two diametrically opposite directions, for a specific search distance, d_b, kept constant. Point T configures a situation of ridge (Fig. 13.7a), which presents a high positive openness, superior to 90°, both for the direction of 45° (in the direction of the right of the figure) and for the direction of 225° (in the direction of the left of the figure). Regarding negative openness (Fig. 13.7b), a ridge point like T has a value less than 90°. The steeper the topographic profile from T, the more reduced the negative openness.

For a valley bottom point such as S in Fig. 13.7c, the positive openness is less than 90° and becomes smaller as the valley is more incised. This point exhibits high negative openness, exceeding 90° and increasing with the degree of valley incision

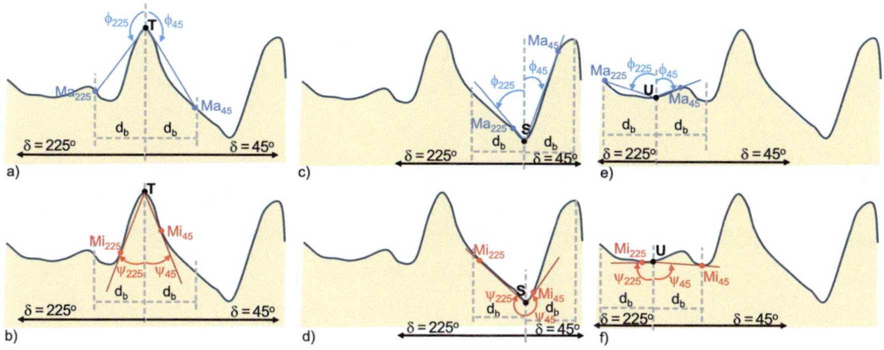

Fig. 13.7 Cross-sectional topographic profile drawn along the azimuthal direction 45° (to the right of the figure) and 225° (to the left of the figure), with indication of the positive (φ) and negative (ψ) openness for a ridge point (T), a valley bottom point (S) and a point of a relatively flat area (U) considering a search distance d_b; points Ma and Mi are those that provide the maximum and minimum elevation angles, respectively, in the corresponding directions (based on Yokoyama et al., 2002)

(Fig. 13.7d). In areas with lower topographic variability within the search radius, such as point U (Fig. 13.7e), positive and negative openness values tend to be close to 90°.

The analyses described from the examples in Fig. 13.7 are for a specific search distance. The variation of this distance changes the degree of surface openness. The smaller the search distance, the more the results refer to the local characteristics of the topography. In contrast, the larger this parameter, the more the analysis corresponds to the larger-scale topography (Prima et al., 2006).

As the search distance increases, the maximum and minimum elevation angles tend to increase in absolute values (they can increase or remain the same value, but not reduce—as described in the item about the topographic index horizon angle in this chapter). As a result, the positive and negative openness tend to decrease with the increase in distance (they can decrease or remain the same value, but not increase).

Determining the surface openness index requires the development of a specific algorithm, such as the one available in the SAGA GIS software and the Whitebox GAT software. Making this determination only by combining standard operations usually available in most GIS software is infeasible. The DEM and the search distance definition are the input information to obtain the surface openness index.

As an example of a real-world case study, the positive and negative surface openness obtained from the SRTM 30 DEM for an area of the municipality of Cristais Paulista (São Paulo state) is shown in Fig. 13.8. It is noticeable how the positive surface openness reaches larger angles for hilltop points, like in A, and smaller values in areas of more embedded topography, like below B. Conversely, the negative openness has smaller values in hilltop areas (like A), and this index reaches the highest values in embedded valley areas (like below B). In areas of more homogeneous relief like C, median values of both positive and negative openness predominate.

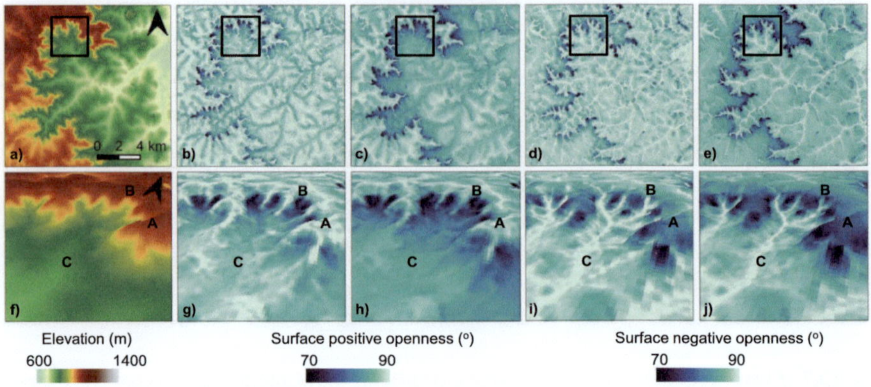

Fig. 13.8 **a** SRTM 30 DEM for an area of the municipality of Cristais Paulista (São Paulo state); **b**, **c** Positive surface openness, for search distances (d_b) of 300 and 2000 m, respectively; **d**, **e** Negative surface openness for d_b equal to 300 and 2000 m, respectively; **f**, **g**, **h**, **i**, **j** 3D view of part of the DEM and the surface openness rasters shown (black rectangle) in **a**, **b**, **c**, **d**, **e**, respectively

By increasing the search distance from 300 to 2000 m in the example, values in the positive and negative openness are reduced, which makes it easier to distinguish between the areas with the largest angles, as can be seen around A and below B, for example.

13.4 Morphometric Protection Index (MPI)

The morphometric protection index (MPI) is related to the positive surface openness index. Although there is confusion in this sense, they are not equivalent attributes (Kaneda & Chiba, 2019).

For the positive openness index at a given point P of the terrain as in the example shown for other indices (Figs. 13.6 and 13.9), the need to identify point Ma, which provides the maximum elevation angle (β), was shown, considering a search along a distance d_b in the chosen direction δ. The positive openness angle (φ) is given by subtracting 90° from β, regardless of whether β is positive (above the horizontal line, making φ less than 90°) or negative (below the horizontal line, making φ higher than 90°).

For the morphometric protection index, the angle β corresponding to the maximum elevation angle is also used, but with a condition: when β is negative, the value zero is taken (Jancewicz & Porębna, 2022; Kaneda & Chiba, 2019).

For obtaining the MPI index, this analysis is done for the eight directions referring to the cardinal and intercardinal directions (0°, 45°, 90°, 135°, 180°, 225°, 270°, and 315°), and the arithmetic mean of these eight values is calculated (Eqs. 13.8 and 13.9).

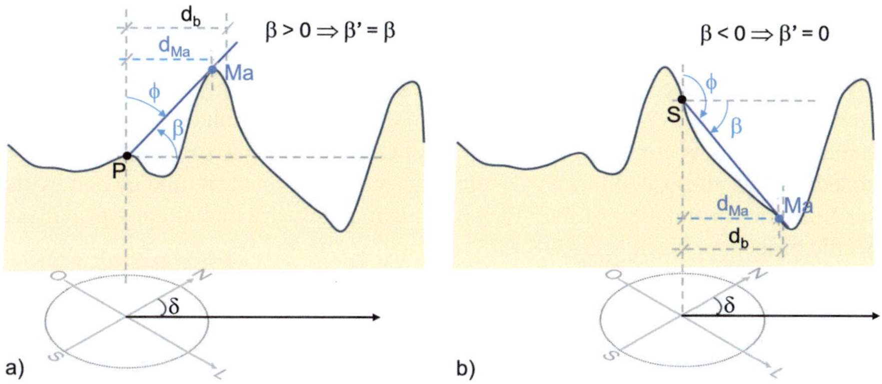

Fig. 13.9 Cross-sectional topographic profile and scheme for determining the angle for calculating the morphometric protection index (MPI) for a point P **a** and S **b** of the terrain, considering a specific analysis direction with azimuth δ and search distance d_b: point Ma, located at a horizontal distance d_{Ma} from the point under analysis, is the one that provides the maximum elevation angle β; for P, $\beta > 0$ and therefore this angle is used for the MPI ($\beta' = \beta$); for S, $\beta < 0$ (below the horizontal line), making $\beta' = 0$ for the MPI

$$\mathrm{MPI_P} = \left(\frac{\beta'_0 + \beta'_{45} + \beta'_{90} + \beta'_{135} + \beta'_{180} + \beta'_{225} + \beta'_{270} + \beta'_{315}}{8} \right) \tag{13.8}$$

$$\beta'_\delta = \begin{Bmatrix} \beta_\delta, \text{ if } \beta_\delta > 0 \\ 0, \quad \text{if } \beta_\delta < 0 \end{Bmatrix} \tag{13.9}$$

where $\mathrm{MPI_P}$ is the morphometric protection index at P; β_δ is the maximum elevation angle in the direction of analysis with azimuth δ; β'_δ is the conversion of β_δ for the calculation of MPI.

This mathematical formulation explains that the MPI at any point on the terrain is an angular measure that is always positive or null. As the MPI considers only the topography above this point within the search area, disregarding lower elevation areas, this index assumes a zero value in flat regions and hilltop areas (Kaneda & Chiba, 2019).

As the name suggests, the MPI is an attribute derived from the DEM used to represent the degree of topographic protection around the point under analysis regarding the wind action. As the MPI is calculated considering the analyses carried out in the eight cardinal and intercardinal directions, it differs by not being a measure referring to a wind with a specific direction of interest, like other indices. The higher the MPI value, the higher the wind protection the surrounding topography offers to the point under analysis, considering a certain distance defined for this analysis. When the MPI is zero, there is a situation of maximum exposure to the wind, which corresponds to flat areas or hilltop areas.

The MPI is an index with various applications related to the influence of wind exposure, such as analyzing ecological characteristics like plant species growth

(Salekin et al., 2021), the spatial variability of soil characteristics (Salekin et al., 2021), and the dispersion of atmospheric pollutants (Pejović et al., 2017).

The DEM and the definition of the search distance for analysis are required to determine the MPI. It is infeasible to perform this determination only by combining standard operations commonly found in most GIS software. A specific algorithm is needed for this purpose, either by developing a computational routine owned by the user from the formulation of the index or by using software that already provides a specific procedure, such as SAGA GIS.

A real-world example of the MPI index derived from the SRTM 30 DEM for an area between the municipalities of Cristais Paulista (São Paulo state) and Claraval (Minas Gerais state) is shown in Fig. 13.10, considering two distinct search distances (300 and 2000 m). In this region, the MPI varies from 0° to 30°. The highest values occur in areas of relatively embedded slopes, like to the right of A, indicating the highest topographic protections against the wind. With the increase of the search distance in the determination of the MPI, from 300 to 2000 m, the areas with higher protection increase.

In the same example, one can note how the areas of more open topography, like around B and, mainly, even further from elevated areas, like in C, represent the areas with the lowest MPI, meaning less protection against the wind. The 3D visualization allows a better understanding of the variation of the MPI according to the relief. Open and relatively flat areas, like on the top of the hills (for example, in D) and in areas of lower elevation (for example, in F), have the lowest morphometric protection indices. The highest protection is in the valley slopes, for example, below, to the left, and to the right of E.

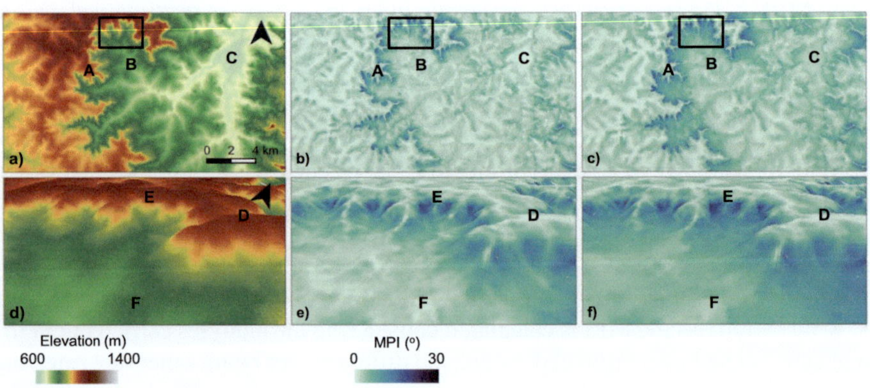

Fig. 13.10 **a** SRTM 30 DEM for an area of the municipality of Cristais Paulista (São Paulo state); **b**, **c** Morphometric protection index (MPI) obtained for search distances of 300 and 2000 m, respectively; **d**, **e**, **f** 3D view of part of the DEM and the surface openness rasters shown (black rectangle) in **a**, **b**, and **c**, respectively

13.5 Terrain Exposure to the Sloped Wind Flux

The terrain exposure to the sloped wind flux was proposed by Antonić and Legović (1999) and combines two topographic variables: the aspect of the terrain relative to the wind, and the horizon angle (Lindsay & Rothwell, 2008). It is an exposure indicator considering the topographic protection effect of areas around the point under analysis within a pre-defined distance. Applications of this index include the analysis of atmospheric pollution, modeling of the topographic effect on solar radiation, and even in ecological studies related to the spatial distribution of plant species (Antonić & Legović, 1999).

For a particular point P of the terrain, the terrain exposure to the sloped wind flux is defined as the cosine of the angle α between two planes: a plane adjusted to represent the local topography at P and a plane orthogonal to the wind flux. This angle is illustrated in Fig. 13.11, which shows the topographic profile in the direction of the considered wind and the tracing of the two mentioned planes. This figure represents a continuity of the scheme shown in Fig. 13.1, when the step-by-step procedure for obtaining the variables related to the determination of the horizon angle and the aspect of the terrain relative to the wind was illustrated.

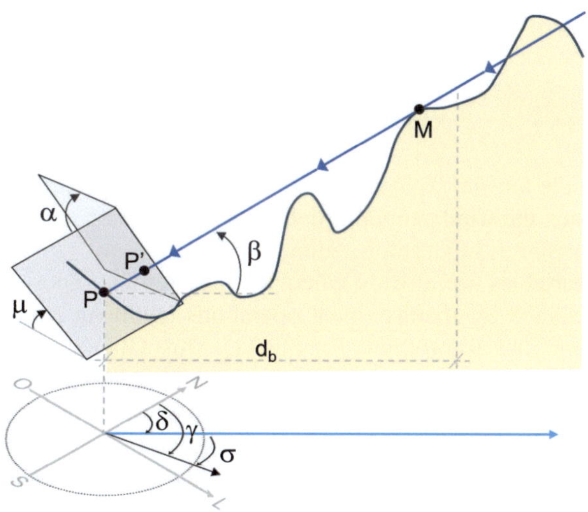

Fig. 13.11 Cross-sectional topographic profile in the direction of the wind flux azimuth (δ), with: indication of a plane adjusted to the local topography at point P, whose local slope is μ and terrain aspect is γ; indication of the terrain aspect relative to the wind (σ), given by the difference between terrain aspect and wind azimuth; determination of the horizon angle β, from the identification of point M in the topographic profile, defined within a search distance d_b in the direction of the wind; identification of the angle α, formed between the plane adjusted to the local topography at P and a plane orthogonal to the wind flux (point P′ is the projection of P on this plane orthogonal to the wind), and whose cosine represents the terrain exposure to the sloped wind flux (*Source* Based on Antonic and Legovic, 1999)

The calculation of the cosine of α is given by Eq. (13.10).

$$\cos \propto = \cos \mu \cdot \sin \beta + \sin \mu \cdot \cos \beta \cdot \cos(\delta - \gamma) \tag{13.10}$$

where α is the angle formed between the plane adjusted to the local topography at the point of interest and a plane orthogonal to the wind flux, β is the horizon angle, μ is the local slope, γ is the terrain aspect, and δ is the azimuth direction of the wind flux.

The maximum value of the terrain exposure to the sloped wind flux index at a given point occurs for cos α equal to 1, that is, α is zero. This means there is a null angle between the adjusted plane and the local topography at this point, and the plane is orthogonal to the wind flux. In this case, the wind flux is orthogonal and frontal to the plane adjusted to the local topography. The minimum terrain exposure to the sloped wind flux occurs when cos α is $-$ 1, that is, α is 180°. This situation occurs when the wind flux is orthogonal to the opposite face of the plane adjusted to the topography at the point under analysis, configuring the case where the topography offers maximum wind protection.

When the wind is horizontal, the horizon angle β is zero. In this case, Eq. (13.10) is simplified to Eq. (13.11), which represents what Antonić and Legović (1999) consider another topographic index, the terrain exposure relative to the horizontal wind flux. The transition from Eq. (13.10) to Eq. (13.11) also occurs if the search distance d_b is considered zero, resulting in β being equal to zero.

$$\cos \propto = \sin \mu \cdot \cos(\delta - \gamma) \tag{13.11}$$

Determining the terrain exposure to the sloped wind flux requires analyzing the local slope's raster, the wind azimuth's definition, the horizon angle's raster, and the terrain aspect's raster as input information.

If all these variables are already calculated, the terrain exposure to the sloped wind flux is obtained by mathematical operations involving rasters, scalars, and trigonometric functions according to Eq. (13.11). Most GIS software presents a tool for this type of calculation. From the DEM, determining the terrain exposure to the sloped wind flux requires obtaining the necessary variables for the calculation and then performing the mathematical operations. In this case, the local slope and terrain aspects are obtained by DEM processing via standard operations available in GIS software. However, as commented in the previous item of this chapter, the horizon angle requires a specific operation, which is not usually available in most software.

An example of a real-world case of determining the terrain exposure to the sloped wind flux is illustrated in Fig. 13.12 for an area of the municipality of Cristais Paulista (São Paulo state). This index was obtained from the SRTM 30 DEM, considering three different configurations: first, for a wind with an azimuth of 45° and two distinct search distances (300 m and 2000 m), and then for a wind with an azimuth of 135° and a search distance of 2000 m. When considering wind from 45°, slope areas such as those to the left of points A and C (the latter being a shaded area in the 3D view) exhibit higher exposure to this wind direction. In contrast, areas to the left of points

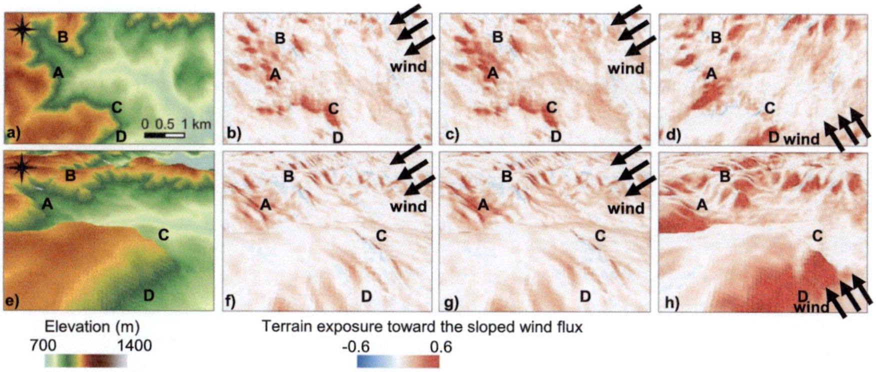

Fig. 13.12 **a** SRTM 30 DEM for an area in the municipality of Cristais Paulista (São Paulo state); **b**, **c** Raster of the terrain exposure to the sloped wind flux, considering wind with azimuth 45° and search distances of 300 m and 2000 m, respectively; **d** Idem, considering wind with azimuth 135° and search distance of 2000 m; **e**, **f**, **g**, **h** 3D view of part of the DEM and the rasters of the exposure index to the wind flow considered in **b**, **c** and **d**, respectively

B and D show lower exposure to wind from 45°. As the search radius increases, the index value also increases, as evidenced by comparing results for a 2000 m search distance and those for 300 m under a 45° wind direction. When considering wind from 135°, areas such as the slope where point D is located, aligned more closely with the wind direction, exhibit the highest levels of wind exposure.

13.6 Wind Fetch

The analysis of wind fetch consists of evaluating the distance from each point of the terrain to topographic obstacles to the incidence of the wind, that is, the distance available for an air flow without obstacles (Lapen & Martz, 1993).

The concept of fetch is widely applied in studies of wind sediment transport (Delgado-Fernandez, 2010; Gillette et al., 1996) and in understanding the formation of waves in water bodies (Ashton et al., 2009). It also has ecological applications, such as explaining the abundance and transition of seedlings to trees in tree lines of forests in alpine regions (McIntire et al., 2016).

The so-called fetch effect, for example, is related to the increase in soil particle movement due to the rise in exposure distance along the wind direction. In water bodies, the higher the extent of the exposed surface in the wind direction (higher fetch), the higher the ease of wave formation due to the action of this wind.

The idea is similar in the context of fetch analysis from a DEM. For a given pixel of the DEM and a specific direction selected for analysis, the DEM is traversed from the pixel along this direction until the first obstacle is identified, which is defined as the pixel that meets the condition established in Eq. (13.12). In other words, a pixel with an elevation higher than the initial pixel's elevation plus an elevation increment is considered an obstacle. This increment gradually increases with the distance from the central pixel under analysis, calculated based on the distance traveled and a slope adopted as constant.

$$Z_{test} \geq Z_P + L \cdot I \tag{13.12}$$

where Z_P is the elevation at the point of interest P; L is the distance from P to the pixel being tested as an obstacle; Z_{test} is the elevation of the pixel tested as an obstacle; I is the constant slope adopted as a calculation parameter in the wind fetch analysis.

As Lapen and Martz (1993) recommended, the user should choose the slope value used to determine the fetch, depending on the area's characteristics and the purpose of the analysis. These authors tested values from 0.025 to 0.01 m/m, observing that the lower slope made fetch patterns more sensitive to smaller-scale topographic variations.

The smaller the value of I in Eq. (13.12), the more gradually the right-hand term increases, making it easier to satisfy the specified condition and to identify an obstacle. On the other hand, steeper slopes result in a pattern that favors identifying larger-scale topographic features, such as valleys within the river drainage network. This occurs because higher I values lead to a more rapid increase in the right-hand term of Eq. (13.12) with distance, making finding a pixel that qualifies as an obstacle more stringent.

The value of the wind fetch in a specific direction is equal to the distance from the central pixel to the identified obstacle. This procedure can be repeated for the eight directions aligned to the eight neighboring pixels in the 3×3 window around the central pixel under analysis: the four cardinal directions (north, south, east, and west) and the four intercardinal directions (northeast, southeast, southwest, and northwest). According to this procedure, the analysis generates eight rasters from the DEM, each having as an attribute the value of the wind fetch in one of the eight mentioned directions.

This approach of working only with the eight cardinal and intercardinal directions was expanded by Lindsay (2016) to perform the fetch analysis for any angle between 0° and 360° from the interpolation of the DEM values. On the one hand, there is a gain in analysis flexibility by allowing any direction to be analyzed. On the other hand, there is the disadvantage of working with interpolated elevation values instead of strictly considering the existing values in the DEM. When interpolating, there are mathematically generated elevation estimates, which may be closer or further from what would be the elevation at that point, but they are still estimates.

Figure 13.13 illustrates the concept of wind fetch analysis for pixel P, considering four directions: north, northwest, west, and east. Despite the schematic example, it is possible to understand the idea of obstacles and wind fetch without values or scales. For instance, in the west direction, from the central pixel P, no pixel before pixel Q could be considered an obstacle, as they all have an elevation lower than P.

Let's assume pixel Q has an elevation that meets the condition of Eq. (13.12). Thus, pixel Q represents the obstacle in this direction, and the wind fetch is given

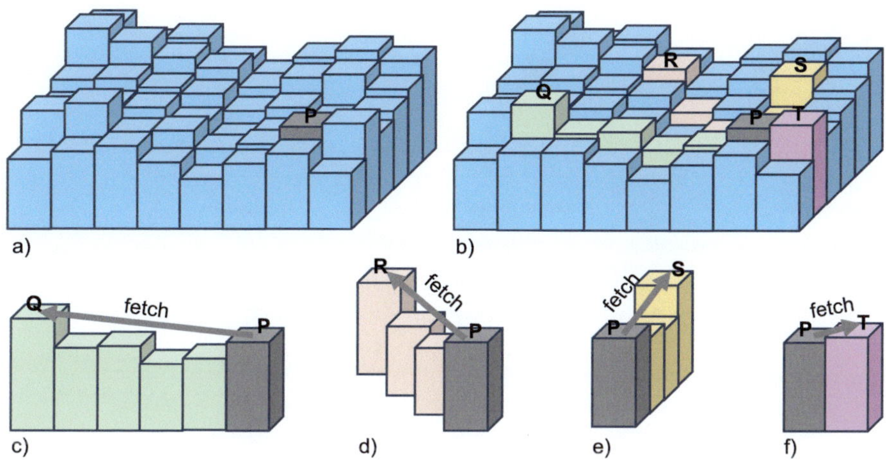

Fig. 13.13 **a** 3D representation of a DEM with emphasis on a pixel P; **b** For P, identification of the pixels located along the fetch according to the west (green), northwest (beige), north (yellow) and east (pink) directions and of the pixels Q, R, S and T, considered as obstacles in these directions, respectively; **c**, **d**, **e**, **f** Detailing of the fetch analysis for the four directions considered in the example

by the distance from P to Q. For the east direction, the first neighboring pixel in this direction already configures an obstacle, representing a reduced wind fetch, of size equal to the side of the pixel. In the northwest direction, the wind fetch has an extension equal to three diagonal steps, from pixel P to pixel R, while the wind fetch in the north direction is slightly smaller, equal to three orthogonal steps.

A quantitative example of the calculation of the wind fetch is illustrated in Fig. 13.14. The DEM shown in Fig. 13.14a has a spatial resolution of 10 m, and a 0.05 m/m slope was adopted for the analysis. For pixel P, whose elevation is 220 m, the study in the north direction indicates that neither pixel N1 nor pixel N2, which are the first two tested, configure an obstacle. This occurs for pixel N3, whose elevation (222 m) is higher than the sum of the elevation of P (220 m) with the elevation increment calculated for a distance of 30 m and a slope of 0.05 m/m. As a result, the wind fetch in the north direction is equal to 30 m.

In Fig. 13.14b, a raster is shown with the values of the sum of pixel P with the increment corresponding to each analyzed direction and spatial position of the pixels tested as obstacles. The colors indicate the pixels tested in each direction, the pixel with the highlighted border being the pixel that configures the obstacle in each direction.

In the northeast direction, the first tested pixel (NE1) has an elevation higher than that of P, but NE1 does not configure an obstacle because it did not meet the condition of Eq. (13.12). No tested pixel met such a condition in the south and southwest directions, with the search reaching the image border. In this case, it is recommended that the fetch corresponds to the distance from the central pixel to the last tested pixel at the edge of the image, but imposing the negative sign for recognition that the actual fetch may be higher than the calculated one (Lapen &

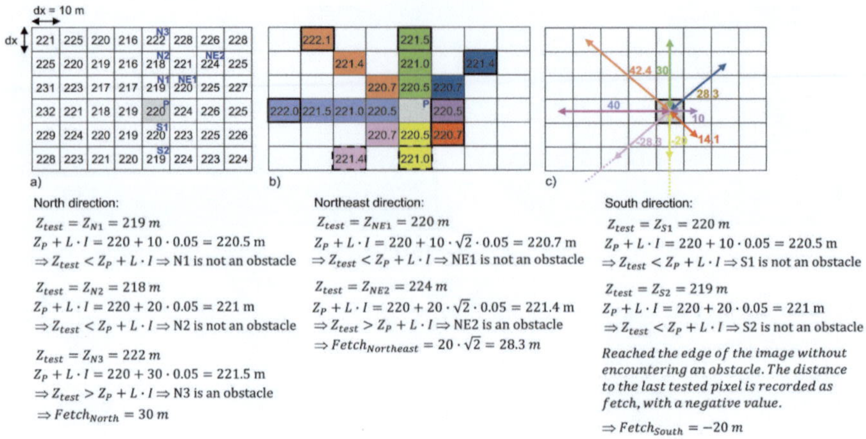

Fig. 13.14 **a** Digital Elevation Model (in meters) with spatial resolution of 10 m; **b** Value of the Zp+*L. I* factor (in meters) for each pixel in the fetch analysis according to the eight directions pointing to the cardinal points (north, south, east, west) and intercardinals (northeast, northwest, southeast, southwest), relative to pixel P; the eight directions for pixel P and considering $I = 0.05$ m/m; **c** Fetch values for point P obtained for the eight directions analyzed (in meters)

Martz, 1993). The values of fetch obtained in the eight directions are shown in Fig. 13.14c.

Obtaining the wind fetch raster from the DEM requires a specific operation, such as the one in the Whitebox GAT software. Determining this topographic index only by combining steps that involve standard operations usually available in GIS software is infeasible.

An example of a real-world case of wind fetch analysis is shown in Fig. 13.15 for an area between the states of São Paulo and Minas Gerais, determined from SRTM 30 DEM data. Specific results are displayed for a wind at 45° and a wind at 135°. A 0.025 m/m slope was considered for analysis in both cases. The maximum value of the wind fetch distances obtained for this region exceeds 100 km. The actual DEM coverage is much more extensive than the area shown in the figure, which allowed for such high distances before reaching the edge of the DEM.

In this example, comparing the results obtained with the two wind directions illustrates the effect of this choice. Lighter areas have the longest wind fetch. These areas have higher topographic exposure to the wind in the specific direction of the analysis. In contrast, the darker the color tone, the shorter the wind fetch until it almost reaches zero—these are the most protected regions from that wind.

Areas where points A, B, C, and D are located have distinct spatial patterns between the two results, but are related to the wind incidence angle. For example, for a wind at 135°, there is an area at northeast of C with two patterns: (i) a light spot (indicated by the orange arrow), with a longer wind fetch, located in a region of higher elevation and that, in the direction under analysis, there is a long distance until finding a topographic obstacle that provides wind protection (assuming the slope of 0.025 m/

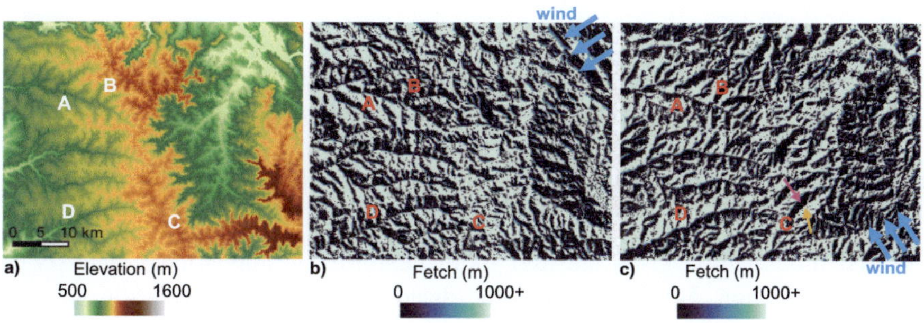

Fig. 13.15 **a** SRTM 30 DEM for an area between the microregion of Franca (SP) and the microregion of Passos (MG); **b**, **c** Wind fetch analysis considering winds at 45° and 135°, respectively

m for the calculation); (ii) a dark spot (indicated by the pink arrow), representing a reduced wind fetch, which is precisely under topographic protection of the previous area. In other words, pixels from the area indicated by the orange arrow represent obstacles for the pixels from the dark area indicated by the pink arrow.

13.7 Directional Relief

Directional relief is a topographic index that denotes how much each point of the terrain has an elevation higher or lower than the average of its neighbors along a specific direction of analysis (Eqs. 13.13 and 13.14).

$$\mathrm{RD_P} = \overline{Z_{\mathrm{dir}}} - Z_{\mathrm{P}} \tag{13.13}$$

$$\overline{Z_{\mathrm{dir}}} = \frac{1}{nd} \cdot \sum_{i=1}^{nd} Z_i \tag{13.14}$$

where $\mathrm{RD_P}$ is the directional relief index at P for a particular direction dir; Z_{P} is the elevation at P; $\overline{Z_{\mathrm{dir}}}$ is the average of the elevations of the pixels neighboring P located along the direction dir; *nd* is the number of pixels neighboring P along this direction; and Z_i is the elevation of pixel i along this direction.

This directional relief index was proposed by Lapen and Martz (1993), who recommend developing this analysis for the eight directions of a pixel's neighborhood formed by the cardinal (north, south, east, and west) and intercardinals (northeast, northwest, southeast, and southwest) directions. In addition, these authors propose computing the index considering all the neighbors along a direction until reaching the edge of the DEM.

Lindsay (2016) proposes two adaptations to this index:

i. Consideration of any direction between 0° and 360° to perform the analysis, which requires linear interpolation to obtain the elevation values at points not coinciding with the DEM pixels;
ii. A maximum search distance is adopted to identify neighboring cells to calculate the mean elevation, with this distance being defined by the user.

The first modification gives the user more flexibility in unrestricted analysis for any wind direction. However, it has the disadvantage of working with elevation values artificially adjusted by interpolation. In the analysis with the eight principal directions, all the elevations worked on are precisely the values present in the DEM. The second modification is interesting because it prevents the analysis from being influenced by the size of the DEM spatial coverage, as it is objectively tied to a predetermined distance.

Positive values obtained for the directional relief indicate areas of elevation lower than the average of the neighbors located along the direction of analysis. These regions represent areas relatively protected from the action of the wind (Lindsay & Rotwell, 2008). In contrast, areas with negative values for the directional relief have elevations higher than the neighbors' average in the analyzed direction, configuring areas with higher exposure to the wind.

In a way, the directional relief index resembles the horizon angle and the wind fetch by analyzing the topographic exposure to the wind influenced by the elevation of neighbors aligned in a specific direction. The main difference is that the horizon angle evaluates this exposure in terms of angle, the wind fetch evaluates regarding the horizontal extension, and the directional relief quantifies the elevation difference.

The analysis of directional relief is helpful for roughly distinguishing the region into uplands (those with a null or negative value for this index) and lowlands (those with a positive value for the index) (Lapen & Martz, 1993).

Figure 13.16 provides a didactic example of determining directional relief for the eight principal directions. For the DEM displayed in Fig. 13.16a and taking pixel P, the pixels considered in the analysis of the neighborhood average in each direction are highlighted in different colors in Fig. 13.16b. To simplify the example, the DEM spatial extent is significantly reduced, and there are only 2 to 4 neighbors in each direction up to the edge of the image.

Figure 13.16c indicates the directional relief values obtained for pixel P. According to the RD index, P is an area with relative wind protection in 5 out of the eight directions (RD values > 0). Point P receives higher topographic protection from northeast, east, and southeast winds. The highest exposure occurs from winds in the south and north directions.

An example with real-world results from determining the directional relief index is shown for part of the municipality of Ibiraci, in Minas Gerais (Fig. 13.17). A wind with an azimuth of 45° and two search distances of 300 m and 2000 m were considered. For the search distance of 300 m, the directional relief varied approximately between − 150 m and 150 m, while for the distance of 2000 m, the variation was

221	225	220	216	222	228	226	228
225	220	219	216	218	221	224	225
231	223	217	217	219	220	225	227
232	221	218	219	220 P	224	226	225
229	224	220	219	220	223	225	226
228	223	221	220	219	224	223	224

a)

North direction:

$$\overline{Z_N} = \frac{219 + 218 + 222}{3} = 219.7\,m$$

$$RD_P^N = 219.7 - 220 = -0.3\,m$$

221	225	220	216	222	228	226	228
225	220	219	216	218	221	224	225
231	223	217	217	219	220	225	227
232	221	218	219	220	224	226	225
229	224	220	219	220	223	225	226
228	223	221	220	219	224	223	224

b)

Northeast direction:

$$\overline{Z_{NE}} = \frac{220 + 224 + 228}{3} = 224\,m$$

$$RD_P^{NE} = 224 - 220 = 4\,m$$

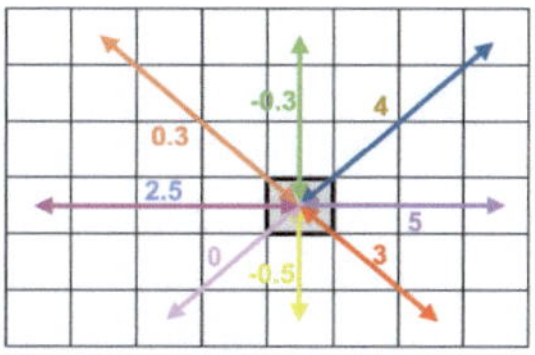

c)

Southwest direction:

$$\overline{Z_{SO}} = \frac{219 + 221}{2} = 220\,m$$

$$RD_P^{SW} = 220 - 220 = 0$$

Fig. 13.16 Example of determination of the directional relief (RD) index: **a** DEM (m); **b** For pixel P, identification in different colors of the neighboring pixels along each of the four cardinal directions (north, south, east, west) and four intercardinal directions (southeast, southwest, northeast, northwest); **c** For pixel P, RD index values (meters) obtained according to the eight directions, with calculation detailing for the north (N), northeast (NE), and southwest (SW) directions

around − 250 to 250 m. However, both results are presented on a scale of − 200 to 200 m for easy comparison.

In general, considering the entire DEM, with the increase in the search distance, there is an increase in the maximum absolute value of the RD index. This is a theoretically expected tendency since the elevations of points close to each other have values closer than between more distant points. Thus, the further away from a point, the set of points taken to calculate the average that composes the RD metric has a higher chance of having a value that is more distinct from that point. It is not

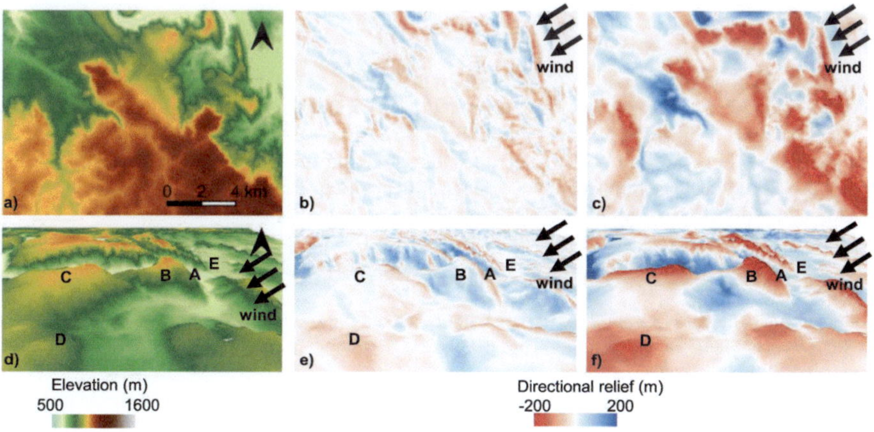

Fig. 13.17 **a** SRTM 30 DEM for an area in the municipality of Ibiraci (Minas Gerais state); **b**, **c** Raster of the directional relief topographic index, considering wind with azimuth 45° and search distances of 300 m and 2000 m, respectively; **d**, **e**, **f** 3D view of part of the DEM and the rasters of the directional relief considered in **b** and **c**, respectively

a rule that will occur when analyzing each point individually, but it is a tendency to predominate in occurrence when all points are taken along a DEM.

In more detail, the effect of the considered wind angle on the RD metric can be understood from the results of Fig. 13.17. Consider, for example, points located in the hill region near A, which form a kind of slope facing the direction of the wind. For this region, the neighbors in such a direction are of lower elevation, making the RD negative. This corroborates the perception by the visual analysis of the 3D scenes that this region around A has high exposure to the wind at 45°. In the search distance of 300 m, the negative RD value reached close to 150 m, and the occurrence was limited to a particular set of pixels. When the distance increased to 2000 m, the RD went to almost – 200 m in this region, and the negative values began to cover a much larger area. The expansion of the search along the direction of analysis made the average of the directional neighborhood cover more pixels of lower elevation than in A and even in B.

The increase in the search distance changed the RD from negative to positive in the region where B is. This also occurred in C: at the distance of 300 m, along 45°, there is an over-elevation in the terrain that makes the average of the elevations of the neighborhood higher than C; but for the distance of 2000 m, predominant elevations are lower than C, mainly referring to the pixels located on the slope covered in the 3D view. Already in the regions around D and E, the increase in the search distance increased the RD in modulus, maintaining the same sign. In D, there is a region of exposure to the wind at 45°, while in E, there is a region like a drainage valley with areas of higher elevation around 45°.

References

Anders, N. S., Seijmonsbergen, A. C., & Bouten, W. (2011). Segmentation optimization and stratified object-based analysis for semi-automated geomorphological mapping. *Remote Sensing of Environment, 115*, 2976–2985. https://doi.org/10.1016/j.rse.2011.05.007

Antonić, O., & Legović, T. (1999). Estimating the direction of an unknown air pollution source using a digital elevation model and a sample of deposition. *Ecological Modelling, 124*, 85–95. https://doi.org/10.1016/S0304-3800(99)00149-0

Ashton, A. D., Murray, A. B., Littlewood, R., Lewis, D. A., & Hong, P. (2009). Fetch-limited self-organization of elongate water bodies. *Geology, 37*, 187–190. https://doi.org/10.1130/G25299A.1

Böhner, J., & Antonić, O. (2009). Chapter 8 land-surface parameters specific to topo-climatology. In: *Developments in soil science* (pp. 195–226). Elsevier. https://doi.org/10.1016/S0166-2481(08)00008-1

Delgado-Fernandez, I. (2010). A review of the application of the fetch effect to modelling sand supply to coastal foredunes. *Aeolian Research, 2*, 61–70. https://doi.org/10.1016/j.aeolia.2010.04.001

Gillette, D. A., Herbert, G., Stockton, P. H., & Owen, P. R. (1996). Causes of the fetch effect in wind erosion. *Earth Surface Processes and Landforms, 21*, 641–659. https://doi.org/10.1002/(SICI)1096-9837(199607)21:7%3c641::AID-ESP662%3e3.0.CO;2-9

Görüm, T. (2019). Landslide recognition and mapping in a mixed forest environment from airborne LiDAR data. *Engineering Geology, 258*, Article 105155. https://doi.org/10.1016/j.enggeo.2019.105155

Jancewicz, K., & Porębna, W. (2022). Point cloud does matter. Selected issues of using airborne LiDAR elevation data in geomorphometric studies of rugged sandstone terrain under forest—Case study from Central Europe. *Geomorphology, 412*, 108316. https://doi.org/10.1016/j.geomorph.2022.108316

Kaneda, H., & Chiba, T. (2019). Stereopaired morphometric protection index red relief image maps (Stereo MPI-RRIMs): Effective visualization of high-resolution digital elevation models for interpreting and mapping small tectonic geomorphic features. *Bulletin of the Seismological Society of America, 109*, 99–109. https://doi.org/10.1785/0120180166

Lapen, D. R., & Martz, L. W. (1993). The measurement of two simple topographic indices of wind sheltering-exposure from raster digital elevation models. *Computers and Geosciences, 19*, 769–779. https://doi.org/10.1016/0098-3004(93)90049-B

Ließ, M., Schmidt, J., & Glaser, B. (2016). Improving the spatial prediction of soil organic carbon stocks in a complex tropical mountain landscape by methodological specifications in machine learning approaches. *PLoS ONE, 11*, Article e0153673. https://doi.org/10.1371/journal.pone.0153673

Lindsay, J. B., & Rothwell, J. J. (2008). Modelling channelling and deflection of wind by topography. *Advances in Digital Terrain Analysis*, 383–406.

Lindsay, J. B. (2016). Whitebox GAT: A case study in geomorphometric analysis. *Computers and Geosciences, 95*, 75–84. https://doi.org/10.1016/j.cageo.2016.07.003

McIntire, E. J. B., Piper, F. I., & Fajardo, A. (2016). Wind exposure and light exposure, more than elevation-related temperature, limit tree line seedling abundance on three continents. *Journal of Ecology, 104*, 1379–1390. https://doi.org/10.1111/1365-2745.12599

Pejović, M., Bajat, B., Gospavić, Z., Saljnikov, E., Kilibarda, M., & Čakmak, D. (2017). Layer-specific spatial prediction of As concentration in copper smelter vicinity considering the terrain exposure. *Journal of Geochemical Exploration, 179*, 25–35. https://doi.org/10.1016/j.gexplo.2017.05.004

Prima, O. D. A., Echigo, A., Yokoyama, R., & Yoshida, T. (2006). Supervised landform classification of Northeast Honshu from DEM-derived thematic maps. *Geomorphology, 78*, 373–386. https://doi.org/10.1016/j.geomorph.2006.02.005

Salekin, S., Mason, E. G., Morgenroth, J., Bloomberg, M., & Meason, D. F. (2021). Hybrid height growth and survival model for juvenile Eucalyptus globoidea (Blakely) and E. bosistoana (F. Muell) in New Zealand. *Forest Ecology and Management*, 490, 119074. https://doi.org/10.1016/j.foreco.2021.119074

Winstral, A., Elder, K., & Davis, R. E. (2002). Spatial snow modeling of wind-redistributed snow using terrain-based parameters. *Journal of Hydrometeorology, 3*, 524–538. https://doi.org/10.1175/1525-7541(2002)003%3c0524:SSMOWR%3e2.0.CO;2

Winstral, A., & Marks, D. (2002). Simulating wind fields and snow redistribution using terrain-based parameters to model snow accumulation and melt over a semi-arid mountain catchment. *Hydrological Processes, 16*, 3585–3603. https://doi.org/10.1002/hyp.1238

Yokoyama, R., Shirasawa, M., & Pike, R. (2002). Visualizing topography by openness: A new application of image processing to digital elevation models. *Photogrammetric Engineering and Remote Sensing, 68*, 257–265.

Chapter 14
Quality, Uncertainties, and Errors Associated with DEM Processing

A DEM can be obtained from different topographic data sources, as described in Chap. 4. Regardless of the data source, the DEM represents the terrain topography and presents deviations from reality, that is, errors. The nature and magnitude of these errors are not fully known and available to users for a given area of interest, making this lack of knowledge of the errors a source of uncertainty (Wechsler, 2007).

It can be considered that the uncertainties present in DEM elevations are primarily related to the uncertainties in the acquisition and processing of the data used to generate the DEM. This chapter addresses these aspects, focusing on DEMs derived from orbital remote sensing. Such uncertainties result in the topographic representation provided by the DEM being an approximation of the real world, never an exact depiction of it.

The uncertainties in the DEM propagate into derived products and studies based on the DEM and its derivatives. This chapter assesses this propagation concerning one of the primary applications of DEMs in environmental studies: the extraction and characterization of river drainage networks and watersheds. The chapter evaluates the DEM characteristics that influence the quality of the extracted river drainage network, with special attention given to the conceptual and methodological aspects of how drainage network quality is defined and measured. Various approaches for assessing river drainage network quality are presented, encompassing qualitative and quantitative methods.

Finally, this chapter provides an illustrative and didactic discussion of the impacts of different errors in the river drainage network and watershed delineation on further hydrological studies.

A. Rolim da Paz, *Digital Elevation Models for Environmental Studies*,
https://doi.org/10.1007/978-3-032-04523-2_14

14.1 Uncertainties in DEM Generation

The various types of observational errors present in DEMs generated from orbital remote sensing are significant and cannot be disregarded in applications within the geosciences and related disciplines (Yamazaki et al., 2017). The uncertainties refer mainly to issues such as the imprecision of the data's georeferencing, the imprecision of the instruments for quantifying the elevations, the contamination of the data acquisition by atmospheric effects such as clouds and dust, and other noise sources in the data.

Four main types of errors in the elevations captured by DEMs derived from orbital remote sensing data can be highlighted (Yamazaki et al., 2017):

- Systematic noises associated with acquisition geometry (stripe noise);
- Speckle noises associated with signal scattering;
- Absolute deviations;
- Effect of vegetation cover.

The contamination of the elevation values represented in the global DEMs due to vegetation cover must be highlighted. The electromagnetic waves that acquire data from some orbital sensors cannot traverse the vegetation canopy completely until they reach the bare ground. As seen in Chap. 4, global DEMs reported in the literature, such as SRTM, ASTER, TANDEM, and AW3D, present this vegetation bias. This justifies why some authors prefer to refer to such global DEMs as DSMs, considering them a representation of the surface and not the terrain itself. However, the elevations present in these DEMs in areas of dense vegetation cover do not refer precisely to the top of the trees but to some intermediate depth between the top and the terrain. This depth varies according to the area's characteristics and other particularities of the data acquisition.

14.2 Uncertainties in DEM Processing

This book works on numerous types and methods for DEM processing. These processes involve, for example, basic geoprocessing operations with raster data, such as cropping, concatenation, reclassification, cross-sectional profiling, and algebraic operations. But the primary focus is the derivation of various products from the DEM, mainly the flow directions, accumulated drainage areas, river drainage network, and watershed, in addition to other products derived from these, such as distances along the river drainage network and dozens of topographic indices.

For most of these processes, the concepts involved and the application of methods to solve them lead to the same results, regardless of the algorithm, command, or software used. For example, making a subset or reclassifying a DEM leads to the same results if different software is used, as long as the same criteria for applying these procedures are maintained.

On the other hand, the treatment of terrain depressions and the determination of flow directions are procedures for which each specific algorithm generates different results. As seen in Chap. 6, considering a single flow direction pointed in the direction of one of the eight neighbors around the 3×3 window and adopting the steepest descent slope rule to determine the flow direction are criteria adopted in most methods and algorithms for deciding flow directions. However, there are situations where additional rules need to be established to define flow directions, mainly involving flat areas, areas of tie regarding the steepest slope, and the presence of complex or single terrain depressions.

Differences in flow directions result in differences in accumulated drainage areas, the river drainage network's layout and characteristics, and even the watershed's delimitation and characteristics. These differences continue propagating to topographic indices and other products based on one or more initial DEM-derived data layers.

Given that a topographic data source is considered for providing a DEM, relatively distinct results from processing this DEM can be obtained by applying different algorithms regarding depression removal and flow direction assignment. In certain situations, variations in results among algorithms can be higher than the variations resulting from the choice of the DEM data source itself.

It is the user's responsibility first to identify which algorithm is employed by the software or command chosen for DEM processing. Subsequently, assessing whether the algorithm has recognized quality and merit in the scientific literature or is considered outdated due to known limitations in handling certain situations is recommended.

14.3 Relationship Between Pixel Size and River Meanders

The ability of a DEM-derived river drainage network to follow the actual course of a river is highly influenced, among other factors, by the relationship between the pixel size of the DEM and the dimensions of the river meanders.

As seen in Chap. 7, when processing the DEM, a raster river drainage network is obtained, with spatial resolution identical to that of the DEM. The pixel size of the river drainage network can limit the network's ability to represent river meanders accurately.

For example, Fig. 14.1a shows a reach of the Piratinim River, flowing into the Uruguay River, in southern Brazil. The river drainage network derived from the SRTM 90 data resampled to a spatial resolution of 100 m (Fig. 14.1b) illustrates how this pixel size is reasonably sufficient to represent the Piratinim River.

In this example, the course of the river identified in the Landsat image is not fully represented in the DEM-derived network due to limitations of the quality of the topographic data and even the DEM processing algorithm. This occurs, for example, for the meanders near points A, B, C, and D. However, it can be considered that there is almost no limitation due to the pixel size itself. This becomes clear when

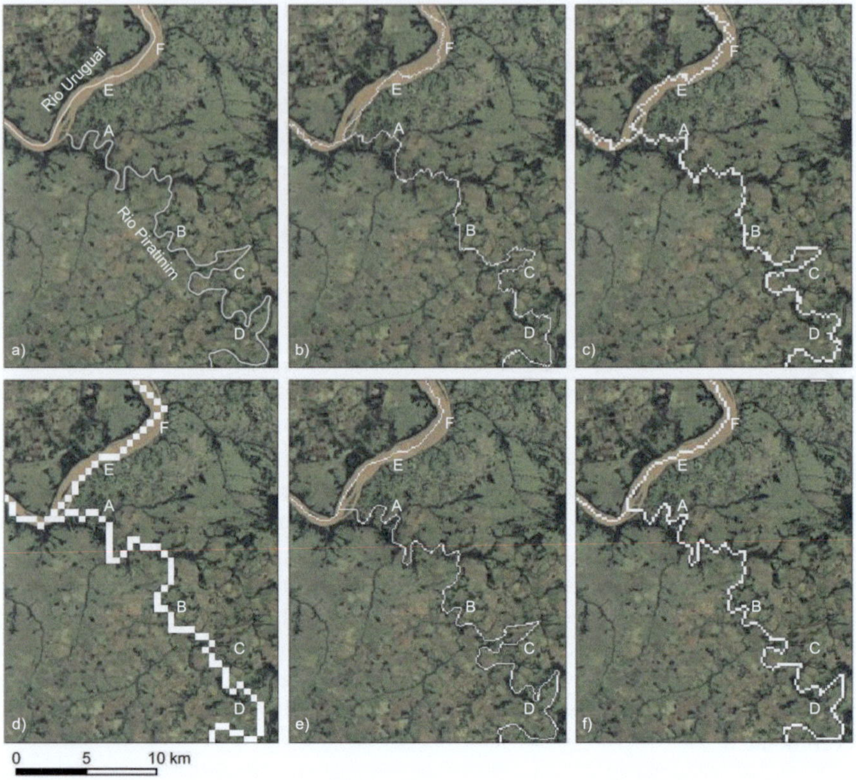

Fig. 14.1 **a** Reaches of the Piratinim and Uruguay rivers, in southern Brazil, visually identified on a Landsat image; **b**, **c**, **d** Corresponding river reaches derived from a DEM with spatial resolution of 100 m, 200 m, and 500 m, respectively, obtained by resampling the SRTM 90 data; **e**, **f** Corresponding river reaches derived from a DEM with spatial resolution of 100 m and 200 m, but having done the preprocessing by stream burning using the vector river network indicated in **a**

applying the preprocessing by stream burning to this 100-m resolution DEM (as seen in Chap. 7). The vector river network digitized from the Landsat image (Fig. 14.1a) was considered for the stream burning. The network derived from this preprocessed DEM can represent almost all the meanders of this river reach, including points A, B, C, and D (Fig. 14.1e).

In this example, when working with a 200 m DEM, there is an explicit limitation related to the pixel size to follow the course of the river (Fig. 14.1c). The preprocessing by stream burning improves the quality of the network derived from the DEM (Fig. 14.1f), but the limitation to reproduce the smaller meanders remains. At a resolution of 500 m, there is a remarkable limitation in reproducing the river's course, with the omission of several small and medium-sized meanders, leaving only the more general pattern of the Piratinim river course (Fig. 14.1d).

Thus, there is a limitation of a network derived from the DEM to follow the course of a river that is independent of the quality of the topographic information present in the DEM and the quality of the algorithm used to process it, being inherent to the relationship between pixel size and the dimensions of the meanders.

In general, the larger the pixel size (i.e., the coarser the spatial resolution), the higher the chance of the river drainage network not precisely following the courses of the rivers. Not only does the absolute spatial resolution matter, but the relationship of this magnitude with the geometric shape of the river curves also matters. A pixel size may be adequate to represent the curves of one river, but it may not be appropriate to follow the flow of another river.

In the example of Fig. 14.1, the reach of the Uruguay River has large meanders, forming much smoother curves than the Piratinim River. Therefore, even at a spatial resolution of 500 m, the pixel size is sufficient to represent the course of this reach of the Uruguay River.

Lower-order and smaller rivers tend to occur in more embedded valleys and have higher sinuosity with tighter meanders. This situation makes it challenging to represent the course of these reaches in the DEM-derived river drainage networks. In these networks, the rivers have a less sinuous representation and are shortened by the incomplete representation of the meanders. In this case, the length of the river reaches derived from the DEM tends to be shorter than the distances of the rivers in the real world. In other words, there is an effect of river length underestimation. A deeper reading of this issue is presented in the works of Paz and Collischonn (2007) and Paz et al. (2008), taking the Uruguay River basin as a case study.

14.4 Relationship Between Pixel Size and River Width

The relationship between the pixel size of the DEM and the width of the rivers is another issue that explains the difficulty of DEM-derived river drainage networks in reproducing the characteristics of the actual river courses. This effect is opposite to the relationship between pixel size and river meanders.

In the case of global DEMs, the general rule is that the collected data does not contain bathymetric information, that is, information about the topography below the water level of the rivers at the time of data acquisition. In these DEMs, the elevations collected in the area covered by water refer roughly to the water surface at the moment of data acquisition. Other factors, such as vegetation along the watercourse's margin, can also interfere with this information.

If a river is wider than the DEM's pixel size, then the area occupied by water will probably form a flat area. This is the case of the Uruguay River section, exemplified in Figs. 14.1 and in 14.2, whose width varies from 600 to 1400 m. The smaller the pixel size relative to the river's width, the more evident the formation of a corresponding flat area is.

Flat areas present in the DEM are one of the situations that require special treatment for the definition of flow directions, as seen in Chap. 6. As a result, there is a

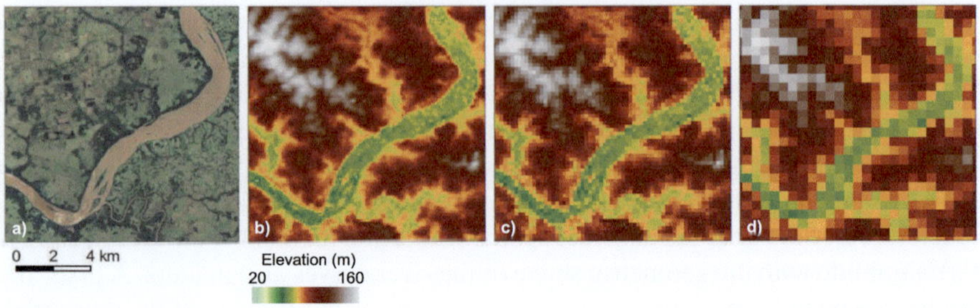

Fig. 14.2 **a** Landsat image showing a reach of the Uruguay River as shown in Fig. 14.1; **b**, **c**, **d** DEMs derived from SRTM 90 data resampled to spatial resolutions of 100 m, 200 m, and 500 m, respectively

high chance that the flow path derived from the DEM corresponding to a river section with a width several times higher than the spatial resolution of the DEM presents a zigzag within the area covered by the river surface.

In the example of Fig. 14.1b, there is a zigzag in the river network derived at the spatial resolution of 100 m, referring to the Uruguay River reach, as illustrated by points E and F. The same effect occurred for the resolution of 200 m (Fig. 14.1c). At the resolution of 500 m, the pixel size is not far from the river's width, and the zigzag effect is minimal. For the Piratinim River, a tributary to the Uruguay River shown in the figure and with a width less than 100 m, this zigzag effect does not exist for any of the three resolutions of the example.

The zigzag effect resulting from wide rivers causes the DEM-derived river reaches to have a higher length than the actual lengths; that is, there is an overestimation error. The preprocessing of the DEM by the stream-burning procedure is a solution to avoid or reduce this, as illustrated in Fig. 14.1d and 14.1e. However, it depends on the availability of a vector river network with sufficient quality for this purpose, as discussed in Chap. 7.

14.5 Improving Distance Calculations in a Raster Environment

To follow a specific path walking between pixels in a raster requires going from pixel to pixel, considering a neighborhood of the 3 × 3 window, independently of the raster attribute. Each step must be taken from one pixel to one of its eight neighbors. For example, to go from pixel 1 to pixel 2 in Fig. 14.3a, there are numerous possible paths if there are no specific data constraints and restrictions. But, regardless of which path is established, to leave pixel 1, one must initially go to one of its eight neighbors, and so on, until reaching pixel 2.

In this example, two paths were determined (Fig. 14.3b and d). Along each path, the distance traveled refers to the sum of the distances of the incremental steps

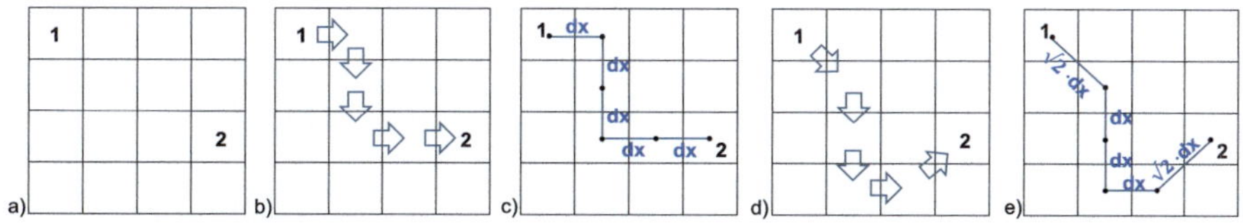

Fig. 14.3 **a** Raster indicating pixels 1 and 2 selected as initial and end points, respectively; **b, d** Two possible paths to follow from pixel 1 to pixel 2; **c, e** Quantifying incremental distances along the defined paths from pixel 1 to pixel 2

between neighboring pixels. In each incremental step, from one pixel to one of its eight neighbors, the distance is established as the one between the centroids of these pixels.

In the case of Fig. 14.3b, the path corresponds to five orthogonal steps, each of them of extension equal to the pixel size (dx). This path's total extension equals $5 \cdot dx$ (Fig. 14.3c). In the situation established in Fig. 14.3d, the path also has five steps between neighboring pixels, but three are orthogonal, and two are diagonal (Fig. 14.3e). The orthogonal steps each have an extension equal to the pixel size, while the diagonal steps have an extension equal to $\sqrt{2} \cdot dx$. The total extension of this second path is equal to $3 \cdot dx + 2 \cdot \sqrt{2} \cdot dx$.

This characteristic of having to walk through a raster always with orthogonal or diagonal incremental steps from one pixel to one of its eight neighbors causes an intrinsic error in calculating the distance traveled by river reaches.

When the path to be followed aligns perfectly and exclusively along the vertical or horizontal direction—that is, aligned with one of the grid axes—this intrinsic error does not exist. Take, for example, the A–B vector in Fig. 14.4a. The path of this vector is perfectly represented by the sequence of three highlighted pixels in the raster shown in Fig. 14.4b. Assuming a pixel size (dx) of 1 m, the length of vector A–B is 2 m, and this same value corresponds to the distance measured along the raster representation of the A–B path. For rasters with finer resolutions, such as $dx/3$ (Fig. 14.4c) and $dx/9$ (Fig. 14.4d), the exact representation occurs, with the total length also being 2 m, assuming $dx = 1$ m: in Fig. 14.4c, the path consists of 6 steps of $dx/3$, and in Fig. 14.4d, 18 steps of $dx/9$ each.

A path aligned precisely at 45° also has an exact representation in a raster. For example, vector C–D in Fig. 14.4e is perfectly represented in the raster of Fig. 14.4f. Assuming dx with 1 m, the length of vector C–D is 2.828 m, and this same value is obtained by the path traced in the raster. The accuracy of the distance traveled remains for the other resolutions exemplified in Fig. 14.4g and h.

Consider a vector aligned at 21.8° with the south direction, like vector E–F from Fig. 14.4i. This vector has an extension of 5.385 m, considering that the spatial scale of dx equals 1 m. The representation of vector E–F in the raster of spatial resolution dx is indicated in Fig. 14.4j, with the corresponding path indicated by the red line, formed by three orthogonal steps and two diagonal steps. Since each step in the pixel-by-pixel traversal can only be made from one centroid to another, the path followed in the raster cannot precisely align with the vector's trajectory. As a result,

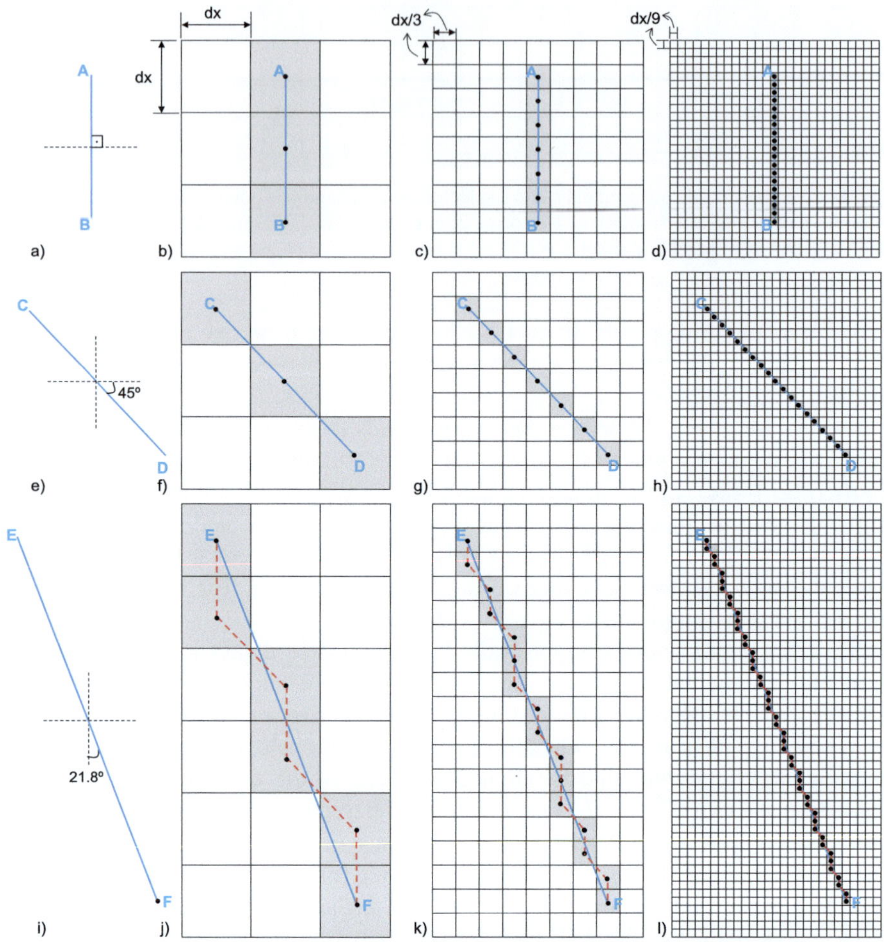

Fig. 14.4 **a**, **b**, **c**, **d** Vector A–B aligned vertically and its equivalent representation along rasters with spatial resolutions *dx*, *dx*/3, and *dx*/9; **e**, **f**, **g**, **h** Vector C–D aligned at 45° to the horizontal in the eastward direction and its corresponding representation in the three rasters; **i**, **j**, **k**, **l** Vector E–F aligned at 21.8° to the vertical in the southward direction and its corresponding representation in the three rasters (in the rasters, the pixels that make up the path representation are shaded in gray and connected by a red line, with centroids indicated by black dots)

the distance traveled in the raster is 5.828 m. This represents an overestimation error of 8.2%, relative to the actual length measured in vector E–F.

In Fig. 14.4k, vector E–F was represented in a raster with spatial resolution three times more refined than the previous raster. Now, the path corresponding to vector E–F is represented by nine orthogonal and six diagonal steps. At first glance, the path in the second raster appears to align more closely with the vector than in the case of the first raster. However, the distance traveled along the second raster is 5.828 m, which is precisely the value obtained for the first raster. That is, the same overestimation error (8.2%) remains. The same error occurs for the third raster, from Fig. 14.4l, whose spatial resolution is 9 times more refined than the first raster and whose path was obtained with 27 orthogonal and 18 diagonal steps. It is evident, therefore, that this error is independent of spatial resolution.

In summary, for paths vertically, horizontally, or diagonally at an angle of 45% and multiples, the representation in a raster is exact, with zero error regarding distance traveled. For paths with an alignment different from these directions, there is an intrinsic overestimation error due to the inability to walk from centroid to centroid at angles other than 90° or 45° and multiples. The maximum overestimation error occurs in the walk at half the angle of 45° (22.5°) or angles added 90° to this value.

Researchers proposed what is called distance transform operators to minimize this problem, adapting the Euclidean way of computing the orthogonal and diagonal distances between each pixel (Butt & Maragos, 1998; De Smith, 2004).

Instead of considering the distance between two neighboring pixels side by side as equal to one dx, one computes a value equal to $OD_{ortho}.dx$, OD_{ortho} being the orthogonal distance operator with a value less than one. For a diagonal step, instead of considering $\sqrt{2} \cdot dx$, which is equal to $1.414.dx$, one computes $OD_{diag}.dx$, where OD_{diag} is the diagonal distance operator, with a value less than 1.414. For example, Butt and Maragos (1998) propose the values of 0.96194 and 1.36039 for OD_{ortho} and OD_{diag}, respectively.

When computing distances using OD_{ortho} and OD_{diag} with values lower than the standard values of Euclidean distances, an intentional underestimation error for orthogonal or diagonal distances at multiple angles of 45° is inserted. However, structures like river reaches do not follow a single direction. They present a significant random alternation of directions and alignments. Therefore, the idea is to introduce an underestimation in the computation of incremental distances between pixels to compensate for the intrinsic effect of overestimating distance in a raster.

In the example of Fig. 14.4, the consideration of the distance transforms of Butt and Maragos (1998) leads to computing a distance of 1.924 m in the raster representation of vector A–B, leading to an underestimation error of 3.8% relative to the vector. For vector C–D, the computed distance in the raster would be 2.721 m, with an underestimation error of 3.8%. These errors are the same for these two vectors, regardless of the spatial resolution. For vector E–F, the distance measured in the raster is 5.607 m when considering the values of OD_{ortho} and OD_{diag} from Butt and Maragos (1998). In this case, the overestimation error of 8.2% was reduced to 4.1% compared to the standard way of calculating Euclidean distances.

As rivers alternate directions between various angles along their course, represented in a raster, adopting distance transform operators leads to the compensation of under and overestimation errors, tending to reduce the total error at the end. In other words, using these operators does not eliminate the intrinsic error of computing distances along a raster. However, it reduces the errors when computing distances for structures that alternate multiple flow directions. This was verified with real-world river reaches in the Uruguay River basin, in the south of Brazil, in the work of Paz et al. (2008).

14.6 How to Evaluate the Quality of a DEM-Derived River Drainage Network?

Several DEM-derived products were described in the previous chapters, particularly emphasizing the river drainage network. Many factors influence the quality of this product, including the quality of the DEM elevations, horizontal and vertical accuracy, and noise. However, the quality of the river drainage network is also affected by the physical characteristics of the study area, the scale of analysis and spatial resolution used, and the influence of the algorithms applied in DEM processing.

Several situations require adequately assessing the quality of the river drainage network derived from a DEM. For example, to analyze the effect of the DEM's spatial resolution, to evaluate new DEM data sources, to assess improvements in existing DEMs, or to evaluate new DEM processing algorithms (Sousa & Paz, 2017).

But how should the quality of a river drainage network be evaluated? There is a prior question: What is meant by the quality of a river drainage network? There is no definitive concept for this, but the river network quality is generally associated with the spatial layout and positioning of river reaches that most closely match those of a reference network. For this reason, various approaches and metrics have been proposed in the literature to quantify the similarity and pattern reproduction of a drainage network relative to a reference network.

As it is expected that a DEM-derived river network follows the layout and positioning of a reference network, it is understandable to expect that direct characteristics of the network, such as lengths and sinuosities of river reaches, also agree with those from the reference network. Thus, the analysis of the characteristics of river reaches has also been used to evaluate the quality of DEM-derived river drainage networks. Similarly, indirect attributes of the network, such as shape and amount of contributing drainage area, are also expected to meet those of the reference network. For this reason, this type of comparison is also used to assess the quality of the river drainage network.

Each approach to evaluating the quality of a river drainage network presents advantages and disadvantages and focuses on one or more specific aspects of judging this quality. Some evaluation methods specifically judge the network's layout and spatial positioning, others judge direct characteristics like river lengths, while another set evaluates indirect characteristics, like the drainage basin size.

What is the intended purpose of the drainage network being evaluated? Will its spatial layout and alignment be essential to its use? Or will only numerical attributes—such as river lengths and drainage areas—be applied in subsequent analyses? These are crucial questions for the user to consider when choosing an appropriate approach for evaluating the river drainage network, as the evaluation method should align with the intended application. Multiple analytical approaches should be combined to achieve a more comprehensive assessment of river drainage network quality.

14.7 What is a Reference River Drainage Network?

A key issue for evaluating DEM-derived river drainage networks is the availability of a reference network and whether this network is suitable to be considered as a reference. However, what is necessary for this suitability, i.e., for a drainage network to be considered a reference for evaluating other networks?

A network taken as a reference does not necessarily represent the layout and spatial positioning in a definitively accurate way concerning the real world. Obtaining a drainage network through any method is subject to imprecision and errors.

Common ways of obtaining a reference network are via a field survey based on topographic instruments, an aerial photogrammetry survey, and manual digitization of satellite images. In all these procedures, there are different sources of error, such as the georeferencing of data and images, difficulty in interpreting the layout of rivers, and imprecision in the manual procedure of digitization, among others.

Moreover, a central issue in determining whether a river drainage network is suitable to serve as a reference for evaluating other networks lies in the purpose of the evaluation itself. What is the goal of such an assessment? For instance, when analyzing the effect of DEM spatial resolution on the drainage network, the network derived from the DEM at the finest resolution may be considered a valid reference. In this case, the objective is to evaluate how well the drainage networks extracted from coarser-resolution DEMs preserve the spatial patterns of the network obtained at the original resolution.

If the purpose is to evaluate the quality of a drainage network derived from a new DEM data source, the interest is to judge how well this network reproduces the real world. In this case, the reference network must have layout, positioning, and characteristics that agree with the real world and are of sufficient quality to serve as a benchmark for judging the other network.

In some cases, the absence of a reference network with an acceptable quality level to serve this purpose makes the comparative analyses between networks solely a relative evaluation, without the association of error. Only the disagreement or divergence between layouts and characteristics is discussed, but there are no elements to infer which river network is wrong.

14.8 Approaches to Evaluating the Quality of the River Drainage Network

The following sub-items describe different approaches used in the scientific literature to assess the quality of a DEM-derived river drainage network.

14.8.1 Visual Inspection of River Drainage Network Patterns

Visual inspection is one of the most used approaches to evaluate the quality of a river drainage network. In this approach, the network layout is visually analyzed in terms of its agreement with the layout of another network, which is taken as a qualitative reference. The network under analysis is overlaid on the reference network, and the observer visually traverses the networks, inferring the spatial correspondence and topological consistency between them.

Following this procedure, there is no absolute judgment of the quality of the network, as the evaluation is merely qualitative and contains subjectivity. In general, a relative inference is made, comparing the ability of the network under analysis to follow the reference network's spatial patterns relative to other networks being analyzed. This analysis provides answers of this type: network A is better at following the layout of the reference network than network B. Identifying critical reaches with gross errors of one network relative to another is also possible. For example, an inference of the type: in the lower part of the basin, the network under analysis showed higher difficulty in following the layout of the reference network than in the upper part.

As a real-world example, the visual inspection of Fig. 14.5 allows one to infer the agreement of two river drainage networks relative to a reference network. The networks derived from the SRTM 30 DEM and the SRTM 90 DEM are under analysis, with the network obtained by manual digitization of Landsat images taken as a reference. By visual inspection, it is perceived that the network originating from the SRTM 30 DEM follows the layout of the reference network more closely than the network generated from the SRTM 90 data.

In this example, areas 1 and 2 show sections with higher network divergence originating from the SRTM 90 DEM relative to the reference network due to the higher distance between them. In this type of figure, it is essential to insert the

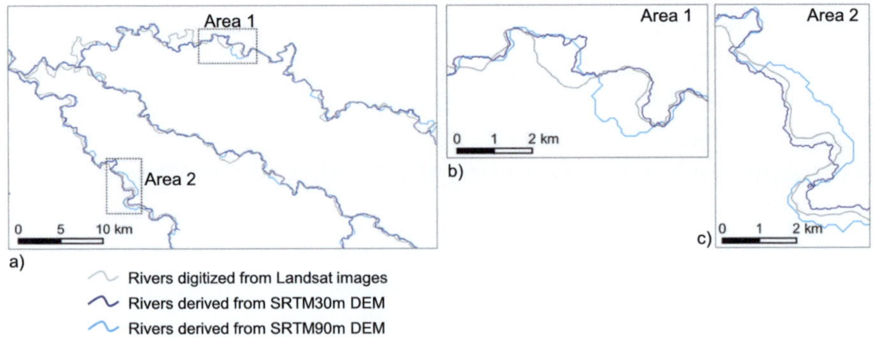

Fig. 14.5 **a** Overlaying river traces from part of the Icamaquã river basin (Rio Grande do Sul state) obtained by different methods, for visual inspection purposes; **b**, **c** Detailing of areas 1 and 2, respectively, indicated in **a**

graphic scale so that the visual inspection has a distance reference to improve the judgment of the layouts of the river drainage networks.

In the visual inspection approach, the ability of the human eye and brain to identify and judge complex spatial patterns that are difficult to measure by other methods is utilized (Saraiva & Paz, 2014). In particular, this type of analysis is very suitable for judging drainage networks obtained at more coarsened spatial resolutions, concerning the maintenance of the spatial drainage patterns present in the more refined drainage network, taken as a reference (Arora & Harrison, 2007; Wu et al., 2011).

The main disadvantage of this type of analysis is the intrinsic subjectivity and the risk of bias, as different observers may arrive at distinct results. In addition, there are two apparent difficulties. The first is to perceive more subtle differences between networks under evaluation, and the second is the chance of incoherence when repeating the analysis procedure (Gatziolis & Fried, 2004). When repeating the visual inspection after a time interval, the same observer may have a different impression from the previous analysis. When there are more discrepant differences between the networks, as in areas 1 and 2 of Fig. 14.5, there is higher convergence of interpretation among different observers and for the same observer repeating the procedure at various times.

14.8.2 Characteristics of the River Drainage Network as a Quality Evaluation Approach

The most common method found in the literature to evaluate a river drainage network is to quantify and analyze the direct and indirect characteristics of the network. Direct river reach characteristics, such as length and sinuosity, are evaluated. The drainage basin's area and perimeter at different network points are assessed as indirect characteristics in addition to geomorphological indices such as shape and compactness coefficients. Other characteristics of rivers and basins, such as the number of headwaters, confluences, drainage density, and stream order, have also been used for this purpose.

The main advantages of this type of analysis lie in the ease of determining such characteristics and the availability of reference values from other studies. Comparing the values obtained for a river drainage network under evaluation with reference values is straightforward and easily interpreted in terms of absolute and relative differences. Moreover, key attributes such as river length and watershed area are highly relevant for many hydrological and environmental studies. This relevance motivates the quantification and analysis of how the delineation of a drainage network impacts these characteristics.

The main limitation of assessing river drainage network quality based solely on river and watershed attributes is that, fundamentally, this type of analysis does not evaluate the quality of the network itself. Two drainage networks may exhibit equal main river length and watershed area, while their spatial layouts differ significantly. In other words, agreement in these attribute values between the evaluated and reference networks does not imply spatial agreement in their configurations.

Another drawback is the potential for error compensation. For instance, suppose the evaluated network underestimates river length in the watershed's upper reaches due to poor representation of smaller meanders. Meanwhile, the network may overestimate the river length in the downstream portion by zigzagging across the floodplain or wider river reaches. These under and overestimations can coincidentally balance each other when computing the total river length, resulting in a value that appears to match the reference closely, but does not reflect the actual spatial accuracy of the network.

14.8.3 *Commission and Omission Errors in the Watershed Delineation*

The processing of a DEM to obtain the river drainage network is almost invariably associated with delineating the watershed corresponding to this network. Identifying which areas have runoff that contributes to the river drainage network is essential for most studies using this data layer.

Suppose an available basin delineation can be considered as a reference. In that case, the DEM-based delineation can be evaluated by overlaying it with the available delineation and quantifying their agreement and disagreement. Two types of errors can be quantified (Fig. 14.6): the areas erroneously included in the delineation under analysis (commission error) and the areas erroneously not included (omission error).

The omission error refers to the relationship between the total areas that failed to be included in the delineation and the total area of the reference basin. The commission error corresponds to the relationship between the amount of area erroneously included in the delineation and the total area of the reference basin. The intersection between the two watershed delineations represents a match, which can be quantified regarding

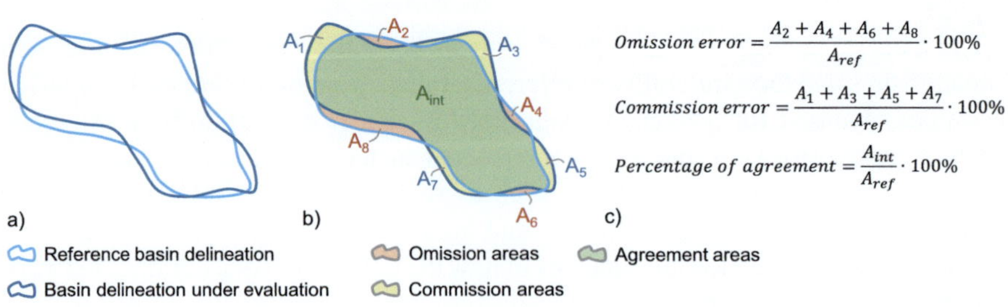

Fig. 14.6 Comparison between the delineation of the basin under evaluation and the reference delineation: **a** Overlap between the delineations; **b** Omission, commission and agreement areas; **c** Calculation of omission error, commission error, and percentage of agreement, where A_{int} is the intersection area and A_{ref} is the total area of the reference basin

the percentage of agreement between the delineations. This metric refers to the ratio between the intersection area and the total area of the reference watershed.

Quantifying omission and commission errors and the percentage of agreement between two watershed delineations in vector format is straightforward, leveraging standard vector manipulation operations available in most GIS software. This process typically involves polygon intersection and union operations and calculating polygon area as a geometric attribute.

Alternatively, this comparative analysis procedure between basin delineations can be easily performed considering these delineations in raster format. A practical and straightforward way of doing this is as follows (Fig. 14.7):

(a) Reclassify the raster basin delineations: the raster of the basin delineation under analysis is reclassified to have attributes 2 and 0 for the pixels belonging or not, respectively, to the basin; the raster of the reference delineation is similarly reclassified to have attributes 1 and 0 for the pixels belonging to the basin or not, respectively;
(b) Perform the sum of the two raster delineations, with the following interpretation: the resulting raster has pixels with values 0, 1, 2, or 3; the value 0 corresponds to pixels not belonging to either of the two delineations; the value 1 refers to pixels that are part of the reference basin but not the basin under analysis; the value 2

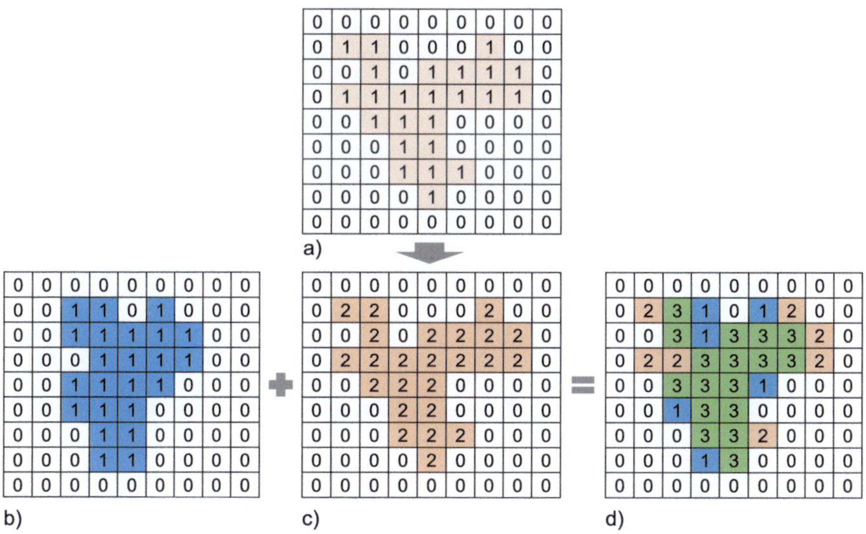

Fig. 14.7 Evaluation scheme of a DEM-derived basin delineation, in raster format: **a** Basin delineation under analysis, obtained from the DEM; **b** Reference basin delineation; **c** Reclassification of the delineation under analysis; **d** Sum of the rasters of the two delineations, resulting in values 0 (pixel not belonging to either basins), 1 (pixel is part of the reference basin but not the basin under analysis), 2 (pixel does not belong to the reference basin but is part of the basin under evaluation); 3 (pixel integrates both basins)

corresponds to pixels that are not part of the reference basin but were included in the basin under evaluation; finally, the value 3 refers to pixels belonging to both delineations;

(c) Quantify the area occupied by each distinct attribute (0, 1, 2, and 3) in the resulting raster;
(d) Quantify the area of the reference basin;
(e) Calculate the metrics of commission error, omission error, and agreement percentage: the commission error is the area occupied by pixels of value 2 in the sum raster, divided by the area of the reference basin and multiplied by 100%; the omission error is the area occupied by pixels of value 1 in the sum raster, divided by the area of the reference basin and multiplied by 100%; the agreement percentage is the ratio between the area occupied by pixels of value 3 in the sum raster and the area of the reference basin, multiplied by 100%.

The example of Fig. 14.7 illustrates the calculation of commission and omission errors and the agreement percentage. In this example, the commission error is 7/23 or 30.4%, while the omission error is 6/23 or 26.1%. The agreement percentage is 73.9% (17/23). It is important to note that omission and commission errors and the percentage of agreement do not add up to 100%, since the reference area used in these calculations is that of the reference watershed. In contrast, the evaluated watershed occupies a different location. In other words, because the total area used as the baseline is fixed (that of the reference), and the evaluated delineation may include overlapping and non-overlapping regions, the combined percentages of agreement, omission, and commission do not necessarily form a complete partition of the evaluated area.

Omission errors have a minimum value of 0%, which occurs when no part of the reference watershed is missing from the evaluated delineation (Fig. 14.8a). The maximum omission error is 100% when the entire reference watershed lies outside the evaluated delineation (Fig. 14.8b).

Fig. 14.8 Extreme cases in the comparison between two watershed delineations: **a** The evaluated delineation entirely contains the reference delineation (omission error = 0%, commission error > 0, agreement percentage = 100%); **b** No intersection between the evaluated and reference delineations (omission error = 100%, commission error > 0, agreement percentage = 0%); **c** Evaluated delineation is entirely contained within the reference delineation (commission error = 0%, omission error < 100%, agreement percentage < 100%)

Commission errors also have a minimum value of 0%, representing the case where no portion of the evaluated watershed lies outside the reference watershed (Fig. 14.8c). However, commission errors do not have a defined upper limit—it can grow indefinitely depending on how much area from the evaluated watershed lies outside the reference watershed (i.e., its conceptual maximum is ∞).

The agreement percentage ranges from 0 to 100% and should be interpreted in the opposite sense of omission and commission errors. For the errors, lower values indicate better delineation accuracy. Higher values indicate better agreement between the evaluated and reference delineations for the agreement percentage.

A minimum agreement value of 0% occurs when the evaluated watershed does not intersect the reference watershed at all (Fig. 14.8b). In such cases, omission error reaches 100%, while commission error depends on the size of the evaluated watershed.

A maximum agreement value of 100% occurs in the opposite case, when the evaluated delineation is entirely contained within the reference watershed (Fig. 14.8c). In this situation, the commission error is 0%, while the omission error depends on how much of the reference watershed was excluded from the evaluated delineation.

Whether as an operation between rasters or vectors, comparing basin delineations is a straightforward determination procedure. Its application quantitatively evaluates how well the basin delineation under analysis reproduces the shape of the reference delineation. It is an approach for analyzing the spatial positioning of the delineation without having a compensation effect, as occurs with the numerical comparison of the total drainage area.

However, the analysis of spatial agreement between delineations by commission and omission errors does not directly infer the quality of the river drainage network itself. Another disadvantage of this analysis is the dependence on the availability of a basin delineation that can be taken as a reference.

Attention should be paid to consider which basin delineation could effectively serve this purpose. For example, basin delineations provided by official databases do not necessarily meet the requirements. They may have been obtained using DEM data sources with more coarsened spatial resolution or less precision than the basin under evaluation, or using flow direction algorithms identified in the literature as of questionable performance.

14.8.4 Mean Distance Metrics Applied to Watershed Delineation

The average distance (d_{mean}) between watershed boundaries is a metric that quantitatively evaluates the spatial displacement between the watershed delineation under analysis and the reference one. It is a typical evaluation approach used when working with watershed boundaries in vector format.

Overlaying the two delineations in vector format, they generally intersect at various points, forming polygons between these points (Fig. 14.9a). At the point of intersection, the distance between the two boundaries is zero. In the sections where polygons are formed, there are variable distances at each point between the two boundaries (Fig. 14.9b). Conceptually, the average of these distances is the metric d_{mean}.

The metric d_{mean} ranges from 0, when the two delineations are completely coincident, to infinity, increasing as much as the distance between the two boundaries increases. The interpretation of this metric is straightforward: the larger the d_{mean}, the higher the spatial discrepancy between the two delineations, and the metric value quantitatively indicates the intensity of this discrepancy. The usual unit of measurement is meters.

A simple and straightforward approach to determine the metric d_{mean} employing standard GIS operations is to quantify the area of each polygon formed by the intersection of the watershed boundaries (Fig. 14.9c). Then, these polygons' areas are added up, and this sum is divided by the perimeter of the reference watershed (Fig. 14.9d).

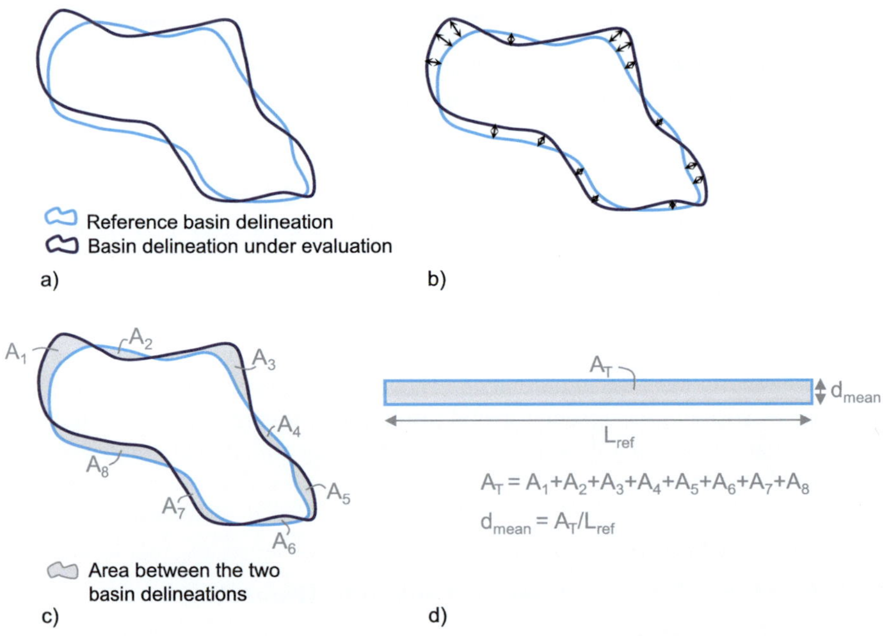

Fig. 14.9 **a** Comparison between two watershed delineations for determining the average distance between them; **b** Indication of the distance between watershed boundaries at some points; **c** Formation of polygons between the intersections of the delineations (eight polygons) and quantification of the areas of these polygons (A_i, $i = 1$ to 8); **d** Determination of the average distance (d_{mean}) between the watershed boundaries by the relationship between the sum of the areas of the polygons (A_T) and the perimeter of the reference watershed (L_{ref})

This procedure is equivalent to transforming these polygons into an equivalent rectangle. The average distance corresponds to the width of a rectangle whose length is the perimeter of the reference watershed and whose area is equal to the sum of the areas of the polygons between the intersections of the delineations in comparison. If the area of the polygons is obtained in square meters and the perimeter of the reference watershed in meters, then d_{mean} is expressed in meters.

The evaluation of the quality of a watershed delineation by the metric d_{mean} measures the spatial discrepancy of the watershed boundary relative to the reference delineation. This metric considers the spatial positioning of the watershed delineation, the shape, and, indirectly, the area. As the metric summarizes an average distance between the boundaries, it becomes essential to overlay them and visually interpret their comparison. This allows the spatial perception of the most significant regions of discrepancy or agreement. A disadvantage of this metric is that it requires the availability of a watershed delineation to serve as a reference, which does not always exist, as pointed out in the previous sub-item of this chapter.

14.8.5 Mean Distance Metrics Applied to the River Drainage Network

The average distance between river drainage networks is a quantitative evaluation of the spatial displacement of a river drainage network under evaluation relative to the trace of a drainage network taken as a reference (Fig. 14.10a).

This metric is conceptually and procedurally analogous to the mean distance between watershed boundaries. Conceptually, it represents the average point-by-point distances between the two river network traces (Fig. 14.10b). In practice, the simplest way to compute this metric is by delineating and measuring the areas of the polygons formed between the network traces, and then dividing the sum of these areas by the total length of the evaluated river network trace. However, two additional steps are required compared to the procedure used for calculating the mean distance between watershed boundaries: isolating segments and closing the endpoints.

The first additional step is isolating the trace for a specific river reach of the network one wants to evaluate, and isolating the corresponding reach of the reference network. The DEM-derived river network comprises the main river and its tributaries, with a ramification of river reaches. However, the computation of distances between corresponding reaches of the reference network and the network under analysis is almost unfeasible when applied between branched networks (Fig. 14.10a).

Multiple river branches create ambiguity when determining which points or segments the distances should be measured. Suppose the goal is to obtain more spatially distributed results. In that case, a river reach can be sub-divided into sub-reaches, and the mean distance for each sub-reach can be calculated individually. The second additional step is to close the endpoints between the two network traces

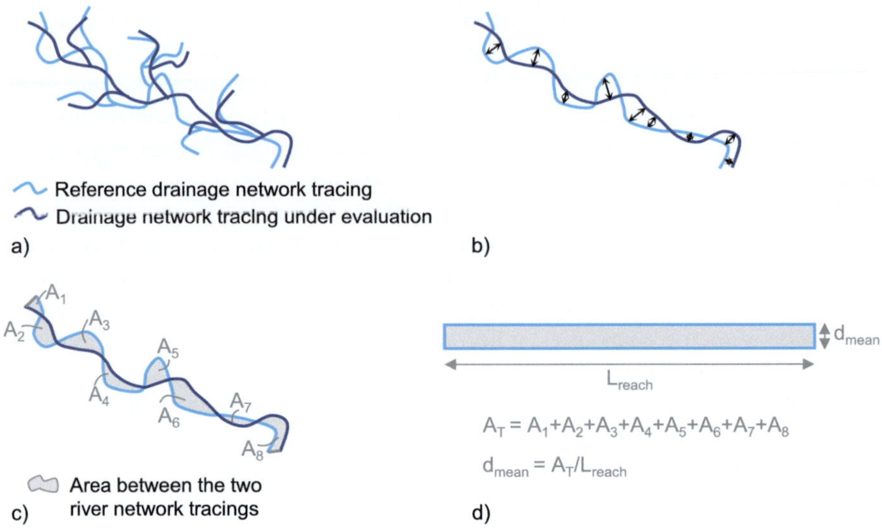

Fig. 14.10 **a** Comparison between two drainage network traces to determine the average distance between them; **b** Simplification of the network and isolation of the section for evaluation, with indication of the distance between the traces at some points; **c** Closure of the ends of the networks, formation of polygons between the crossings of the traces (eight polygons) and quantification of the areas of these polygons (A_i, $i = 1$ to 8); **d** Determination of the average distance (d_{mean}) between the traces by the relationship between the sum of the areas of the polygons (A_T) and the length of the trace of the reference network (L_{ref})

with a straight line segment, allowing for the formation of polygons at the network extremities (Fig. 14.10c).

The complete step-by-step guide for obtaining the average distance between two drainage networks is as follows:

(i) Edit the two vector networks that you want to compare to leave them with only the trace of the section selected for analysis (Fig. 14.10b);
(ii) Closure of the ends between these two sections (Fig. 14.10c);
(iii) Quantification of the areas of all the polygons formed (Fig. 14.10c);
(iv) Sum of the areas of the polygons (Fig. 14.10d);
(v) Division of the sum of the areas of the polygons by the length of the trace of the reference river (Fig. 14.10d).
(vi) Repeat this set of previous steps for each section of interest, performing the analysis for each section individually.

If the areas of the polygons between the traces are quantified in square meters and the length of the trace of the reference river in meters, the average distance is obtained in meters. The larger the d_{mean}, the higher the spatial distance between the two traces. The minimum value of d_{mean} is zero when the two traces coincide and can increase indefinitely.

This metric assumes positional correctness in the delineation of the reference network and considers any deviation from that delineation as an error in the evaluated network. This can be seen as a limitation of the d_{mean} metric. Considering that the river reach may be wider than the pixel size of the DEM used to derive the drainage network, it may be overly strict to assume that a correct delineation only occurs when it coincides with the reference network. Any delineation within the riverbanks would be conceptually acceptable, especially given that DEMs lack bathymetric information, i.e., they do not contain topographic data for the riverbed below the water surface.

To illustrate, Fig. 14.11a shows a section of the lower course of the Uruguay River in southern Brazil, where the river's width ranges from approximately 1 to 3 km. A centrally aligned channel trace was manually digitized over a color composite satellite image (Fig. 14.11b). Any DEM-derived delineation falling within the area between the riverbanks should be considered entirely correct (Fig. 14.11c).

Suppose the trace presented in Fig. 14.11d is compared to the reference trace obtained from the Landsat image. When computing the average distance between these traces, the result is $d_{mean} = 431$ m. There is a spatial displacement between the two traces, and its average value is 431 m. However, the DEM-derived river trace is entirely contained in the area between the river banks, highlighting how the isolated interpretation of the metric d_{mean} can be problematic.

The limitation of this metric regarding the assumption of positional correctness in the reference network should not be forgotten, as there are errors associated with the georeferencing of the Landsat image and the manual digitization of the river trace. In this example, there are two points where the river bifurcates and goes around an internal island (points A and B in Fig. 14.11b). In manual digitization, the corresponding network arbitrarily followed the trace on one side of the island at each

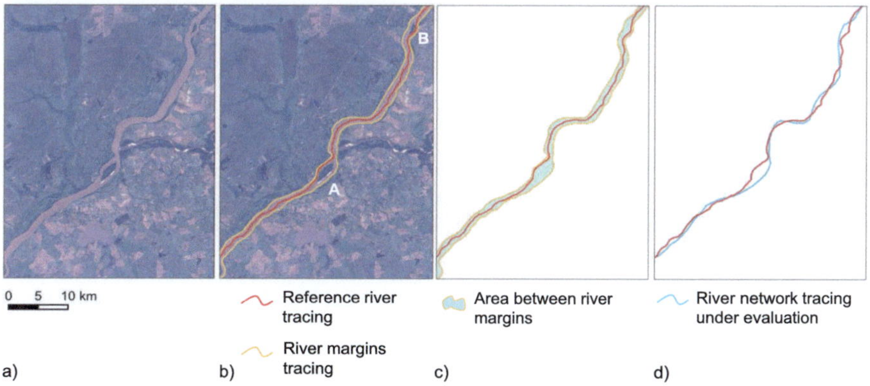

Fig. 14.11 **a** RGB colored composition of Landsat satellite image of a reach of the lower course of the Uruguay River (border between Brazil and Argentina); **b** River drainage network of the Uruguay River and the river banks obtained by manual digitization over the satellite image; **c** Identification of the area between the river banks; **d** DEM-derived river drainage network for comparative evaluation against the trace obtained from the satellite image

point. However, the network obtained from the DEM followed the opposite side of the island at both bifurcations, which tends to increase the value of the metric d_{mean} without necessarily indicating an error in the DEM-derived river flow path.

14.8.6 Percentage Within Buffer

One way to evaluate the spatial positioning of a DEM-derived river drainage network relative to that of a drainage network taken as a reference is by determining the percentage within buffer (PWB). This metric quantifies the fraction of the network's trace evaluated within a specific marginal area of the reference trace, which is the buffer.

This metric aims to evaluate the discrepancy of a network's layout regarding spatially lying within a specific tolerance range around the reference network. It is assumed that the reference network has some inaccuracy in its layout and that it is acceptable for any other network to follow it at a distance up to a specific limit, which is the width of the buffer.

The application of this metric is most recommended to be performed with vectors of the networks under comparison, using standard vector operations usually available in GIS software. However, like the average distance, determining the percentage within the buffer is problematic or even unfeasible when done with branched reaches of the river drainage network. It is recommended that each reach of interest be analyzed and isolated to quantify the corresponding PWB.

The following sequence of operations is indicated to obtain the PWB metric between two river drainage networks taken in vector format:

i. The first step is to select the river reach to be evaluated and exclude the other reaches, isolating the selected reach from both networks to be compared (Fig. 14.12a and b).
ii. A buffer is applied around the reach of the reference river network (Fig. 14.12c), defining the marginal area considered acceptable for the network under evaluation.
iii. A vector intersection operation determines the reference river network sections inside and outside the buffer (Fig. 14.12d).
iv. The individual extension of the sections within the buffer is quantified;
v. The lengths of these sections within the buffer are added up;
vi. This sum is divided by the length of the river reach under analysis and multiplied by 100%. This result is the PWB metric, meaning the percentage fraction of the river reach under evaluation located within the buffer area.
vii. Repeat the previous steps for each river selected for individual analysis.

A key issue when applying the PWB metric is the definition of the buffer width. One option for this definition is to adopt the buffer width as equal to the river width (Collischonn et al., 2010) if this information is available or can be estimated from satellite images, for example. A difficulty with this is the non-uniformity of the

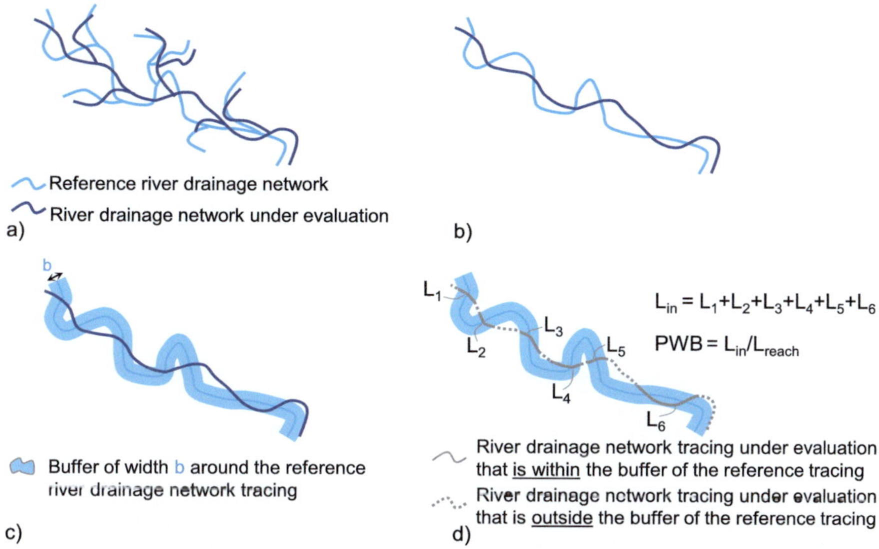

Fig. 14.12 **a** Comparison between two river drainage network layouts for determining the percentage within buffer metric (PWB); **b** Simplification of the network and isolation of the reach for evaluation; **c** Definition of width *b* and formation of the buffer around the reference river drainage network; **d** Determination of the sections of the network under evaluation that are inserted or not in the buffer of the reference layout, and obtaining the PWB metric, where $L_{section}$ is the length of the reference river reach

river width. To consider the variable width of the river along its length, the analysis procedure could separate smaller river reaches and adopt different buffer widths for each one according to the specific average width of each reach.

Another way to define the buffer width in calculating the PWB metric is to consider the limitation that a DEM-derived river drainage network has to follow a path along the centroids of the pixels, forcibly. The larger the pixel size (i.e., the coarser the spatial resolution) of the DEM, the higher the spatial limitation for tracing the river network. Therefore, some authors define the buffer width based on the pixel dimension, taking precisely the size of one pixel, two pixels, or even more (Sousa & Paz, 2017).

The smaller the buffer width used to obtain the PWB, the more rigorous the evaluation is considered when evaluating a river network. In contrast, the larger the buffer width, the higher the tolerance for the non-exact agreement between the layout of the network under evaluation and that of the reference network.

When you have a set of networks for comparative evaluation relative to the same reference, adopting a relatively large buffer width is not recommended. This can result in all networks under evaluation being almost entirely contained within the buffer, reducing the metric's usefulness in distinguishing the different networks' quality.

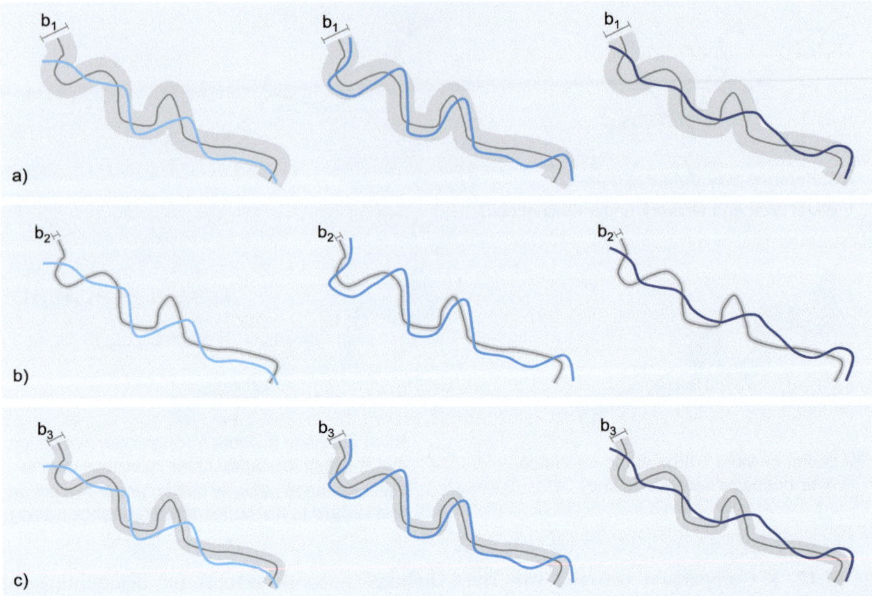

Fig. 14.13 Comparison between the layouts of three river drainage networks (shades of blue) relative to a reference network (in dark gray) by the percentage within the buffer approach (gray area around the reference network), considering buffer widths b_1 **a**, b_2 **b**, and b_3 **c**

For example, in the comparison between the three networks in Fig. 14.13a with the corresponding reference network, a buffer of width b_1 results in almost 100% of the reaches of the three networks within the buffer. By this metric, all three networks have nearly the same layout quality.

Similarly, a too-narrow buffer width may represent an exaggerated rigor of requirement for the accuracy of the network layouts. This leads to so little agreement measured by the metric that it cannot make any helpful network quality distinctions. This situation is exemplified by adopting width b_2 in Fig. 14.13b. In this case, the value obtained for the metric would be very low for the three networks, which would be interpreted as a sign of equally low quality for them.

A buffer width value with an intermediate effect should be sought without being too strict or too tolerant. In the example, a buffer of width b_3 (Fig. 14.13c) would be recommended to increase the chance of differentiating the quality between the three networks.

For a deeper discussion on the definition of buffer width in applying the PWB metric, it is recommended to read the work of Sousa and Paz (2017).

14.9 Impacts of River Drainage Network Errors on Hydrological Studies

Errors in the DEM-derived river drainage network and its direct and indirect attributes—such as river lengths or drainage areas—propagate through the processing of these data for various types of studies, whether in hydrology, ecology, environmental management, or any other field. In short, all studies and activities that rely on the spatial information of the river drainage network can be affected by errors in the location and layout of the network. For example, this includes ecological studies on species occurrence and distribution based on proximity to watercourses, or studies aimed at selecting suitable sites for landfills, which must meet minimum distance requirements from rivers.

In hydrology, the river drainage network is one of the fundamental inputs for applying hydrological models. Errors in spatially representing the river drainage networks and their attributes propagate through the modeling structure. The propagation of errors can hinder, for example, the representation of the flow regime in rivers, the simulation of the flood wave routing, and the reproduction of flooded areas. As a result, there is a predisposition to compensate for these errors' effects by calibrating the hydrological model's parameters. One can force parameter values that counterbalance errors in the river drainage's representation within the model, helping the model reproduce expected hydrological patterns. This compensation through parameter adjustment is undesirable, according to best practices recommended for hydrological modeling (Semenova & Beven, 2015). As a result, there is a tendency to overemphasize parameter fitting and to prioritize achieving model accuracy at any cost, rather than striving for model accuracy for the right reasons (Kirchner, 2006).

The following sub-items illustrate more concretely the effects of river drainage network errors on hydrological studies, focusing on the issue of errors in the layout and spatial positioning of the network, errors in river lengths, and errors in the drainage basin.

14.9.1 Errors in the Layout and Spatial Positioning of the River Drainage Network

As already seen in this chapter, several factors lead to errors in the layout and spatial positioning of the DEM-derived river drainage network relative to the actual layout of rivers. As examples of this type of error, river reaches may be spatially displaced or present a drainage network with more or fewer meanders or meanders of different geometry. In addition, confluences may be positioned spatially displaced.

One of the impacts of the error in the layout and spatial positioning of a drainage network is on the estimate of flooded areas resulting from extreme events, with the application of hydrological models. For example, consider the drainage network formed by rivers A, B, and C, as shown in Fig. 14.14a. Suppose that, in this region,

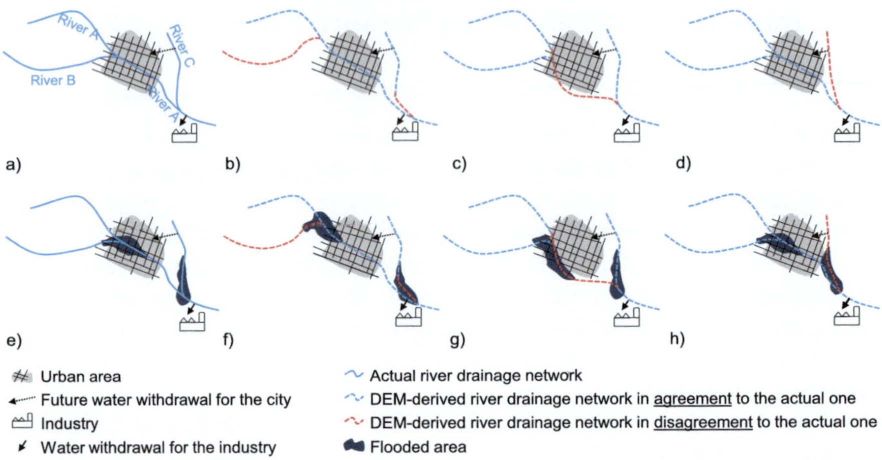

Fig. 14.14 **a**, **e** Actual layout of the river drainage network and flood extension occurred for an extreme event; **b**, **f** DEM-derived river drainage network with error in the spatial positioning of the two confluences and corresponding flood extent simulated with a hydrodynamic model; **c**, **g** DEM-derived river drainage network with error in the layout of river A between the two confluences and the corresponding flood extent simulated with a hydrodynamic model; **d**, **h** DEM-derived river drainage network with the river C flow trace spatially wrong.

there is an urban area cut by river reaches, as indicated, and that there are prospective studies to expand the water supply for this urban area from a catchment of river C. There is also an industry downstream of the confluence between rivers A and C, which withdraws water at the indicated point.

Figure 14.14b–d show drainage networks derived from different DEM processing methods. Each of these networks contains segments that match and segments that diverge from the actual network. As a result, the flood extents obtained from the hydrodynamic simulation of an extreme rainfall event, when using each of these networks, differ from the simulation that uses the actual drainage network (Fig. 14.14e).

In the case of the drainage network in Fig. 14.14b, both confluences are spatially displaced relative to the actual network layout. The confluence between rivers A and B occurs upstream of the location where it should appear. As a result, the river overflow points and the corresponding flood extent are also shifted upstream (Fig. 14.14f), assuming that the contribution of river B to river A is a key factor in causing the overflow of river A in this reach. With the flood area shifted upstream in the hydrodynamic simulation results, the flood-related issue in the urban area appears less severe. If this study were used to support decision-making, it would lead to planned actions based on incorrect information, ultimately underestimating the actual flood risk in that area.

In the simulation using actual data, another flood extent is associated with the overflow of river C, upstream of its confluence with river A. This could be caused by the backwater effect from river A, which is a larger river, blocking or slowing the flow of river C. However, when using the network from Fig. 14.14b in the hydrodynamic model simulation, the spatial error in positioning the confluence between rivers A and C causes the flood area to shift downstream. As a result, the industry in that region appears to be in an area with a much higher flood risk than it is, due to the errors in river tracing and spatial positioning.

In the case of the drainage network shown in Fig. 14.14c, the error lies in the course of river A between the confluences with rivers B and C, while the locations of those confluences are spatially well-represented. In the DEM-derived network, river A runs along the edge of the urban area, which leads the hydrodynamic simulation of the extreme event to show a reduced urban area subject to flooding when compared to the simulation using the actual network. In contrast, the error in the course of river C in the drainage network from Fig. 14.14d results in simulated flood risks being falsely increased in the southeastern end of the urban area.

Another example of the impact of errors in the drainage network's layout and positioning is in water availability studies. In the actual drainage network illustrated in Fig. 14.14a, the industry withdraws water from river A, in a section downstream of the confluence with river C. In this section of the river where the intake is, the water availability is influenced by the flow of rivers A and C upstream of the confluence.

However, in the DEM-derived river drainage network illustrated in Fig. 14.14b, there is a spatial misalignment in the confluence of rivers A and C, resulting in the industry's water intake point on river A being located downstream of the confluence. Suppose a water availability study and a comparison between water supply and demand in this basin are carried out using the drainage network shown in Fig. 14.14b. In that case, the estimated water availability at the industry's intake point will be lower than the actual value, due to the error of not accounting for the contribution of river C. Consequently, decision-making processes such as granting water use permits in this section may be based on an underestimated water availability, potentially leading to a false scenario of water scarcity.

A third example of the impact of drainage network errors is in assessing water sources for increasing water supply. In addition to the quantitative effect on contributing flows, as previously mentioned, there is also an impact on the preliminary feasibility analysis regarding the routing alternatives for water conveyance systems. In the example shown in Fig. 14.14a, river C is a certain distance from the urban area, with a corresponding cost associated with conveying water for consumption. Due to the river misalignment presented in Fig. 14.14d, the conveyance distance and related cost are falsely reduced, which could lead to misguided decision-making and a waste of time, resources, and energy until the error is detected, typically during the detailed planning of the conveyance system based on more accurate data.

14.9.2 Errors in the Definition of River Drainage Network Initiation

Chapter 7 explicitly addresses the issue of defining the beginning of the river drainage network when this network is extracted by processing a DEM. The extracted network can present two types of errors relative to the actual network regarding the river initiation.

The first type of error concerns the actual starting point of each river segment. This starting point may be either anticipated or delayed concerning the actual drainage network, meaning that the river course begins further upstream or downstream than it should. For example, in Fig. 14.15a, the river segment that starts at point A in the actual drainage network begins further upstream, at point B, in the DEM-derived network (Fig. 14.15b).

This type of error impacts hydrological studies that use the drainage network in two main ways:

(i) it alters the length of the river segment, resulting in consequences associated with this change, as described in the next section;
(ii) it modifies the extent of overland flow required to reach the drainage network. Three clear examples of studies affected by both factors include travel time estimation of surface runoff within the watershed, rainfall-runoff transformation modeling, and studies on erosion and sediment transport processes.

For instance, consider the simulation of erosion processes triggered by rainfall and overland flow originating from point C (Fig. 14.15c). In the actual network, from C to the drainage network at point A, sediment transport occurs as overland flow across the terrain, while from point A downstream, sediment is conveyed through channel flow along river segments. Between C and A, sediment movement is hindered by shallow flow depth and high surface resistance, such as in forested or pasture areas.

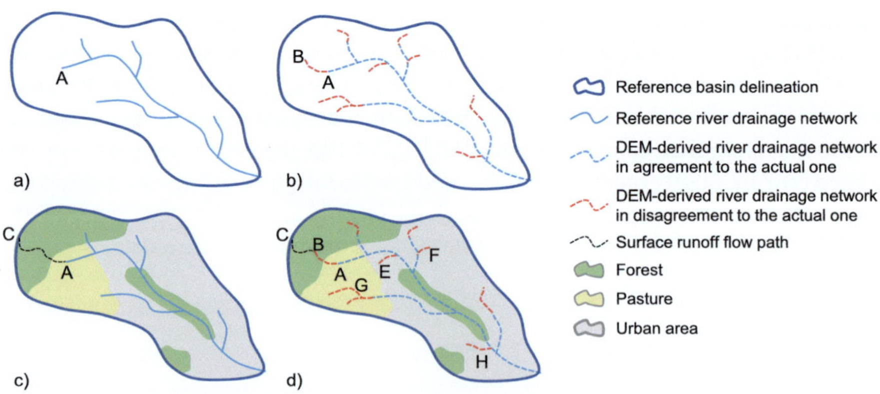

Fig. 14.15 **a**, **b** Actual and DEM-derived river drainage network, respectively; **c**, **d** Overlay of the actual and DEM-derived networks on land use and land cover mapping

Once the drainage network is reached (from A downstream), sediment transport becomes more efficient.

In the DEM-derived drainage network shown in Fig. 14.15d, the beginning of the river reach is anticipated to point B instead of A. Consequently, the overland flow segment downstream of point C is shortened to the path between C and B. As a result, a hydro-sedimentological simulation incorporating this DEM-derived network would tend to overestimate sediment transport velocity due to the premature initiation of channel flow segments.

The second type of error involves the over or underestimation of the number of river branches as a consequence of misidentifying the beginning points of the network. Depending on the method used, tributaries may be erroneously included in or omitted from the DEM-derived network compared to the actual drainage network. For instance, in Fig. 14.15d, the river branches labeled E, F, G, and H do not exist in the actual network but appear in the DEM-derived network, representing errors originating from the stream initiation definition in DEM processing.

When the network includes more river branches than it should, hydrological studies using this network will overestimate flow conveyance speed, flood potential in the basin, and sediment transport capacity. In contrast, if river branches are omitted, the simulations will underestimate the basin's flood susceptibility and the likelihood of surface runoff generation and sediment transport.

14.9.3 Errors in the River Reaches' Lengths

River length is one of the most valuable information extracted from DEMs for hydrological studies. Errors in the alignment and spatial positioning of the drainage network often lead to errors in river reach lengths as well.

Even studies that do not explicitly incorporate the drainage network layout often require information on river lengths within the study area. This variable represents the distance water must travel along the watercourse before exiting the watershed.

River length is essential for estimating flow travel time within the basin, simulating flood wave routing, modeling sediment transport, estimating pollutant attenuation and dilution processes, and other water quality-related phenomena, as well as a wide range of hydrological studies.

If errors occur in the river length extracted from DEM processing, and this information is used in hydrological analyses, all such studies will struggle to accurately represent the actual conditions of the water body, producing results that deviate from reality.

Consider, for example, the A–B river reach shown in Fig. 14.16a, represented by the drainage network layout derived from the DEM. The actual and DEM-derived layouts connect points A and B; however, the latter path covers a shorter distance due to reduced sinuosity and the incorrect representation of meanders. The DEM-extracted river length is equivalent to a network connecting point A to point C, which is located upstream of point B (Fig. 14.16b).

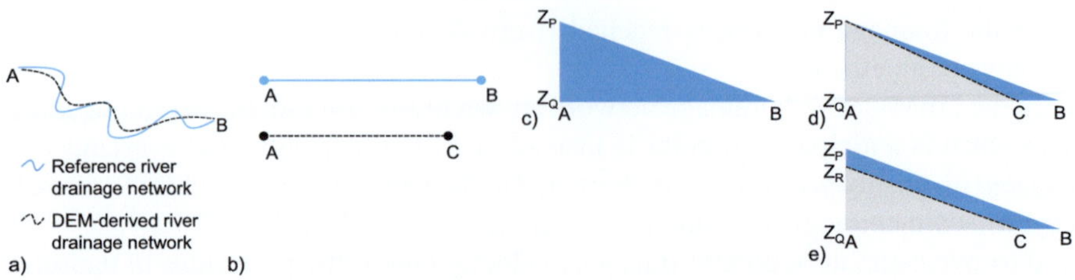

Fig. 14.16 **a** Actual and DEM-derived layout for river reach between points A and B; **b** Conversion of the layout of the river reach A–B into a straight segment adjusting the extension according to the actual layout and the layout derived from the DEM; **c** Schematic representation of the longitudinal topographic profile of the river reach A–B assuming constant slope, going from elevation Z_P to Z_Q; **d** Simplified longitudinal topographic profile extracted from the DEM, overcoming the same topographic difference Z_P to Z_Q over a shortened distance; **e** Simplified longitudinal topographic profile considering the same slope of the actual reach but over a shortened distance

With a shorter path to travel, the hydrological simulation of flood wave routing tends to underestimate the arrival time and attenuation, since a longer flow path results in more time for wave travel and a higher reduction in peak discharge. Similarly, a water quality study would likely overestimate the pollutant concentration at the end of the reach, since the calculation would be based on a travel distance shorter than the actual one.

14.9.4 Errors in the River Reaches' Slopes

River slope is another highly relevant variable for hydrological studies, closely related to river length, since slope is defined as the ratio between the elevation drop along the river course and the length of that course. When processing DEM-derived data, errors in river lengths directly affect slope estimates, further compounding the impact on hydrological studies that rely on such information.

The slope of a river reach is a key input for simulating flow routing. For instance, overestimated river slopes result in underestimated travel times of flood waves and overestimated conveyance capacity, which may falsely reduce the evaluated flood risk. In contrast, underestimated slopes lead to longer simulated travel times, underestimated conveyance, and falsely inflated flood risk estimates.

In the example shown in Fig. 14.16, assume that the topographic profile of the A–B river reach is as depicted in Fig. 14.16c, descending from elevation Z_P to Z_Q with a constant slope. Since the DEM-derived network shortens the A–B reach, a steeper slope is required to overcome the same elevation drop (Z_P–Z_Q) over a shorter horizontal distance. The topographic profile of this river reach would then follow the dashed line and gray-shaded area in Fig. 14.16d. In practical terms, this implies that the DEM-derived river is equivalent to flowing from point A to point C, located upstream of point B.

If the hydrological study uses the slope value from the actual river profile but applies it to the shortened length extracted from the DEM, the resulting topographic profile would resemble the dashed line and gray-shaded area in Fig. 14.16e, assuming a constant slope simplification. To maintain the same slope over a shorter distance, the upstream elevation would have to be artificially reduced from Z_P to Z_R (equivalently, the downstream elevation would need to be increased).

14.9.5 Errors in Watershed Delineation

Watershed delineation is one of the primary objectives of DEM processing, as watersheds define the area of interest in most environmental and hydrological studies.

Errors in watershed delineation can be analyzed in two main aspects:

(i) the numerical value of the drainage area;
(ii) the spatial area included within the delineation.

Like river length, drainage area is a fundamental variable in many hydrological studies. In some cases, only the area value is required as input, regardless of the actual spatial boundaries of the watershed.

An error in the total drainage area directly affects hydrological simulations. When combined with rainfall depth, this area determines the volume of incoming water to the watershed (Fig. 14.17a). This water volume exits the basin through evaporation, transpiration, deep percolation, or streamflow. Underestimating the watershed area makes it challenging to conduct a hydrological study of the water balance between inputs and outputs.

If the delineated watershed is smaller than it should be (Fig. 14.17b), the hydrological study will consider a lower water input. To reproduce the observed outflow adequately, the hydrological model must artificially reduce water losses during rainfall–runoff transformation, such as decreasing evapotranspiration.

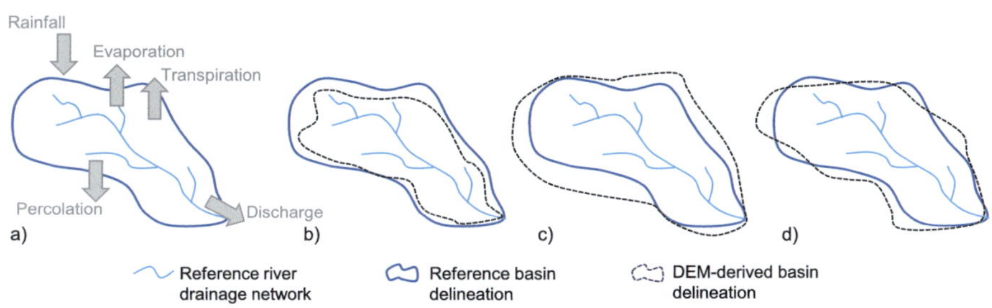

Fig. 14.17 **a** Diagram of water inputs and outputs in a watershed water balance; **b** DEM-derived watershed delineation with an area smaller than the reference basin and entirely contained within it; **c** DEM-derived delineation with an area larger than the reference basin, fully enclosing it; **d** DEM-derived delineation intersecting the reference at multiple points

If the area is overestimated (Fig. 14.17c), the model assumes a larger catchment than reality. In this case, it compensates for the inflated water input by increasing losses (e.g., evapotranspiration or deep percolation) during calibration.

Such compensations are commonly done through parameter adjustment and calibration in hydrological models. However, this approach is undesirable, as it can result in a model that is less suitable for simulating future scenarios and less reliable for decision-making support.

Moreover, calibration often cannot fully compensate for drainage area errors, which leads to models that tend to overestimate or underestimate streamflow. Underestimating watershed areas may produce artificially lower water availability and flood risk estimates. In contrast, overestimating drainage areas may result in a wrongly exaggerated water supply and inflated flood risks.

Figure 14.17b and c illustrate extreme cases where the DEM-derived watershed is either entirely within or encompassing the reference watershed. In practice, most cases fall somewhere in between, with delineations that partially intersect the reference watershed (Fig. 14.17d), leading to partial over or underestimation.

A second aspect of watershed delineation errors concerns the spatial inclusion of specific areas. Incorrectly included areas (commission errors) and omitted areas (omission errors) may occur.

Commission and omission errors can offset each other numerically, resulting in a total area value close to the actual one. Nevertheless, they can still impact hydrological modeling, particularly due to differences in soil type and land use/land cover within the included or omitted zones.

Consider the watershed shown in Fig. 14.18a and the mapping of land use/land cover. The DEM-derived delineation (Fig. 14.18b) includes erroneously added and omitted areas (Fig. 14.18c). Forest and pasture areas were wrongly added in the west and north, while urban and pasture areas were omitted in the south and east.

This leads to a distorted land cover distribution, with more forested and fewer urbanized areas than reality. Hydrological models using this data will tend to underestimate surface runoff for two reasons: (i) forested areas, which promote higher

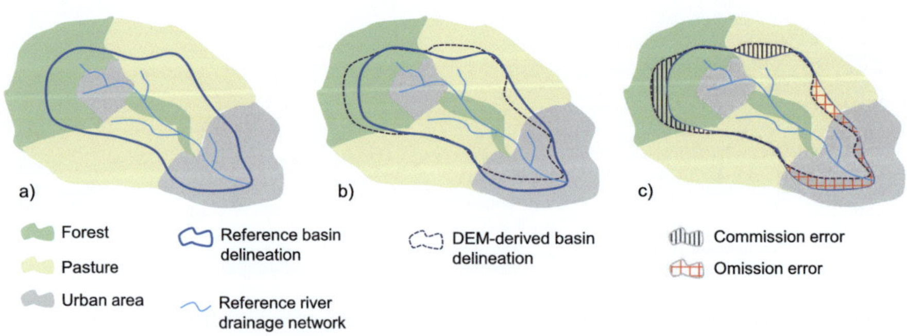

Fig. 14.18 **a** Reference delineation of a watershed with land use/land cover mapping; **b** DEM-derived delineation overlaid on the reference; **c** Identification of areas with commission and omission errors in the DEM-derived delineation

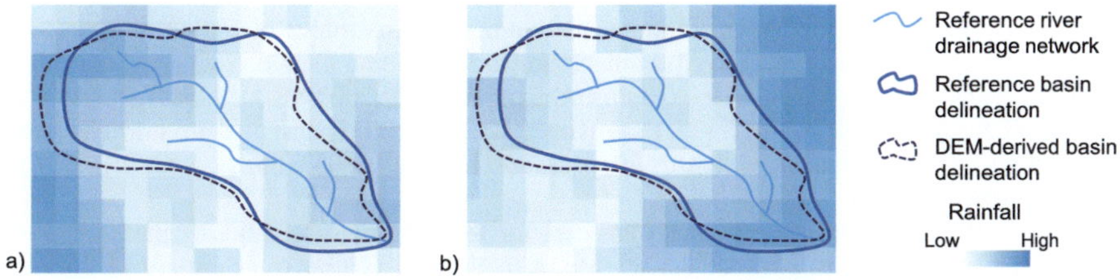

Fig. 14.19 **a**, **b** Spatial rainfall fields at two different time steps overlaid by the reference watershed and DEM-derived delineation

infiltration and evapotranspiration losses, are overrepresented; (ii) urban areas, which produce higher runoff due to impervious surfaces, are underrepresented.

If a rainfall–runoff model is calibrated based on this flawed representation, parameter values are again forced to compensate for the input error.

Incorrect watershed delineation may also impair hydrological modeling when spatial variability in precipitation is considered. Several studies demonstrate the benefits of incorporating spatial rainfall variability (Ghimire et al., 2021; Loritz et al., 2021). Assuming spatially uniform precipitation can induce calibration compensation effects (Schuurmans & Bierkens, 2007), and similar effects, albeit less pronounced, can arise from watershed delineation errors.

For instance, in Fig. 14.19a, the DEM-derived watershed includes additional areas to the west (commission error), which experience higher rainfall than those omitted to the south and east (omission error). In Fig. 14.19b, the opposite occurs—commission errors include drier regions, while wetter zones are omitted.

In both cases, watershed delineation errors lead to two key consequences. Firstly, there is an error in the total precipitation input caused by misrepresenting the catchment's spatial extent, even if the total area is approximately correct. Secondly, there is an error in the spatial distribution of rainfall input. Depending on the hydrological modeling approach, this error can impact the shape, timing, and magnitude of simulated hydrographs.

References

Arora, V. K., & Harrison, S. (2007). Upscaling river networks for use in climate models. *Geophysical Research Letters, 34*, L21407. https://doi.org/10.1029/2007GL031865

Butt, M. A., & Maragos, P. (1998). Optimum design of chamfer distance transforms. *IEEE Transactions on Image Processing, 7*, 1477–1484.

Collischonn, W., Buarque, D. C., Paz, A. R., Mendes, C. A. B., & Fan, F. (2010). *Impact of pit removal methods on DEM derived drainage lines in flat regions*. Presented at the AWRA 2010 Spring Specialty Conference, Orlando (EUA).

De Smith, M. J. (2004). Distance transforms as a new tool in spatial analysis, urban planning, and GIS. *Environmental and Planning b: Planning and Design, 31*, 85–104. https://doi.org/10.1068/b29123

Gatziolis, D., & Fried, J. S. (2004). Adding Gaussian noise to inaccurate digital elevation models improves spatial fidelity of derived drainage networks: Adding Gaussian noise to inaccurate DEM. *Water Resources Research, 40*. https://doi.org/10.1029/2002WR001735

Ghimire, G. R., Jadidoleslam, N., Goska, R., & Krajewski, W. F. (2021). Insights into storm direction effect on flood response. *Journal of Hydrology, 600*, Article 126683. https://doi.org/10.1016/j.jhydrol.2021.126683

Kirchner, J. W. (2006). Getting the right answers for the right reasons: Linking measurements, analyses, and models to advance the science of hydrology: Getting the right answers for the right reasons. *Water Resources Research, 42*. https://doi.org/10.1029/2005WR004362

Loritz, R., Hrachowitz, M., Neuper, M., & Zehe, E. (2021). The role and value of distributed precipitation data in hydrological models. *Hydrology and Earth Systems Science, 25*, 147–167. https://doi.org/10.5194/hess-25-147-2021

Paz, A. R., & Collischonn, W. (2007). River reach length and slope estimates for large-scale hydrological models based on a relatively high-resolution digital elevation model. *Journal of Hydrology, 343*, 127–139. https://doi.org/10.1016/j.jhydrol.2007.06.006

Paz, A. R., Collischonn, W., Risso, A., & Mendes, C. A. B. (2008). Errors in river lengths derived from raster digital elevation models. *Computers and Geosciences, 34*, 1584–1596. https://doi.org/10.1016/j.cageo.2007.10.009

Saraiva, A. G. S., & Paz, A. R. (2014). Multi-step change of scale approach for deriving coarse-resolution flow directions. *Computers and Geosciences, 68*, 53–63. https://doi.org/10.1016/j.cageo.2014.04.002

Schuurmans, J. M., & Bierkens, M. F. P. (2007). Effect of spatial distribution of daily rainfall on interior catchment response of a distributed hydrological model. Hydrology and Earth System Sciences.

Semenova, O., & Beven, K. (2015). Barriers to progress in distributed hydrological modelling. *Hydrological Processes, 29*, 2074–2078. https://doi.org/10.1002/hyp.10434

Sousa, T. M. I., & Paz, A. R. (2017). How to evaluate the quality of coarse-resolution DEM-derived drainage networks. *Hydrological Processes, 31*, 3379–3395. https://doi.org/10.1002/hyp.11262

Wechsler, S. P. (2007). Uncertainties associated with digital elevation models for hydrologic applications: A review. *Hydrology and Earth Systems Science, 11*, 1481–1500. https://doi.org/10.5194/hess-11-1481-2007

Wu, H., Kimball, J. S., Mantua, N., & Stanford, J. (2011). Automated upscaling of river networks for macroscale hydrological modeling: Upscaling of global river networks. *Water Resources Research, 47*. https://doi.org/10.1029/2009WR008871

Yamazaki, D., Ikeshima, D., Tawatari, R., Yamaguchi, T., O'Loughlin, F., Neal, J. C., Sampson, C. C., Kanae, S., & Bates, P. D. (2017). A high-accuracy map of global terrain elevations: Accurate global terrain elevation map. *Geophysical Research Letters, 44*, 5844–5853. https://doi.org/10.1002/2017GL072874

MIX
Papier aus verantwortungsvollen Quellen
Paper from responsible sources
FSC® C105338

If you have any concerns about our products,
you can contact us on
ProductSafety@springernature.com

In case Publisher is established outside the EU,
the EU authorized representative is:
Springer Nature Customer Service Center GmbH
Europaplatz 3, 69115 Heidelberg, Germany

Printed by Libri Plureos GmbH
in Hamburg, Germany